Methods in Enzymology

Volume 258

REDOX-ACTIVE AMINO ACIDS IN BIOLOGY

METHODS IN ENZYMOLOGY

EDITORS-IN-CHIEF

John N. Abelson Melvin I. Simon

DIVISION OF BIOLOGY
CALIFORNIA INSTITUTE OF TECHNOLOGY
PASADENA, CALIFORNIA

FOUNDING EDITORS

Sidney P. Colowick and Nathan O. Kaplan

Methods in Enzymology

Volume 258

Redox-Active Amino Acids in Biology

EDITED BY

Judith P. Klinman

DEPARTMENT OF CHEMISTRY
UNIVERSITY OF CALIFORNIA, BERKELEY
BERKELEY, CALIFORNIA

ACADEMIC PRESS
San Diego New York Boston London Sydney Tokyo Toronto

This book is printed on acid-free paper. ♾

Academic Press, Inc.
A Division of Harcourt Brace & Company
525 B Street, Suite 1900, San Diego, California 92101-4495

United Kingdom Edition published by
Academic Press Limited
24-28 Oval Road, London NW1 7DX

International Standard Serial Number: 0076-6879

International Standard Book Number: 0-12-182159-5

PRINTED IN THE UNITED STATES OF AMERICA
95 96 97 98 99 00 MM 9 8 7 6 5 4 3 2 1

Table of Contents

Contributors to Volume 258 vii

Preface xi

Volumes in Series xiii

1. Precursors of Quinone-Tanning: Dopa-Containing Proteins — J. Herbert Waite — 1

2. Isolation of 2,4,5-Trihydroxyphenylalanine Quinone (Topa Quinone) from Copper Amine Oxidases — Susan M. Janes and Judith P. Klinman — 20

3. Spectrophotometric Detection of Topa Quinone — Monica Palcic and Susan M. Janes — 34

4. Model Studies of Topa Quinone: Synthesis and Characterization of Topa Quinone Derivatives — Minae Mure and Judith P. Klinman — 39

5. Catalytic Aerobic Deamination of Activated Primary Amines by a Model for the Quinone Cofactor of Mammalian Copper Amine Oxidases — Lawrence M. Sayre and Younghee Lee — 53

6. Detection of Reaction Intermediates in Topa Quinone Enzymes — Christa Hartmann and David M. Dooley — 69

7. Mass Spectrometric Studies of the Primary Sequence and Structure of Bovine Liver and Serum Amine Oxidase — Gregory W. Adams, Petra Mayer, Katalin F. Medzihradszky, and Alma L. Burlingame — 90

8. Cloning of Mammalian Topa Quinone-Containing Enzymes — David Mu and Judith P. Klinman — 114

9. Isolation of Active Site Peptides of Lysyl Oxidase — Herbert M. Kagan and Ping Cai — 122

10. Resonance Raman Spectroscopy of Quinoproteins — David M. Dooley and Dorren E. Brown — 132

11. Redox-Cycling Detection of Dialyzable Pyrroloquinoline Quinone (PQQ) and Quinoproteins — Rudolf Flückiger, Mercedes A. Paz, and Paul M. Gallop — 140

12. Tryptophan Tryptophylquinone (TTQ) in Bacterial Amine Dehydrogenases — William S. McIntire — 149

13. Model Studies of Cofactor Tryptophan Tryptophylquinone (TTQ) — SHINOBU ITOH AND YOSHIKI OHSHIRO — 164

14. Detection of Intermediates in Tryptophan Tryptophylquinone (TTQ) Enzymes — VICTOR L. DAVIDSON, HAROLD B. BROOKS, M. ELIZABETH GRAICHEN, LIMEI H. JONES, AND YOUNG-LAN HYUN — 176

15. X-Ray Studies of Quinoproteins — F. SCOTT MATHEWS — 191

16. Genetics of Bacterial Quinoproteins — MARY E. LIDSTROM — 217

17. Biogenesis of Pyrrolquinone Quinone from ^{13}C-Labeled Tyrosine — CLIFFORD J. UNKEFER, DAVID R. HOUCK, B. MARK BRITT, TOBIN R. SOSNICK, AND JOHN L. HANNERS — 227

18. X-Ray Crystallographic Studies of Cofactors in Galactose Oxidase — NOBUTOSHI ITO, PETER F. KNOWLES, AND SIMON E. V. PHILLIPS — 235

19. Spectroscopic Studies of Galactose Oxidase — JAMES W. WHITTAKER — 262

20. Use of Rapid Kinetics Methods to Study the Assembly of the Diferric-Tyrosyl Radical Cofactor of *E. coli* Ribonucleotide Reductase — J. MARTIN BOLLINGER, JR., WING HANG TONG, NATARAJAN RAVI, BOI HANH HUYNH, DALE E. EDMONDSON, AND JOANNE STUBBE — 278

21. Tyrosyl Radicals in Photosystem II — BRIDGETTE A. BARRY — 303

22. Role of Tryptophans in Substrate Binding and Catalysis by DNA Photolyase — SANG-TAE KIM, PAUL F. HEELIS, AND AZIZ SANCAR — 319

23. Glycyl Free Radical in Pyruvate Formate-Lyase: Synthesis, Structure Characteristics, and Involvement in Catalysis — JOACHIM KNAPPE AND A. F. VOLKER WAGNER — 343

24. Characterization of a Radical Intermediate in the Lysine 2,3-Aminomutase Reaction — GEORGE H. REED AND MARCUS D. BALLINGER — 362

25. Role of Oxidized Amino Acids in Protein Breakdown and Stability — EARL R. STADTMAN — 379

AUTHOR INDEX 395
SUBJECT INDEX 405

Contributors to Volume 258

Article numbers are in parentheses following the names of contributors.
Affiliations listed are current.

GREGORY W. ADAMS (7), *Department of Pharmaceutical Chemistry, University of California, San Francisco, San Francisco, California 94143*

MARCUS D. BALLINGER (24), *Department of Protein Engineering, Genentech, Inc., South San Francisco, California 94080*

BRIDGETTE A. BARRY (21), *Department of Biochemistry, University of Minnesota, St. Paul, Minnesota 55108*

J. MARTIN BOLLINGER, JR. (20), *Department of Chemistry, Massachusetts Institute of Technology, Cambridge, Massachusetts 02139*

B. MARK BRITT (17), *10 Los Alamos National Laboratory, Inc., Los Alamos, New Mexico 87545*

HAROLD B. BROOKS (14), *Department of Biochemistry, University of Mississippi Medical Center, Jackson, Mississippi 39216*

DOREEN BROWN (10), *Department of Chemistry, Montana State University, Bozeman, Montana 59717*

ALMA L. BURLINGAME (7), *Department of Pharmaceutical Chemistry, University of California, San Francisco, San Francisco, California 94143*

PING CAI (9), *Department of Chemistry, Boston University School of Medicine, Boston, Massachusetts 02118*

VICTOR L. DAVIDSON (14), *Department of Biochemistry, University of Mississippi Medical Center, Jackson, Mississippi 39216*

DAVID M. DOOLEY (6, 10), *Departments of Chemistry and Biochemistry, Montana State University, Bozeman, Montana 59717*

DALE E. EDMONDSON (20), *Department of Biochemistry, Emory University, Atlanta, Georgia 30322*

RUDOLF FLÜCKIGER (11), *Children's Hospital Boston, Harvard School of Dental Medicine, Boston, Massachusetts 02115*

PAUL M. GALLOP (11), *Laboratory of Human Biochemistry, Children's Hospital Corporation, and Departments of Biological Chemistry, Oral Biology, and Pathophysiology, Harvard Schools of Medicine and Dental Medicine, Boston, Massachusetts 02115*

M. ELIZABETH GRAICHEN (14), *Department of Biochemistry, University of Mississippi Medical Center, Jackson, Mississippi 39216*

JOHN L. HANNERS (17), *10 Los Alamos Laboratory, Inc., Los Alamos, New Mexico 87545*

CHRISTA HARTMANN (6), *Khepri Pharmaceuticals, Inc., South San Francisco, California 94080*

PAUL F. HEELIS (22), *Department of Biochemistry, University of North Carolina at Chapel Hill, Chapel Hill, North Carolina 27599*

DAVID R. HOUCK (17), *10 Los Alamos National Laboratory, Inc., Los Alamos, New Mexico 87545*

BOI HANH HUYNH (20), *Department of Physics, Emory University, Atlanta, Georgia 30322*

YOUNG-LAN HYUN (14), *Department of Biochemistry, University of Mississippi Medical Center, Jackson, Mississippi 39216*

NOBUTOSHI ITO (18), *Department of Biochemistry and Molecular Biology, University of Leeds, Leeds LS2 9JT, England*

SHINOBU ITOH (13), *Department of Applied Chemistry, Faculty of Engineering, Osaka University, Suita, Osaka 565, Japan*

SUSAN M. JANES (2, 3), *Amylin Pharmaceuticals, San Diego, California 92121*

LIMEI H. JONES (14), *Department of Biochemistry, University of Mississippi Medical Center, Jackson, Mississippi 39216*

HERBERT M. KAGAN (9), *Department of Chemistry, Boston University School of Medicine, Boston, Massachusetts 02118*

SANG-TAE KIM (22), *Department of Biochemistry, University of North Carolina at Chapel Hill, Chapel Hill, North Carolina 27599*

JUDITH P. KLINMAN (2, 4, 8), *Department of Chemistry, University of California, Berkeley, Berkeley, California 94720*

JOACHIM KNAPPE (23), *Institut für Biologische Chemie, Ruprecht-Karls-Universität Heidelberg, D-69120 Heidelberg, Germany*

PETER F. KNOWLES (18), *Department of Biochemistry and Molecular Biology, University of Leeds, Leeds LS2 9JT, England*

YOUNGHEE LEE (5), *Department of Chemistry, Case Western Reserve University, Cleveland, Ohio 44106*

MARY E. LIDSTROM (16), *Environmental and Engineering Science, California Institute of Technology, Pasadena, California 91125*

F. SCOTT MATHEWS (15), *Department of Biochemistry and Molecular Biophysics, Washington University School of Medicine, St. Louis, Missouri 63110*

PETRA MAYER (7), *Department of Pharmaceutical Chemistry, University of California, San Francisco, San Francisco, California 94143*

WILLIAM S. MCINTIRE (12), *Department of Veterans Affairs Medical Center, Molecular Biology Division, San Francisco, California 94121 and Departments of Biochemistry and Biophysics, and Anesthesia, University of California, San Francisco, San Francisco, California 94143*

KATALIN F. MEDZIHRADSZKY (7), *Department of Pharmaceutical Chemistry, University of California, San Francisco, San Francisco, California 94143*

DAVID MU (8), *Departments of Biochemistry and Biophysics, School of Medicine, University of North Carolina at Chapel Hill, Chapel Hill, North Carolina 27599*

MINAE MURE (4), *Department of Chemistry, University of California, Berkeley, Berkeley, California 94720*

YOSHIKI OHSHIRO (13), *Department of Applied Chemistry, Faculty of Engineering, Osaka University, Suita, Osaka 565, Japan*

MONICA PALCIC (3), *Department of Chemistry, University of Alberta, Edmonton, Alberta, Canada T6G 2G2*

MERCEDES A. PAZ (11), *Children's Hospital Boston, Harvard School of Dental Medicine, Boston, Massachusetts 02115*

SIMON E. V. PHILLIPS (18), *Department of Biochemistry and Molecular Biology, University of Leeds, Leeds, LS2 9JT, England*

NATARAJAN RAVI (20), *Department of Physics, Emory University, Atlanta, Georgia 30322*

GEORGE H. REED (24), *Institute for Enzyme Research, University of Wisconsin—Madison, Madison, Wisconsin 53706*

AZIZ SANCAR (22), *Department of Biochemistry, University of North Carolina at Chapel Hill, Chapel Hill, North Carolina 27599*

LAWRENCE M. SAYRE (5), *Department of Chemistry, Case Western Reserve University, Cleveland, Ohio 44106*

TOBIN R. SOSNICK (17), *10 Los Alamos National Laboratory, Inc., Los Alamos, New Mexico 87545*

EARL R. STADTMAN (25), *NHLBI, Laboratory of Biochemistry, National Institutes of Health, Bethesda, Maryland 20892*

JOANNE STUBBE (20), *Departments of Chemistry and Biology, Massachusetts Institute of Technology, Cambridge, Massachusetts 02139*

WING HANG TONG (20), *Department of Chemistry, Massachusetts Institute of Technology, Cambridge, Massachusetts 02139*

CLIFFORD J. UNKEFER (17), *10 Los Alamos National Laboratory, Inc., Los Alamos, New Mexico 87545*

A. F. VOLKER WAGNER (23), *Institut für Biologische Chemie, Ruprecht-Karls-Universität Heidelberg, D-69120 Heidelberg, Germany*

J. HERBERT WAITE (1), *Department of Chemistry/Biochemistry, University of Delaware, Newark, Delaware 19716*

JAMES W. WHITTAKER (19), *Department of Chemistry, Carnegie-Mellon University, Pittsburgh, Pennsylvania 15213*

Preface

Recent advances in enzymology have dramatically altered our knowledge of cofactor catalysis. In contrast to the classical picture, whereby cofactors are viewed as freely dissociating low-molecular-weight compounds, an increasing number of enzyme systems have been documented as using peptide-bound amino acid side chains as catalytic units in redox chemistry. As described here, these new prosthetic groups may be either radicals derived from glycyl, tyrosyl, and trypyophanyl residues or quinones derived from the posttranslational oxidation of tyrosyl or tryptophanyl precursors. Although the term *cofactor* has traditionally been reserved for small molecules, I propose that we extend this terminology to include the new functional groups derived from amino acid side chains.

The goal of this volume is to introduce the researcher to the methods available for the detection and characterization of amino acid side chains functioning as redox catalysts. Although the field is moving rapidly and new structures may emerge in the near future, the chapters contained herein should provide the groundwork for anyone wishing to enter this exciting field. I am indebted to all of the contributors for their willingness to submit their chapters in a timely manner; to John Abelson and Mel Simon for their original interest in the volume; and to Shirley Light for her skill and persistence in producing a final book.

Judith Klinman

METHODS IN ENZYMOLOGY

VOLUME I. Preparation and Assay of Enzymes
Edited by SIDNEY P. COLOWICK AND NATHAN O. KAPLAN

VOLUME II. Preparation and Assay of Enzymes
Edited by SIDNEY P. COLOWICK AND NATHAN O. KAPLAN

VOLUME III. Preparation and Assay of Substrates
Edited by SIDNEY P. COLOWICK AND NATHAN O. KAPLAN

VOLUME IV. Special Techniques for the Enzymologist
Edited by SIDNEY P. COLOWICK AND NATHAN O. KAPLAN

VOLUME V. Preparation and Assay of Enzymes
Edited by SIDNEY P. COLOWICK AND NATHAN O. KAPLAN

VOLUME VI. Preparation and Assay of Enzymes (*Continued*)
Preparation and Assay of Substrates
Special Techniques
Edited by SIDNEY P. COLOWICK AND NATHAN O. KAPLAN

VOLUME VII. Cumulative Subject Index
Edited by SIDNEY P. COLOWICK AND NATHAN O. KAPLAN

VOLUME VIII. Complex Carbohydrates
Edited by ELIZABETH F. NEUFELD AND VICTOR GINSBURG

VOLUME IX. Carbohydrate Metabolism
Edited by WILLIS A. WOOD

VOLUME X. Oxidation and Phosphorylation
Edited by RONALD W. ESTABROOK AND MAYNARD E. PULLMAN

VOLUME XI. Enzyme Structure
Edited by C. H. W. HIRS

VOLUME XII. Nucleic Acids (Parts A and B)
Edited by LAWRENCE GROSSMAN AND KIVIE MOLDAVE

VOLUME XIII. Citric Acid Cycle
Edited by J. M. LOWENSTEIN

VOLUME XIV. Lipids
Edited by J. M. LOWENSTEIN

VOLUME XV. Steroids and Terpenoids
Edited by RAYMOND B. CLAYTON

VOLUME XVI. Fast Reactions
Edited by KENNETH KUSTIN

VOLUME XVII. Metabolism of Amino Acids and Amines (Parts A and B)
Edited by HERBERT TABOR AND CELIA WHITE TABOR

VOLUME XVIII. Vitamins and Coenzymes (Parts A, B, and C)
Edited by DONALD B. MCCORMICK AND LEMUEL D. WRIGHT

VOLUME XIX. Proteolytic Enzymes
Edited by GERTRUDE E. PERLMANN AND LASZLO LORAND

VOLUME XX. Nucleic Acids and Protein Synthesis (Part C)
Edited by KIVIE MOLDAVE AND LAWRENCE GROSSMAN

VOLUME XXI. Nucleic Acids (Part D)
Edited by LAWRENCE GROSSMAN AND KIVIE MOLDAVE

VOLUME XXII. Enzyme Purification and Related Techniques
Edited by WILLIAM B. JAKOBY

VOLUME XXIII. Photosynthesis (Part A)
Edited by ANTHONY SAN PIETRO

VOLUME XXIV. Photosynthesis and Nitrogen Fixation (Part B)
Edited by ANTHONY SAN PIETRO

VOLUME XXV. Enzyme Structure (Part B)
Edited by C. H. W. HIRS AND SERGE N. TIMASHEFF

VOLUME XXVI. Enzyme Structure (Part C)
Edited by C. H. W. HIRS AND SERGE N. TIMASHEFF

VOLUME XXVII. Enzyme Structure (Part D)
Edited by C. H. W. HIRS AND SERGE N. TIMASHEFF

VOLUME XXVIII. Complex Carbohydrates (Part B)
Edited by VICTOR GINSBURG

VOLUME XXIX. Nucleic Acids and Protein Synthesis (Part E)
Edited by LAWRENCE GROSSMAN AND KIVIE MOLDAVE

VOLUME XXX. Nucleic Acids and Protein Synthesis (Part F)
Edited by KIVIE MOLDAVE AND LAWRENCE GROSSMAN

VOLUME XXXI. Biomembranes (Part A)
Edited by SIDNEY FLEISCHER AND LESTER PACKER

VOLUME XXXII. Biomembranes (Part B)
Edited by SIDNEY FLEISCHER AND LESTER PACKER

VOLUME XXXIII. Cumulative Subject Index Volumes I–XXX
Edited by MARTHA G. DENNIS AND EDWARD A. DENNIS

VOLUME XXXIV. Affinity Techniques (Enzyme Purification: Part B)
Edited by WILLIAM B. JAKOBY AND MEIR WILCHEK

VOLUME XXXV. Lipids (Part B)
Edited by JOHN M. LOWENSTEIN

VOLUME XXXVI. Hormone Action (Part A: Steroid Hormones)
Edited by BERT W. O'MALLEY AND JOEL G. HARDMAN

VOLUME XXXVII. Hormone Action (Part B: Peptide Hormones)
Edited by BERT W. O'MALLEY AND JOEL G. HARDMAN

VOLUME XXXVIII. Hormone Action (Part C: Cyclic Nucleotides)
Edited by JOEL G. HARDMAN AND BERT W. O'MALLEY

VOLUME XXXIX. Hormone Action (Part D: Isolated Cells, Tissues, and Organ Systems)
Edited by JOEL G. HARDMAN AND BERT W. O'MALLEY

VOLUME XL. Hormone Action (Part E: Nuclear Structure and Function)
Edited by BERT W. O'MALLEY AND JOEL G. HARDMAN

VOLUME XLI. Carbohydrate Metabolism (Part B)
Edited by W. A. WOOD

VOLUME XLII. Carbohydrate Metabolism (Part C)
Edited by W. A. WOOD

VOLUME XLIII. Antibiotics
Edited by JOHN H. HASH

VOLUME XLIV. Immobilized Enzymes
Edited by KLAUS MOSBACH

VOLUME XLV. Proteolytic Enzymes (Part B)
Edited by LASZLO LORAND

VOLUME XLVI. Affinity Labeling
Edited by WILLIAM B. JAKOBY AND MEIR WILCHEK

VOLUME XLVII. Enzyme Structure (Part E)
Edited by C. H. W. HIRS AND SERGE N. TIMASHEFF

VOLUME XLVIII. Enzyme Structure (Part F)
Edited by C. H. W. HIRS AND SERGE N. TIMASHEFF

VOLUME XLIX. Enzyme Structure (Part G)
Edited by C. H. W. HIRS AND SERGE N. TIMASHEFF

VOLUME L. Complex Carbohydrates (Part C)
Edited by VICTOR GINSBURG

VOLUME LI. Purine and Pyrimidine Nucleotide Metabolism
Edited by PATRICIA A. HOFFEE AND MARY ELLEN JONES

VOLUME LII. Biomembranes (Part C: Biological Oxidations)
Edited by SIDNEY FLEISCHER AND LESTER PACKER

VOLUME LIII. Biomembranes (Part D: Biological Oxidations)
Edited by SIDNEY FLEISCHER AND LESTER PACKER

VOLUME LIV. Biomembranes (Part E: Biological Oxidations)
Edited by SIDNEY FLEISCHER AND LESTER PACKER

VOLUME LV. Biomembranes (Part F: Bioenergetics)
Edited by SIDNEY FLEISCHER AND LESTER PACKER

VOLUME LVI. Biomembranes (Part G: Bioenergetics)
Edited by SIDNEY FLEISCHER AND LESTER PACKER

VOLUME LVII. Bioluminescence and Chemiluminescence
Edited by MARLENE A. DELUCA

VOLUME LVIII. Cell Culture
Edited by WILLIAM B. JAKOBY AND IRA PASTAN

VOLUME LIX. Nucleic Acids and Protein Synthesis (Part G)
Edited by KIVIE MOLDAVE AND LAWRENCE GROSSMAN

VOLUME LX. Nucleic Acids and Protein Synthesis (Part H)
Edited by KIVIE MOLDAVE AND LAWRENCE GROSSMAN

VOLUME 61. Enzyme Structure (Part H)
Edited by C. H. W. HIRS AND SERGE N. TIMASHEFF

VOLUME 62. Vitamins and Coenzymes (Part D)
Edited by DONALD B. MCCORMICK AND LEMUEL D. WRIGHT

VOLUME 63. Enzyme Kinetics and Mechanism (Part A: Initial Rate and Inhibitor Methods)
Edited by DANIEL L. PURICH

VOLUME 64. Enzyme Kinetics and Mechanism (Part B: Isotopic Probes and Complex Enzyme Systems)
Edited by DANIEL L. PURICH

VOLUME 65. Nucleic Acids (Part I)
Edited by LAWRENCE GROSSMAN AND KIVIE MOLDAVE

VOLUME 66. Vitamins and Coenzymes (Part E)
Edited by DONALD B. MCCORMICK AND LEMUEL D. WRIGHT

VOLUME 67. Vitamins and Coenzymes (Part F)
Edited by DONALD B. MCCORMICK AND LEMUEL D. WRIGHT

VOLUME 68. Recombinant DNA
Edited by RAY WU

VOLUME 69. Photosynthesis and Nitrogen Fixation (Part C)
Edited by ANTHONY SAN PIETRO

VOLUME 70. Immunochemical Techniques (Part A)
Edited by HELEN VAN VUNAKIS AND JOHN J. LANGONE

VOLUME 71. Lipids (Part C)
Edited by JOHN M. LOWENSTEIN

VOLUME 72. Lipids (Part D)
Edited by JOHN M. LOWENSTEIN

VOLUME 73. Immunochemical Techniques (Part B)
Edited by JOHN J. LANGONE AND HELEN VAN VUNAKIS

VOLUME 74. Immunochemical Techniques (Part C)
Edited by JOHN J. LANGONE AND HELEN VAN VUNAKIS

VOLUME 75. Cumulative Subject Index Volumes XXXI, XXXII, XXXIV–LX
Edited by EDWARD A. DENNIS AND MARTHA G. DENNIS

VOLUME 76. Hemoglobins
Edited by ERALDO ANTONINI, LUIGI ROSSI-BERNARDI, AND EMILIA CHIANCONE

VOLUME 77. Detoxication and Drug Metabolism
Edited by WILLIAM B. JAKOBY

VOLUME 78. Interferons (Part A)
Edited by SIDNEY PESTKA

VOLUME 79. Interferons (Part B)
Edited by SIDNEY PESTKA

VOLUME 80. Proteolytic Enzymes (Part C)
Edited by LASZLO LORAND

VOLUME 81. Biomembranes (Part H: Visual Pigments and Purple Membranes, I)
Edited by LESTER PACKER

VOLUME 82. Structural and Contractile Proteins (Part A: Extracellular Matrix)
Edited by LEON W. CUNNINGHAM AND DIXIE W. FREDERIKSEN

VOLUME 83. Complex Carbohydrates (Part D)
Edited by VICTOR GINSBURG

VOLUME 84. Immunochemical Techniques (Part D: Selected Immunoassays)
Edited by JOHN J. LANGONE AND HELEN VAN VUNAKIS

VOLUME 85. Structural and Contractile Proteins (Part B: The Contractile Apparatus and the Cytoskeleton)
Edited by DIXIE W. FREDERIKSEN AND LEON W. CUNNINGHAM

VOLUME 86. Prostaglandins and Arachidonate Metabolites
Edited by WILLIAM E. M. LANDS AND WILLIAM L. SMITH

VOLUME 87. Enzyme Kinetics and Mechanism (Part C: Intermediates, Stereochemistry, and Rate Studies)
Edited by DANIEL L. PURICH

VOLUME 88. Biomembranes (Part I: Visual Pigments and Purple Membranes, II)
Edited by LESTER PACKER

VOLUME 89. Carbohydrate Metabolism (Part D)
Edited by WILLIS A. WOOD

VOLUME 90. Carbohydrate Metabolism (Part E)
Edited by WILLIS A. WOOD

VOLUME 91. Enzyme Structure (Part I)
Edited by C. H. W. HIRS AND SERGE N. TIMASHEFF

VOLUME 92. Immunochemical Techniques (Part E: Monoclonal Antibodies and General Immunoassay Methods)
Edited by JOHN J. LANGONE AND HELEN VAN VUNAKIS

VOLUME 93. Immunochemical Techniques (Part F: Conventional Antibodies, Fc Receptors, and Cytotoxicity)
Edited by JOHN J. LANGONE AND HELEN VAN VUNAKIS

Volume 94. Polyamines
Edited by Herbert Tabor and Celia White Tabor

Volume 95. Cumulative Subject Index Volumes 61–74, 76–80
Edited by Edward A. Dennis and Martha G. Dennis

Volume 96. Biomembranes [Part J: Membrane Biogenesis: Assembly and Targeting (General Methods; Eukaryotes)]
Edited by Sidney Fleischer and Becca Fleischer

Volume 97. Biomembranes [Part K: Membrane Biogenesis: Assembly and Targeting (Prokaryotes, Mitochondria, and Chloroplasts)]
Edited by Sidney Fleischer and Becca Fleischer

Volume 98. Biomembranes (Part L: Membrane Biogenesis: Processing and Recycling)
Edited by Sidney Fleischer and Becca Fleischer

Volume 99. Hormone Action (Part F: Protein Kinases)
Edited by Jackie D. Corbin and Joel G. Hardman

Volume 100. Recombinant DNA (Part B)
Edited by Ray Wu, Lawrence Grossman, and Kivie Moldave

Volume 101. Recombinant DNA (Part C)
Edited by Ray Wu, Lawrence Grossman, and Kivie Moldave

Volume 102. Hormone Action (Part G: Calmodulin and Calcium-Binding Proteins)
Edited by Anthony R. Means and Bert W. O'Malley

Volume 103. Hormone Action (Part H: Neuroendocrine Peptides)
Edited by P. Michael Conn

Volume 104. Enzyme Purification and Related Techniques (Part C)
Edited by William B. Jakoby

Volume 105. Oxygen Radicals in Biological Systems
Edited by Lester Packer

Volume 106. Posttranslational Modifications (Part A)
Edited by Finn Wold and Kivie Moldave

Volume 107. Posttranslational Modifications (Part B)
Edited by Finn Wold and Kivie Moldave

Volume 108. Immunochemical Techniques (Part G: Separation and Characterization of Lymphoid Cells)
Edited by Giovanni Di Sabato, John J. Langone, and Helen Van Vunakis

Volume 109. Hormone Action (Part I: Peptide Hormones)
Edited by Lutz Birnbaumer and Bert W. O'Malley

Volume 110. Steroids and Isoprenoids (Part A)
Edited by John H. Law and Hans C. Rilling

VOLUME 111. Steroids and Isoprenoids (Part B)
Edited by JOHN H. LAW AND HANS C. RILLING

VOLUME 112. Drug and Enzyme Targeting (Part A)
Edited by KENNETH J. WIDDER AND RALPH GREEN

VOLUME 113. Glutamate, Glutamine, Glutathione, and Related Compounds
Edited by ALTON MEISTER

VOLUME 114. Diffraction Methods for Biological Macromolecules (Part A)
Edited by HAROLD W. WYCKOFF, C. H. W. HIRS, AND SERGE N. TIMASHEFF

VOLUME 115. Diffraction Methods for Biological Macromolecules (Part B)
Edited by HAROLD W. WYCKOFF, C. H. W. HIRS, AND SERGE N. TIMASHEFF

VOLUME 116. Immunochemical Techniques (Part H: Effectors and Mediators of Lymphoid Cell Functions)
Edited by GIOVANNI DI SABATO, JOHN J. LANGONE, AND HELEN VAN VUNAKIS

VOLUME 117. Enzyme Structure (Part J)
Edited by C. H. W. HIRS AND SERGE N. TIMASHEFF

VOLUME 118. Plant Molecular Biology
Edited by ARTHUR WEISSBACH AND HERBERT WEISSBACH

VOLUME 119. Interferons (Part C)
Edited by SIDNEY PESTKA

VOLUME 120. Cumulative Subject Index Volumes 81–94, 96–101

VOLUME 121. Immunochemical Techniques (Part I: Hybridoma Technology and Monoclonal Antibodies)
Edited by JOHN J. LANGONE AND HELEN VAN VUNAKIS

VOLUME 122. Vitamins and Coenzymes (Part G)
Edited by FRANK CHYTIL AND DONALD B. MCCORMICK

VOLUME 123. Vitamins and Coenzymes (Part H)
Edited by FRANK CHYTIL AND DONALD B. MCCORMICK

VOLUME 124. Hormone Action (Part J: Neuroendocrine Peptides)
Edited by P. MICHAEL CONN

VOLUME 125. Biomembranes (Part M: Transport in Bacteria, Mitochondria, and Chloroplasts: General Approaches and Transport Systems)
Edited by SIDNEY FLEISCHER AND BECCA FLEISCHER

VOLUME 126. Biomembranes (Part N: Transport in Bacteria, Mitochondria, and Chloroplasts: Protonmotive Force)
Edited by SIDNEY FLEISCHER AND BECCA FLEISCHER

VOLUME 127. Biomembranes (Part O: Protons and Water: Structure and Translocation)
Edited by LESTER PACKER

Volume 128. Plasma Lipoproteins (Part A: Preparation, Structure, and Molecular Biology)
Edited by JERE P. SEGREST AND JOHN J. ALBERS

Volume 129. Plasma Lipoproteins (Part B: Characterization, Cell Biology, and Metabolism)
Edited by JOHN J. ALBERS AND JERE P. SEGREST

Volume 130. Enzyme Structure (Part K)
Edited by C. H. W. HIRS AND SERGE N. TIMASHEFF

Volume 131. Enzyme Structure (Part L)
Edited by C. H. W. HIRS AND SERGE N. TIMASHEFF

Volume 132. Immunochemical Techniques (Part J: Phagocytosis and Cell-Mediated Cytotoxicity)
Edited by GIOVANNI DI SABATO AND JOHANNES EVERSE

Volume 133. Bioluminescence and Chemiluminescence (Part B)
Edited by MARLENE DELUCA AND WILLIAM D. MCELROY

VOLUME 134. Structural and Contractile Proteins (Part C: The Contractile Apparatus and the Cytoskeleton)
Edited by RICHARD B. VALLEE

Volume 135. Immobilized Enzymes and Cells (Part B)
Edited by KLAUS MOSBACH

Volume 136. Immobilized Enzymes and Cells (Part C)
Edited by KLAUS MOSBACH

Volume 137. Immobilized Enzymes and Cells (Part D)
Edited by KLAUS MOSBACH

Volume 138. Complex Carbohydrates (Part E)
Edited by VICTOR GINSBURG

Volume 139. Cellular Regulators (Part A: Calcium- and Calmodulin-Binding Proteins)
Edited by ANTHONY R. MEANS AND P. MICHAEL CONN

Volume 140. Cumulative Subject Index Volumes 102–119, 121–134

VOLUME 141. Cellular Regulators (Part B: Calcium and Lipids)
Edited by P. MICHAEL CONN AND ANTHONY R. MEANS

Volume 142. Metabolism of Aromatic Amino Acids and Amines
Edited by SEYMOUR KAUFMAN

Volume 143. Sulfur and Sulfur Amino Acids
Edited by WILLIAM B. JAKOBY AND OWEN GRIFFITH

Volume 144. Structural and Contractile Proteins (Part D: Extracellular Matrix)
Edited by LEON W. CUNNINGHAM

Volume 145. Structural and Contractile Proteins (Part E: Extracellular Matrix)
Edited by LEON W. CUNNINGHAM

Volume 146. Peptide Growth Factors (Part A)
Edited by DAVID BARNES AND DAVID A. SIRBASKU

Volume 147. Peptide Growth Factors (Part B)
Edited by DAVID BARNES AND DAVID A. SIRBASKU

Volume 148. Plant Cell Membranes
Edited by LESTER PACKER AND ROLAND DOUCE

Volume 149. Drug and Enzyme Targeting (Part B)
Edited by RALPH GREEN AND KENNETH J. WIDDER

Volume 150. Immunochemical Techniques (Part K: *In Vitro* Models of B and T Cell Functions and Lymphoid Cell Receptors)
Edited by GIOVANNI DI SABATO

Volume 151. Molecular Genetics of Mammalian Cells
Edited by MICHAEL M. GOTTESMAN

Volume 152. Guide to Molecular Cloning Techniques
Edited by SHELBY L. BERGER AND ALAN R. KIMMEL

Volume 153. Recombinant DNA (Part D)
Edited by RAY WU AND LAWRENCE GROSSMAN

Volume 154. Recombinant DNA (Part E)
Edited by RAY WU AND LAWRENCE GROSSMAN

Volume 155. Recombinant DNA (Part F)
Edited by RAY WU

Volume 156. Biomembranes (Part P: ATP-Driven Pumps and Related Transport: The Na,K-Pump)
Edited by SIDNEY FLEISCHER AND BECCA FLEISCHER

Volume 157. Biomembranes (Part Q: ATP-Driven Pumps and Related Transport: Calcium, Proton, and Potassium Pumps)
Edited by SIDNEY FLEISCHER AND BECCA FLEISCHER

Volume 158. Metalloproteins (Part A)
Edited by JAMES F. RIORDAN AND BERT L. VALLEE

Volume 159. Initiation and Termination of Cyclic Nucleotide Action
Edited by JACKIE D. CORBIN AND ROGER A. JOHNSON

Volume 160. Biomass (Part A: Cellulose and Hemicellulose)
Edited by WILLIS A. WOOD AND SCOTT T. KELLOGG

Volume 161. Biomass (Part B: Lignin, Pectin, and Chitin)
Edited by WILLIS A. WOOD AND SCOTT T. KELLOGG

Volume 162. Immunochemical Techniques (Part L: Chemotaxis and Inflammation)
Edited by GIOVANNI DI SABATO

Volume 163. Immunochemical Techniques (Part M: Chemotaxis and Inflammation)
Edited by GIOVANNI DI SABATO

Volume 164. Ribosomes
Edited by HARRY F. NOLLER, JR., AND KIVIE MOLDAVE

Volume 165. Microbial Toxins: Tools for Enzymology
Edited by SIDNEY HARSHMAN

Volume 166. Branched-Chain Amino Acids
Edited by ROBERT HARRIS AND JOHN R. SOKATCH

Volume 167. Cyanobacteria
Edited by LESTER PACKER AND ALEXANDER N. GLAZER

Volume 168. Hormone Action (Part K: Neuroendocrine Peptides)
Edited by P. MICHAEL CONN

Volume 169. Platelets: Receptors, Adhesion, Secretion (Part A)
Edited by JACEK HAWIGER

Volume 170. Nucleosomes
Edited by PAUL M. WASSARMAN AND ROGER D. KORNBERG

Volume 171. Biomembranes (Part R: Transport Theory: Cells and Model Membranes)
Edited by SIDNEY FLEISCHER AND BECCA FLEISCHER

Volume 172. Biomembranes (Part S: Transport: Membrane Isolation and Characterization)
Edited by SIDNEY FLEISCHER AND BECCA FLEISCHER

Volume 173. Biomembranes [Part T: Cellular and Subcellular Transport: Eukaryotic (Nonepithelial) Cells]
Edited by SIDNEY FLEISCHER AND BECCA FLEISCHER

Volume 174. Biomembranes [Part U: Cellular and Subcellular Transport: Eukaryotic (Nonepithelial) Cells]
Edited by SIDNEY FLEISCHER AND BECCA FLEISCHER

Volume 175. Cumulative Subject Index Volumes 135–139, 141–167

VOLUME 176. Nuclear Magnetic Resonance (Part A: Spectral Techniques and Dynamics)
Edited by NORMAN J. OPPENHEIMER AND THOMAS L. JAMES

Volume 177. Nuclear Magnetic Resonance (Part B: Structure and Mechanism)
Edited by NORMAN J. OPPENHEIMER AND THOMAS L. JAMES

Volume 178. Antibodies, Antigens, and Molecular Mimicry
Edited by JOHN J. LANGONE

Volume 179. Complex Carbohydrates (Part F)
Edited by VICTOR GINSBURG

Volume 180. RNA Processing (Part A: General Methods)
Edited by JAMES E. DAHLBERG AND JOHN N. ABELSON

Volume 181. RNA Processing (Part B: Specific Methods)
Edited by JAMES E. DAHLBERG AND JOHN N. ABELSON

Volume 182. Guide to Protein Purification
Edited by Murray P. Deutscher

Volume 183. Molecular Evolution: Computer Analysis of Protein and Nucleic Acid Sequences
Edited by Russell F. Doolittle

Volume 184. Avidin–Biotin Technology
Edited by Meir Wilchek and Edward A. Bayer

Volume 185. Gene Expression Technology
Edited by David V. Goeddel

Volume 186. Oxygen Radicals in Biological Systems (Part B: Oxygen Radicals and Antioxidants)
Edited by Lester Packer and Alexander N. Glazer

Volume 187. Arachidonate Related Lipid Mediators
Edited by Robert C. Murphy and Frank A. Fitzpatrick

Volume 188. Hydrocarbons and Methylotrophy
Edited by Mary E. Lidstrom

Volume 189. Retinoids (Part A: Molecular and Metabolic Aspects)
Edited by Lester Packer

Volume 190. Retinoids (Part B: Cell Differentiation and Clinical Applications)
Edited by Lester Packer

Volume 191. Biomembranes (Part V: Cellular and Subcellular Transport: Epithelial Cells)
Edited by Sidney Fleischer and Becca Fleischer

Volume 192. Biomembranes (Part W: Cellular and Subcellular Transport: Epithelial Cells)
Edited by Sidney Fleischer and Becca Fleischer

Volume 193. Mass Spectrometry
Edited by James A. McCloskey

Volume 194. Guide to Yeast Genetics and Molecular Biology
Edited by Christine Guthrie and Gerald R. Fink

Volume 195. Adenylyl Cyclase, G Proteins, and Guanylyl Cyclase
Edited by Roger A. Johnson and Jackie D. Corbin

Volume 196. Molecular Motors and the Cytoskeleton
Edited by Richard B. Vallee

Volume 197. Phospholipases
Edited by Edward A. Dennis

Volume 198. Peptide Growth Factors (Part C)
Edited by David Barnes, J. P. Mather, and Gordon H. Sato

Volume 199. Cumulative Subject Index Volumes 168–174, 176–194

VOLUME 200. Protein Phosphorylation (Part A: Protein Kinases: Assays, Purification, Antibodies, Functional Analysis, Cloning, and Expression)
Edited by TONY HUNTER AND BARTHOLOMEW M. SEFTON

Volume 201. Protein Phosphorylation (Part B: Analysis of Protein Phosphorylation, Protein Kinase Inhibitors, and Protein Phosphatases)
Edited by TONY HUNTER AND BARTHOLOMEW M. SEFTON

VOLUME 202. Molecular Design and Modeling: Concepts and Applications (Part A: Proteins, Peptides, and Enzymes)
Edited by JOHN J. LANGONE

VOLUME 203. Molecular Design and Modeling: Concepts and Applications (Part B: Antibodies and Antigens, Nucleic Acids, Polysaccharides, and Drugs)
Edited by JOHN J. LANGONE

VOLUME 204. Bacterial Genetic Systems
Edited by JEFFREY H. MILLER

VOLUME 205. Metallobiochemistry (Part B: Metallothionein and Related Molecules)
Edited by JAMES F. RIORDAN AND BERT L. VALLEE

VOLUME 206. Cytochrome P450
Edited by MICHAEL R. WATERMAN AND ERIC F. JOHNSON

VOLUME 207. Ion Channels
Edited by BERNARDO RUDY AND LINDA E. IVERSON

VOLUME 208. Protein–DNA Interactions
Edited by ROBERT T. SAUER

VOLUME 209. Phospholipid Biosynthesis
Edited by EDWARD A. DENNIS AND DENNIS E. VANCE

VOLUME 210. Numerical Computer Methods
Edited by LUDWIG BRAND AND MICHAEL L. JOHNSON

VOLUME 211. DNA Structures (Part A: Synthesis and Physical Analysis of DNA)
Edited by DAVID M. J. LILLEY AND JAMES E. DAHLBERG

VOLUME 212. DNA Structures (Part B: Chemical and Electrophoretic Analysis of DNA)
Edited by DAVID M. J. LILLEY AND JAMES E. DAHLBERG

VOLUME 213. Carotenoids (Part A: Chemistry, Separation, Quantitation, and Antioxidation)
Edited by LESTER PACKER

VOLUME 214. Carotenoids (Part B: Metabolism, Genetics, and Biosynthesis)
Edited by LESTER PACKER

VOLUME 215. Platelets: Receptors, Adhesion, Secretion (Part B)
Edited by JACEK J. HAWIGER

VOLUME 216. Recombinant DNA (Part G)
Edited by RAY WU

VOLUME 217. Recombinant DNA (Part H)
Edited by RAY WU

VOLUME 218. Recombinant DNA (Part I)
Edited by RAY WU

VOLUME 219. Reconstitution of Intracellular Transport
Edited by JAMES E. ROTHMAN

VOLUME 220. Membrane Fusion Techniques (Part A)
Edited by NEJAT DÜZGÜNEŞ

VOLUME 221. Membrane Fusion Techniques (Part B)
Edited by NEJAT DÜZGÜNEŞ

VOLUME 222. Proteolytic Enzymes in Coagulation, Fibrinolysis, and Complement Activation (Part A: Mammalian Blood Coagulation Factors and Inhibitors)
Edited by LASZLO LORAND AND KENNETH G. MANN

VOLUME 223. Proteolytic Enzymes in Coagulation, Fibrinolysis, and Complement Activation (Part B: Complement Activation, Fibrinolysis, and Nonmammalian Blood Coagulation Factors)
Edited by LASZLO LORAND AND KENNETH G. MANN

VOLUME 224. Molecular Evolution: Producing the Biochemical Data
Edited by ELIZABETH ANNE ZIMMER, THOMAS J. WHITE, REBECCA L. CANN, AND ALLAN C. WILSON

VOLUME 225. Guide to Techniques in Mouse Development
Edited by PAUL M. WASSARMAN AND MELVIN L. DEPAMPHILIS

VOLUME 226. Metallobiochemistry (Part C: Spectroscopic and Physical Methods for Probing Metal Ion Environments in Metalloenzymes and Metalloproteins)
Edited by JAMES F. RIORDAN AND BERT L. VALLEE

VOLUME 227. Metallobiochemistry (Part D: Physical and Spectroscopic Methods for Probing Metal Ion Environments in Metalloproteins)
Edited by JAMES F. RIORDAN AND BERT L. VALLEE

VOLUME 228. Aqueous Two-Phase Systems
Edited by HARRY WALTER AND GÖTE JOHANSSON

VOLUME 229. Cumulative Subject Index Volumes 195–198, 200–227

VOLUME 230. Guide to Techniques in Glycobiology
Edited by WILLIAM J. LENNARZ AND GERALD W. HART

VOLUME 231. Hemoglobins (Part B: Biochemical and Analytical Methods)
Edited by JOHANNES EVERSE, KIM D. VANDEGRIFF, AND ROBERT M. WINSLOW

VOLUME 232. Hemoglobins (Part C: Biophysical Methods)
Edited by JOHANNES EVERSE, KIM D. VANDEGRIFF, AND ROBERT M. WINSLOW

VOLUME 233. Oxygen Radicals in Biological Systems (Part C)
Edited by LESTER PACKER

VOLUME 234. Oxygen Radicals in Biological Systems (Part D)
Edited by LESTER PACKER

VOLUME 235. Bacterial Pathogenesis (Part A: Identification and Regulation of Virulence Factors)
Edited by VIRGINIA L. CLARK AND PATRIK M. BAVOIL

VOLUME 236. Bacterial Pathogenesis (Part B: Integration of Pathogenic Bacteria with Host Cells)
Edited by VIRGINIA L. CLARK AND PATRIK M. BAVOIL

VOLUME 237. Heterotrimeric G Proteins
Edited by RAVI IYENGAR

VOLUME 238. Heterotrimeric G-Protein Effectors
Edited by RAVI IYENGAR

VOLUME 239. Nuclear Magnetic Resonance (Part C)
Edited by THOMAS L. JAMES AND NORMAN J. OPPENHEIMER

VOLUME 240. Numerical Computer Methods (Part B)
Edited by MICHAEL L. JOHNSON AND LUDWIG BRAND

VOLUME 241. Retroviral Proteases
Edited by LAWRENCE C. KUO AND JULES A. SHAFER

VOLUME 242. Neoglycoconjugates (Part A)
Edited by Y. C. LEE AND REIKO T. LEE

VOLUME 243. Inorganic Microbial Sulfur Metabolism
Edited by HARRY D. PECK, JR., AND JEAN LEGALL

VOLUME 244. Proteolytic Enzymes: Serine and Cysteine Peptidases
Edited by ALAN J. BARRETT

VOLUME 245. Extracellular Matrix Components
Edited by E. RUOSLAHTI AND E. ENGVALL

VOLUME 246. Biochemical Spectroscopy
Edited by KENNETH SAUER

VOLUME 247. Neoglycoconjugates (Part B: Biomedical Applications)
Edited by Y. C. LEE AND REIKO T. LEE

VOLUME 248. Proteolytic Enzymes: Aspartic and Metallo Peptidases
Edited by ALAN J. BARRETT

VOLUME 249. Enzyme Kinetics and Mechanism (Part D: Developments in Enzyme Dynamics)
Edited by DANIEL L. PURICH

VOLUME 250. Lipid Modifications of Proteins
Edited by PATRICK J. CASEY AND JANICE E. BUSS

VOLUME 251. Biothiols (Part A: Monothiols and Dithiols, Protein Thiols, and Thiyl Radicals)
Edited by LESTER PACKER

VOLUME 252. Biothiols (Part B: Glutathione and Thioredoxin; Thiols in Signal Transduction and Gene Regulation)
Edited by LESTER PACKER

VOLUME 253. Adhesion of Microbial Pathogens
Edited by RON J. DOYLE AND ITZHAK OFEK

VOLUME 254. Oncogene Techniques
Edited by PETER K. VOGT AND INDER M. VERMA

VOLUME 255. Small GTPases and Their Regulators (Part A: Ras Family)
Edited by W. E. BALCH, CHANNING J. DER, AND ALAN HALL

VOLUME 256. Small GTPases and Their Regulators (Part B: Rho Family)
Edited by W. E. BALCH, CHANNING J. DER, AND ALAN HALL

VOLUME 257. Small GTPases and Their Regulators (Part C: Proteins Involved in Transport)
Edited by W. E. BALCH, CHANNING J. DER, AND ALAN HALL

VOLUME 258. Redox-Active Amino Acids in Biology
Edited by JUDITH P. KLINMAN

VOLUME 259. Energetics of Biological Macromolecules (in preparation)
Edited by MICHAEL L. JOHNSON AND GARY K. ACKERS

VOLUME 260. Mitochondrial Biogenesis and Genetics, Part A (in preparation)
Edited by GIUSEPPE M. ATTARDI AND ANNE CHOMYN

VOLUME 261. Nuclear Magnetic Resonance and Nucleic Acids (in preparation)
Edited by THOMAS L. JAMES

VOLUME 262. DNA Replication (in preparation)
Edited by JUDITH L. CAMPBELL

VOLUME 263. Plasma Lipoproteins (Part C: Quantitation) (in preparation)
Edited by WILLIAM A. BRADLEY, SANDRA H. GIANTURCO, AND JERE P. SEGREST

[1] Precursors of Quinone Tanning: Dopa-Containing Proteins

By J. HERBERT WAITE

Introduction

Dopa-containing proteins are proteins that contain the amino acid 3,4-dihydroxyphenyl-L-alanine in their primary sequence. They appear to be rather widely distributed in nature and have been reported in at least five different animal phyla.[1] Although the functions attributed to dopa-containing proteins are still speculative, they are commonly associated with extraorganismic structural materials such as helminth egg cases,[2] molluscan byssus,[3,4] annelid cement,[5] and ascidian tunic.[6] Dopa proteins are usually cosecreted with stoichiometric quantities of the enzyme catechol oxidase (EC 1.10.3.1)[1] which is known to catalyze the oxidative dehydrogenation of peptidyl-dopa groups to peptidyl-dopaquinones. There is little agreement about what follows the generation of the quinones. Suffice it to say for the present that the appearance of the quinones leads to a rapid discoloration and sclerotization of the proteins, rendering them intractable to all but hydrolytic treatments. The term "quinone tanning" has been attached to describe this chemical and physical transformation. Many have argued that quinone tanning results from the formation of quinone-derived intermolecular cross-links,[7] but this point has so far eluded biochemical verification.[8]

Comparison of dopa protein sequences with their corresponding cDNA-derived sequences suggests that peptidyl-dopa owes its existence to the action of putative co- or posttranslational modifications on peptidyl tyrosines.[9–11] Nothing is known about these putative hydroxylases except that

[1] J. H. Waite, *Comp. Biochem. Physiol. B* **97B,** 19 (1990).
[2] J. H. Waite and A. M. Rice-Ficht, *Biochemistry* **26,** 7819 (1987).
[3] J. H. Waite, *J. Biol. Chem.* **258,** 2911 (1983).
[4] L. M. Rzepecki, K. M. Mueller, and J. H. Waite, *Biol. Bull.* (*Woods Hole, Mass.*) **183,** 123 (1992).
[5] J. H. Waite, R. A. Jensen, and D. E. Morse, *Biochemistry* **31,** 5733 (1992).
[6] L. C. Dorsett, C. J. Hawkins, J. A. Grice, M. F. Lavin, P. M. Merefield, D. L. Parry, and I. L. Ross, *Biochemistry* **26,** 8078 (1986).
[7] R. Davies and J. L. Frahn, *J. Chem. Soc., Perkin Trans. 1*, p. 2295 (1977).
[8] S. M. Holl, D. Hansen, J. H. Waite, and J. Schaefer, *Arch. Biochem. Biophys.* **302,** 255 (1993).
[9] D. R. Filpula, S.-M. Lee, R. P. Link, S. L. Strausberg, and R. L. Strausberg, *Biotechnol. Prog.* **6,** 171 (1990).
[10] R. A. Laursen, *Results Prob. Cell Differ.* **19,** 55 (1992).

the hydroxylation may be autocatalytic.[12] Because of the wide variety of compositions, primary sequences, and molecular weights in the dopa-containing proteins, there is no single purification strategy that applies to all such proteins. There are, however, two caveats that apply generally to working with these proteins:

1. Dopa is oxidatively unstable at physiological pH and above.[13] It is therefore critical to work at low pH whenever possible; when a pH > 7 is necessary, use borate (which complexes dopa) or an antioxidant such as ascorbate.[14]

2. The diphenolic moiety of dopa has an extremely high affinity for metals, particularly iron(III) and aluminum(III). This applies whether the metal atom is present as a cation, oxide, mineral, or metal.[15–17] Avoid exposing the protein to metals if dopa recovery is desired.

The purpose of this chapter is to outline some useful methodology for the isolation and characterization of two different groups of dopa proteins: (A) proteins with tandemly repeated sequences from the common mussel *Mytilus edulis*, and (B) protein precursors of the eggshells of the liver fluke *Fasciola hepatica*. Some techniques with regard to dopa-containing proteins have been previously described in this series[18] and will not be repeated here, but rather referred to whenever pertinent.

Procedures for Extracting and Isolating Dopa-Containing Proteins

Extraction of M. edulis Foot Proteins

Dopa-containing byssal precursor proteins are stockpiled in the mussel foot, a small pigmented appendage in the center of the animal. The location of the proteins has led to the name "*Mytilus edulis* foot proteins" and the acronym Mefp. Any plans for the preparation of dopa-containing Mefps must include a mussel shucking party for the purpose of accumulating a 100 or more mussel feet which can be stored indefinitely at −80°. *M. edulis*,

[11] A. C. Rice-Ficht, K. A. Dusek, G. J. Kochevar, and J. H. Waite, *Mol. Biochem. Parasitol.* **54,** 129 (1992).

[12] A. Åberg, M. Ormö, P. Norlund, and B.-M. Sjöberg, *Biochemistry* **32,** 9845 (1993).

[13] T. Sarna, A. Duleba, W. Korytowski, and H. Swartz, *Arch. Biochem. Biophys.* **200,** 140 (1980).

[14] J. H. Waite, *Anal. Chem.* **56,** 1935 (1984).

[15] M. B. McBride and L. G. Wesselink, *Environ. Sci. Technol.* **22,** 703 (1988).

[16] S. W. Taylor, C. J. Hawkins, and D. J. Winzor, *Inorg. Chem.* **32,** 422 (1993).

[17] V. K. F. Chia, M. P. Soriaga, A. T. Hubbard, and S. E. Anderson, *J. Phys. Chem.* **87,** 232 (1983).

[18] J. H. Waite and C. V. Benedict, this series, Vol. 107, p. 397.

colloquially called "blue," "common," or "sea" mussels, should have a shell length of at least 4 cm and can be obtained either by collection from the rocky intertidal zone or from seafood wholesalers. The latter course is recommended for neophytes. Mussels obtained from a wholesaler should be examined for vigor before use. The "litmus test" for mussel vigor is a simple one based on gaping. When mussels are maintained out of sea water, they gape, i.e., their valves are ajar. If they close their valves promptly when tapped or pinched, then they are in fine fettle. If not, they are moribund and should be discarded. If mussels are to be obtained by collection from the field, some care is required to sort the different species that can coexist in natural clusters. This is particularly true of the U.S. west coast where as many as three species can occur in coaggregated clusters.

The following procedure has been jointly developed by laboratories in the United States and Chile.[19,20] It appears to be applicable to the extraction of homologous proteins from a variety of marine mussel species.[19] On occasion, however, it has been necessary to "tweek" the acetone concentration used to precipitate efficiently the target protein(s). This was the case for dopa protein from the hairy mussel *Trichomya hirsuta*.[19] The procedure does not work for byssal precursor proteins from the zebra mussel *Dreissena polymorpha*.[21]

Materials A

0.7% (w/v) perchloric acid (PCA) cooled to 4°
Ethanol–water ice bath
Concentrated sulfuric acid
Acetone cooled to −80°
5% acetic acid cooled to 4°
Equipment: Waring blender, deep freezer, refrigerated preparative centrifuge, magnetic stirrer, small tissue grinder (Kontes, Vineland, NJ)

Procedure A. The following protocol for isolating Mepf-1 and -2 can be scaled according to the wet weight of tissue to be extracted. About 20 to 30 g of feet (wet weight) can be obtained from 100 mussels depending on size. About 30 g feet are placed in 300 ml ice-cold 0.7% PCA and homogenized for 1 min to a smooth puree using a small Waring blender at top speed. The puree is centrifuged at 30,000*g* for 30 min, and the supernatant (S-1) is carefully decanted and its volume recorded. S-1 is

[19] L. M. Rzepecki, X.-X. Qin, J. H. Waite, and M. F. Lavin, *Mar. Mol. Biol. Biotechnol.* **1,** 123 (1991).
[20] J. Pardo, E. Gutierrez, C. Saez, M. Brito, and L. O. Burzio, *Protein Express. Purif.* **1,** 147 (1990).
[21] L. M. Rzepecki and J. H. Waite, *Mar. Mol. Biol. Biotechnol.* **2,** 267 (1993).

transferred to a 1-liter beaker, chilled in an ethanol–water ice bath (>0°), and vigorously stirred on a magnetic stirrer. While stirring, concentrated sulfuric acid is added (vol = S-1 × 0.0168) to give a final concentration of 0.3 *M*. This is followed by the dropwise addition of prechilled acetone (2 × S-1 vol) using a large separatory funnel. When all the acetone has been added, the suspension is left to sit for 15–20 min. After redispersing the precipitate with gentle stirring, it is centrifuged for 30 min at 30,000*g* as before. Do not use polycarbonate centrifuge tubes with the acetone-containing supernatants. The second supernatant is decanted and can be discarded in most cases. The pellets are left to drain upside down on ice and are then resuspended by 0.5 ml vol of 5% acetic acid (S-3) and rehomogenized in small (2 ml) tissue grinders (Kontes, Vineland, NJ). The samples are spun at 15,000*g* for 5 min in a microfuge, and are then ready for further purification by gel filtration and high-performance liquid chromatography (HPLC).

Extraction of Vitelline Proteins from the Liver Fluke

Liver flukes (*F. hepatica*) are hermaphroditic, continually active sexually, and prolific egg layers producing some 30,000 eggs per day in sheep.[22] The eggs are protected by an organic shell or varnish that is derived from proteins produced in the vitelline gland network. For obvious reasons, the vitelline gland-extracted proteins are called "vitelline proteins" or Vps.[22] In some parts of the world, e.g., southwest United States, liver fluke infection of livestock can reach epidemic proportions. Public slaughterhouses in these regions are the best places to collect liver flukes. Freshness of material is absolutely essential when working with trematodes. This is in part due to the profusion of proteolytic enzymes in the flukes. Even a few minutes at room temperature can lead to a significant autolysis of animal tissues.

Materials B

Neutral salt buffer: 0.15 *M* NaCl, 0.05 *M* Tris(hydroxymethyl)aminomethane hydrochloride (Tris–HCl), pH 7.5, with 1 m*M* phenylmethylsulfonyl fluoride, 10 m*M* *N*-ethylmaleimide, 25 m*M* ethylenediaminetetracetic acid (EDTA), and 1 m*M* potassium cyanide
5% (v/v) acetic acid with 4 *M* urea, 10 μM leupeptin, and 1 m*M* iodoacetamide chilled to 4°
5% acetic acid with 8 *M* urea chilled to 4°
2.5% acetic acid chilled to 4°
Equipment: Deep freezer, refrigerated preparative centrifuge, large tissue grinders (Kontes), magnetic stirrer

[22] J. H. Waite and A. M. Rice-Ficht, *Biochemistry* **28,** 6104 (1989).

Procedure B. The following can be used to purify dopa-containing vitelline proteins B and C: Worms collected from condemned livers at the abattoir are immediately dropped into liquid nitrogen for transport and storage. Prior to extraction, 15- to 20-g (wet weight) worms are briefly thawed, spread out on 5-mm-thick glass plates, and deep frozen at −80°. The plate strategy allows for the selective dissection of the vitellaria from the rest of the animal.[22] The vitellaria contain the eggshell precursors and are located in the posterior third and peripheral 3–5 mm of the animal. The dissective separation of the vitellaria from the more interior digestive diverticulum is advisable to minimize protein degradation (Fig. 1). Once dissected, the vitellaria are homogenized in large frosted glass tissue grinders using the neutral salt buffer (10 ml/g wet weight). The homogenate is centrifuged at 5000*g* at 4° for 10 min, and pellets are collected, rehomogenized in neutral salt buffer, and centrifuged as before. The second pellet is carefully collected and rehomogenized in cold 5% acetic acid with 4 *M* urea and inhibitors (5 ml/g wet weight). The inhibitors, particularly leupeptin, are indispensible for the efficacy of this step.[22] If, in time, yield of the dopa protein decreases, it is quite possible that the leupeptin has begun to deteriorate. Homogenates are centrifuged at 35,000*g* at 4° for 30 min. On collection, supernatants are either stored frozen at −20° or brought to 10% (w/v) ammonium sulfate at room temperature, stirred an additional 40 min at 10°, and allowed to sit without stirring for 10 min. The precipitate is removed by centrifugation at 5000*g* for 30 min. The ammonium sulfate concentration is then raised to 20% (w/v) at room temperature and stirred as before. The precipitate is removed again by centrifugation. The supernatant is collected and dialyzed against at least 200 vol of 2.5% acetic acid at

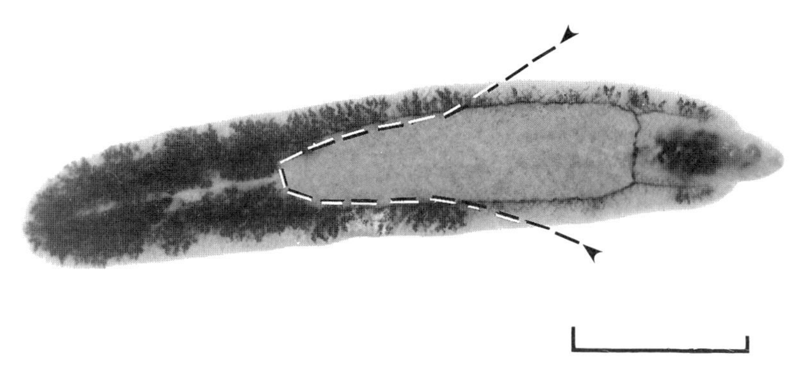

FIG. 1. A whole mount of *Fasciola* stained with Arnow's reagent[23] to visualize the vitellaria. Dashes denote the approximate line of incision. Only the vitelline-rich lower third is retained. The scale bar is equal to 1 cm.

TABLE I
PURIFICATION TABLE FOR *Mytilus edulis* FOOT PROTEIN NO. 1 (Mefp-1) BASED ON THE ARNOW AND BRADFORD DETECTION METHODS FOR DOPA AND PROTEIN, RESPECTIVELY

Step	Total protein (mg)	Total dopa (μmol)	Dopa/protein (μmol/mg)	Yield (%)	Fold purification
From 35 g of feet					
S-1, perchloric acid	221.0	75.0	0.34	100	1.0
S-3, acetone ppt	70.2	43.8	0.62	58	1.2
Sephadex G-200	30.9	32.1	1.04	43	3.1
C-8 HPLC	13.8	16.2	1.17	21	3.5

4° using dialysis tubing with a molecular weight cutoff of 1000 (Spectrum Industries, Los Angeles, CA). Upon equilibration, protein in the tubing flocculates and can be harvested by gentle centrifugation at 10,000*g* for 30 min in a swinging bucket rotor. Centrifuge tubes can be drained upside down after discarding the supernatant, and the precipitate at the bottom can be redissolved in a small volume of 5% acetic acid with 4–8 *M* urea. It should be vigorously microfuged (5 min at 15,000*g*) before purification by HPLC.

Methods for Detection of Dopa Proteins

Enzyme purifications are commonly evaluated by changes in specific activity to ascertain yields and folds of purification. This can be applied to dopa-containing proteins using specific assays for dopa and protein and determining a dopa/protein ratio.[5,19] The assay of choice for dopa is that of Arnow[23] described previously in this series.[18] For protein, it is the Bradford assay[24] available as a kit from Bio-Rad (Richmond, CA). Table I itemizes typical dopa/protein ratios for the purification of Mefp-1. Two features are notable: the apparent yield is low, and the fold purification is little more than three times that of the PCA extract. The apparent yield is low because there are several different dopa-containing proteins extracted by PCA. Mefp-2, in fact, is much more abundant than Mefp-1. When it is removed from the latter by chromatography on Sephadex G-200, the total dopa level decreases sharply. Fold purification is low because PCA is very selective in that only a few proteins are extracted to begin with. If the dopa/protein ratio of the intact mussel tissue could be estimated, it would likely be a small fraction of the PCA "specific activity." Finally, it is important to

[23] L. E. Arnow, *J. Biol. Chem.* **118,** 531 (1937).
[24] M. M. Bradford, *Anal. Biochem.* **72,** 248 (1976).

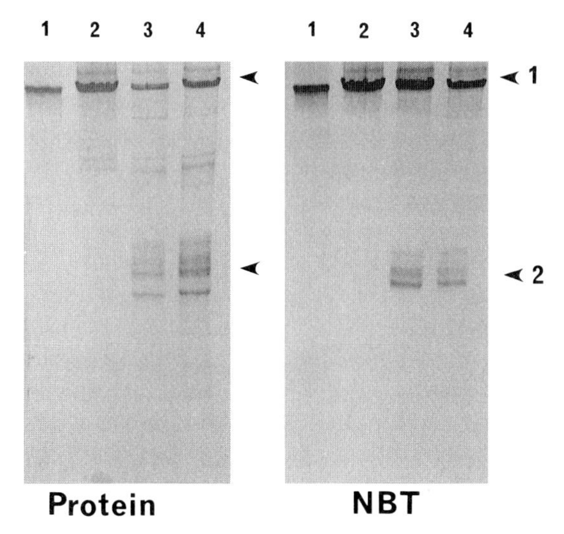

FIG. 2. Acetic acid-urea polyacrylamide gel electrophoresis of samples taken from different steps in the purification of Mefp-1. The protein panel was stained with Coomassie blue R-250, whereas the NBT panel was stained with nitroblue tetratrazolium in glycinate. Lane 1, 10 μg HPLC-purified protein; lane 2, 10 μg first peak Sephadex G-200; lane 3, 10 μg of S-3; and lane 4, 10 μg of S-1. See Table I for dopa/protein ratios. Arrows denote positions of Mefp-1 and -2.

point out that since there are often several different kinds of dopa-containing proteins in the same tissues, the dopa/protein ratio, while useful, cannot be relied upon per se to reflect the degree of purity. Purity can, however, be assessed when dopa/protein ratios are coupled to results following electrophoresis on acid-urea polyacrylamide gels. The application of gel electrophoresis to dopa-containing proteins has been described earlier in this series.[18] In short, a purification strategy would entail sampling each critical step, e.g., those listed in Table I, by acid-urea polyacrylamide gel electrophoresis. Proteins can be detected by staining with Coomassie blue R-250 and dopa, by a modification of the Arnow assay,[18] or by redox cycling[25] (Fig. 2). Redox cycling reactions involving nitroblue tetrazolium (NBT) and glycinate are not as specific for *o*-diphenols as the Arnow reaction, but they are much more sensitive and produce a more stable color.[26] Acid-urea gels can be quickly and directly stained with NBT. (Sigma Chemical Co.) as follows: Prepare 1 liter of 2 *M* glycinate (sodium salt),

[25] M. A. Paz, R. Flückinger, A. Boak, H. Kagan, and P. M. Gallop, *J. Biol. Chem.* **266,** 689 (1991).

[26] R. Flückiger, M. A. Paz, and P. M. Gallop, this volume [11].

pH 10. Dissolve 5 mg of NBT in 1 ml of methanol in a 25-ml graduated cylinder. Bring the volume up to 25 ml with glycinate and mix. After electrophoresis, place the acid-urea gel (volume $\leq$7 ml) in a small pan of transparent material and add 25 ml NBT–glycinate cocktail to cover the gel. Place the pan on an orbital shaker at room temperature for up to 1 hr for color to develop. Wash excess glycinate from the gel with deionized water. The reaction can be stopped by keeping the gel in 1 *M* sodium borate, pH 10.[4]

Chromatography

Affinity Chromatography

Use of phenyl boronate immobilized to agarose (PBA; Glyco-Gel B, Pierce Chemical Co., and Matrix PBA, Amicon Corp.) has been reported as an affinity matrix for the purification of ferreascidin, a dopa-containing protein from hemocytes of the tunicate *Pyura stolonifera.*[27] Binding to phenylboronate would seem to be dictated by the same condition that leads to formation of a soluble complex between *o*-diphenols and borate, that is, at pH $>$ 8. In fact, phenylboronate is considerably less predictable. Protein must be soluble and sufficiently stable in a nonborate-containing buffer at pH $\geq$ 7.5 to allow loading and binding to the column. This is not fulfilled by Mefp-1 and -2 or VpB and C, which precipitate or adsorb irreversibly at or above neutral pH. Ferreascidin is loaded and bound to Matrix PBA-10 in 0.25 *M* ammonium acetate, pH 8.5. After nonbinding proteins are washed out with the same buffer, those coupled by diols other than *o*-diphenols are eluted with 0.1 *M* Tris with 0.2 *M* sorbitol and 10 m*M* EDTA, pH 8.5. After a brief wash with more of the first buffer to remove sorbitol, ferreascidin is recovered by a pH jump, i.e., eluting PBA at lower pH, using 1 *M* sodium acetate, pH 3.0. PBA is more broadly applicable to the isolation of dopa peptides. However, even with these there have been unexpected results. After loading trypsin-digested VpB to Glyco-Gel B, for example, Waite and Rice-Ficht[2] report on the acid elution of a mere handful of dopa-containing peptides. More than 20 or so other dopa-containing peptides do not bind.[28] Since most of the latter are acidic, nonbinding may be a consequence of the mutual repulsion between boronate and acidic groups on the peptides.

[27] C. J. Hawkins, M. F. Lavin, D. L. Parry, and I. L. Ross, *Anal. Biochem.* **159,** 187 (1986).
[28] J. H. Waite and A. M. Rice-Ficht, *Mol. Biochem. Parasitol.* **54,** 143 (1992).

Gel Filtration

Reversed-phase HPLC is a godsend for the purification of many dopa-containing proteins that are otherwise too adsorptive to be recoverable from other chromatographic media. Indeed, C-8 reversed-phase HPLC is the only chromatography necessary for *Fasciola* vitelline proteins B and C. For mussel proteins, however, a prior application to Sephadex G-200 or -150 is helpful to improve separation between Mefp-1 and -2.[4] For an extraction from 30 g mussel feet, be prepared to fill a large column (3 × 100 cm) with Sephadex G-200 (20–30 ml/g dry gel) preswollen in an excess volume of 5% acetic acid for at least 72 hr at room temperature. Degas the swollen resin for at least 30 min with a vacuum aspirator. After pouring the resin into the column, elute with at least two column volumes of 5% acetic acid at room temperature before loading the protein. Flow should be between 1 and 3 ml/hr. Eluant is monitored at 280 nm, and Mefp-1 can be expected to elute immediately following the void volume. The resin is best used only once. Column fractions absorbing at 280 nm and reacting with Arnow's method should be subjected to acid-urea polyacrylamide gel electrophoresis. Those with protein staining with Coomassie blue R-250, NBT, and comigrating with Mefp-1 should be pooled, frozen at −80°, and lyophilized prior to HPLC.

High Performance Liquid Chromatography

Materials

HPLC gradient system with UV detector

C-8 analytical or preparative reversed-phase column such as Brownlee Aquapore RP-300 (always include in-line prefilter and C-8 guard column)

Solvent A [0.1% (v/v) trifluoroacetic acid made with Milli-Q water that has been previously filtered through 0.45-μm nylon-66 filter and degassed]

Solvent B (HPLC-grade acetonitrile with 0.1% trifluoroacetic acid)

Sample loading syringes (100 and 2000 μl)

Fraction collector

Procedure. Reversed-phase HPLC has played a critical role in the purification of more than 10 different dopa-containing proteins.[2,4,5,19,22,29] Running conditions in all of these are quite similar. At the pre-HPLC stage, dopa-containing proteins, regardless of source, are usually found in a small volume (1–2 ml) of 5% acetic acid with or without urea. This solution

[29] J. H. Waite, T. J. Housley, and M. L. Tanzer, *Biochemistry* **24,** 5010 (1985).

should be spun for 5–10 min at 15,000g on a microfuge such as the Eppendorf before loading to HPLC. If the sample contains 1 or more mg protein/ml, an aliquot of 10–50 μg should be sacrificed for a pilot run. Most elution programs for dopa-containing proteins consist of three successive continuous gradients of solvent B: 0 to 15–19% B in the first 10 min, to 30–40% B in the following 60 min, and finishing at 90–100% B before returning to 0% B. The proteins generally elute in the middle shallow gradient. Figure 3 illustrates the relative elution position of Mefp-1 and -2 (mussel-derived) and VpB/VpC (liver fluke-derived) by the samc program. Results will vary somewhat from one column manufacturer to another and from even one lot to another. As before, fractions absorbing at 280 nm are quick-frozen in a methanol–dry ice bath, and then lyophilized overnight on a −80° freeze drier. The freeze-dried residue can be dissolved in a small volume of distilled or deionized water and stored at −20°. At this stage, the protein is highly purified. It can now be sequenced, digested, hydrolyzed, or electrophoresed. Acid-urea polyacrylamide gel electrophoresis is the method of choice for assessing the number of impurities in each fraction (Fig. 1). Amino acid analysis of freeze-dried hydrolyzed HPLC fractions can, however, point to compositional disparities not evident by electrophoresis. This is particularly true of dopa-containing proteins. In VpB,[2] as in several of the mytilid foot proteins,[19,21] there are complementary dopa–tyrosine gradients across the HPLC peak, with dopa highest in fractions from the leading edge of the HPLC peak and tyrosine highest in fractions of the trailing edge (Table III). The meaning of this with respect to protein modification is unclear.

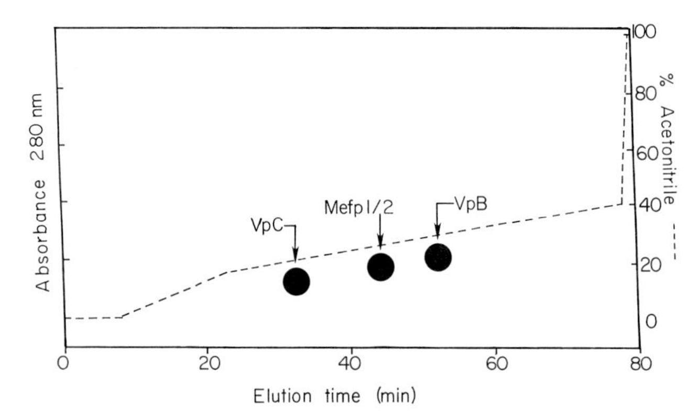

FIG. 3. Elution positions of various dopa-containing proteins by preparative reversed-phase C-8 HPLC. The elution program consisted of 0–16% acetonitrile in 16 min then increasing to 33% acetonitrile at 53 min. The column was stripped with 100% acetonitrile at 58 min. Flow rate is 1 ml/min. Column is at ambient temperature.

Protein Digestion and Peptide Purification

Materials

1–5 mg pure dopa protein
Pure protease (1:50 by weight with the sample protein)
5 ml Reacti-Vial with stir bar (Pierce Chemicals, Rockford, IL)
Magnetic stirrer
Buffers (see below)
Reversed-phase HPLC (as described earlier)

Comment. Most dopa-containing proteins can be digested by commercially available pure enzymes. Those from *Mytilus* and *Fasciola* are rich in basic amino acids such as lysine and are particularly labile to enzymes such as trypsin and lysine C-endopeptidase. Chymotrypsin also digests some of the proteins and can cleave on the C-side of dopa under some conditions.[22] Since dopa residues tend to become oxidized at pH values >6, some special precautions must be observed to minimize oxidative losses during digestion. The following recipes have been used to do this:

1. 0.1 *M* Tris–HCl at pH 7.5 to 8.0 under 40 psi N_2. This attempts to reduce dopa oxidation by imposing a nitrogen-rich environment. There are problems when the protein is not soluble at this pH.
2. 0.10 to 0.20 *M* sodium borate at pH 8.0–8.5. This reduces dopa oxidation by the latter's complexation with borate which occurs favorably between pH 8 and 10.[14] Another consequence of complexation is that the protein charge changes by virtue of the borate, conferring a negative charge on previously uncharged dopa. Complications arise if complexation in itself modifies the target site, e.g., the sequence Lys–dopa is cleaved by trypsin in the absence but not presence of borate.[28] Moreover, chymotrypsin will not cleave dopa modified by borate (J. H. Waite, unpublished observations).
3. 0.1 *M* Tris–ascorbate at pH 7.5 to 8.0. This is made by titrating 0.2 *M* Tris base with an equal volume of 0.2 *M* ascorbic acid, thus using ascorbate as a counterion and antioxidant. It is basically the same approach as No. 1.

The options then are as follows: If the protein is not soluble in Tris (No. 1 or No. 3) at pH 7.5–8.5, try adding up to 4 *M* urea; if it is still not soluble, switch to borate. If the protein is soluble but not digested in borate at pH 7.5–8.5, try adding up to 4 *M* urea or switch to Tris with or without 4 *M* urea. Some dopa proteins are insoluble at all but acid pH; in such cases, longer digestion times may be necessary. Mefp-1 is equally labile to trypsin digestion in Tris or borate buffers. VpB is digested by trypsin in either Tris or borate buffers, but some of the peptides produced are different. For example, four peptides produced from VpB digested by trypsin in 0.15 *M* borate contain the sequence -Lys–dopa- which does not survive digestion in Tris.[28]

A typical digestion protocol is as follows: Add 1–5 mg of the dopa-containing protein to be digested dissolved in about 0.5 ml deionized water to the reaction vial; add trypsin (Boehringer-Mannheim, Indianapolis), at most a 1:50 (enzyme:protein weight/weight ratio) concentration, and set astir. Finally, add 0.5 ml of a buffer such as 0.2 *M* Tris or 0.3 *M* borate both with 0.02 m*M* $CaCl_2$ (for trypsin). Continue stirring magnetically at room temperature for 8 to 12 hr. Aliquots can be periodically removed to assess the progress of digestion by polyacrylamide gel electrophoresis. When digestion is complete, the reaction is stopped by adding 10 μl glacial acetic acid, frozen at $-80°$, and lyophilized. Before chromatography, the freeze-dried residue will need to be dissolved in 0.2–0.5 ml deionized water or 0.5% acetic acid and microfuged or filtered through a 0.46-μm HPLC sample filter.

Peptides are best separated by reversed-phase HPLC using either C-8 or -18 columns. Here too, gradients are constructed of Solvents A and B listed earlier for HPLC, but, as a rule, they are much shallower. Figure 4 illustrates the elution profile of six tryptic consensus peptides derived from Mefp-1. Although they are of similar mass and sequence, the more hydrophilic ones elute first from C-8 (Brownlee Aquapore RP-300) in a shallow gradient running from 5 to 11% acetonitrile in 40 min.[29] If peptides are

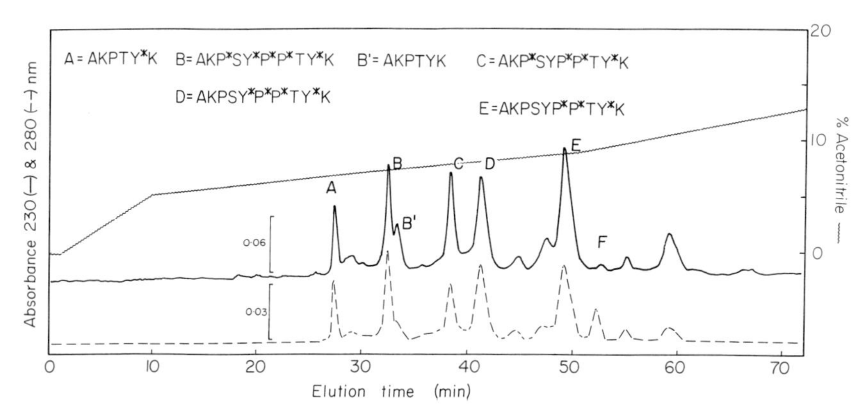

FIG. 4. Preparative reversed-phase C-8 HPLC of trypsin-digested Mefp-1. Gradient was made from buffers described in the text under HPLC. The elution program here consisted of 0% B for 1 min increasing to 5% at 10 min; second gradient to 9% B ending at 50 min, and a third gradient to 13% B ending at 74 min followed by a 100% B wash at 80 min. The one-letter representation of amino acids is standard with the exception of P* which is hydroxyproline and Y* which is dopa. P* is *trans*-4-hydroxyproline in all but the sixth position where it is probably *trans*-3-hydroxyproline.[29]

small or extremely polar, better retention can generally be achieved on C-18 (e.g., Rainin Microsorb C-18). Thus, Y*-D-Q-Y*-G-K and G-Y*-G-G-S-S-A-A-S-K typically are not bound to C-8 but are well-resolved on C-18.[28] Peak containing fractions collected from the eluate are fast-frozen in a methanol–dry ice bath and lyophilized at −80° after which they can be analyzed for composition and sequence.

Hydrolysis of Dopa Proteins and Peptides

Materials

6 *M* HCl
Recrystallized phenol (liquified)
Trifluoroacetic acid
2-ml long-stemmed hydrolysis vials (Wheaton, Vineland, NJ)
Vacuum pump with −50° trap
Oxygen/propane torch
Dedicated dry baths at 110 and 155°
Flash evaporation manifold (Büchler)

Procedure. When a protein or peptide sample reaches a stage of purification where it is free of all contaminating substances save water, it is ready for amino acid analysis. Ideally, 5–30 μg protein is required for hydrolysis. The samples are added to long-stemmed 2-ml hydrolysis vials that have been previously "vacuumed" to remove packaging lint particles. Vacuuming can be done with a 20-cm Pasteur pipette attached by a rubber hose to a vacuum aspirator. Two rather similar methods of hydrolysis are used that we routinely rely on for preparing dopa-containing proteins for amino acid analysis. These are the "short" and "long" methods. The long method is hydrolysis *in vacuo* in 6 *M* HCl with or without 5% (v/v) phenol as antioxidant for at least 24 hr at 110°.[30] The short method also involves 6 *M* HCl with 5–10% phenol and 10% trifluoroacetic acid for at least 22 min *in vacuo* at 155° in vapor or liquid phase.[31] The results are comparable, although yields of serine and dopa are marginally better with the long method. To quantitate exactly how much dopa is present in the hydrolysate, the extrapolation of dopa recovery following a series of timed hydrolyses (e.g., 24, 48, and 72 hr) to zero time is recommended. The hydrolysates are flash evaporated at 60° under vacuum and washed to dryness in two steps: with 1 ml deionized water followed by 1 ml absolute alcohol (chromatography grade). Residues are taken up in 100 μl "NaS" (Beckman Instruments, Fullerton, CA), of which half or more is applied to an autoanalyzer. It

[30] E. L. Malin, H. J. Dower, and E. G. Piotrowski, *Anal. Biochem.* **181,** 315 (1989).
[31] T. Tsugita, H. Uchida, W. Mewes, and T. Ataka, *J. Biochem.* (*Tokyo*) **102,** 1593 (1987).

is noteworthy, perhaps, that, of the other methods available for protein hydrolysis, 2 *M* NaOH[32] completely destroys dopa (J. H. Waite, unpublished observations) and that the use of methanesulfonic acid[33] places dopa at some risk. The reason for the latter seems to hinge on the oxidation of dopa during pH adjustment following hydrolysis (unpublished results). Even the recommended hydrolysis in HCl can lead to diminished dopa yields when contaminants such as iron are present: traces of Fe(III) will lead to a one-electron oxidation of dopa to semiquinone.[34] Reagents should be of the highest purity and glassware should be clean.

Amino Acid Analysis

Materials and Equipment

Complete HPLC system for the separation, postcolumn detection, and analysis of amino acids
Standard amino acid mixture (500 pmol each amino acid/10 μl)
Standard dopa mixture (500 pmol/10 μl)

Procedure. The detection of dopa in protein hydrolysates by amino acid analysis has been described before in this series.[18] However, since amino acid microanalysis has become more rapid and sensitive, it may be useful to recapitulate some strategies. In the case of dopa, only postcolumn derivatization by ninhydrin has been studied in any detail. We typically use a Beckman System 6300 autoanalyzer which relies on cation-exchange chromatography and canned elution programs (Beckman Na high performance column No. 338076) to separate the common amino acids. In these, as in previous elution programs recommended by the manufacturer, leucine and dopa coelute. An 88-min elution program for the separation of the 22 amino acids typically encountered in dopa-containing proteins has been developed.[35] Critical run parameters are listed in Table II. Dopa/leucine separation appears to be particularly sensitive to column temperature t_2 (Fig. 5). Decreasing the temperature from 55 to 50° can lead to complete separation of the two. However, since this ultimately leads to the merger of other amino acids later in the run, a compromise of 52° is selected. It is noteworthy that the method of hydrolysis can affect the analysis. For example, hydrolyzing a Cys-rich protein with phenol can lead to Cys/phenol-derived artifacts obscuring histidine.[4] Table III lists the amino acid composition of Mefp-1 and -2 as well as VpB and C.

[32] S. P. Robins, *in* "The Methodology of Connective Tissue Research" (D. A. Hall, ed.), pp. 37–52. Joynson Bruvvers Ltd., Oxford, 1976.

[33] R. J. Simpson, M. R. Neuberger, and T.-Y. Liu, *J. Biol. Chem.* **251,** 1936 (1976).

[34] E. Mentasti, E. Pelizzetti, and G. Saini, *J. Chem. Soc.*, p. 2609 (1973).

[35] J. H. Waite, *Anal. Biochem.* **192,** 429 (1991).

TABLE II
CRITICAL RUN PARAMETERS FOR AMINO ACID ANALYSIS ON A BECKMAN AUTOANALYZER 6300/7300[35,a,b]

Parameter	Column temperature (°C)	Start time (min)
Buffer A, pH 3.15 (b_1)	—	0.0
Buffer B, pH 3.90 (b_2)	—	29.0
Buffer D, pH 6.26 (b_3)	—	52.5
Buffer R (regeneration)	—	89.0
Temperature 1 (t_1)	43.0	0.0
Temperature 2 (t_2)	52.0	4.0
Temperature 3 (t_3)	70.0	40.0

[a] *Note:* Flow rate is 16.5 ml/hr (11 ml/hr for buffers; 5.5 ml/hr for ninhydrin reagent).
[b] Reprinted with permission from Academic Press.

Sequencing

Materials and Equipment

Automated system for Edman microsequencing chemistry, e.g., Porton or Applied Biosciences Inc.
Dedicated on-line reversed-phase HPLC such as Hewlett-Packard Model 1090
C-18 reversed-phase column (200 × 2.1 mm, e.g., Hewlett-Packard No. 79916 opt 572)
Solvent A [made by mixing the following Beckman sequencing grade reagents: 3 *M* sodium acetate, pH 4.0 (24 ml), 3.5% (v/v) tetrahydrofuran in water (900 ml), triethylamine (90 μl), and adjusting final pH to 4.15 with 6 *N* NaOH]
Solvent B acetonitrile (HPLC grade) filtered through a 0.45-μm nylon-66 filter
L-Dopa standard, 1 m*M*

Procedure. While there is no point in rehashing Edman chemistry here, it is salient to remark that peptidyl-dopa behavior during Edman degradation is unremarkable. A little "fine-tuning" is necessary to resolve and detect the phenylthiohydantoin (PTH) derivative of dopa. Most of this laboratory's sequencing experience has been with a Porton Instruments Model 2090 sequencer (Beckman Instruments, Fullerton, CA). However, given that most microsequencing technology is evolving along similar lines, our experience should be generally valid. Readers are referred in particular to Porton Programs No. 40 (purified peptides and proteins) and 4A (proteins electroblotted onto PDVF membranes) for the automated stepwise

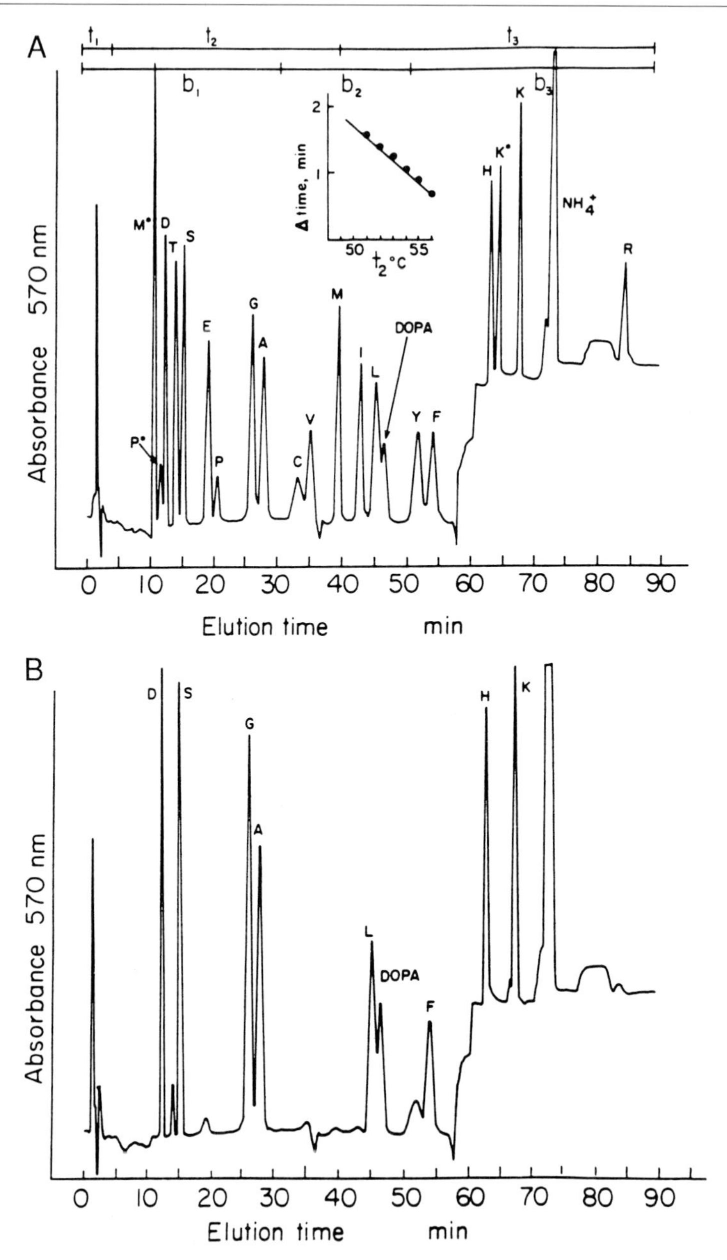

FIG. 5. (A) Chromatogram of a mixture of standard amino acids (500 pmol except where otherwise indicated) including methionine sulfoxide (M*), 4-hydroxyproline (P*, 2500 pmol), aspartate (D), threonine (T), serine (S), glutamate (G), proline (P, 2500 pmol), glycine (G),

TABLE III
AMINO ACID COMPOSITION OF Mefp-1, -2, AND VpB AND C IN RESIDUES/1000 RESIDUES[a]

Amino acid	Mefp-1[4]	Mefp-2[4]	VpB[28]	VpC[22]
3-Hyp	42-27	0	0	0
4-Hyp	161-120	0	0	0
Asp	12	127	150	34
Thr	113	42	17	1
Ser	93	74	51	77
Glu	5	49	87	4
Pro	41-82	112	18	22
Gly	10	142	177	414
Ala	81	39	70	3
Cys/2	0	69	0	0
Val	5	43	5	1
Met	2	2	22	1
Ile	8	9	3	0
Leu	Trace	13	40	0
Dopa	181-110	29	120-67	208-182
Tyr	31-73	56	13-52	2-11
Phe	0	10	31	1
His	3	8	40	209
Lys	214	135	113	0
Arg	3	44	52	14

[a] Amino acids are listed in order of elution. Ranges given for 3-Hyp, 4-Hyp, Pro, Dopa, and Tyr represent relative concentrations of these amino acids at the leading and trailing edges respectively of C-8 HPLC peaks.

Edman degradation of proteins to PTH derivatives. Following transfer of the PTH amino acid (100 μl) from the reaction flask to the HPLC sample flask, the sample is automatically loaded onto a 75-μl sample coil and injected. While the elution position of the common PTH derivatives is unique using the HPLC gradient conditions provided by the manufacturer, this is not always the case with rarer, modified amino acids. Thus, when

alanine (A), cystine (C, 250 pmol), valine (V), methionine (M), isoleucine (I), leucine (L), dopa (Y*), tyrosine (Y), phenylalanine (F), histidine (H), hydroxylysine (K*), lysine (K), and arginine (R). Detection is at 570 nm following postcolumn derivatization with ninhydrin. Full-scale absorbance deflection is 0.5. The functions t_n and b_n above the elution profile correspond to temperature and buffer changes in the program. The inset plots the time interval between the leucine and dopa peaks as a function of t_2. (B) Chromatogram of VpB tryptic peptide AY*LHGSFDK following acid hydrolysis.[28] (Reprinted from ref. 35 with permission from Academic Press).

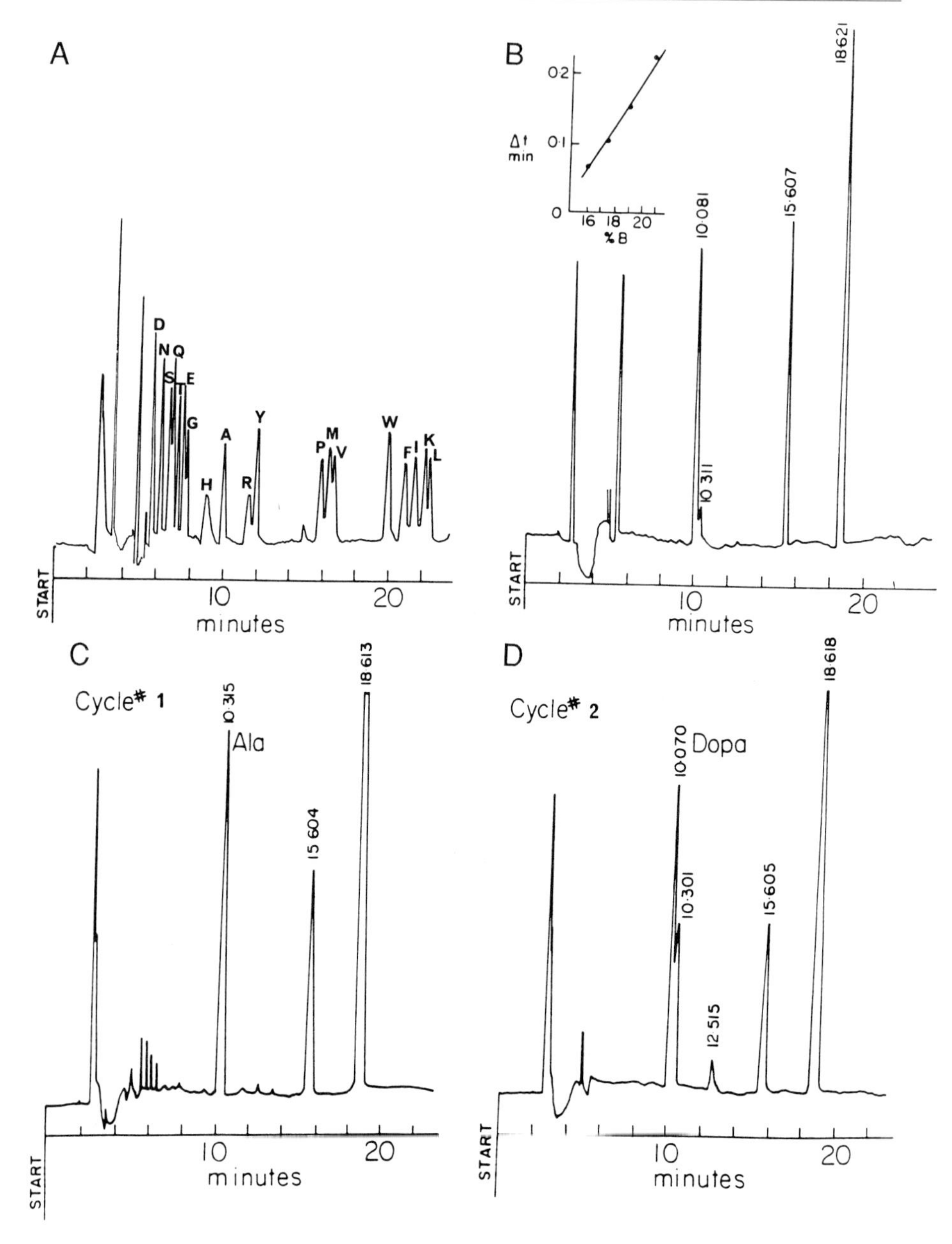

FIG. 6. (A) Chromatogram of a mixture of standard PTH amino acids (100 pmol) including aspartate (D), aspargine (N), serine (S), glutamine (Q), threonine (T), glutamate (E), glycine (G), histidine (H), alanine (A), arginine (R), tyrosine (Y), proline (P), methionine (M), valine

10 μl of dopa standard was blotted onto the sample filter, converted to PTH-dopa by program No. 40, transferred to HPLC, and combined with a mixture of PTH amino acid standards, it tended to coelute with PTH-alanine on HPLC.[35] To prevent this, the following critical gradient conditions were developed at a column temperature of 42° and a flow rate of 0.2 ml/min: 9% B (0 min), 21% B (1.5 min), 38% B (18 min), 60% B (22 min), 80% B (25 min), 90% B (27 min), 9% B (28 min), and 9% B (38 min). The eluant is continuously monitored at 269 nm. As indicated in Fig. 6A, common PTH amino acid standards are still well separated by this program. Note that Δt separation between PTH-dopa and PTH-Ala increases linearly with the percentage of B at 1.5 min. At 21% B, it is approaching 0.25 min which is adequate for routine resolution of the two (Figs. 6B–6D).[35] A more extensive listing of peptide sequences from dopa-containing proteins can be found in the pertinent literature about *Fasciola* vitelline proteins[2,22,28] and mussel byssal precursors.[3,4,10,19,21,29]

Postscript on Dopa-Containing Proteins

Many of the methodologies presented here will lead, if attempted, to products that are homogeneous by some measures of assessment (N-terminal analysis) and heterogeneous by others (electrophoresis, amino acid analysis of fractions under HPLC peaks). In some proteins, heterogeneity may arise primarily from variations in the extent of the hydroxylation of tyrosine to dopa. In others, e.g., VpB, however, there is increasing evidence for the existence of gene families in addition to variations in hydroxylation.[11] We are far from understanding the processing details or function of this exotic genre of proteins. Mefp-1 has found considerable application as a nonspecific attachment factor for cell and tissue culture.[36]

[36] M. F. D. Notter, *Exp. Cell Res.* **177,** 237 (1988).

(V), tryptophan (W), phenylalnine (F), isoleucine (I), lysine (K), and leucine (L). Absorbance is monitored at 269 nm. The percentage of B at 1.5 min is 21 as indicated in text. (B) Chromatogram of PTH-dopa prepared by blotting 1000 pmol of free dopa onto a peptide filter disk in the sample holder and running two cycles on the microsequencer. Peaks at 15.607 and 18.6231 min represent an ethyl acetate impurity and *N,N'*-diphenylthiourea, respectively. These peaks recur in every cycle of the automated Edman degradation. The inset shows the dependence of the time interval between the elution times of PTH-alanine and PTH-dopa as a function of the percentage of B at 1.5 min in the elution program. (C) PTH-alanine (500 pmol) of cycle No. 1 of VpB tryptic peptide AY*LHGSFDK.[28] (D) PTH-dopa (450 pmol) of cycle No. 2 of the same peptide. Note carryover of residual PTH-alanine at 10.301 min. Full-scale deflection is 16 mV in all panels. (Reprinted from ref. 35 with permission from Academic Press).

Acknowledgments

I am grateful to Dr. Leszek Rzepecki for trailblazing much of the purification strategy for Mefp-1 and -2. Thanks are due to the following graduate students for helping me troubleshoot various methods over the past 5 years: T. V. Diamond, D. C. Hansen, L. G. Huggins, K. Mueller Hansen, X.-X. Qin, and S. J. Samulewicz. The Office of Naval Research and the National Institutes of Health (NIDR) supported the research.

[2] Isolation of 2,4,5-Trihydroxyphenylalanine Quinone (Topa Quinone) from Copper Amine Oxidases

By SUSAN M. JANES and JUDITH P. KLINMAN

Introduction

Copper-containing amine oxidases (EC 1.4.3.6) are ubiquitous, occurring in bacteria, yeast, plants, and animals. These enzymes are proposed to catalyze physiologically important transformations: for example, mammalian plasma amine oxidases are involved in the oxidative elimination of biogenic amines from blood[1] and tissue diamine oxidases appear important to the regulation of levels of spermine and spermidine via their oxidation of intracellular putrescine.[2] The sensitivity of these enzymes to carbonyl reagents, coupled with the presence of an intense 450- to 500-nm chromophore in the oxidized enzymes, has been used as evidence for the presence of a carbonyl-containing active site cofactor.[3] However, elucidation of the structure of this chromophoric moiety eluded researchers for many years.

Previous Attempts at Prosthetic Group Identification

The failure of attempts to remove the carbonyl cofactor through the use of denaturing conditions led to the realization that there was a stable covalent linkage between the organic moiety and the enzyme.[4] Subsequent efforts focused on chemical or enzymatic methods of protein degradation that could release the organic chromophore from the enzyme backbone. For example, one of the earliest attempts[5] used pronase digestion of boiled native enzyme to release the enzyme chromophore. Although data obtained

[1] F. Buffoni, *Pharmacol. Rev.* **18,** 1163 (1966).
[2] U. Bachrach, *in* "Structure and Functions of Amine Oxidases" (B. Mondovi, ed.), p. 5. CRC Press, Boca Raton, FL, 1985.
[3] K. T. Yasunobu, H. Ishizaki, and N. Minamiura, *Mol. Cell. Biochem.* **13,** 2 (1976).
[4] C. W. Tabor, H. Tabor, and S. M. Rosenthal, *J. Biol. Chem.* **208,** 645 (1954).
[5] H. Yamada and K. T. Yasunobu, *J. Biol. Chem.* **238,** 2669 (1963).

from ultraviolet-visible (UV/VIS) and fluorescence analysis of the isolated chromophore were insufficient to identify the structure, they were used to eliminate flavin from consideration.

In the 1970s the focus shifted to using phenylhydrazine derivatives of the enzymes in an attempt to improve sensitivity and cofactor stability. Reaction of bovine serum amine oxidase (BSAO) with phenylhydrazine resulted in the formation of an intense yellow chromophore with an absorbance maximum at 430 nm.[6] Subsequent reduction, aminoethylation, and proteolytic digestion resulted in the isolation of a small amount of a N-terminally blocked peptide that was characterized by amino acid analysis and carboxypeptidase A treatment. Although the results were not conclusive, marked spectral differences were observed between the spectra of the isolated peptide and the phenylhydrazine derivative of pyridoxal phosphate, implying that the cofactor structure was distinct from pyridoxal phosphate.

In 1984, two groups published independent reports identifying pyrroloquinoline quinone (PQQ) as the covalently linked carbonyl cofactor in bovine serum amine oxidase.[7,8] Lobenstein-Verbeek *et al.*[7] used a modification of the previous protocol. They prepared a reduced and aminoethylated enzyme that had been labeled with 2,4-dinitrophenylhydrazine, which was then proteolyzed by chymotrypsin, pepsin, and pronase. The digested material was purified on a Sep-Pak C-18 cartridge to afford a yellow complex in 6.2% yield that coeluted with the 2,4-dinitrophenylhydrazine derivative of PQQ on high performance liquid chromatography (HPLC). Although the absorbance spectra of the isolated enzyme chromophore and the PQQ derivative were shown to be identical, both were significantly different from the spectrum of the intact derivatized enzyme. Despite the noted spectral differences, the similarity of the isolated adduct to the PQQ model was used to identify PQQ as the carbonyl cofactor in bovine serum amine oxidase.[7]

Ameyama *et al.*[8] used acid hydrolysis to obtain yellow chromophores from the unlabeled, native forms of methylamine dehydrogenase, bovine serum amine oxidase, pig kidney diamine oxidase, and *Aspergillus niger* amine oxidase. The isolated chromophores were examined using fluorescence and were shown to have spectra similar to PQQ. Additionally, the chromophores were reported to give partial reconstitution of activity in a PQQ-deficient mutant. This evidence was used to support the presence of PQQ in copper amine oxidases.[8]

[6] K. Watanabe, R. A. Smith, M. Inamasu, and K. T. Yasunobu, *Adv. Biochem. Psychopharmacol.* **5,** 107 (1972).

[7] C. L. Lobenstein-Verbeek, J. A. Jongejan, J. Frank, and J. A. Duine, *FEBS Lett.* **170,** 305 (1984).

[8] M. Ameyama, M. Hayashi, K. Matsushita, E. Shinagawa, and O. Adachi, *Agric. Biol. Chem.* **48,** 561 (1984).

Over the next 5 years these methodologies were extended and used to identify PQQ in enzymes as diverse as lysyl oxidase,[9] galactose oxidase,[10] dopa decarboxylase,[11] soybean lipoxygenase,[12] and dopamine β-hydroxylase.[13] However, none of these claims can now be substantiated.[14]

General Strategy and Rationale

The strategy discussed in this chapter involves a mild, nondestructive technique that can be used to obtain a labeled cofactor-containing peptide in high yield from copper amine oxidases. Reducing agents or harsh conditions are avoided to minimize unwanted side reactions and unintentional modification of the prosthetic group. The reaction of enzymes with phenylhydrazine or closely related compounds provides a derivative of the carbonyl cofactor which is stable to limited proteolytic digestion and HPLC purification. In this manner, labeled cofactor-containing peptides have been obtained from a variety of copper amine oxidases. Through a combination of sequencing, mass spectral analysis, nuclear magnetic resonance (NMR), UV/VIS, and resonance Raman spectroscopy, the prosthetic group has been unambiguously identified as 2,4,5-trihydroxyphenylalanine quinone (topa quinone) (Scheme 1,A).[15–17]

Methods

Preparation of Active Site-Labeled Enzymes

A solution of copper amine oxidase is freshly prepared in pH 7–8 phosphate buffer using enzymes of known purity and activity. Because most copper amine oxidases are stored as concentrated aliquots in dilute phosphate buffer, it is convenient to take the protein aliquots and dilute

[9] P. R. Williamson, R. S. Moog, D. M. Dooley, and H. M. Hagan, *J. Biol. Chem.* **261,** 16302 (1986).

[10] R. A. van der Meer, J. A. Jongejan, and J. A. Duine, *J. Biol. Chem.* **264,** 7792 (1989).

[11] B. W. Groen, R. A. van der Meer, and J. A. Duine, *FEBS Lett.* **237,** 98 (1988).

[12] R. A. van der Meer and J. A. Duine, *FEBS Lett.* **235,** 194 (1988).

[13] R. A. van der Meer, J. A. Jongejan, and J. A. Duine, *FEBS Lett.* **231,** 303 (1988).

[14] J. P. Klinman and D. Mu, *Annu. Rev. Biochem.* **63,** 299 (1994).

[15] S. M. Janes, D. Mu, D. Wemmer, A. J. Smith, S. Kaur, D. Maltby, A. L. Burlingame, and J. P. Klinman, *Science* **248,** 981 (1990).

[16] D. Mu, S. M. Janes, A. J. Smith, D. E. Brown, D. M. Dooley, and J. P. Klinman, *J. Biol. Chem.* **267,** 7979 (1992).

[17] S. M. Janes, M. M. Palcic, C. H. Scaman, A. J. Smith, D. E. Brown, D. M. Dooley, M. Mure, and J. P. Klinman, *Biochemistry* **31,** 12147 (1992).

[18] Deleted in proof.

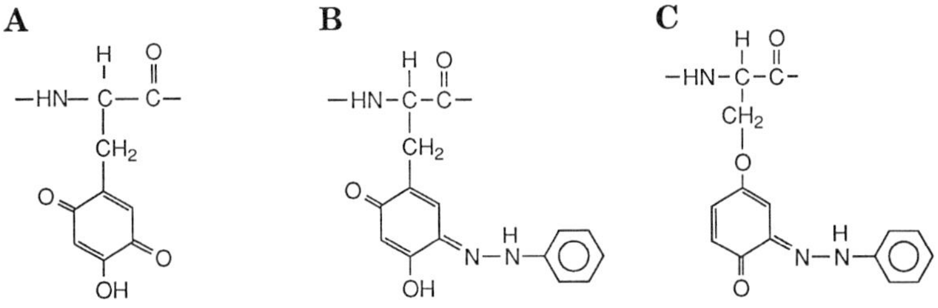

SCHEME 1

them to 0.1–20 mg/ml with 10–100 m*M* potassium phosphate, pH 7.2–7.5, with the final solution volume not to exceed 3 ml. The protein solution is then transferred to a disposable container and is labeled by the addition of a slight molar excess of phenylhydrazine in at least four portions over a 30-min period. This can be performed with unlabeled or radiolabeled phenylhydrazine (initially [U-^{14}C]phenylhydrazine, 21,200 dpm/nmol, from ICN was used; it is no longer available from ICN, and is being purchased from California Bionuclear Corp. at lower specific activity). If the labeling reaction is done with less than 2 mg/ml protein in a 1-cm path disposable cuvette, the reaction progress can be monitoring directly using absorbance at 450 nm; otherwise reaction aliquots should be diluted to 0.1–1.0 mg/ml and examined for absorbance at 450 nm. The reaction is complete when no further color development is observed on phenylhydrazine addition. Excess phenylhydrazine can be quenched by the addition of spectral grade acetone to 1% (v/v) and the reaction is desalted using a Bio-Rad 10DG disposable desalting column equilibrated with 50–100 m*M* NH_4HCO_3, pH 8. This procedure has been successfully used with 0.1–20 mg of copper amine oxidase. Either fluorophenylhydrazine or nitrophenylhydrazine can be used in place of phenylhydrazine. However, the use of 2,4-dinitrophenylhydrazine is not recommended because of the long incubation times normally required to label amine oxidases with this reagent.

For example, 3.4 mg of pea seedling amine oxidase in 200 μl of dilute phosphate buffer was added to 2 ml of 100 m*M* potassium phosphate, pH 7.2, in a UV grade disposable cuvette and titrated using eight sequential 2-μl aliquots of [U-^{14}C]phenylhydrazine HCl (7.4 m*M*) to give the spectra shown in Fig. 1. A 20-μl aliquot of the resulting yellow solution was counted to determine the total amount of radioactivity added and the remaining solution was loaded onto a Bio-Rad 10 DG column that had been previously equilibrated with 2 column volumes of 100 m*M* ammonium bicarbonate, pH 8. The labeled enzyme solution was carefully loaded onto the column

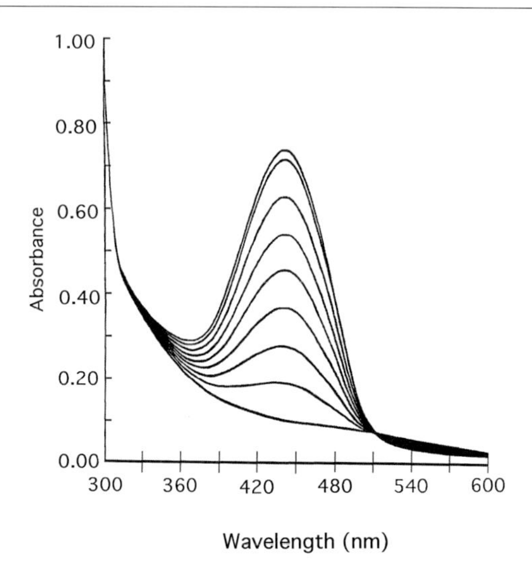

FIG. 1. Phenylhydrazine titration of pea seedling amine oxidase. Pea seedling amine oxidase (3.4 mg) in 200 μl of dilute phosphate buffer was added to 2 ml of 100 m*M* potassium phosphate, pH 7.2, in a UV grade disposable cuvette and titrated using eight sequential 2-μl aliquots of [U-^{14}C]phenylhydrazine HCl (7.4 *m*M). Spectra were recorded on a Varian DMS-200 spectrophotometer using a scan speed of 100 nm/min with a 2-nm slit against a baseline spectra of 100 m*M* potassium phosphate, pH 7.2.

using a 1-ml pipettor and the loading eluate was collected into a 20-ml scintillation vial. The column was then eluted using the equilibration buffer. The yellow inactivated enzyme band was collected into a preweighed 15-ml conical centrifuge tube, and a 20-μl aliquot was counted to determine the yield of labeled enzyme. The remaining eluate was collected using 2-min fractions into 20-ml scintillation vials. Analysis of the elution profile indicated a clean separation of radiolabeled peptide from excess reagent. A total of 38.5 nmol of phenylhydrazine-labeled pea seedling amine oxidase was obtained in quantitative yield using this method.

Proteolytic Digestion of Labeled Enzymes

The labeled enzyme solution in ammonium bicarbonate buffer is generally brought to 2–4 *M* urea and is proteolyzed with 2–5% (w/w) thermolysin or trypsin for up to 2 days at 20–37° in a shaker bath. The enzymatic digestions are monitored by small-scale HPLC injections, and the reactions are stopped when the amount of labeled peptide observed in the HPLC profiles reaches a maximum. A C-4 or C-8 column (4.6 × 250 mm, 5 μm, 300 Å) eluted with a gradient from 0.3% triethylamine acetate, pH 7, to 0.3% triethylamine acetate in 60% acetonitrile works quite well for this purpose. The elution of peptides is monitored at 214 nm (peptide bond absorbance), at 350 nm (topa quinone-phenylhydrazone absorbance), or by the radioactivity in HPLC fractions. The HPLC solvent is prepared in the following manner: a 3% solution of triethylamine in water is adjusted to pH 7 by the careful addition of glacial acetic acid; this stock solution is diluted with water to give 0.3% triethylamine acetate, pH 7, and the organic HPLC phase is made by mixing 100 ml of the 3% stock solution with 300 ml of water and 600 ml of acetonitrile.

The elution of the labeled chromophore can also be monitored at 450 nm when using a solvent with pH 6–8. Although a 0.1% trifluoroacetic acid (TFA) in water/acetonitrile solvent system will successfully fractionate peptides in the initial purification, it is not recommended because this solvent system leads to a bleaching of the phenylhydrazine-labeled prosthetic group. This bleaching effect is variable and is diminished when high concentrations of peptide are present. In general, it is found that UV monitoring at 350 nm (a wavelength where both the acid and neutral pH forms of the phenylhydrazone adduct of topa quinone absorb) avoids most of these spectral shift problems, and can be used to detect topa quinone-phenylhydrazone peptides during HPLC. When working with a new protein, radiolabeled phenylhydrazine is used to aid protocol development and determination of labeling and digestion yields.

For example, a thermolytic digest of 2 mg/ml of labeled BSAO, using 5% (w/w) thermolysin in 4 *M* urea, 50 m*M* ammonium bicarbonate, pH 8, was essentially complete after 2 days of incubation.

Isolation of Labeled TPQ Peptides from Thermolytic Digests

Thermolytic digests are injected onto a C-8 column equilibrated with 0.3% (v/v) triethylamine acetate, pH 6.8 (solvent A), and are eluted using a 10-min gradient to 20% solvent B [0.3% (v/v) triethylamine acetate, pH 6.8, 60% acetonitrile], followed by a 60-min gradient to 45–60% solvent B (depending on protein derivatized), with fractions collected at 1-min intervals. Phenylhydrazine-labeled TPQ peptides typically elute at 30 to 50 min into the profile as bright yellow droplets from the HPLC column and can be detected using either absorbance at 350 nm or by determining the radioactivity in the collected fractions. Sample chromatograms from injection of 60 nmol of a BSAO thermolytic digest are shown in Figs. 2 and 3. Peptide-containing fractions are generally rechromatographed by

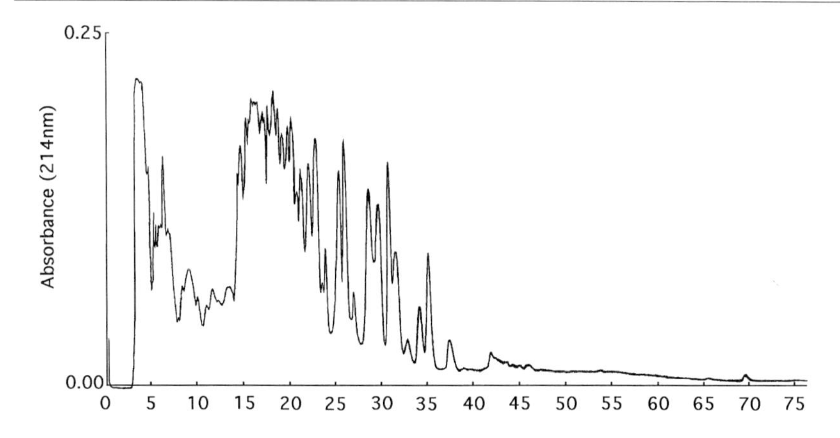

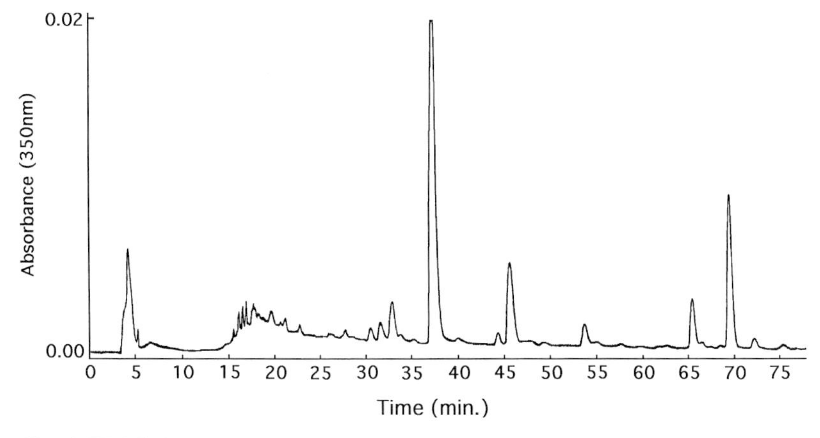

FIG. 2. HPLC chromatograms from thermolytic digest of labeled bovine serum amine oxidase. The HPLC column was eluted using the following gradient at a flow rate of 1 ml/min: 0% B at 1 min to 20% B at 10 min to 45% B at 70 min to 100% B at 80 min, where the aqueous phase is 0.3% triethylamine acetate, pH 7, and the organic phase is 0.3% triethylamine acetate in 60% acetonitrile.

HPLC using a trifluoroacetic acid/acetonitrile solvent system to ensure purity prior to sequencing or mass spectral analysis.

Isolation of Labeled TPQ Peptides from Tryptic Digests

Tryptic digests are injected onto a C-4 column equilibrated with 0.11% (v/v) TFA containing 5% (v/v) acetonitrile (solvent A) and are eluted using

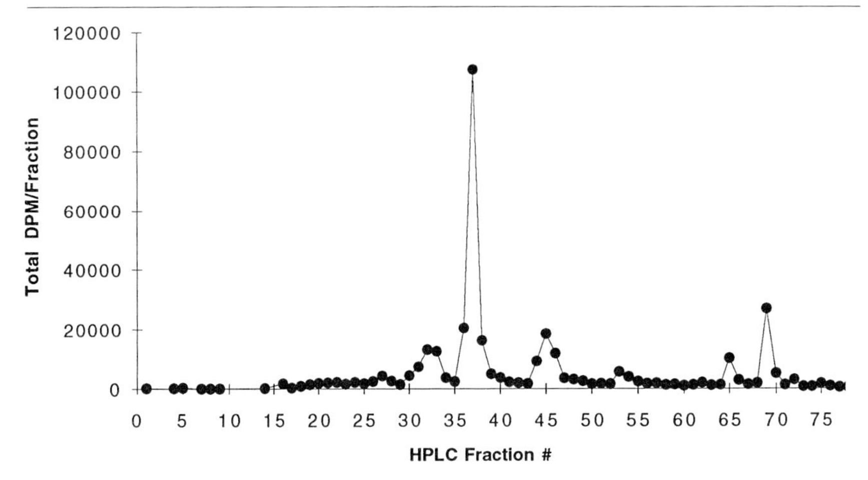

FIG. 3. Radioactivity profile for HPLC chromatogram from the thermolytic digest of labeled bovine serum amine oxidase.

a 70-min gradient to 75% solvent B [0.1% (v/v) TFA, 80% acetonitrile], with fractions collected at 1-min intervals. Phenylhydrazine-labeled TPQ peptides can be detected using either absorbance at 350 nm or by determining the radioactivity in the collected fractions. The fractions containing TPQ peptides are generally colorless using this solvent system; a yellow color is only observed in those instances when the peptide is present at a high enough concentration to buffer the HPLC solvent. TPQ peptide-containing fractions can be rechromatographed by HPLC using the triethylamine acetate solvent system to ensure purity prior to sequencing or mass spectral analysis.

Characterization of Labeled TPQ-Containing Peptides

Peptides containing the phenylhydrazine label are initially characterized by sequencing and mass spectrometry where possible. These peptides are sometimes blocked toward N-terminal sequencing and often exhibit poor mass spectral characteristics. In particular, peptides derived from nitrophenylhydrazine-derivatized proteins are especially prone to poor ionization. Resonance Raman spectroscopy and UV/VIS spectroscopy can also be used to identify the TPQ-phenylhydrazone in these peptides.

Initial characterization of this first thermolytic peptide revealed the following: the sequence of the peptide was Leu-Asn-X-Asp-Tyr, the radioactivity from the labeled phenylhydrazine was associated with the unknown residue in the sequence, and the absorption spectrum of the peptide was very similar to that of the labeled protein. In general, the method described herein yields unblocked peptides for sequencing about 70% of the time. A sample of sequence data for the purified peak seen at 45.5 min in the example from Figs. 2 and 3 is given in Table I.

Mass Spectral and NMR Analyses

The pentapeptide was then exhaustively examined by mass spectral analysis. Positive and negative liquid secondary ion mass spectrometry established that the neutral mass of the pentapeptide was 806.5. Subtraction of the mass for Leu, Asn, Asp, and Tyr leaves a mass of 283 for the cofactor phenylhydrazone residue. Tandem mass spectral analysis of this pentapeptide confirmed the amino acid sequencing results and hexylation of the peptide demonstrated a lack of free carboxylate groups in the unknown residue. This lack of free carboxylates and the low mass associated with the unknown residue provided the first conclusive evidence that PQQ (MW 330.2) was not the prosthetic group in BSAO.

Computer analysis of an exact mass generated from the dihexyl derivative of the pentapeptide gave rise to a list of seven empirical formulae for the unknown residue: $C_9H_{19}N_2O_6S$, $C_{17}H_{15}O_4$, $C_{13}H_{18}NO_4P$, $C_{11}H_{16}N_4O_3P$, $C_{15}H_{13}N_3O_3$, $C_7H_{17}N_5O_5S$, and $C_{18}H_{11}N_4$. The first three and last two formulae were not reasonable because they lack enough carbons and nitrogens to form both a peptide backbone and a phenylhydrazine ring. The fourth formula does not contain enough heteroatoms to satisfy the valence requirements of phosphorus and is too saturated to build an aromatic structure. Of the formulae, only $C_{15}H_{13}N_3O_3$ was concluded to be consistent with the physical and chemical characteristics of the unknown residue.

Given the evidence from both mass spectrometry and Edman sequenc-

TABLE I
SEQUENCE DATA FROM A BSAO THERMOLYTIC PEPTIDE SAMPLE

Residue No.	Amino acid	Yield (pmol)	^{14}C cpm (background correction)
1	Leu	70	25
2	Asn	54	17
3	Blank cycle	—	172
4	Asp	25	86
5	Tyr	20	33
6	Val	20	18
7	Trp	3	11

ing that the unknown residue resided in the polypeptide chain, two possible structures were proposed to be consistent with an empirical formula of $C_{15}H_{13}N_3O_3$. These are shown as structures B and C in Scheme I where B is the phenylhydrazone of the oxidized, quino form of a 2,4,5-trihydroxyphenylalanine side chain and C is the phenylhydrazone of an oxidized catechol ring linked to the hydroxyl group of a serine side chain. The presence of a stable phenylhydrazone moiety was confirmed following examination of [1-^{15}N]phenylhydrazine-labeled pentapeptide by mass spectrometry. The observed mass of 808.3 (MH+) was strong evidence that the phenylhydrazine remained attached to the oxidized cofactor. Mass spectrometry also provided an important first step in resolving whether the unknown structure was B or C since these structures were expected to fragment quite differently (with B giving rise to cleavage between the α and β carbons and C giving rise to cleavage between the β carbon and phenolic oxygen). In fact, only the former pattern of cleavage was observed, presenting strong initial support for the presence of 2,4,5-trihydroxyphenylalanine as the missing cofactor.

The next phase involved preparation of enough pentapeptide from BSAO for NMR experiments and the synthesis of a stable analog of B for structural proof. The hydantoin of TPQ (2,4,5-trihydroxyphenylalanine quinone) was chosen as a model because it provided a peptide-like backbone for the prosthetic group and was much simpler to synthesize and analyze than a pentapeptide. A series of NMR experiments were first conducted in D_2O. Examination of the aromatic region of the spectrum indicated some disparity between the active site-derived peptide and the model compound. As shown in Fig. 4, a singlet at 5.6 ppm, seen clearly in the model compound (Frame A), was absent from the peptide spectrum (Frame B). Given the limited amount of cofactor containing peptide available for NMR studies (we used 30–50 nmol dissolved in 400 μl for characterization), a considerably larger number of scans were averaged (128 to 1024) than was necessary for the model compound. Referring to structure B in Scheme I, it can be seen that one of the two cofactor ring protons lies between an enol and a ketone functional group and, as such, is expected to be chemically labile. It was therefore postulated that prolonged incubation of the cofactor containing peptide in D_2O had led to replacement of protium by deuterium at this position. Reinvestigation of the NMR spectrum of the peptide in H_2O confirmed this hypothesis, with the missing peak at 5.6 ppm becoming plainly visible (Frame C, Fig. 4). In this classical manner, the structure of TPQ was firmly established in BSAO. Subsequent investigations of a broad spectrum of copper amine oxidases have employed less labor-intensive approaches which include UV/VIS spectroscopy ([3], this volume) and resonance Raman spectroscopy ([10], this volume).

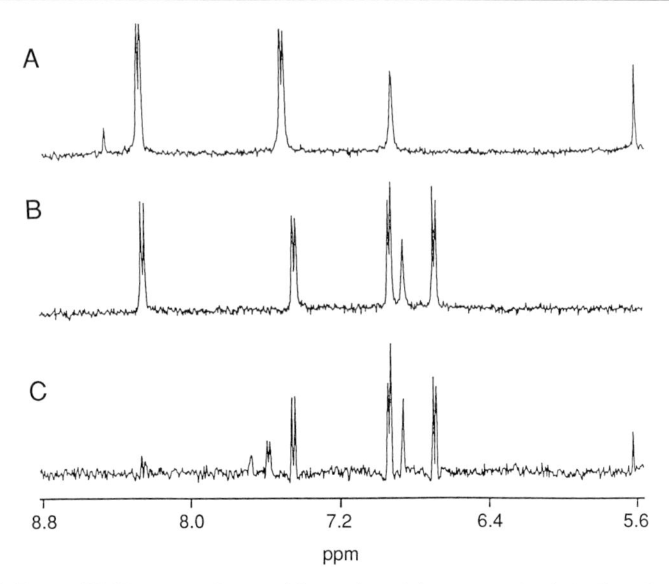

FIG. 4. Proton NMR spectra for peptides and model compounds. (A) Nitrophenylhydrazone-derivatized topa quinone hydantoin. (B and C) Nitrophenyldrazone-containing peptide in D_2O and H_2O, respectively. For the spectrum in H_2O (containing 10% D_2O as a lock for the spectrometer), the solvent signal was suppressed with the use of a $1:\bar{3}:3:\bar{1}$ selective excitation sequence with a delay of 750 μsec between pulses.

Discussion

Proteolytic Digestion of Phenylhydrazine-Labeled Amine Oxidases

Initial efforts to obtain labeled peptides from copper amine oxidases were fraught with difficulties, and it is useful to document some of the "pitfalls." As previously noted, topa quinone-phenylhydrazone chromophores readily bleach in trifluoroacetic acid containing HPLC solvents. Thus attempts to monitor at 450 nm for chromophore-containing peptides from proteolyzed amine oxidases failed when standard TFA-containing solvent systems were used. The use of radiolabeled phenylhydrazine for cofactor derivitization made rigorous tracking of labeled peptides possible and eliminated prior difficulties in assessing protocol success.

Protease screening with trypsin, thrombin, *Staphylococcus aureus* V8 protease, chymotrypsin, elastase, and thermolysin was initially done using reduced and carboxymethylated [U-^{14}C]phenylhydrazine-labeled bovine serum amine oxidase under denaturing conditions. Thermolysin gave the most promising result: 4–8% yield of a thermolytic peptide from the labeled BSAO. Subsequent use of thermolysin with native, labeled BSAO increased the yield to 16%, and the addition of 2 *M* urea to the digestion buffer generated a dramatic increase in yield to the 30–50% range. In more recent results, it has also been shown that reduced and carboxymethylated BSAO works just as well as native protein, as long as 2 *M* urea is included in the digestion buffer (S. Wang and J. P. Klinman, unpublished results). Overall, the methods presented in this chapter have worked well and reproducibly for a number of copper amine oxidases, providing the thermolytic peptides and longer tryptic peptides summarized in Table II. From time to time, N-terminally blocked peptides are obtained in the course of these studies. Mass spectral analyses of some of these products suggest a cyclic peptide formed by condensation of the amino-terminal residue to the topa quinone phenylhydrazone.

Discovery of TPQ

Bovine serum amine oxidase was used as a prototype for cofactor identification as it was available in gram quantities. Isolation of a short thermolytic peptide in 50% yield provided ample peptide for spectroscopic characterization. Significantly, comparison of the resonance Raman spectra of the isolated peptide to the intact protein indicated similar properties, ruling out the generation of breakdown products during the course of peptide isolation. The UV/VIS absorbance spectrum of the peptide was also very similar to that of the intact protein. An important early observation was that the unknown residue was released during Edman degradation (as demonstrated by specific release of radioactivity from the sequencer during that cycle of Edman degradation), indicating a normal amino acid backbone structure.

TABLE II
TPQ–PHENYLHYDRAZONE-CONTAINING PEPTIDES FROM COPPER AMINE OXIDASES

Copper amine oxidase	Peptide sequence[a]	Protease
Bovine serum	LN**X**DY[15]	Thermolysin
Bovine serum	LN**X**DYVW	Thermolysin
Bovine serum	SVSTMLN**X**DYVWDMVFYPNGAIE[16]	Trypsin
Porcine plasma	LN**X**DY[17]	Thermolysin
Porcine plasma	SVSTMLN**X**DYVWDMIFHP	Trypsin
Pig kidney	VYN**X**DY[17]	Thermolysin
Pig kidney	tTSTVYN**X**DYIWDFIFYYn[b,17]	Trypsin
Pea seedling	VGN**X**DNV[17]	Thermolysin
Pea seedling	VGN**X**DNVIDWE[17]	Thermolysin
Yeast	VAN**X**EYV[16]	Thermolysin

[a] **X**, TPQ derivative.

[b] Lowercase letters in sequence indicate ambiguity in the assignments for that residue.

TABLE III
SUMMARY FOR TOPA QUINONE-CONTAINING ENZYMES

Source	Ref.
Prokaryotic	
Arthobacter P1 and *E. coli* amine oxidases	19, 20
Yeast	
Hansenula polymorpha amine oxidase	16
Plant	
Pea and chick pea seedling amine oxidases	17
Mammalian	
Serum amine oxidases	15, 17
Intracellular amine oxidases	17

Following the availability of an empirical formula determined from mass spectroscopy for the missing amino acid, it was clear that PQQ could not be the correct structure. However, PQQ had become a very plausible and entrenched cofactor model and it was difficult to discard the preconception that the prosthetic group in BSAO was a heterocycle. Given the lines of evidence against a (PQQ) breakdown product, it took only a short time to hypothesize that the phenylhydrazone derivative of topa quinone was the unknown residue in the BSAO pentapeptide. In contrast, many months and the efforts of a large number of people were necessary to derive absolute structure proof using analog synthesis, NMR, and mass spectroscopy.[15]

UV/VIS and resonance Raman methods have now largely supplanted NMR and mass spectroscopy for topa quinone detection due to their ability to employ intact proteins. Both methods have been shown to be specific for topa quinone-containing peptides, and the UV/VIS method in particular is straightforward and can be done using a simple desktop spectrophotometer. The list of topa-containing proteins has grown substantially (Table III),[15–20] showing topa quinone to be a ubiquitous cofactor.

Generality of Method

Although we were able to generate enough TPQ peptide from 1 g of BSAO for exhaustive mass spectral and NMR experiments, this cannot be done for most copper amine oxidases due to limitations in the amounts of enzyme available. Consequently, we have since used a combination of

[19] D. M. Dooley, W. S. McIntire, M. A. McGuirl, C. E. Cote, and J. L. Bates, *J. Am. Chem. Soc.* **112,** 2782 (1990).

[20] R. A. Cooper, P. F. Knowles, D. E. Brown, M. A. McGuirl, and D. M. Dooley, *Biochem. J.* **288,** 337 (1992).

TABLE IV
SUMMARY FOR PHENYLHYDRAZINE, THERMOLYSIN PROTOCOL

Copper amine oxidase	Phenylhydrazine bound (%)	λ_{max} of enzymatic phenylhydrazone (nm)	TPQ-peptide yield (%)
Bovine serum	87	448	34
Porcine plasma	59	450	32
Pig kidney	38	443	15
Pea seedling	100	443	26

sequencing, mass spectral analysis, UV/VIS spectrophotometry, and resonance Raman spectra to establish the presence of TPQ in other copper amine oxidase peptides.[16,17]

The overall yield of derivatized topa quinone-containing peptides can vary according to copper amine oxidase and the particular phenylhydrazone used for labeling. A summary of results for thermolytic digestion of phenylhydrazone-labeled copper amine oxidases is given in Table IV. In general, the techniques described herein have been successful in yielding active site-labeled peptides from most of the enzymes examined. In some cases, it is preferable to use either trypsin or thermolysin to obtain peptides that could be sequenced. In others, we achieved more satisfactory results using *p*-nitrophenylhydrazine as a TPQ label. The general applicability of this method to TPQ-containing copper amine oxidases is proven by the number of cofactor-containing peptides that have been successfully isolated and characterized.

Relationship to Earlier Studies

By examining the unique structure and reactivity of topa quinone, it is possible to gain a valuable perspective on the identification of novel redox cofactors. It can now be postulated that many of the early attempts at obtaining this moiety by extensive proteolysis failed because of the unique reactivity of this unusual amino acid. Topa quinone as the free amino acid and related molecules have been shown to undergo internal cyclization due to Michael addition of the amine nitrogen on the amino acid backbone to the double bond in the quinone ring.[21] Thus the yellow chromophore that was isolated from extensive proteolysis of copper amine oxidases can be postulated to be a ring-closed form of topa quinone.[5] In an analogous way,

[21] S. Senob and B. Witkop, *J. Am. Chem. Soc.* **81,** 6231 (1959).

pronase digestion of the phenylhydrazine-labeled enzyme would result in the formation of a ring-closed form of topa quinone–phenylhydrazone.[7,8] These ring-closed derivatives closely resemble PQQ precursors and metabolites and would be expected to have similar spectral properties. Topa quinone could only be correctly identified when it was retained in a long enough peptide to prevent this ring closure. The major drawback of previous studies was the assumption that the chromophore isolated from the protein backbone by rigorous treatment would be identical to the prosthetic group in the native enzyme.

Acknowledgment

This work was supported in part by Grant GM39296 from the National Institutes of Health.

[3] Spectrophotometric Detection of Topa Quinone

By MONICA M. PALCIC and SUSAN M. JANES

Introduction

Trihydroxyphenylalanine quinone (topa quinone) was first identified as the redox-active cofactor in the copper amine oxidase isolated from bovine plasma.[1] The structural determination was achieved by nuclear magnetic resonance (NMR) and mass spectrometric characterization of the phenylhydrazine adduct of an isolated active-site peptide. Subsequently, a variety of copper amine oxidases obtained from mammalian,[1,2] plant,[2] bacterial,[3,4] and yeast[5] sources were demonstrated to contain topa quinone. Identification was based on resonance Raman spectroscopy of phenylhydrazine or *p*-nitrophenylhydrazine adducts and on the topa quinone consensus sequence Asn–topa–Asp/Glu. The visible absorbance spectral properties of the *p*-nitrophenylhydrazine adducts of both native enzymes or isolated active site peptides are also unique to topa quinone and can thus be used

[1] S. M. Janes, D. Mu, D. Wemmer, A. J. Smith, S. Kaur, D. Maltby, A. L. Burlingame, and J. P. Klinman, *Science* **248,** 981 (1990).

[2] S. M. Janes, M. M. Palcic, C. H. Scaman, A. J. Smith, D. E. Brown, D. M. Dooley, M. Mure, and J. P. Klinman, *Biochemistry* **31,** 12147 (1992).

[3] R. A. Cooper, P. F. Knowles, D. E. Brown, M. A. McGuirl, and D. M. Dooley, *Biochem. J.* **288,** 337 (1992).

[4] X. Zhang, J. H. Fuller, and W. S. McIntire, *J. Bacteriol.* **175,** 5617 (1993).

[5] D. Mu, S. M. Janes, A. J. Smith, D. E. Brown, D. M. Dooley, and J. P. Klinman, *J. Biol. Chem.* **267,** 7979 (1992).

to identify the cofactor.[2] At neutral pH (pH 7.2) the derivatives are bright yellow in color with absorbance peak maxima at 457–472 nm. In basic solution (1–2 *M* KOH), there is a dramatic red shifting of the absorbance spectrum with the generation of a new purple chromophore at 577–587 nm. Advantages of this method are the ease of detection of topa quinone, use of a simple desktop spectrophotometer, and the fact that spectra obtained with intact proteins do not appear to differ significantly from active-site-derived peptides.

Reagents

- 1 ml of freshly prepared 10 m*M* aqueous *p*-nitrophenylhydrazine hydrochloride
- Enzyme (0.5–8.8 mg) in 1–2 ml of sodium or potassium phosphate buffer, pH 7.0–7.5
- One Bio-Rad DG 10 column equilibrated with 50 m*M* NH_4HCO_3, pH 7.8
- 200 ml of 50 m*M* NH_4HCO_3, pH 7.8

The protein solution is titrated in a disposable 1-ml cuvette with the addition of 5-μl aliquots of the 10 m*M* *p*-nitrophenylhydrazine solution. After the addition of each aliquot, the increase in absorbance is monitored at 457 nm. When there is no further increase in absorbance the next 5-μl aliquot of hydrazine solution is added. There is considerable variability in the accessibility of the cofactor, which is reflected in the time required for complete reaction after *p*-nitrophenylhydrazine addition. This varies from 5 min between additions for pea seedling amine oxidase to 45 min for horse plasma amine oxidase. Titration is continued until the addition of *p*-nitrophenylhydrazine gives no further increase in absorbance.

Excess titrant is removed from derivatized protein by applying the sample to a Bio-Rad DG 10 column equilibrated with 50 m*M* NH_4HCO_3, pH 7.8. The column is developed with equilibration buffer and 20 drop fractions are collected in 10 × 75-mm disposable glass test tubes. For most solutions, both the bright yellow adduct and the unreacted *p*-nitrophenylhydrazine are visible on the column. The volume of the desalted protein adduct is typically 2–3 ml. Solutions that are cloudy after desalting can be clarified by passage through a 0.8-μm Nalgene CA filter.

For visible spectroscopy, 0.1–0.5 ml of the desalted enzyme is added to 0.5–0.9 ml of the potassium phosphate buffer, pH 7.2 (final concentration 0.1–0.2 M), or to 0.5–0.9 ml of KOH (final concentration 2 *M*). Figure 1A shows the visible spectra of the *p*-nitrophenylhydrazine adduct of 0.15 mg of pea seedling amine oxidase in 0.1 *M* potassium phosphate buffer, pH 7.2, and in 2 *M* KOH solution. The λ_{max} at neutral pH is 463 nm with a

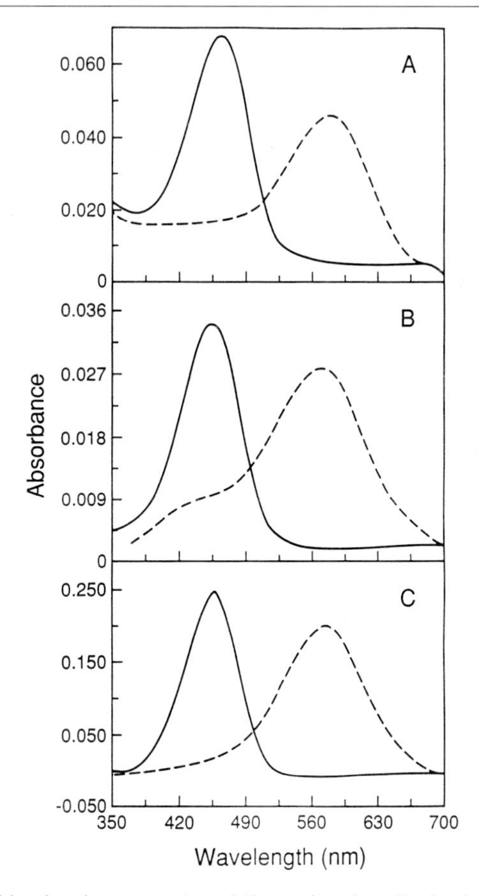

FIG. 1. The visible absorbance spectra of the *p*-nitrophenylhydrazine adducts of native pea seedling amine oxidase (A), an isolated active site peptide obtained from thermolytic digests of derivatized pea seedling amine oxidase (B), and a topa hydantoin model (C). Spectra are obtained in 0.1 *M* potassium phosphate buffer, pH 7.2 (——), and 2 *M* KOH (----). Reproduced with permission from Janes *et al.*[2]

shift of this peak to 581 nm in basic solution. This is essentially identical to the spectra of a peptide obtained from the thermolytic digestion of *p*-nitrophenylhydrazine-derivatized pea seedling enzyme[2]; as shown in Fig. 1B, λ_{max} of the peptide adduct is 457 nm in pH 7.2 buffer and 567 nm in 1 *M* KOH. The preparation of active-site peptides of derivatized enzymes is discussed elsewhere (see [2], this volume). Figure 1C illustrates the absorbance spectra of the *p*-nitrophenylhydrazine derivative of a model topa aquinone hydantoin,[2] with absorbance maxima at 455 nm in neutral solution and 575 nm in 1 *M* KOH.

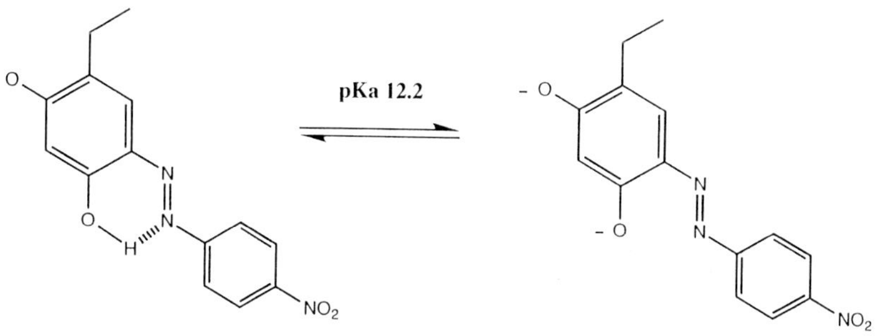

FIG. 2. Proposed structures for the *p*-nitrophenylhydrazine derivative of topa aquinone at pH 7 (left) and pH 14 (right).

The absorbance shift in base is attributed to the ionization of the azo group of the derivatized cofactor (Fig. 2). A pK_a of 12.2 has been measured by titration of the model topa aquinone hydantoin adduct.[2] An elevated pK_a in native-derivatized enzymes, as evident from the requirement to utilize 2 *M* KOH to complete the spectral shift in base, is ascribed to the buffering capacity of the protein.

Table I lists the absorbance maxima of a range of derivatized amine oxidases from mammalian, plant, and bacterial sources. At neutral pH all

TABLE I
VISIBLE SPECTRAL PROPERTIES OF *p*-NITROPHENYLHYDRAZONES OF AMINE OXIDASES

	λ_{max} (nm)	
Enzyme source	pH 7.2	2 *M* KOH
Bovine serum[a,b,c]	457	585
Porcine kidney[a,b]	460	578
Pea seedling[a,b]	463	581
Chick pea seedling	462	581
Sheep serum	462	587
Porcine serum[a,b]	462	578
Horse serum	464	580
Rabbit serum	462	584
Arthrobacter species	472	580

[a] Confirmation by resonance Raman spectroscopy.[1,2,6]
[b] Confirmation by topa quinone consensus sequence.[1,2]
[c] Confirmation by NMR and mass spectrometry.[1]

of the proteins exhibit an absorbance peak at 457–472 nm, which is shifted about 120 nm in base to 578–587 nm. Additional confirmation of topa quinone as a cofactor in many of these enzymes has been carried out using a variety of techniques annotated in Table I. These include resonance Raman spectroscopy of phenylhydrazine and/or *p*-nitrophenylhydrazine adducts,[1,2,6] the demonstration of the active-site consensus sequence Asn–topa–Asp/Glu,[1,2] and mass spectrometry and NMR spectroscopy.[1]

Phenylhydrazine is frequently used to derivatize topa aquinone enzymes. Although the visible spectral properties of these adducts at neutral pH are all similar with absorbance maxima from 437 to 447 nm, there is only about a 50-nm spectral shift in 2 *M* KOH solution to 480–490 nm. However, a significant absorbance remains at 437–447 nm, indicating an incomplete conversion to the high pH form with base. This suggests that the pK_a for protein phenylhydrazine adducts is greater than for *p*-nitrophenylhydrazine adducts, and precludes the use of visible spectroscopy for a straightforward identification of topa aquinone by titration with phenylhydrazine.

The spectral properties of the *p*-nitrophenylhydrazones of pyridoxal and pyrroloquinoline quinone are quite different from those of derivatized topa aquinone. The pyridoxal adduct has absorbance maxima at 422 nm at neutral pH and 541 nm in basic solution. The 541-nm band is also unstable and decays in a few minutes. The *p*-nitrophenylhydrazone of pyrroloquinoline quinone has an absorbance maximum at 443 nm at pH 7.2; however, the adduct exhibits only a minor red shift in base to 456 nm.

In summary, the characteristic visible spectral properties of *p*-nitrophenylhydrazine-derivatized topa quinone provide the basis for the spectrophotometric determination of this cofactor in as little as a 0.5-mg sample of enzyme. The adducts of native enzymes are yellow colored with absorbance maxima between 457 and 472 nm at pH 7.2. The absorbance spectrum shifts to 578–587 nm in 2 *M* KOH solution with conversion to a purple chromophoric form.

Acknowledgment

This work was funded by an operating grant from the Natural Sciences and Engineering Research Council of Canada to M. M. P.

NOTE ADDED IN PROOF. Recent data suggest that tyrosine derived quinones with different functional groups from topa may exist in proteins (J. P. Klinman, personal communication). Further experiments will be needed to determine whether quinones of this type can be distinguished spectrophotometrically from topa quinone.

[6] D. E. Brown, M. A. McGuirl, D. M. Dooley, S. M. Janes, D. Mu, and J. P. Klinman, *J. Biol. Chem.* **266,** 4049 (1991).

[4] Model Studies of Topa Quinone: Synthesis and Characterization of Topa Quinone Derivatives

By MINAE MURE and JUDITH P. KLINMAN

Introduction

6-Hydroxydopaquinone (topa quinone, **A** in Scheme 1) is the covalently bound prosthetic group of the copper amine oxidases.[1] A transamination mechanism involving substrate and product Schiff base intermediates (**B** and **C**, respectively, in Scheme 1) has been proposed for the anaerobic portion of the enzyme reaction.[1] This chapter describes the preparation and characterization of model compounds for the quinone, reaction intermediates, and the reduced form of the cofactor (**A**–**D** in Scheme 1). These structural analogs of topa quinone provide a frame of reference for interpretation of studies of topa in both enzymatic and model systems.

Synthesis of Topa Quinone Analogs

Topa Hydantoin Quinone ($\mathbf{1_{ox}H}$)

Topa hydantoin quinone ($\mathbf{1_{ox}H}$) was first synthesized as a model compound of the covalently bound prosthetic group in order to determine its spectroscopic and electrochemical properties.[2] It was prepared from 5-(2,4,5-tribenzyloxybenzylidene) hydantoin[3] which was synthesized following the method of Lee and Dickson,[4] with some modifications. The quinone was isolated as a very unstable solid both in its protonated form ($\mathbf{1_{ox}H}$) and as a salt ($\mathbf{1_{ox}^{-}Na^{+}}$) after silica gel chromatography. It is strongly recommended that this compound be stored in its reduced form ($\mathbf{1_{red}H_3}$)[5]; quinone can be generated by air-oxidation in the solution just before usage.

Synthesis of 5-(2,4,5-Tribenzyloxybenzilidine) Hydantoin.[3] 2,4,5-Tribenzyloxybenzaldehyde (29.4 mmol, 12.5 g) is slowly added to a stirred mixture of hydantoin (64 mmol, 6.4 g) and sodium acetate (78 mmol, 6.4 g) in 16 ml of glacial acetic acid and 1 ml of acetic anhydride at 100° in a three-necked round-bottomed flask equipped with a thermometer. After the

[1] J. P. Klinman and D. Mu, *Annu. Rev. Biochem.* **63,** 299 (1994).
[2] M. Mure and J. P. Klinman, *J. Am. Chem. Soc.* **155,** 7117 (1993).
[3] S. M. Janes, Ph.D. Thesis, University of California, Berkeley (1990).
[4] F. G. H. Lee and D. E. Dickson, *J. Med. Chem.* **14,** 266 (1971).
[5] The quinol ($\mathbf{1_{red}H_3}$) is stable in the solid state and can be stored in a dessicator.

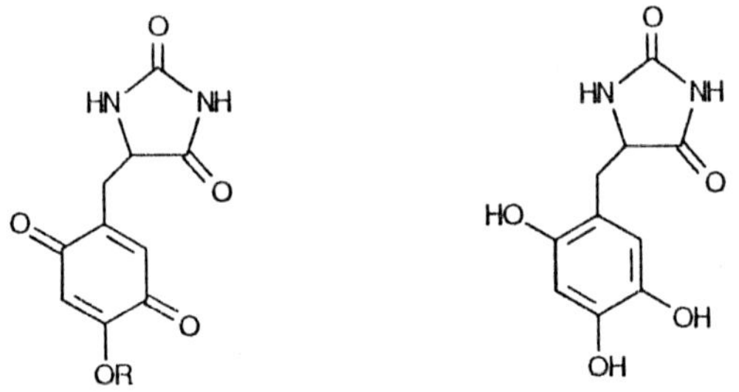

R = H $1_{ox}H$
= Na $1_{ox}^{-}Na^{+}$

$1_{red}H_3$

addition is complete, the reaction is refluxed for 3.5 hr, during which the internal temperature increases from 150 to 160°. The hot, viscous reaction mixture is poured into a 1-liter heavy-walled beaker and is slowly cooled, yielding a crystalline mass covered with a thin layer of red material, which is removed by rinsing with copious amounts of water and then several times

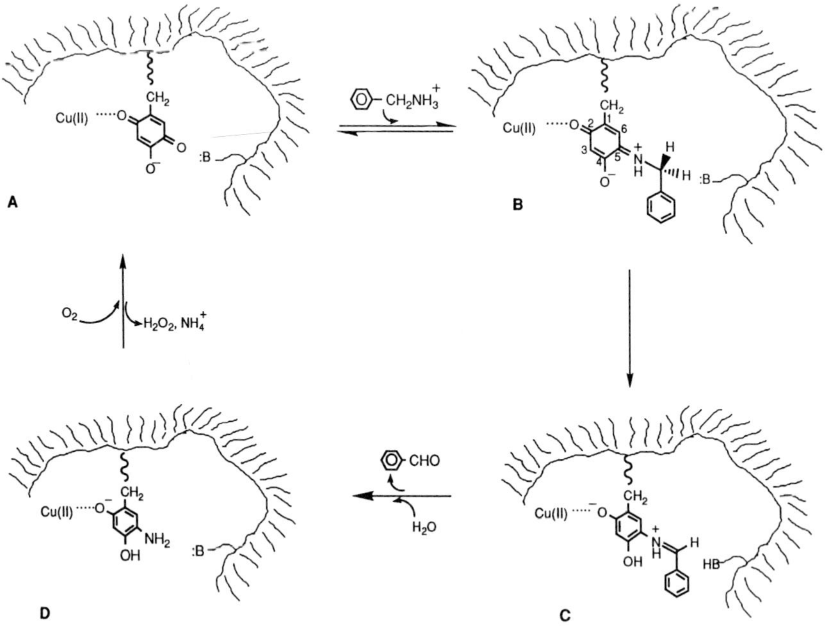

SCHEME 1. Proposed reaction mechanism for the oxidative half reaction catalyzed by bovine serum amine oxidase.[1] (A) Topa quinone, (B) substrate Schiff base intermediate, (C) product Schiff base intermediate, and (D) aminoresorcinol.

with a small amount of cold ethanol. The remaining solid is recrystallized from chloroform to yield 5-(2,4,5-tribenzyloxybenzylidine) hydantoin (6.95 g, 54%) as yellow needles: mp 189–190°; TLC (silica) $R_f = 0.67$ (chloroform), MS, *m/e* 506 (M^+); ^{1}H-NMR (acetone-d_6): δ 5.15 (2H, s), 5.19 (2H, s), 5.22 (2H, s), 6.74 (1H, s), 7.02 (1H, s), 7.28 (1H, s), 7.4 (15H, m). Anal. Calcd. for $C_{31}H_{26}N_2O_5$: C, 73.5: H, 5.1: N, 5.5. Found: C, 73.61: H, 4.80: N, 5.14.

Synthesis of 5-(2,4,5-Tribenzyloxybenzyl) Hydantoin.[3] Sodium amalgam (3% w/w, 100 g) is added over 15 min to a stirred suspension of 5-(2,4,5-tribenzyloxybenzylidine) hydantoin (9 g, 18 mmol) in dioxane/water (130 ml/20 ml). The reaction is vigorously stirred until the color changes to pale yellow (ca. 12 hr). The solution is decanted from sodium amalgam, diluted with 100 ml of water, cooled on ice, and acidified to pH 6.5 with HCl. The precipitated product is extracted with ethylacetate, dried with $MgSO_4$, and concentrated under reduced pressure to give a yellow oil. A solid is obtained after treatment with benzene and hexane, which recrystallized from acetone/H_2O to give 5-(2,4,5-tribenzyloxybenzyl) hydantoin (7 g, 77%) as a white needle. mp 177–178°; TLC (silica) $R_f = 0.8$ (ethylacetate/pet. ether, 9/1); MS, *m/e* 508 (M^+); ^{1}H-NMR (acetone-d_6): δ 2.73 (1H, dd), 3.29 (1H, dd), 4.36(1H, dd), 5.06(2H, s), 5.12 (2H, s), 5.14 (2H, s), 6.92 (1H, s), 7.00 (1H, s), 7.4 (15H, m). Anal. Calcd. for $C_{31}H_{28}N_2O_5$: C, 73.2: H, 5.5: N, 5.5. Found: C, 72.91: H, 5.25: N, 5.42.

Synthesis of 5-(2,4,5-Trihydroxybenzyl) Hydantoin ($\mathbf{1_{red}H_3}$). A suspension of 5-(2,4,5-tribenzyloxybenzyl) hydantoin (200 mg, 0.4 mmol) and palladized charcoal (5%) in 13 ml of 95% ethanol containing 3 drops of 3.0 *N* HCl is hydrogenated at 45 psi in a Parr apparatus for 5 hr. The catalyst is removed by filtration and rinsed with a small portion of acidic ethanol for several times in an Ar atmosphere. The solvent is removed under vacuum and the remaining solid is washed with ether for several times and dried under vacuum to give 5-(2,4,5-trihydroxybenzyl) hydantoin ($\mathbf{1_{red}H_3}$) as a white solid (82.5 mg, 87%). HRMS $C_{10}H_{11}N_2O_5$ (MH^+) Calcd. 239.0668, Obsd. 239.0666. ^{1}H-NMR (DMSO-d_6): δ 2.46 (dd, 1H, J = 13.8, 7.5 Hz), 2.90 (dd, 1H, J = 13.8, 4.7 Hz), 4.14 (dd, 1H, J = 7.5, 4.7 Hz), 6.28 (s, 1H), 6.42 (s, 1H), 7.59 (s, 1H, exchangeable with D_2O), 8.03 (br s, exchangeable with D_2O), 8.57 (s, 1H, exchangeable with D_2O), 8.61 (br s, exchangeable with D_2O), 10.49 (s, 1H, exchangeable with D_2O).

Synthesis of 2-Hydroxy-5-(5′-hydantoin methyl)-1,4-benzoquinone ($\mathbf{1_{ox}H}$). One hundred milligrams (0.42 mmol) of 5-(2,4,5-trihydroxybenzyl)-hydantoin ($\mathbf{1_{red}H_3}$) is oxidized to the quinone with a catalytic amount of triethylamine in 10 ml of dry methanol. The reaction is stirred for 1 hr, and the solvent is removed by rotary evaporation. The remaining sample is purified on a silica gel column. First, the small amount of the quinone ($\mathbf{1_{ox}H}$) is eluted with ethanol (a pale yellow fraction). The deprotonated form of the quinone ($\mathbf{1_{ox}^-}$) is then eluted as a Na^+ salt (Na^+ is counter ion in neutral silica gel) with ethanol/methanol (5/2, v/v) (a red fraction). Total yield, 91%. $\mathbf{1_{ox}H}$: ^{1}H-NMR (DMSO-d_6): δ 2.46 (dd, 1H, J = 13.58, 9.10 Hz), 2.83 (dd, 1H, J = 13.58, 4.64 Hz), 4.15 (dd, 1H, J = 9.10, 4.63 Hz), 6.07 (s, 1H, exchangeable with DCl), 6.63 (s, 1H), 8.00 (br s, exchangeable with D_2O), 10.73 (br s, exchangeable with D_2O), λ_{max} 350 nm (MeOH). $\mathbf{1_{ox}^-Na^+}$: ^{1}H-NMR (DMSO-d_6): δ 2.32 (dd, 1H, J = 13.52, 9.01 Hz), 2.85 (dd, 1H, J = 13.52, 4.51 Hz), 4.18 (dd, 1H, J = 9.01, 4.51 Hz), 5.00 (s, 1H, exchangeable with D_2O), 6.11 (s, 1H), 7.93 (s, 1H, exchangeable with D_2O), 10.62 (s, 1H, exchangeable with D_2O).

2-Hydroxy-5-alkyl-1,4-benzoquinones **(2a–2d)**

Several topa quinone analogs bearing alkyl substituents **(2a–2d)** were easily prepared from the corresponding 4-alkylresorcinols[6] by oxidation with Fremy's salt in one step according to the reported method.[7] These quinones were purified by rapid sublimation under reduced pressure and were stable and could be stored in a desiccator under dark for a few months. Studies on the catalytic oxidation of benzylamine by topa analogs indicate that the *tert*-butyl quinone **(2a)** functions as well as topa hydantoin quinone **($\mathbf{1_{ox}H}$)**, whereas ethyl **(2c)** and methyl quinone **(2d)** are less efficient catalysts.[8] A bulky substituent appears necessary to prevent side reaction in these model systems.

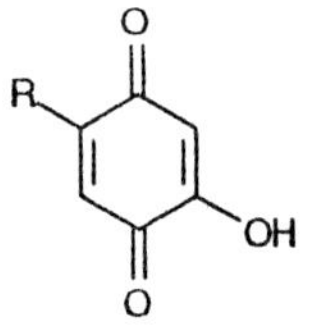

O
R
OH
O

R = $C(CH_3)_3$ **2a**
= $CH(CH_3)_2$ **2b**
= C_2H_5 **2c**
= CH_3 **2d**

Following is the method to prepare the quinone (**2a**). To 1.0 g (0.006 mol) of 4-*tert*-butyl resorcinol and 1.048 g (0.0046 mol) of $K_2HPO_4 \cdot 3H_2O$ in 15 ml of water is added 120 ml of water containing 4.0 g (0.015 mol) of Fremy's salt [$(KSO_3)_2NO$] and 1.048 g of $K_2HPO_4 \cdot 3H_2O$. After 2 min, the mixture is acidified with 2 N H_2SO_4 and extracted with ether. The ether

[6] A. E. Tchitchibabime, *Bull. Soc. Chim. Fr.* **5**(2), 497 (1935).
[7] H. Musso and D. Maassen, *Justus Liebigs Ann. Chem.* **689,** 93 (1965).
[8] M. Mure and J. P. Klinman, unpublished results.

layer is dried with Na_2SO_4 and concentrated to give a dark yellow solid. The solid is recrystallized from cyclohexane and is further purified by rapid sublimation under reduced pressure. 2-Hydroxy-5-*tert*-butyl-1,4-benzoquinone **(2a)** is obtained as a golden yellow needle (980 mg, 91%). Anal Calcd. ($C_{10}H_{12}O_3$): C, 66.67; H, 6.67; O, 26.66. Found: C, 66.93; H, 6.85; O, 26.22. ^{1}H-NMR (d_6-acetone) δ1.271 (9H, S, $-^t$Bu), 5.896 (1H, s, exchangeable with D), 6.523 (1H, s).; (CD_3CN) δ 1.249 (s, 9H, $-^t$Bu), 5.907 (s, 1H, exchangeable with D), 6.546 (s, 1H), 7.80 (br s, 1H, $-$OH).

Substrate Schiff Base Analogs

2-Hydroxy-5-*tert*-butyl-1,4-benzoquinone **(2a)** formed Schiff base complexes with various amines.[8] Complexes with cyclohexylamine **(3a)** and *n*-propylamine **(3b)** were isolable as the amine salts. α-Methylbenzylamine also formed the substrate Schiff base **(3c)** but its high solubility hampered the isolation.[9] The substrate Schiff base with benzylamine **(3d)** was very unstable and rapidly transformed to the product Schiff base **(5)**.

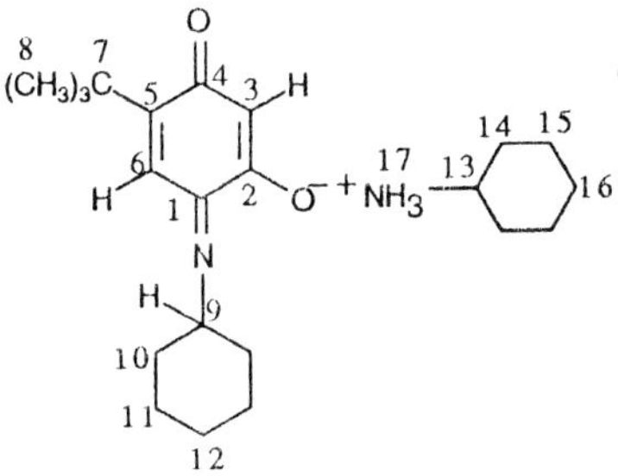

8 7 (CH3)3C 5 4 3 H O 6 1 2 O⁻ +NH3 17 13 14 15 16 H N H 9 10 11 12

3a

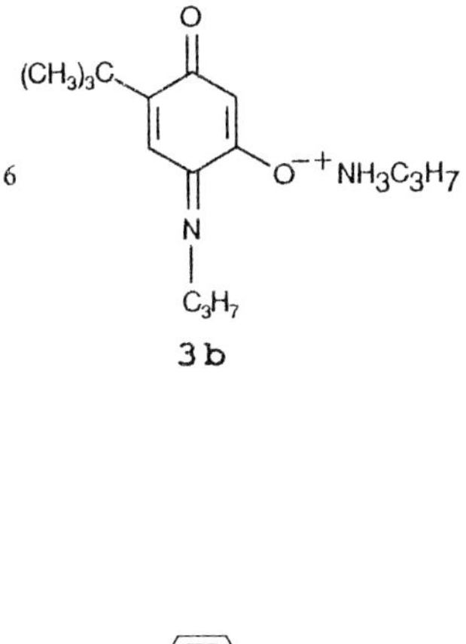

(CH3)3C O O⁻ +NH3C3H7 N C3H7

3b

(CH3)3C O O⁻ + NH3CH(R) N RCH

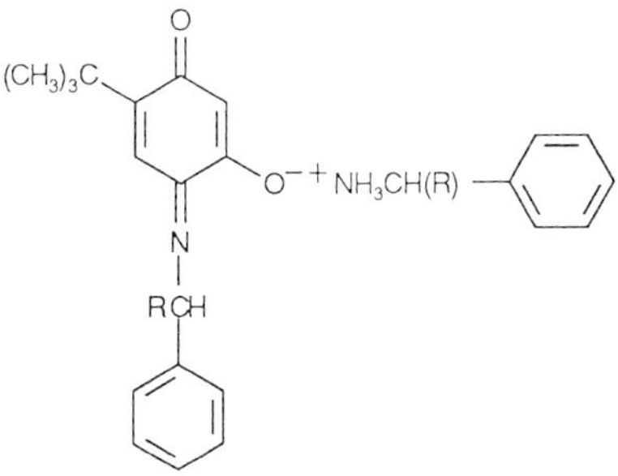

R = CH3 3c
= H 3d

[9] The formation of **3c** was detected by either ^{1}H-NMR or UV/VIS spectroscopy. ^{1}H-NMR (CD_3CN) δ 1.254, (9H, s), 1.547 (3H, d, J = 6.54 Hz), 5.337 (1H, q, J = 6.54 Hz), 5.738 (1H, s), 7.103 (1H, s). UV/VIS (CH_3CN) λ_{max} 352 nm, (pH 10.0) λ_{max} at 454 nm.

The following method is used to prepare the substrate Schiff base analog **(3a)**. **2a** (30.3 mg, 0.168 mmol) is treated with a five-fold excess of cyclohexylamine in 10 ml of anhydrous acetonitrile under anaerobic conditions. After 1 hr, an orange red precipitate is collected by centrifugation, washed with a small amount of hexane, and dried *in vacuo* to give the substrate Schiff base **(3a)** as an orange red solid (52.3 mg, 86%). NMR data are given below under spectroscopic characterizations.

Aminoresorcinol Analogs

The aminoresorcinol **($4aH_2$, $4bH_2$)** was prepared from the corresponding 4-alkylresorcinol via nitrosation and reduction in two steps and isolated as the hydrochloride salt.[2,8] The following method is used to prepare the aminoresorcinol **($4aH_2$)**.

R, HO, NH_2, OH

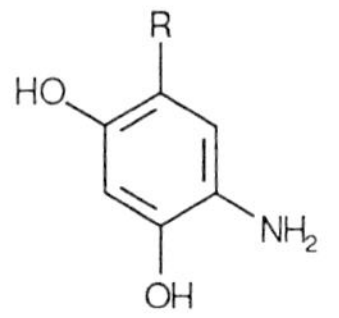

R = $C(CH_3)_3$ $4aH_2$
= C_2H_5 $4bH_2$

A solution of $NaNO_2$ (0.831 g, 0.0120 mol) in 4 ml of water is added dropwise to a stirred solution of 2 g (0.0120 mol) of 4-*tert*-butylresorcinol in 42 ml of 95% ethanol containing a 10-fold excess of HCl. The mixture is stirred at −5 to 5° for 15 min. The solution is concentrated to a small volume under vacuum, diluted with water, and extracted with CH_2Cl_2. The CH_2Cl_2 layer is dried over Na_2SO_4 and concentrated to give a red oil. The oil is treated with water to yield 4-*tert*-butyl-6-nitrosoresorcinol as an orange red solid (1.638 g, 70%). ^{1}H-NMR (DMSO-d_6), δ 1.208 (9H, s, ^{t}Bu), 5.595 (1H, s), 7.303 (1H, s). Anhydrous ethanol (Aldrich) is deoxygenated by bubbling O_2-free Ar[10] through it for 30 min. A 3-ml aliquot of the O_2-free EtOH was added to the reaction vessel containing $SnCl_2 \cdot 2H_2O$ (1.1 g, 4.88 mmol) and 4-*tert*-butyl-6-nitrosoresorcinol (326.2 mg, 1.67 mmol). After the solution is stirred for 30 min at 70° under the stream of O_2-free Ar, a half volume of the solvent is removed under vacuum followed by the addition of 1.67 mmol of concentrated HCl. Removing the solvent under vacuum leaves a greenish white solid. This is dissolved in a small amount of H_2S-saturated 0.1 *N* HCl (ca. 3 ml) and H_2S gas is bubbled through it until all Sn^{2+} is converted to SnS. The solution is kept under an H_2S

[10] The O_2-free Ar was generated by passing Ar through an alkaline pyrogallol solution.

atmosphere overnight. SnS is removed by filtration and washed with H_2S-saturated 0.1 *N* HCl several times. The filtrate and the HCl solutions are combined and concentrated under O_2-free conditions to give the hydrochloride salt of 4-amino-6-*tert*-butylresorcinol **($4aH_2$)** as a white solid (120 mg, 52%). ^{1}H-NMR (CD_3CN) δ 1.272 (9H, s, $-^t$Bu), 6.717 (1H, s), 7.261 (1H, s), 7.8 (br s, exchangeable with D_2O), 8.9 (br s, exchangeable with D_2O). HRMS (EI) ($M^+ -$ HCl) Calcd. for $C_{10}H_{15}NO_2$ 181.1103; Obsd. 181.1103.

Product Schiff Base Analog

The product Schiff base **(5)** was prepared from the condensation reaction of the aminoresorcinol **($4aH_2$)** and benzaldehyde.[8]

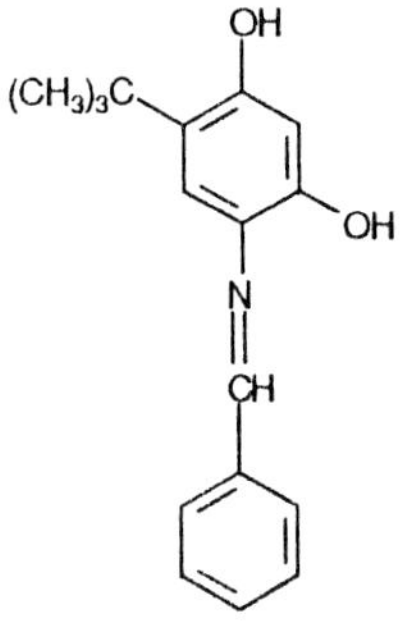

OH
$(CH_3)_3C$
OH
N
CH

5

A suspension of the hydrochloride salt of **$4aH_2$** (29.3 mg, 0.135 mmol) in anhydrous acetonitrile is treated with a slight excess (1.5 equiv) of benzaldehyde. The reaction mixture is stirred at room temperature for 3 hr followed by the addition of a small amount of anhydrous Et_2O to precipitate the product. A yellow solid **(5)** is collected by centrifugation, washed with anhydrous Et_2O, and dried *in vacuo* (29.1 mg, 80%). ^{1}H-NMR (CD_3CN) δ 1.380 (9H, s, $-^t$Bu), 6.653 (1H, s), 7.471 (1H, s), 7.662 (2H, m), 7.793 (1H, m), 8.436 (2H, m), 8.923 (1H, s) (in the presence of $PhCH_2NH_2$) δ 1.374 (9H, s, ^{t}Bu), 6.392 (1H, s), 7.260 (1H, s), 7.316 ~ 7.471 (5H, m), 7.984 (2H, m), 8.727 (1H, s). HRMS (EI) $C_{17}H_{19}NO_2$ Calcd. 269.1416. Obsd. 269.1412.

Spectroscopic Characterizations

*Two-Dimensional NMR Analysis of the Substrate Schiff Base Analog **(3a)***

Method. All NMR spectra were acquired on a Bruker AM-500 spectrometer fitted with an inverse ^{1}H-^{13}C probe operating at a proton frequency of 500.13 MHz. ^{1}H chemical shifts are relative to a residual $CHCl_3$ signal

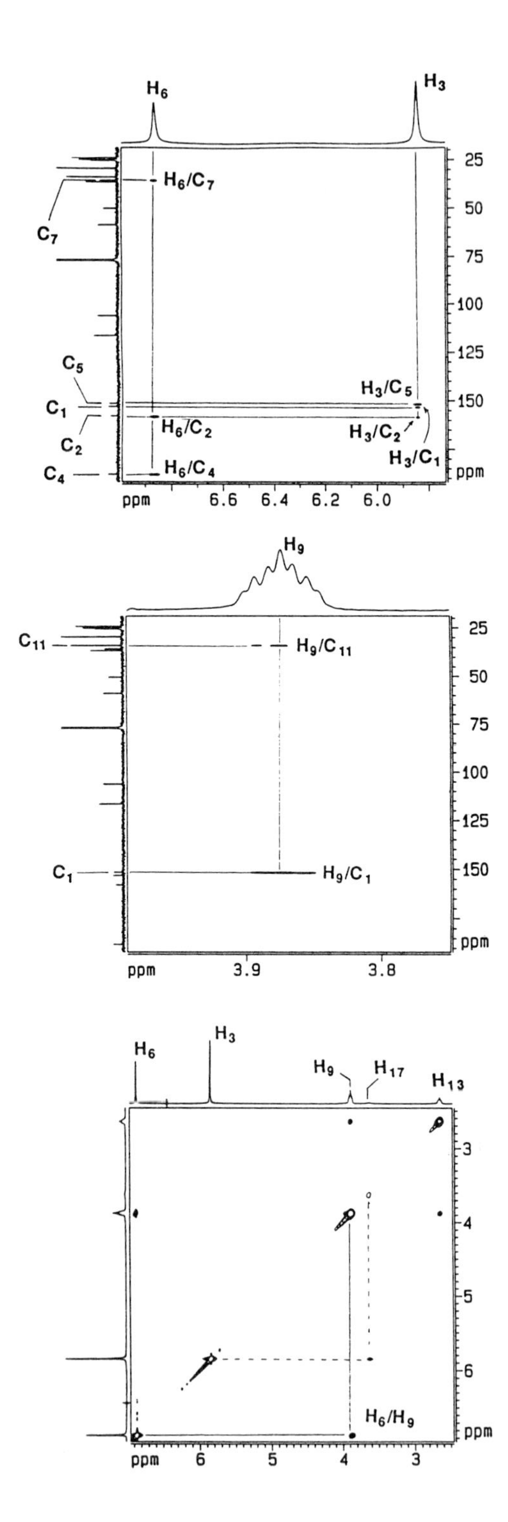

H6
H3
H6/C7
C7
C5
C1
C2
C4
H3/C5
H6/C2
H3/C2
H3/C1
H6/C4
ppm
6.6
6.4
6.2
6.0
25
50
75
100
125
150
H9
C11
H9/C11
C1
H9/C1
3.9
3.8
H17
H13
H6/H9
6
5
4
3

set to 7.250 ppm. ^{13}C chemical shifts are relative to the ^{13}C signal of $CDCl_3$ set to 77.0 ppm. The long-range inverse proton carbon correlation (HMBC) spectra are processed in magnitude mode; 512 t_1 increments of 2K data points are corrected.[11] A delay of 60 msec is used in the pulse sequence to allow for the evolution of long range couplings. The NOESY spectrum is obtained using a 100-msec mixing time. All two-dimensional spectra are zero fitted to twice their size in both dimensions prior to processing.

Spectral Assignments. The ^{13}C data of **3a** in $CDCl_3$ showed a typical carbonyl carbon signal at 188.1 ppm, which was assigned to C_4. The ^1H-NMR spectrum showed two ring protons with δ 5.821 and 6.838 chemical shifts. The proton at 5.821 ppm was assigned to H_3 on the basis of its up-field chemical shift and smaller intensity in CD_3OD due to the proton–deuterium exchange as seen in the quinone.[2] There were two 1H multiplets at 2.612 and 3.847 ppm for two cyclohexyl α-protons. The signal at 3.847 ppm was assigned to the cyclohexyl α-proton on the imine moiety because of its down-field chemical shift of 1.235 ppm, reflecting the electronic environment, adjacent to the quinone imine ring. To assign the carbons, a multiple-bond heteronuclear correlation experiment (HMBC) was performed. Figure 1 (top, middle) shows two two-dimensional cross-sections with the ^1H and ^{13}C one-dimensional spectra plotted along the horizontal and vertical axes, respectively. The imine carbon (C_1) was identified as the signal at 151.6 ppm by its strong $^3J_{C\text{-}H}$ correlation with H_3 (Fig. 1, top) and H_9 (Fig. 1, middle). In Fig. 1 (top), the strong correlations between H_6 (6.838 ppm) with C_4 (188.1 ppm), C_2 (158.0 ppm), and C_7 (35.7 ppm), and between H_3 (5.821 ppm) with C_1 (151.6 ppm) and C_5 (153.2 ppm) are shown. In addition, a small $^2J_{C\text{-}H}$ correlation between C_2 (158.0 ppm) and H_3 (5.821 ppm) was observed. A ^{13}C signal at 106.3 ppm was assigned to C_3 since in CD_3OD it became a multiplet with small intensity due to the proton–deuterium exchange of H_3. Thus, the signal at 116.6 ppm was assigned to C_6. In a NOESY experiment (nuclear Overhauser enhancement spectroscopy), positive cross peaks (NOE effect) between H_6 and H_9, supporting the substitution at C_1 with the cyclohexyl-imino group, were observed (Fig. 1, bottom). On the other hand, negative cross peaks indicating exchangeable

[11] A. Bax and M. F. Summers, *J. Am. Chem. Soc.* **109,** 2093 (1986).

FIG. 1. Two-dimensional NMR data of the substrate Schiff base **(3a)** in $CDCl_3$. (Top) Parts of HMBC spectrum of **3a** covering correlations between the ring protons (H_3 and H_6) and carbons. (Middle) Parts of HMBC spectrum of **3a** covering correlations between the α-proton of the cyclohexyl imino group (H_9) and carbons. (Bottom) Parts of NOESY spectrum of **3a** covering the ring protons (H_3 and H_6) and cyclohexyl α-protons (H_9 and H_{13}).

TABLE I
UV/VIS SPECTRAL DATA OF TPQ ANALOGS

	Protonated form at pH 1.91	Anionic form at pH 6.73
1$_{ox}$	266[a] (13300)[b]	274 (9590)
	388 (710)	488 (1850)
2a	268 (17720)	274 (13070)
	388 (610)	488 (1870)
2d	266 (19540)	272 (13540)
	388 (730)	490 (2110)

[a] λ_{max}, nm.
[b] ε, M^{-1} cm^{-1}.

protons with either water or the amine (v < 10 sec) were detected between H_9 and H_{13}, and H_{17} and H_3. These results show that the position of the nucleophilic addition was at C_1, the carbonyl carbon next to the 2-hydroxyl group.

UV/VIS Properties

Topa Quinone Analogs. Solutions of topa quinone analogs (**1**$_{ox}$**H, 2a–2d**) show a pale yellow color at pH values below 4. At pH values above

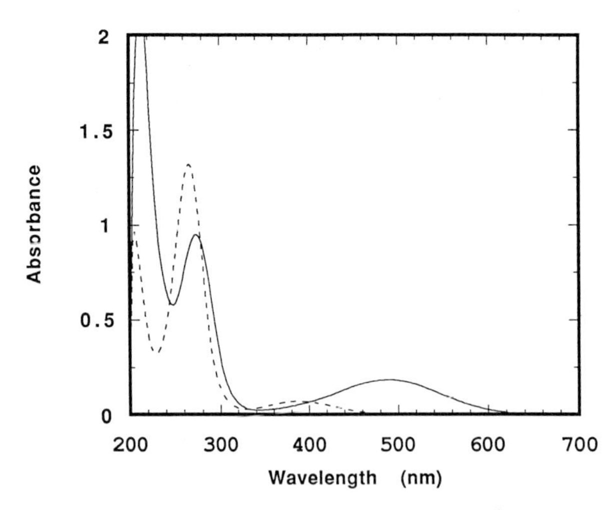

FIG. 2. Absorption spectra of the quinone. [**1**$_{ox}$**H**] = 9.91 × 10^{-5} M. (· · ·) **1**$_{ox}$**H** at pH 1.91, (——) **1**$_{ox}$$^-$ at pH 6.73.

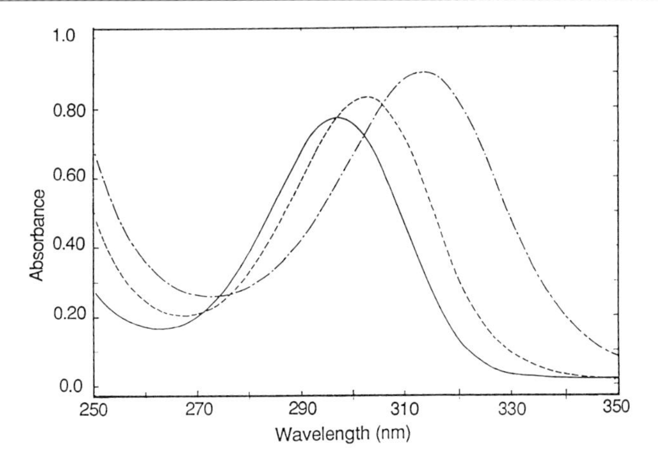

FIG. 3. Absorption spectra of the quinol. $[\mathbf{1_{red}H_3}] = 1.97 \times 10^{-4}$ *M*. (——) $\mathbf{1_{red}H_3}$ at pH 7.5, (· · ·) $\mathbf{1_{red}H_2}^-$ at pH 11.0, (– · –) $\mathbf{1_{red}H}^{2-}$ at pH 12.9.

4, solutions show a pink color. This is due to the appearance of an absorbance at around 488 nm. Spectroscopic titration determined the pK_a value of the hydroxyl group of the quinone **($\mathbf{1_{ox}H}$)** to be 4.13 ± 0.01 [Eq. (1)].[2] The resonance effect involving delocalization of electrons is

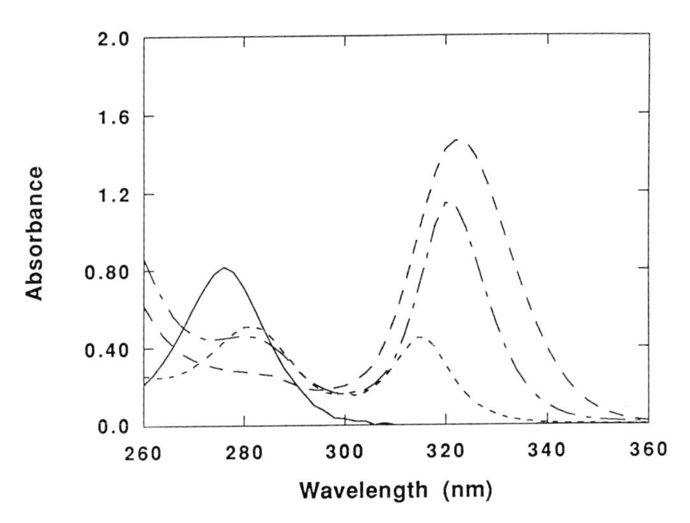

FIG. 4. Absorption spectra of the aminoresorcinol. $[\mathbf{4bH_2}] = 3.48 \times 10^{-4}$ *M*. (——) $\mathbf{4bH_3}^+$ at pH 3.9, (· · ·) $\mathbf{4bH_2}$ at pH 7.1, (– · –) $\mathbf{4bH}^-$ at pH 10.4, (---), $\mathbf{4b}^{2-}$ at pH 13.6.

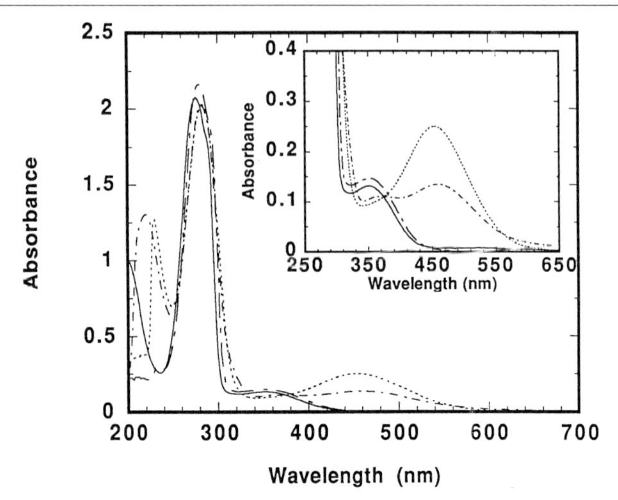

FIG. 5. Solvent effect for the UV/VIS spectra of the substrate Schiff base **(3a)**. **[3a]** = 1.0×10^{-4} *M*. (——) CH_2Cl_2, (- - -) CH_3CN, (- · -) MeOH, (· · ·) pH 9.9, 0.1 *M* carbonate buffer. (Insert) Expanded spectra.

$1_{ox}H$ $\underset{+H^+}{\overset{-H^+}{\rightleftharpoons}}$ 1_{ox}^- (1)

expected to stabilize the anionic structure **(1_{ox}^-)**. Table I summarizes values for λ_{max} and ε for species **1**, **2a**, and **2d**. The spectra of the protonated form

TABLE II
UV/VIS SPECTRAL DATA OF SUBSTRATE SCHIFF BASE **(3a)** AND PRODUCT SCHIFF BASE **(5)**

	CH_3CN	pH 10.0
3a	276[a] (20750)[b]	284 (20290)
	352 (1320)	454 (2500)
5	368 (13450)	ND

[a] λ_{max}, nm.
[b] ε, M^{-1} cm^{-1}.

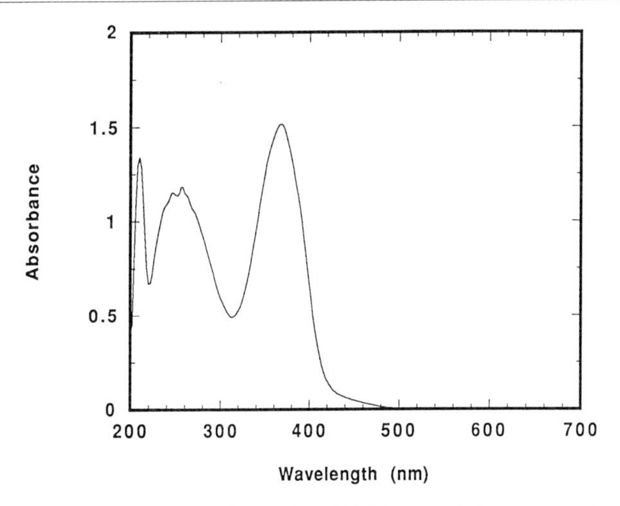

FIG. 6. Absorption spectrum of the product Schiff base **(5)** in acetonitrile. **[5]** = 1.0×10^{-4} *M*.

(1_{ox}H) and anionic form **(1_{ox}^-)** are shown in Fig. 2. Comparison of the enzyme to the model compound indicated an even lower pK_a of 3.0,[2] which may reflect the proximity of cupric ion or a positively charged amino acid side chain.

Quinol and Aminoresorcinol Analogs. Solutions of the quinol **($1_{ox}H_3$)** and the aminoresorcinols **($4aH_2$, $4bH_2$)** are colorless.[2] These compounds are highly air-sensitive and oxygen has to be removed from the solution by bubbling oxygen-free Ar through it.[10] The UV/VIS spectra of the quinol and aminoresorcinol at different pH are shown in Figs. 3 and 4, respectively. The quinol and the aminoresorcinol have no absorption at around 488 nm, comparable to the reduced form of enzymes. The pK_as of the quinol **($1_{ox}H_3$)** were determined to be 9.17, 11.66, and 13.0 [Eq. (2)] and those of the aminoresorcinol **($4bH_2$)** were determined to be 5.88, 9.59, and 11.62 [Eq. (3)].[2]

$$1_{red}H_3 \underset{+H^+}{\overset{-H^+}{\rightleftharpoons}} 1_{red}H_2^- \underset{+H^+}{\overset{-H^+}{\rightleftharpoons}} 1_{red}H^{2-} \underset{+H^+}{\overset{-H^+}{\rightleftharpoons}} 1_{red}^{3-} \qquad (2)$$

$$\mathbf{4bH_3^+} \underset{+H^+}{\overset{-H^+}{\rightleftharpoons}} \mathbf{4bH_2} \underset{+H^+}{\overset{-H^+}{\rightleftharpoons}} \mathbf{4bH^-} \underset{+H^+}{\overset{-H^+}{\rightleftharpoons}} \mathbf{5b^{2-}} \qquad (3)$$

Substrate Schiff Base Analogs. Figure 5 shows the spectra of the substrate Schiff base analog **(3a)** in different solvents. In a nonpolar aprotic solvent (CH_2Cl_2), **3a** has a λ_{max} at 352 nm. Changing the solvent to a polar aprotic one (CH_3CN) results in a slight difference of the spectrum. On the other hand, in a polar protic solvent (MeOH), a new band at λ_{max} 454 nm appears with a corresponding loss of the band at 352 nm. In water (pH 10.0), the band at 352 nm is completely gone and **3a** shows only a 454-nm absorption band. In an aprotic solvent, the amine salt of the substrate Schiff base **(3a)** exists as a contact ion pair with λ_{max} 352 nm, whereas in a polar protic solvent, **3a** is expected to be strongly solvated by hydrogen bonding and to exist as a solvent-separated ion pair (with λ_{max} at 454 nm). The origin of the 100-nm red shift of the λ_{max} is presumably the resonance effect involving electron delocalization of the anionic species between oxygens at carbons 2 and 4.

Comparison of the enzyme to model compound indicates that there is no such resonance effect in the substrate Schiff base complex of the enzyme. Species **B** (Scheme 1) in bovine serum amine oxidase has been assigned a λ_{max} at 340 nm[12] which can be rationalized by a strong electrostatic interaction of the oxyanion at C-4 with the protonated Schiff base complex on the neighboring carbon.

Product Schiff Base Analogs. The absorption spectrum of the product Schiff base **(5)** in CH_3CN is shown in Fig. 6. It has a λ_{max} at 368 nm and an ε value about 10 times larger than that of the substrate Schiff base (Table II), reflecting a double bond in conjugation with the benzene rings. The product Schiff base was very unstable in an aqueous solution and rapidly hydrolyzed to the aminoresorcinol and benzaldehyde.

[12] C. Hartmann, P. Brzovic, and J. P. Klinman, *Biochemistry* **32,** 2234 (1993).

[5] Catalytic Aerobic Deamination of Activated Primary Amines by a Model for the Quinone Cofactor of Mammalian Copper Amine Oxidases

By LAWRENCE M. SAYRE and YOUNGHEE LEE

Introduction

The copper amine oxidases utilize a covalently bound "active carbonyl" cofactor to achieve a pyridoxal-like transamination of primary amines to aldehydes, with the role of copper being to mediate the O_2-dependent reoxidation of the reductively aminated cofactor (H_2O_2 and NH_3 are by-products) subsequent to hydrolytic release of the aldehyde product. The cofactor for the copper amine oxidases has been shown to be the quinone form of an active site based 2,4,5-trihydroxyphenylalanine residue.[1,2] Quinone-mediated transamination of amines has precedent in the synthetically useful stoichiometric deamination of branched primary amines (giving ketones) by 3,5-di-*tert*-butyl-1,2-benzoquinone (DTBQ),[3] but a catalytic process has not been described, and DTBQ cannot be used for the conversion of unbranched primary amines (e.g., benzylamine) to aldehydes on account of its propensity toward a benzoxazole-forming side reaction. On the other hand, benzylamine is successfully deaminated by pyrroloquinoline quinone (PQQ), and the reaction can be made catalytic, although both transamination and addition-elimination mechanisms can occur.[4,5] Based on this past work, the course of deamination by the unique hydroxyquinone nucleus of topa quinone (TPQ) models could not be predicted with certainty. This chapter describes a TPQ model system which achieves catalytic deamination of activated amines. The synthesis and characterization of various forms of the catalyst implicated in the catalytic cycle are also reported.

In choosing an appropriate TPQ model, it should be appreciated that the simplest model, 2-hydroxy-1,4-benzoquinone, is itself commercially available in the reduced form (1,2,4-benzenetriol), but the absence of a C-5 alkyl substituent is known to result in an oxidative intermolecular

[1] S. M. Janes, S. Mu, D. Wemmer, A. J. Smith, S. Kaur, D. Maltby, A. L. Burlingame, and J. P. Klinman, *Science* **248,** 981 (1990).
[2] D. E. Brown, M. A. McGuirl, D. M. Dooley, S. M. Janes, D. Mu, and J. P. Klinman, *J. Biol. Chem.* **266,** 4049 (1991).
[3] E. J. Corey and K. Achiwa, *J. Am. Chem. Soc.* **91,** 1429 (1969).
[4] Y. Ohshiro and S. Itoh, *Bioorg. Chem.* **19,** 169 (1991).
[5] E. J. Rodriguez and T. C. Bruice, *J. Am. Chem. Soc.* **111,** 7947 (1989).

C–C coupling, with ultimate formation of 2,2′-bi-*p*-quinones.[6] Also, some workers have utilized the commercially available 2,4,5-trihydroxyphenylalanine as a cocatalyst with Cu(II) in the O_2-dependent deamination of benzylamine.[7,8] An alternative possibility is the descarboxy version, the well-known neurotoxin 6-hydroxydopamine.[9] However, neither of these two compounds is suitable as a robust catalyst because the α-amino group undergoes cyclocondensation at the quinone stage.[10,11] One TPQ model which closely reproduces the context of TPQ as it exists in the protein and which is not subject to cyclocondensation complications is the pivalamide **2** of 6-hydroxydopamine quinone.[12]

Results and Discussion

Catalytic deaminative turnover. Quinone **2** is prepared via $FeCl_3$ oxidation of the triol **1** generated by hydrogenolysis of the *N*-pivaloyl derivative of the known tri(benzyloxy)phenethylamine.[13] As is the case for the quinone forms of 2,4,5-trihydroxyphenylalanine and 6-hydroxydopamine, **2** gives weak yellow solutions at low pH ($\lambda_{max} \sim 370$ nm), but forms a deep red conjugate base anion ($\lambda_{max} \sim 490$ nm) at pH ≥ 5.

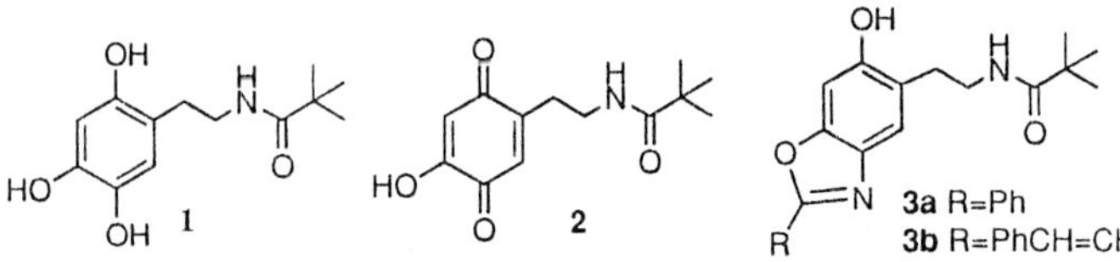

1 **2**

3a R=Ph
3b R=PhCH=CH

Table I lists data for the successful catalytic action of **2** in the O_2-dependent deamination of benzylamine (present in large excess) in buffered aqueous CH_3CN at 26°. Although the amine concentration is higher than the buffer concentration, the latter need only be in excess of the reaction stoichiometry to minimize the pH drop which otherwise occurs during the course of the reaction according to Eq. (1):

$$2\ PhCH_2NH_3^+ + O_2 \rightarrow PhCH_2N{=}CHPh + H_2O_2 + H^+ + NH_4^+. \quad (1)$$

[6] J. F. Corbett, *J. Chem. Soc.* C, p. 2101 (1970).
[7] N. Nakamura, T. Kohzuma, H. Kuma, and S. Suzuki, *J. Am. Chem. Soc.* **114,** 6550 (1992).
[8] M. A. Shah, P. R. Bergethon, A. M. Boak, P. M. Gallop, and H. M. Kagan, *Biochim. Biophys. Acta* **1159,** 311 (1992).
[9] R. N. Adams, E. Murrill, R. McCreery, L. Blank, and M. Karolczak, *Eur. J. Pharmacol.* **17,** 287 (1972).
[10] R. F. Chapman, A. Percival, and G. A. Swan, *J. Chem. Soc.*, C, p. 1664 (1970).
[11] S. Senoh, and B. Witkop, *J. Am. Chem. Soc.* **81,** 6231 (1959).
[12] F. Wang, J.-Y. Bae, A. R. Jacobson, Y. Lee, and L. M. Sayre, *J. Org. Chem.* **59,** 2409 (1994).
[13] F. G. H. Lee, D. E. Dickson, and A. A. Manian, *J. Med. Chem.* **14,** 266 (1971).

TABLE I
CATALYTIC AEROBIC DEAMINATIONS MEDIATED BY TPQ MODEL 2[a]

Substrate	pH	Buffer, conc. (cat.)	Time (hr)	Yield[a] (%)
Benzylamine	8.5	Boric acid, 50 mM	5	250
			18	294
Benzylamine	9.0	Phosphate, 5 mM	4	290
			23	420
Benzylamine	9.0	Phosphate, 5 mM (Cu^{2+})[b]	4	180
			23	240
Benzylamine	10.0	Boric acid, 8 mM[c]	3	330
			10	410
Benzylamine	10.0	Phosphate, 5 mM[c]	3	350
			4	385
			9	470
			23	610
			48	630
Benzylamine-d_2	10.0	Phosphate, 5 mM[c]	4	30
			23	39
Benzylamine	10.0	Phosphate, 5 mM (Cu^{2+})[b]	4	86
			23	99
Cinnamylamine[d]	10.0	Phosphate, 5 mM	14	180
Dibenzylamine[e]	8.0	Phosphate, 5 mM[c]	4	37
			23	146
Dibenzylamine[e]	9.0	Phosphate, 5 mM[c]	4	53
			23	147

[a] The reactions are 50 mM in amine and 1 mM in **2**, conducted at 26° in buffer–CH_3CN (7:3). Yields represent mol % of isolated 2,4-dinitrophenylhydrazine derivative of corresponding aldehyde on the basis of **2** present, and numbers shown are an average of two to five determinations.

[b] Presence of 1 mM cu$(ClO_4)_2$ and 1 mM 2,2′-bipyridine.

[c] The yields shown are corrected for the amount of aldehyde formed as a consequence of cofactor-independent autoxidation. Such autoxidation for benzylamine itself was negligible at pH 9 and below.

[d] The amine concentration was 25 mM.

[e] The solvent used was buffer–CH_3CN (1:1) (the dibenzylamine was incompletely soluble in the 7:3 system). The corrected yields do not account for **2**-mediated production of PhCH=O which may arise from benzylamine released in the non-**2**-dependent autoxidation of dibenzylamine.

Excessive buffer concentrations preclude the use of the mixed solvent system. NMR analysis of organic extracts reveals that the PhCH=O product is generated in the form of the hydrolytically stable Schiff base PhCH=NCH_2Ph under these reactions conditions. Although the reaction can be quantified on this basis, the yield of PhCH=O is most easily determined by weighing the precipitated 2,4-dinitrophenylhydrazone of

PhCH=O obtained upon workup of the reaction with the acidified 2,4-dinitrophenylhydrazine reagent.

In preliminary studies, we determined that the pH was the major factor controlling the deamination yield, not the identity of the buffering agent. Control studies showed that the amount of PhCH=O produced in the absence of **2** from simple autoxidation was negligible below pH 10, but required a correction to be made in the pH 10 yields. Table I shows that the 2-dependent deamination yield increases with increasing pH over the range of 8.5–10.0 and that deamination slows considerably over a period of several hours. The yields shown at the longest times represent essentially the t = infinity yields at each pH. The catalysis by **2** ceases on account of its irreversible conversion into forms which are incapable of turnover. The progress toward the ultimate fate of the catalyst is accompanied by a loss of the TPQ anion chromophore at 486 nm and growth of a strong, broad absorption in the 310- to 370-nm range, as evidenced by a color change of the reaction mixture from deep red to weak yellow.

α-Deuteration is known to slow the deamination of primary amines by copper amine oxidases according to a primary kinetic isotope effect.[14–16] Data in Table I show that deamination of benzylamine-d_2 is slowed considerably relative to its d_0 partner. Although relative yield data of this type cannot be translated into an assessment of the intrinsic deuterium kinetic isotope effect, it is clear that α-carbon–hydrogen bond cleavage is substantially rate limiting under our model deamination conditions. An increase in the difference between PhCH=O and PhCD=O yields at longer reaction times is probably an artifact of kinetic partitioning effects[17] associated with the side-reaction(s) which removes the active form of the catalyst.

Compared to benzylamine, cinnamylamine also undergoes deamination (quantified in terms of the 2,4-dinitrophenylhydrazine derivative of cinnamaldehyde), although the turnover yield is much lower on account of a more rapid deterioration of the catalyst (Table I). In contrast to these two activated amines, neither *trans*-2-phenylcyclopropylamine nor butylamine undergo any detectable deamination. This is in marked contrast to DTBQ, which, being a stronger transaminating agent, affects reaction of nonbenzylic as well as benzylic amines and induces oxidative ring cleavage of *trans*-2-phenylcyclopropylamine.[18]

[14] M. M. Palcic and J. P. Klinman, *Biochemistry* **22,** 5957 (1983); M. Farnum, M. Palcic, and J. P. Klinman, *ibid.* **25,** 1898 (1986); K. L. Grant and J. P. Klinman, *ibid.* **28,** 6597 (1989).

[15] P. H. Yu, *Biochem. Cell. Biol.* **66,** 853 (1988).

[16] B. Olsson, J. Olsson, and G. Pettersson, *Eur. J. Biochem.* **64,** 327 (1976).

[17] A. Thiblin and P. Ahlberg, *Chem. Soc. Rev.* **18,** 209 (1989).

[18] L. M. Sayre, M. P. Singh, P. K. Kokil, and F. Wang, *J. Org. Chem.* **56,** 1353 (1991).

Curiously, the activated secondary amine dibenzylamine appears to undergo some deamination under our model conditions. However, this amine is more susceptible to catalyst-independent autoxidation than benzylamine, and yield corrections (PhCH=O equivalents) were required at both pH 8 and 9. Moreover, these yield corrections cannot account for the **2**-mediated generation of PhCH=O from any benzylamine liberated in the autoxidation of dibenzylamine. Thus, the values given in Table I must be considered as upper estimate limits for the PhCH=O arising specifically from the reaction of **2** with dibenzylamine—the actual values may be very small. In this regard, it should be recalled that secondary amines are not substrates for the enzymes themselves.

In the previously studied catalytic aerobic deamination of benzylamine by 2,4,5-trihydroxyphenylalanine itself, the presence of (bipy)Cu(II) was claimed to be an essential ingredient.[7] However, data in Table I show that (bipy)Cu(II) is inhibitory in our catalytic system, with the inhibition being greater at higher pH. Since Cu(II) would be expected to aid in the autoxidative recycling of the reduced catalyst, the different effect of Cu(II) in our case compared to the 2,4,5-trihydroxyphenylalanine case must arise from differences in the nature of coordination of Cu(II) at the amine processing stage. For our model, one would expect to observe coordination of Cu(II) in a five-membered chelate ring with the oxo-enolate portion of the anionic **2** nucleus. We suspect that such coordination inhibits initial condensation of **2** with the amine, and that any aid by Cu(II) in autoxidative recycling must be of little significance. In the case of 2,4,5-trihydroxyphenylalanine, the strongest coordination of Cu(II) must be to the free amino carboxylate end of the molecule [strong chelation of amino acids by Cu(II) is well known]. The favorable effect of Cu(II) in this case probably arises mainly because coordination of Cu(II) to the free amino group prevents its cyclization into the quinone nucleus, which otherwise results in a loss of deaminating capability. By itself, 2,4,5-trihydroxyphenylalanine has been reported to be an ineffective catalyst for the aerobic deamination of benzylamine.[19]

Catalyst Destruction

We endeavored to determine the nature of irreversible loss of turnover competence of our catalyst. One possible factor we considered was decomposition of the catalyst by the H_2O_2 generated on autoxidative recycling of the reduced catalyst. We found that the addition of H_2O_2 to **2** anion at pH 8–10 caused a time-dependent but substoichiometric loss of the 486-nm chromophore in a manner which could be prevented by the presence

[19] K. Kano, T. Mori, B. Uno, M. Goto, and T. Ikeda, *Biochim. Biophys. Acta* **1157,** 324 (1993).

of catalase in the reaction mixture. However, since catalase did not enhance the turnover yields with benzylamine at any time point in the reaction (data not shown), H_2O_2 could be ruled out as the major cause of catalyst deterioration.

A second possibility is related to the side reaction that occurs in the case of DTBQ-mediated transamination of unbranched primary amines, wherein the product Schiff base generated on transamination is in equilibrium with a cyclized tautomer (dihydrobenzoxazole) which is oxidized to a benzoxazole[3] by starting DTBQ[20,21] more rapidly than DTBQ is consumed by condensation with amine. In our catalytic system using **2**, an analysis of the neutral organic materials present at the point where deamination turnover for benzylamine and cinnamylamine had ceased resulted in the isolation of benzoxazole products **3**. NMR spectral analysis of the crude organic extract indicated that these benzoxazoles are the major end products accounting for 30–40% of the catalyst **2**, although several other species are also present as evidenced by the number of *t*-butyl singlets appearing in the high field ^1H NMR spectra. The more rapid catalyst deterioration that occurs in the case of cinnamylamine (thus resulting in inferior turnover yields) may reflect the greater ease of oxidation at the dihydrobenzoxazole stage in this case (relative to the productive turnover pathway). In contrast, the persistence of catalyst in the dibenzylamine reactions is readily explained by the impossibility of benzoxazole formation in this case.

Based on the realization that turnover yield is limited by eventual benzoxazole formation, we guessed that improved deamination yields might be seen for the branched primary amine, α-methylbenzylamine (1-phenethylamine), which gives excellent yields of acetophenone using DTBQ.[3] In the case of **2**, however, only poor turnover yields of the 2,4-dinitrophenylhydrazine derivative were seen in 24 hr (less than 100%). As was observed for dibenzylamine, analysis of the reaction mixture at 24 hr revealed the persistence of unconverted **2**, suggesting that the limiting factor in this case may be steric interference by the α-methyl group with one or more steps in the deamination mechanism rather than catalyst deterioration. It is interesting to note that branched primary amines are also poor substrates for the copper amine oxidase enzymes, presumably for steric reasons.

[20] DTBQ rather than O_2 is the oxidant: M. P. Singh, P. B. Kokil, B. Venkataraman, M. Klein, and L. M. Sayre, *Abstr. Pap., 199th Natl. Meet., Am. Chem. Soc.*, Boston, *1990*, Abstr. ORGN 429 (1990).

[21] A similar reaction occurs with amino acids: M. C. Vander Zwan, F. W. Hartner, R. A. Reamer, and R. Tull, *J. Org. Chem.* **43,** 509 (1978).

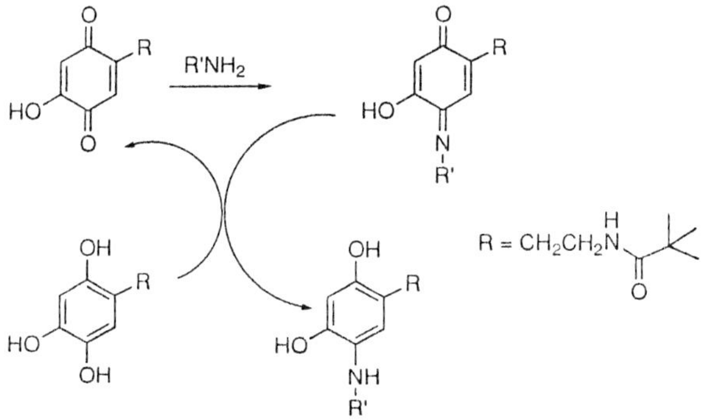

SCHEME I

*Independent Preparation of Benzoxazole **3a** from Aminoresorcinol **4**, and General Conversion of Triol **1** to (Alkylamino)resorcinols*

The benzoxazole structure **3** implicates condensation of amine at the more electrophilic C-5 carbonyl of **2** (using the numbering scheme for 2,4,5-trihydroxyphenylalanine). The other possible benzoxazole isomer, which would have the N and O reversed, could not be excluded by any of the spectral data and would arise from some type of transamination mechanism initiated by reaction of amines at C-4. In order to prove that the benzoxazole structure is **3**, we carried out an independent synthesis in the case of the benzylamine-derived material **3a**. We have shown that the reduced benzenetriol forms of TPQ models can be readily converted to C-5 alkylamino-substituted derivatives by reaction of the triol with alkyl amines.[12] The reaction involves the catalytic action of a trace of quinone present in the triol preparation, wherein the quinoneimine formed from the trace of quinone is reduced by triol to give (alkylamino)resorcinol and regenerate the quinone catalyst (Scheme I). The regiochemistry of this substitution was verified by a combination of NOE difference and long-range ^{13}C–^{1}H coupling data.[12] In fact, our observation confirmed the proposal that amine condensation is directed to the electrophilic C-5 carbonyl of TPQ models.[22] According to Scheme I, bubbling of NH_3 through a solution of triol **1** in CH_3CN gave the aminoresorcinol **4**. Reaction of **4** with a twofold excess of PhCH=O in CH_3CN, followed by removal of the solvent (and excess PhCH=O), yielded the Schiff base **5** (by NMR and HRMS), which upon exposure to

[22] M. Mure and J. P. Klinman, *J. Am. Chem. Soc.* **115,** 7117 (1993).

air was observed to be predominantly converted, presumably via oxidation of its cyclic tautomer **6**, to benzoxazole **3a** (Eq. 2). We do not know whether

$$1 \xrightarrow[\text{Scheme I}]{NH_3} \mathbf{4} \overset{PhCHO}{\rightleftharpoons} \mathbf{5} \rightleftharpoons \mathbf{6} \xrightarrow{[O]} \mathbf{3a} \qquad R = CH_2CH_2NHC(O)C(CH_3)_3 \tag{2}$$

this oxidation is effected directly by O_2 or by a catalytic amount of quinone **2** present in the reaction mixture (in which case the reduced form **1** would be reconverted to **2** in air[12]).

Mechanisms for Catalytic Deamination

There are basically two distinct mechanisms for quinone-catalyzed deamination of primary amines that have been discussed, most recently for pyrroloquinoline quinone (PQQ) and its analogs: (i) transamination operating via a quinoneimine intermediate, and (ii) addition elimination involving β-elimination of aldimine at the carbinolamine preceding the quinoneimine. The copper amine oxidases appear to utilize the transamination mechanism,[22] since the aldehyde product can be observed anaerobically in single turnover studies, with NH_3 being observed only upon O_2-mediated regeneration of the quinone cofactor.

Application of these mechanisms to our present model in the case of benzylamine is shown in Scheme II. The transamination mechanism (Path A) would involve tautomerization of the quinoneimine **8** to give the PhCH=O Schiff base **5**. In the presence of high [$PhCH_2NH_2$], the PhCH=NCH_2Ph product would undoubtedly arise from direct interception of **5** by $PhCH_2NH_2$ rather than via hydrolysis of **5** and recondensation of PhCH=O with $PhCH_2NH_2$. The reductively aminated catalyst **4** thereby released would be oxidized by O_2, generating the quinoneimine **9**, in turn being converted to quinone **2** by hydrolysis and/or directly to **8** by transimination with $PhCH_2NH_2$ (not shown). Alternatively, product Schiff base **5** could, in the form of its tautomer **6**, undergo oxidation to benzoxazole **3a**. According to this scenario, a key factor governing catalyst lifetime would be the partitioning of **5** between direct oxidation to **3a** and transimination with $PhCH_2NH_2$ prior to oxidation, so that an increased amount of PhCH=O product would be expected at higher [$PhCH_2NH_2$] (all other factors remaining unchanged).

If our model deaminations followed the alternative addition-elimination mechanism (Scheme II, Path B), the triol **1** formed on β-elimination of PhCH=NH from carbinolamine **7** would be autoxidized back to quinone

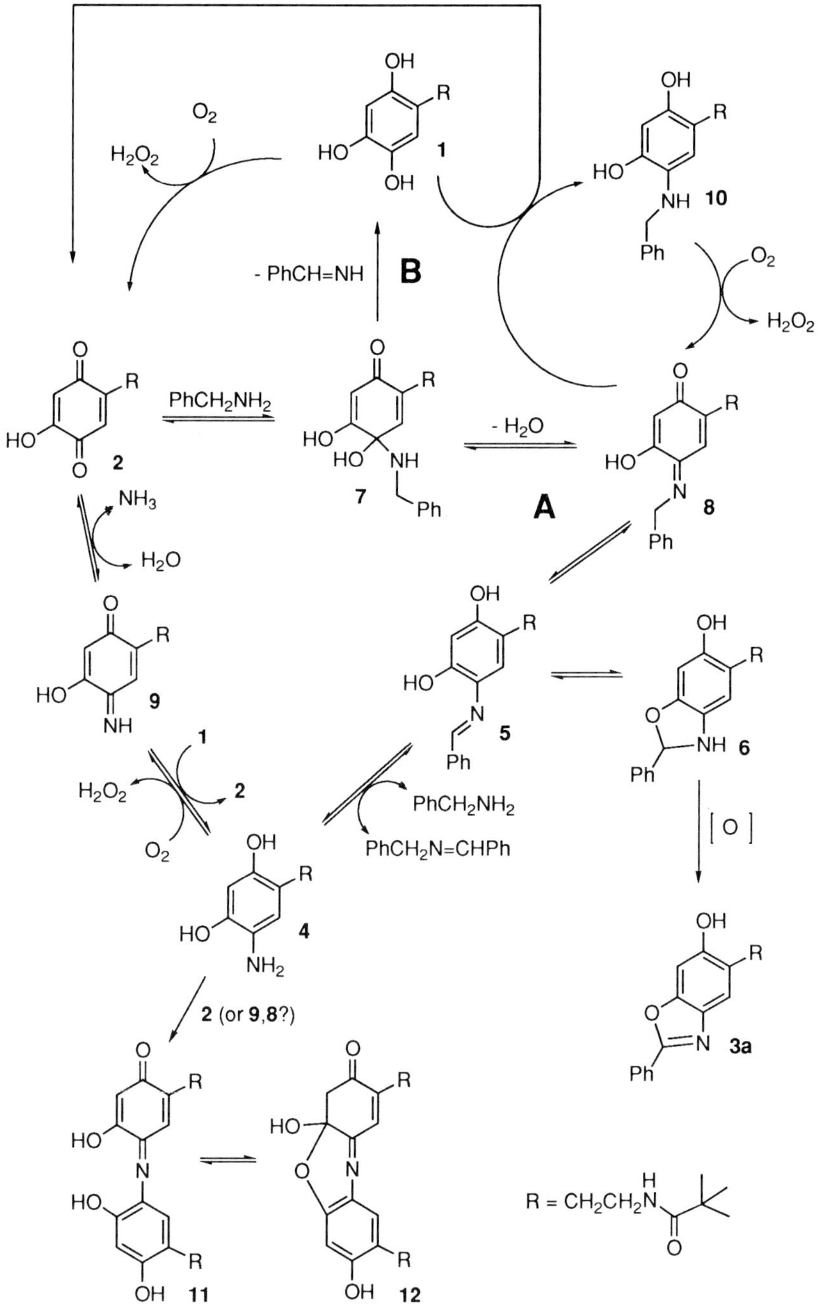

O_2
H_2O_2
OH
R
HO
OH
1
NH 10
Ph
O_2
H_2O_2
- PhCH=NH
B
O
R
$PhCH_2NH_2$
HO
O 2
- H_2O
HO NH
7 Ph
A
N 8
NH_3
H_2O
9
NH
5
Ph
O
NH 6
Ph
1
H_2O_2
2
O_2
$PhCH_2NH_2$
$PhCH_2N=CHPh$
[O]
4
NH_2
3a
N
2 (or 9,8?)
11
12
R = CH_2CH_2N(H)C(O)C(CH_3)$_3$

SCHEME II

2 under the reaction conditions.[12] In this case, the product PhCH=NCH_2Ph would undoubtedly arise from the direct reaction of PhCH=NH with $PhCH_2NH_2$. If Path B is the mechanism for deaminative turnover, dehydration of **7** to quinoneimine **8** would constitute a side reaction, presumably responsible for the inevitable "draining off" of active catalyst in the form of the isolated benzoxazole side product **3a**.

Reaction Analysis under Anaerobic Conditions

Theoretically, the two mechanisms can be distinguished through determination of the product of quinone reduction when the reaction is conducted under anaerobic (single turnover) conditions; transamination (Path A) results in aminoresorcinol **4**, whereas carbinolamine elimination (Path B) results in triol **1**. Since both products are easily autooxidized, a more practical approach than direct identification is to characterize them in the form of air-stable derivatives. In this manner, the final mixture of anaerobic deamination is evaporated *in vacuo*, and the residue is treated under argon with an excess of acetic anhydride to afford the triacetylated derivative. The triacetyl forms of **1** and **4** were prepared independently for comparison purposes. The isolated product from the anaerobic benzylamine reaction was shown to be neither **1** nor **4**, and instead was exclusively the triacetylated derivative of (benzylamino)resorcinol **10**. The finding of **10** as the product of quinone reduction can arise if the mechanism follows Path B, since we already showed that triol **1** is converted to **10** in the presence of excess benzylamine (Scheme I),[12] catalyzed by quinone **2**. However, we also found that aminoresorcinol **4** is converted to **10** in the presence of excess benzylamine, catalyzed by either quinone **2** or quinoneimine **9**,[23] so that **10** would be the expected product of Path A as well. The ambiguity of this product study highlights the difficulty of establishing mechanisms in this model system.

Another approach is to follow the anaerobic reactions simultaneously by ^1H NMR and UV/VIS spectroscopy in an effort to distinguish pathway-specific intermediates. For this purpose, the large excess of amine used in the aerobic turnover experiments of Table I is not amenable to NMR studies. Using instead a 3-fold excess of benzylamine in CH_3CN–H_2O (9:1) (with or without pH 10 phosphate buffer), the **2** anion absorption at 490 nm was seen to convert to the 300-nm absorption characteristic of **10** (~2.5-fold greater extinction), with isosbestic points at 290 and 350 nm. The conversion of **2** to **10** is confirmed by ^1H NMR in CD_3CN–H_2O (no buffer added in this case). At a long reaction time, a broad absorption at

[23] Y. Lee and L. M. Sayre, *J. Am. Chem. Soc.* **117,** 3096 (1995).

590 nm appears, accompanied by a blue-violet coloration of the solution. The indophenol **11** was suspected as the most likely species exhibiting this unusual spectrum, an assignment supported by our ability to recreate the spectrum through independent synthesis of **11** by condensing **4** with **2**.

Curiously, the reaction of $PhCD_2NH_2$ under the same conditions follows a very different reaction course, where the **2** anion chromophore at 490 nm is replaced by a strong dual 335/345-nm absorption with a single isosbestic point at 395 nm. The same behavior is seen for "unactivated" primary amines (though not for $PhCH_2NH_2$). We previously attributed[12] the 335/345 absorption to the quinoneimine (e.g., **8**) on the basis that a 340-nm absorption seen in rapid-scan studies on the anaerobic reaction of bovine serum amine oxidase with benzylamine was proposed to represent the substrate-derived quinoneimine.[24] However, isolation of our 335/345-absorbing compounds, exhibiting two vinyl $^1H/^{13}C$ signals that could be mistaken as the quinoneimines when following reactions by NMR, actually afforded a fragmentation product derived from the C-1 Michael addition of amine to the quinoneimine and subsequent elimination of the pivalamidoethyl side chain.[23] Thus, the 335/345 absorption generated in our case corresponds to a side reaction peculiar to our model which predominates in the absence of high amine deaminative reactivity.

Although a detailed discussion of the mechanism is beyond the scope of this chapter, any complete mechanistic assignment requires an evaluation of all potential intermediates that can be directly accessed by independent synthesis. In this regard, at least one example of each species shown in Scheme II other than the unstable carbinolamine **7** and dihydrobenzoxazole **6** has been independently characterized. We have not characterized the benzylamine-derived quinoneimine **8**, but the NH_3-derived quinoneimine **9** could be generated by exposure of **2** to aqueous CH_3CN solutions of NH_3. Increasing concentrations of NH_3 resulted in an increasingly more rapid and more complete isosbestic shift of the 486-nm (red) absorption for **2** anion to a λ_{max} 448-nm absorption (orange) with essentially the same extinction. The 448-nm absorption and chemical shift positions seen in the ^1H NMR spectrum suggest to us the existence of quinoneimine **9** as an *anion.*

Experimental Section

General Methods

NMR spectra were obtained on a Varian Gemini 300 instrument (^{13}C NMR at 75 MHz), with chemical shifts being referenced to TMS or the

[24] C. Hartmann, P. Brzovic, and J. P. Klinman, *Biochemistry* **32,** 2234 (1993).

solvent peak. For spectra taken in CD_3CN-H_2O, the CD_3CN ^1H and ^{13}C signals served as reference, and the water signal was suppressed. High-resolution mass spectra (HRMS, electron impact) were obtained at 20–40 eV on a Kratos MS-25A instrument. UV-visible spectra were obtained using a jacketed (temperature-controlled) cell compartment. For anaerobic UV/VIS and NMR experiments, the solvent was thoroughly degassed by freeze–pump–thaw prior to the admission of argon. Doubly distilled water was used for all experiments. Melting points are uncorrected. Thin-layer and preparative-layer chromatography were run on Merck silica gel 60 plates with a 254-nm indicator. All solvents, reagents, and organic fine chemicals were the most pure available from commercial sources. Amines were freshly fractionally distilled under N_2 from NaOH pellets. Benzylamine-d_2 was prepared from $LiAlD_4$ (98% atom D) reduction of benzamide. 2,4,5-Tribenzyloxybenzaldehyde was purchased from Regis Chemical Company. All evaporations were conducted at reduced pressure using a rotary evaporator.

2,4,5-Tribenzyloxyphenethylamine

Essentially as described,[13] a mixture of 5 g (11.8 mmol) of 2,4,5-tribenzyloxybenzaldehyde, 0.45 g of NH_4OAc, and 60 ml of nitromethane was heated at reflux for 5 hr under N_2. The yellow-orange precipitate obtained on cooling was collected by filtration, washed with water, cold methanol, and hexane, and dried in a vacuum oven overnight, affording 4.6 g (88%) of 2,4,5-tribenzyloxy-β-nitrostyrene: mp 139–140° (lit.[13] 138–140°). The latter material (4.6 g, 9.9 mmol) was reduced with $LiAlH_4$ in THF as described.[13] The solid obtained after the aqueous workup was extracted twice with hot THF, and the combined THF extract was dried (Na_2SO_4) and evaporated to give 3.0 g (70%) of the white amine: mp 75–77° (lit.[13] 70–77°).

N-(2,4,5-Trihydroxyphenethyl)pivalamide ***(1)***

A solution of 0.7 g (5.8 mmol) of pivaloyl chloride in 10 ml of dry ether was added dropwise with stirring to a solution of 2.4 g (5.8 mmol) of 2,4,5-tribenzyloxyphenethylamine and 0.88 g (8.7 mmol) of triethylamine in 125 ml of dry ether. After 1 hr at room temperature, the white solid that had precipitated was filtered and extracted with ether. The combined ether layers were washed with 40 ml of water, dried (Na_2SO_4), and evaporated to give 2.8 g (97%) of *N*-(2,4,5-tribenzyloxyphenethyl)pivalamide as a white solid: mp 79–80°. A solution of 2.4 g (4.6 mmol) of the latter material in 70 ml of EtOH containing 0.70 g of 10% Pd/C was subjected to H_2 at 60 psi in a Parr apparatus at 40° for 3 hr. The catalyst was removed by filtration, the filtrate was evaporated, and the residue was washed with $CHCl_3$ (3 ×

15 ml), yielding 1.3 g (81%) of **1** as a light brown solid: mp 58–60°. Spectroscopic data for **1** and its tri(benzyloxy) precursor are given in Wang *et al.*[12]

2-Hydroxy-5-(2-pivalamidoethyl)-1,4-benzoquinone **(2)**

To a solution of 0.6 g (2.4 mmol) of **1** in 6 ml CH_3CN was added 1.6 g (6.0 mmol) of $FeCl_3 \cdot H_2O$ in 6 ml H_2O with stirring and cooling in an ice bath. After the addition was complete, the mixture was extracted with CH_2Cl_2 (3 × 60 ml). The combined organic layer was dried (Na_2SO_4) and evaporated, leaving a dark brown residue that was recrystallized from 5 ml of CH_3CN to give 0.33 g (55%) of a yellow solid: mp 138–139° (dec). Spectroscopic data are given in Wang *et al.*[12]

General Method for Aerobic Catalytic Deamination

A mixture of 5 mmol of benzylamine, 0.5 mmol of KH_2PO_4 (or 0.5 to 5 mmol boric acid), and 0.1 mmol of **2** in 100 ml of 30% aqueous CH_3CN was adjusted to pH 10.0 (or some other pH, see Table I) with KOH. The solution was magnetically stirred in an open 500-ml Erlenmeyer flask at 26° as vigorously as possible without splashing. The reaction volume was maintained by periodic addition of CH_3CN, which evaporates somewhat at long reaction times. Two 50-ml aliquots were worked up at different reaction times by addition to each of 18 ml of standard 2,4-dinitrophenylhydrazine reagent (142 m*M*, in 2.7 *N* H_2SO_4 in 70% aqueous EtOH). After cooling to 0° for 1 hr, the solution was filtered, and the precipitate was dried to constant weight to obtain the yield of the 2,4-DNP derivative of PhCH=O, the identity and purity of which were confirmed by TLC and ^1H NMR. Deamination of cinnamylamine was conducted in the same manner except that 2.5 mmol of the amine · HCl and 0.5 mmol of K_2HPO_4 were used. Deamination of dibenzylamine required the use of 50% aqueous CH_3CN as solvent. The 2,4-DNP derivatives of the corresponding aldehydes (benzaldehyde and cinnamaldehyde) were confirmed by TLC and ^1H NMR. Similar experiments on *n*-butylamine and *trans*-2-phenylcyclopropylamine · HCl did not yield any amine-derived carbonyl product.

Isolation of ***2****-Derived Benzoxazoles* ***3***

For the benzylamine and cinnamylamine reactions described earlier, aliquots after 24 hr were concentrated to half the original volume and extracted with EtOAc (2 × 40 ml). The organic layer was dried and evaporated, and the residue was subjected to silica gel chromatography (eluent EtOAc) to give, after recrystallization, analytical samples of the respective benzoxazoles: 6-Hydroxy-2-phenyl-5-[2-(pivalamido)ethyl]benzoxazole

(3a) (7% isolated yield): mp 208°; λ_{max} (CH_3CN-H_2O) 314 nm (sh 330 nm); ^1H NMR (DMSO-d_6) δ 1.05 (s, 9H), 2.77 (t, 2H, J = 7.1 Hz), 3.29 (m, 2H), 7.11 (s, 1H), 7.41 (s, 1H), 7.49 (bt, 1H, NH), 7.56–7.58 (m, 3H), 8.09–8.13 (m, 2H), 9.98 (s, 1H, OH); ^{13}C NMR (DMSO-d_6) δ 27.4, 30.1, 37.9, 39.0, 96.4, 120.2, 124.3, 126.6, 126.8, 129.2, 131.1, 133.7, 149.6, 154.2, 160.3, 177.2; HRMS calcd for $C_{20}H_{22}N_2O_3$ 338.1632, found 338.1661 (M^+, 28%). 6-Hydroxy-2-styryl-5-[2-(pivalamido)ethyl]benzoxazole **(3b)** (15% isolated yield): mp 159°; λ_{max} (CH_3CN-H_2O) 342 nm; ^1H NMR (DMSO-d_6) δ 1.05 (s, 9H), 2.50 (t, 2H, J = 6.9 Hz), 3.28 (m, 2H), 7.06 (s, 1H), 7.23 and 7.66 (2d, 1H each, J = 16.4 Hz), 7.34 (s, 1H), 7.40 (m, 1H), 7.42 (m, 2H), 7.47 (bt, 1H, NH), 7.75 (m, 2H), 9.97 (bs, 1H, OH); ^{13}C NMR (DMSO-d_6) δ 27.4, 30.0, 37.9, 39.0, 96.2, 114.1, 120.1, 124.1, 127.5, 128.8, 129.4, 134.1, 135.1, 137.2, 149.2, 154.2, 160.5, 177.2; HRMS calcd for $C_{22}H_{24}N_2O_3$ 364.1788, found 364.1768 (M^+, 7%).

General Method for Preparation of 4-(Alkylamino)-6-[2-(pivalamido)ethyl]resorcinols (e.g., ***10****)*

To a solution of triol **1** (containing a trace of quinone **2** as evidenced by its slight yellow coloration) in dry degassed CH_3CN was added 1.05 equiv of amine. Evaporation of the solvent after several hours affords the desired product, for which NMR spectra can be obtained under argon without any noticeable decomposition. Solvents other than dry CH_3CN (e.g., wet CH_3CN, DMSO) can be used for this reaction except in the case of $PhCH_2NH_2$, for which some deamination accompanies the generation of **10**. Spectroscopic data are as follows: Using benzylamine gives 4-(benzylamino)-6-[2-(pivalamido)ethyl]resorcinol **(10)**: ^1H NMR (DMSO-d_6) δ 1.07 (s, 9H), 2.48 (t, 2H, J = 7.3 Hz), 3.12 (app q, 2H), 4.21 (s, 2H), 6.21 (s, 1H), 6.35 (s, 1H), 7.19–7.34 (m, 5H), 8.38 (br s, 1H, NH), 9.0 (br, 1H, OH). ^{13}C NMR (DMSO-d_6) δ 27.3, 29.5, 37.8, 39.9, 47.8, 102.7, 112.9, 115.3, 126.5, 127.2, 128.1, 129.7, 140.8, 143.2, 145.9, 177.2. HRMS calcd for $C_{20}H_{26}N_2O_3$ 342.1944, found 342.1935 (15.9%). Using neopentylamine gives 4-(neopentylamino)-6-[2-(pivalamido)ethyl]resorcinol: ^1H NMR (CD_3CN) δ 0.96 (s, 9H), 1.11 (s, 9H), 2.64 (t, 2H, J = 6.9 Hz), 2.79 (s, 2H), 3.26 (app q, 2H, CH_2NH), 6.38 (app s, 2H), 6.78 (br t, 1H, NH); ^{13}C NMR (CD_3CN) δ 27.81, 27.97, 30.54, 32.45, 39.15, 41.60, 58.09, 103.98, 115.20, 117.21, 132.45, 144.90, 147.23, 180.11. HRMS calcd for $C_{18}H_{30}N_2O_3$ 322.2258, found 322.2246 (21.8%).

4-Amino-6-[2-(pivalamido)ethyl]resorcinol ***(4)***

Through a solution of triol **1** (50.6 mg, 0.2 mmol, containing a trace of quinone **2**) in 0.5 ml of degassed CH_3CN was bubbled NH_3 gas for 2 min.

The mixture was allowed to stand for 2 hr in a sealed system, and the solvent (and NH_3) was removed under high vacuum to leave an air-sensitive resin: ^{1}H NMR (CD_3CN) δ 1.09 (s, 9H), 2.57 (t, 2H, J = 6.7 Hz), 3.22 (m, 2H), 6.36 (s, 1H), 6.38 (s, 1H), 6.89 (b, >1H, NH); ^{13}C NMR (CD_3CN) δ 27.8, 30.2, 39.1, 41.7, 104.4, 117.3, 118.8, 128.7, 145.3, 148.5, 179.9; HRMS calcd for $C_{13}H_{20}N_2O_3$ 252.1469, found 252.1473 (M^+, 17.1%).

General Method for Spectral Monitoring and Product Identification for Anaerobic Reactions of Amines with Quinone Cofactor Model **2**

For NMR analysis of anaerobic reactions, a 20 m*M* solution of **2** in 0.5 ml of CD_3CN–H_2O (9 : 1) was degassed, and 3–5 equiv of amine was added under argon via a microliter syringe. For UV/VIS studies, we use CH_3CN–H_2O (9 : 1) solutions of **2** (5–10 m*M*) and 3–5 equiv of amine (in some cases the solvent H_2O component is 0.3 *M* phosphate buffer). The use of 0.1-cm cuvettes is needed at these concentrations of **2** (1-cm cuvettes require the use of concentrations of **2** which are too dilute for the conduct of parallel NMR studies).

Since the three possible products of benzylamine reduction of **2**, triol **1**, aminoresorcinol **4**, and (alkylamino)resorcinol **10**, are all unstable in air, a quantitative product analysis is best accomplished after a per-acetylation derivatization step. At the end of the observation period, all volatiles are removed under high vacuum. To the resulting residue (or, to the independently generated **1**, **4**, or **10**) dissolved in a minimal amount of DMSO, 50 equiv of Et_3N and then 50 equiv of acetic anhydride are added under argon. Utilization of a smaller excess of the reagent risks isolation of some diacetylated derivative. After 24 hr, the mixture is taken to dryness at high vacuum, and the residue is partitioned between water and EtOAc. Drying of the organic extract (Na_2SO_4) and evaporation afford the triacetylated derivatives, which can be purified by silica gel chromatography (EtOAc eluant) if necessary.

The *O,O,O*-triacetylated derivative of **1** displays ^{1}H NMR ($CDCl_3$) δ 1.11 (s, 9H), 2.239/2.245/2.31 (3s, 3H each, $CH_3C{=}O$), 2.70 (t, 2H, J = 6.8 Hz), 3.41 (app q, 2H, CH_2NH), 5.84 (br t, 1H, NH), 6.97 (s, 1H), 7.02 (s, 1H). The *N,O,O*-triacetylated derivative of **4** displays ^{1}H NMR ($CDCl_3$) δ 1.12 (s, 9H), 2.22/2.28/2.36 (3s, 3H each, $CH_3C{=}O$), 2.76 (t, 2H, J = 6.9 Hz), 3.45 (app q, 2H, CH_2NH), 5.83 (br t, 1H, NH), 7.09 (s, 1H), 7.11 (s, 1H). The *N,O,O*-triacetylated derivative of **10** exhibits large diastereotopic splittings on account of restricted rotation: ^{1}H NMR ($CDCl_3$) δ 1.12 (s, 9H), 1.88/2.16/2.36 (3s, 3H each, $CH_3C{=}O$), 2.61 (m, 2H), 3.19 and 3.37 (2m of symmetric ABXY, 1H each), 4.30 and 5.21 (2d, 1H each, J = 14.3 Hz, $PhCH_2$), 5.62 (br t, 1H, NH), 6.76 (s, 1H), 7.01 (s, 1H), 7.17–7.27 (m,

5H). ^{13}C NMR ($CDCl_3$) δ 20.69, 20.98, 22.26, 27.54, 29.49, 38.63, 39.38, 51.69, 118.30, 127.62, 128.47, 129.10, 129.81, 131.33, 132.49, 137.27, 145.67, 148.75, 168.30, 169.04, 170.79, 178.49. HRMS calcd for $C_{26}H_{32}N_2O_6$ 468.2262, found 468.2284.

*Independent Synthesis of **3a** from **4** and Benzaldehyde via **5***

The aminoresorcinol **4** obtained from 0.2 mmol of **1** as described earlier was dissolved in 1 ml of CH_3CN, to which benzaldehyde (40 μl, 0.4 mmol) was added. After 30 min, the solvent and excess PhCHO were removed under high vacuum. Characterization of the residue was consistent with the Schiff base **5**: λ_{max} ($CH_3CN–H_2O$) 370 nm. ^1H NMR (CD_3CN) δ 1.12 (s, 9H), 2.71 (t, 2H, J = 7.1 Hz), 3.32 (m, 2H), 6.47 (s, 1H), 6.75 (br t, 1H, NH), 7.18 (s, 1H), 7.44–7.47 (m, 3H), 7.92–7.96 (m, 2H), 8.66 (s, 1H, N=CH). ^{13}C 27.8, 30.6, 39.2, 40.9, 102.8, 118.5, 119.1, 128.8, 129.4, 129.7, 131.8, 137.6, 153.4, 154.9, 157.3, 180.2. HRMS calcd for $C_{20}H_{24}N_2O_3$ 340.1788, found 340.1788 (23.5%).

The residue **(5)** was taken up in 50 ml of $CH_3CN–H_2O$ (1:1), which was adjusted to pH 10 and stirred for 16 hr open to the air. The reaction mixture was concentrated to remove CH_3CN and extracted with EtOAc. The organic extract was dried (Na_2SO_4) and evaporated. The residue was shown by TLC (EtOAc) and NMR (DMSO-d_6) to contain benzoxazole **3a** as a major component.

*Independent Synthesis of **11** from **4** and **2***

The aminoresorcinol **4** was prepared as described earlier from 0.23 mmol of **1** in degassed CH_3CN. After removal of the solvent, quinone **2** (0.23 mmol) was added to the residue dissolved in 0.5 ml of degassed DMSO-d_6. The NMR spectrum at this time indicated the presence of a new 1:1 adduct (coinciding with appearance of an absorption at 580 nm) along with trace amounts of **4** and triol **1**. Purification of **11** was accomplished by concentration at high vacuum and application of the residue to a preparative TLC plate, eluting with EtOAc–acetone (3:1). Extraction of the fast-moving yellow band (λ_{max} 365 nm) with EtOAc–acetone affords the neutral material, whereas extraction with MeOH–acetone affords the sodium salt of the anionic form (silica contains sodium ions) responsible for the 590-nm absorption. The NMR spectra of the neutral form in DMSO-d_6 are most consistent with **11** existing as the C-4/C-4′-bridged cyclic hemiketal **12**,[23] resulting in large diastereotopic ^1H NMR splittings of one of the CH_2CH_2 side chains and conversion of one sp^2 ^{13}C signal to a sp^3 ^{13}C signal: ^1H NMR (DMSO-d_6) δ 1.00 (s, 9H), 1.06 (s, 9H), 2.30 (m, 1H), 2.37 (m, 1H), 2.64 (t, 2H), 3.12 (AB, 2H), 3.22 (m, 4H), 6.46 (s, 1H), 6.94

(s, 1H), 7.11 (s, 1H), 7.3 (variable, br s, 1H, slowly exchanged OH), 7.48 (m, 2H, NH). ^{13}C NMR (DMSO-d_6) δ 27.36 (CH_3), 27.42 (CH_3), 28.99 (CH_2), 29.80 (CH_2), 36.87 (CH_2), 37.89 (C), 38.96 (CH_2), 49.90 (CH_2), 88.65 (C), 102.63 (CH), 120.72 (C), 126.29 (C), 129.44 (CH), 138.72 (CH), 142.15 (C), 143.66 (C), 151.54 (C), 157.00 (C), 177.17 (C), 177.43 (C), 194.31 (C). FAB HRMS calcd for $C_{26}H_{35}N_3O_6 \cdot$ 485.2527, found 485.2488 (5%).

*Conversion of **2** to Quinoneimine **9***

Exposure of solutions of **2** in CH_3CN–H_2O to excess NH_4OH results in a shift of the **2** anion chromophore at 490 nm to a 448-nm absorption of equal extinction, interpreted in terms of generation of the anion of **9**. For NMR spectral confirmation of this conversion, anhydrous NH_3 was bubbled for 1 min through a solution of **2** (0.1 mmol) in 0.5 ml DMSO-d_6. After 2 hr, excess NH_3 was removed through freeze–pump–thaw cycling at high vacuum. NMR indicated quantitative conversion to **9**. ^{1}H NMR (DMSO-d_6) δ 1.05 (s, 9H), 2.42 (t, 2H, J = 6.9 Hz), 3.13 (m, 2H), 4.93 (s, 1H), 6.50 (s, 1H), 7.60 (br t, 1H, J = 4.9 Hz, NH); ^{13}C NMR (DMSO-d_6) δ 27.4, 29.4, 37.8, 38.9, 100.0, 130.2, 145.1, 167.2, 169.1, 177.1, 184.3. Addition of a slight excess of $NaBH_4$ to samples of **9** results in an instantaneous appearance of the spectral properties exhibited by aminoresorcinol **4**.

Acknowledgments

We are grateful to the National Institutes of Health (GM 48812) for support of this research and to Dr. Fengjiang Wang for some preliminary studies.

[6] Detection of Reaction Intermediates in Topa Quinone Enzymes

By CHRISTA HARTMANN and DAVID M. DOOLEY

Introduction

Copper-containing amine oxidases are ubiquitous in nature, having been isolated from bacteria, fungi, yeast, plants, and mammalian sera and tissues,[1] and are responsible for the oxidative deamination of biogenic amines. These

[1] W. S. McIntire and C. Hartmann, *in* "Principles and Applications of Quinoproteins" (V. L. Davidson, ed.), p. 97. Dekker, New York, 1992.

enzymes consist of two identical subunits, each of which contains an organic cofactor and one copper atom.[1] The role of the copper atoms in amine oxidases has been the subject of much debate. Copper has been implicated in the reoxidation of reduced enzyme,[2–4] but numerous electron paramagnetic resonance (EPR) experiments failed to detect changes in the oxidation state of the copper in the presence of substrates.[5–10] Therefore, proposals were made implicating that Cu(II) acted as a Lewis acid,[11] had an indirect role in catalysis,[12] or that it served a structural role.[9] Stopped-flow studies by Bellelli *et al.*[13,14] and EPR, circular dichroism, and temperature jump experiments by Dooley *et al.*[15] and Turowski *et al.*[16] have provided significant insight into the role of the copper cofactor in amine oxidases.

The cofactor in the oxidized enzyme has been identified as the quinone of 3-(2,4,5-trihydroxyphenyl)-L-alanine (topa quinone),[17–19] which is produced by the post-translational modification of a tyrosine residue.[18] The native, oxidized form of these enzymes contains Cu(II) and topa quinone, with the quinone being responsible for the characteristic peach color.[1] Primary substrate amines are oxidatively deaminated in a series of two half

[2] D. M. Dooley, W. S. McIntire, M. A. McGuirl, C. E. Coté, and J. L. Bates, *J. Am. Chem. Soc.* **112,** 2782 (1990).
[3] J. M. Hall and P. J. G. Mann, *Biochem. J.* **91,** 171 (1964).
[4] B. Mondovi, G. Rotilio, A. Finazzi-Agró, and E. Antonini, *in* "Magnetic Resonance in Biological Research" (G. Franconi, ed.), p. 233. Gordon & Breach, London, 1971.
[5] H. Yamada, K. Yasunobu, T. Yamano, and H. S. Mason, *Nature* (*London*) **198,** 1092 (1963).
[6] B. Mondovi, G. Rotilio, M. T. Costa, A. Finazzi-Agró, E. Chiancone, R. E. Hansen, and H. Beinert, *J. Biol. Chem.* **242,** 1160 (1967).
[7] H. Yamada, O. Adachi, and T. Yamano, *Biochim. Biophys. Acta* **191,** 751 (1969).
[8] A. Finazzi-Agró, A. Rinaldi, G. Floris, and G. Rotilio, *FEBS Lett.* **176,** 378 (1984).
[9] L. Morpurgo, E. Agostinelli, O. Befani, and B. Mondovi, *Biochem. J.* **260,** 19 (1989).
[10] J. Grant, I. Kelly, P. Knowles, J. Olsson, and G. Pettersson, *Biochem. Biophys. Res. Commun.* **83,** 1216 (1978).
[11] K. P. S. Yadav and P. Knowles, *Eur. J. Biochem.* **114,** 139 (1981).
[12] S. Suzuki, T. Sakurai, A. Nakahara, T. Manabe, and T. Okuyama, *Biochemistry* **25,** 338 (1986).
[13] A. Bellelli, M. Brunori, A. Finazzi-Agró, G. Floris, A. Giartosi, and A. Rinaldi, *Biochem. J.* **232,** 923 (1985).
[14] A. Bellelli, A. Finazzi-Agró, G. Floris, and M. Brunori, *J. Biol. Chem.* **266,** 20654 (1992).
[15] D. M. Dooley, M. A. McGuirl, D. E. Brown, P. N. Turowski, W. S. McIntire, and P. F. Knowles, *Nature* (*London*) **349,** 6306 (1991).
[16] P. N. Turowski, M. A. McGuirl, and D. M. Dooley, *J. Biol. Chem.* **268,** 17680 (1993).
[17] S. M. Janes, D. Mu, D. Wemmer, A. J. Smith, S. Kaur, D. Maltby, A. L. Burlingame, and J. P. Klinman, *Science* **248,** 981 (1990).
[18] D. E. Brown, M. A. McGuirl, D. M. Dooley, S. M. Janes, D. Mu, and J. P. Klinman, *J. Biol. Chem.* **266,** 4049 (1991).
[19] D. Mu, S. M. Janes, A. J. Smith, D. E. Brown, D. M. Dooley, and J. P. Klinman, *J. Biol. Chem.* **267,** 7979 (1992).

reactions, referred to as the reductive and oxidative half reactions. As summarized in Schemes I and II, reaction intermediates in each half reaction have been detected by a variety of chemical, kinetic, and spectroscopic methods. The methodology employed for the detection of each of these intermediates is described herein.

Assay Method

Principle

Two basic methods can be used to assay amine oxidases. Method 1 involves the direct polarographic measurement of O_2 concentrations using a Yellow Springs oxygen electrode as described by Hartmann and Klinman.[20] Method 2 utilizes the absorbance change that occurs on enzymatic oxidation of benzylamine to benzaldehyde. Method 1 is convenient to use with substrates where the extinction coefficients of the product aldehydes are not known. Method 2 is convenient because it only requires an ultraviolet-visible (UV/VIS) spectrophotometer.

Reagents for Methods 1 and 2

Sodium phosphate buffer, 100 m*M*, pH 7.2
Benzylamine hydrochloride, 0.1 *M* (for spectrophotometric assay)
Primary amine hydrochloride 0.1 *M* (for oxygen electrode)

Procedure for Method 1

The buffer is thermostated at 25° and is saturated with air to ensure a constant $[O_2]$ for each assay. Buffer (0.885 ml) is put into the Yellow Springs electrode (1-ml capacity) followed by 100 μl of primary amine hydrochloride. The cell is stirred with the magnetic stir bar as supplied. Oxygen consumption is monitored after syringe injection of enzyme (0.01–0.1 units) via the groove on the side of the electrode.

Procedure for Method 2

Buffer (0.900 ml) saturated with air at the assay temperature of 25° is pipetted into a 1-cm pathlength, 1-ml quartz cuvette. To this is added 100 μl of the benzylamine solution, also air-saturated at the assay temperature. After the addition of microliter quantities of enzyme, the absorbance change

[20] C. Hartmann and J. P. Klinman, *Biochemistry* **30,** 4605 (1991).

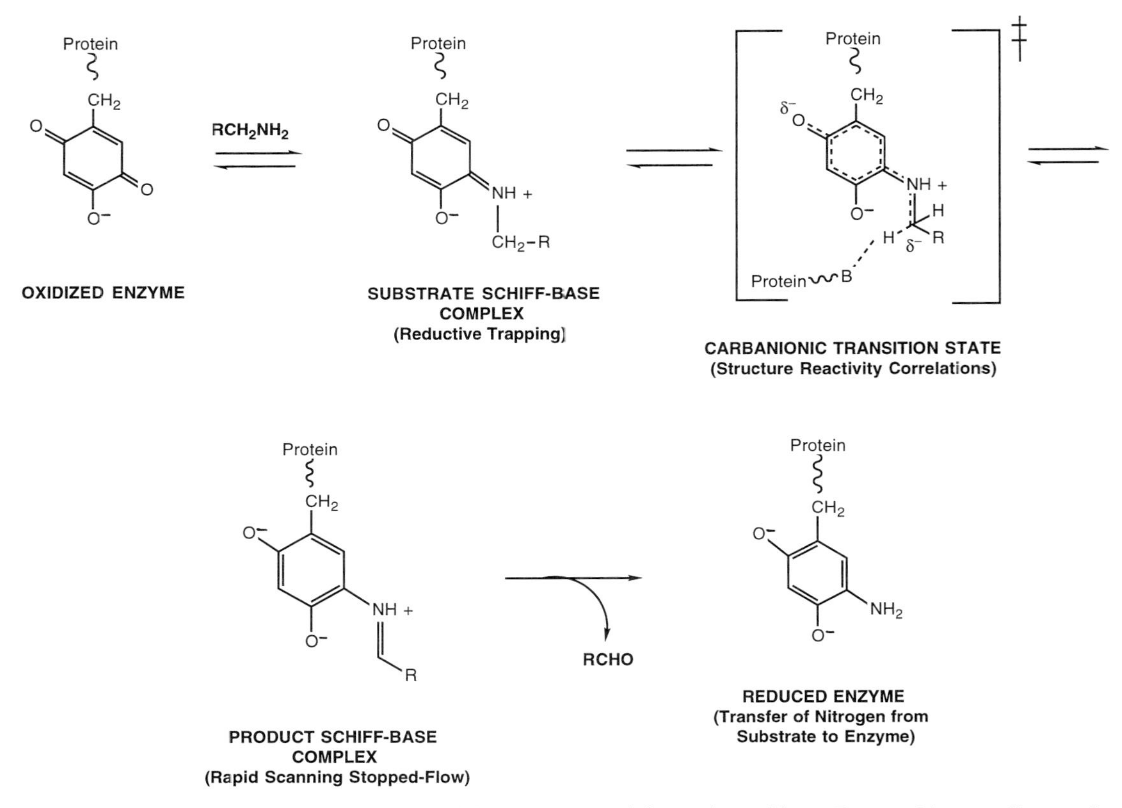

SCHEME 1. Reductive half reaction postulated for copper-containing amine oxidases. Proposed intermediates and method of detection for the conversion of amine substrate and oxidized enzyme to aldehyde product and reduced enzyme (reviewed by McIntire and Hartmann[1]).

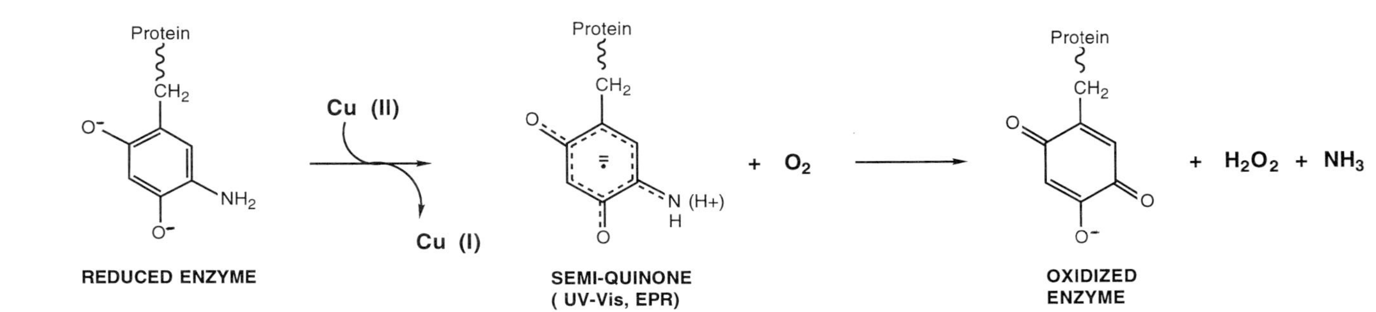

SCHEME 2. Oxidative half reaction postulated for copper-containing amine oxidases. Proposed intermediate and method of detection for conversion of reduced to oxidized enzyme. One of the possible structures for the topa aminosemiquinone radical is shown. The species in brackets is a hypothetical intermediate proposed by Turowski *et al.*[16]

associated with the formation of benzaldehyde is monitored at 250 nm. A $\Delta\varepsilon_{250}$ of 12,500 $M^{-1}cm^{-1}$ is used to calculate the rate of substrate oxidation.[21]

Specific Activity and Unit Definitions

Method 1: One unit of activity is defined as the amount of amine oxidase needed to reduce 1 μmol of O_2 per minute at 25°. Method 2: One unit is defined as the amount of enzyme required to produce 1 μmol of benzaldehyde per minute at 25°. Specific activity is expressed in units per milligram of protein. Protein concentrations for amine oxidases are determined from their individual ε_{280} values.[1]

Detection of Intermediates during the Reductive Half Reaction

Reductive Trapping of Substrate to Bovine Serum Amine Oxidase (BSAO): Evidence for the Presence of an Enzyme–Schiff Base Complex

Materials

BSAO was purified to homogeneity by a modification of the procedure of Summers *et al.*[22] [^{14}C]Benzylamine was obtained from Sigma (specific activity 12–16 mCi/mmol) and [^{3}H]$NaCNBH_3$ was from Amersham Corp. (3.8–11 Ci/mmol). All other reagents were obtained from commercial sources and were reagent grade or better. Radioactivity was determined using a Beckman LS-8000 liquid scintillation spectrometer in a mixture composed of either Chemfluor or Crystalfluor with Surfactol-100 (E&K Scientific Products) in toluene (30:70, v/v).

Method

Inactivation of Enzyme with $NaCNBH_3$ and [^{14}C]Benzylamine. Typical reaction mixtures contained 2–3 mg/ml amine oxidase, 20 m*M* [^{14}C]benzylamine (final concentration; added in 5- to 10-μl aliquots during the course of the experiment), and 20 m*M* $NaCNBH_3$ in 50 m*M* phosphate buffer, pH 6.0, 25°. Substrate was allowed to react for 15–30 sec prior to the addition of reductant. At time points varying from 5 to 180 min, 100- to 200-μl aliquots were removed and applied to a G-25 Sephadex column (0.5 × 12.5 cm) preequilibrated using 100 m*M* phosphate buffer, pH 6.0.

[21] R. Neumann, R. Hevey, and R. H. Abeles, *J. Biol. Chem.* **250,** 6362 (1975).

[22] M. C. Summers, R. Markovic, and J. P. Klinman, *Biochemistry* **18,** 1969 (1979).

These conditions led to the separation of the radiolabeled protein from the large pool of benzylamine, reductant, and products. In cases where a clean separation was not achieved, protein peaks were subjected to dialysis in a Centricon-30 apparatus (Amicon Corp.).

It is possible that the lysine side chains in the enzymes could be labeled by [^{14}C]benzaldehyde. To determine the amount of labeling by the product aldehyde, a coupled assay procedure was used, containing liver alcohol dehydrogenase (LADH) in a 10-fold excess in units over amine oxidase, NADH (2–5 mg), amine oxidase (in amounts described earlier), 20 m*M* [^{14}C]benzylamine, and 20 m*M* $NaCNBH_3$. This reaction was run side by side with a typical reaction described earlier without LADH. In the analysis for the stoichiometry of protein labeling, the amine oxidase was assumed to be recovered in the same ratio as in the initial reaction mixture without LADH. No significant difference between reaction mixtures incubated with or without LADH was observed, thereby ruling out labeling of the lysyl side chains by [^{14}C]benzaldehyde.

The time course for inactivation for BSAO was compared to that for radiolabeling, indicating that incorporation of C-14 label from substrate amine occurred concomitantly with the loss of enzyme activity. A plot of counts incorporated versus enzyme inactivated extrapolated to 1 mol of radiolabeled substrate/mol of enzyme inactivated.[23] A similar analysis of methylamine oxidase from *Arthrobacter* P1[24] gave nearly identical results.[25]

Inactivation of Enzyme with [^{3}H]NaCNBH$_3$. Reaction conditions were identical to those described earlier with the exception of using unlabeled benzylamine and 20 m*M* [^{3}H]$NaCNBH_3$ ($1.1–4.1 \times 10^9$ cpm/μmol) as reductant. Specific activity of [^{3}H]$NaCNBH_3$ was measured by the method of Jentoft and Dearborn[26] and was found to be within 5% of the reported value. Incubations of amine oxidase to 93% inactivation with [^{3}H]$NaCNBH_3$ alone or in the presence of benzylamine yielded <0.6 and $<1.2\%$ incorporation of tritium, respectively. This failure to trap tritium was not due to instability of the [^{3}H]$NaCNBH_3$ stock, which was assayed for chemical and radiochemical integrity.[26]

A mechanistic explanation that accounts for the trapping of C-14 from the substrate without tritium from the reductant is outlined in Scheme III. As shown, substrate forms a Schiff's base with topa quinone (1), which undergoes reduction by $NaCNBH_3$. Although tritiated reductant will initially label the enzyme (2), this intermediate is expected to undergo rapid

[23] C. Hartmann and J. P. Klinman, *J. Biol. Chem.* **262,** 962 (1987).

[24] W. S. McIntire, this series, Vol. 188, p. 227.

[25] C. Hartmann and J. P. Klinman, *FEBS Lett.* **261,** 441 (1990).

[26] N. Jentoff and D. G. Dearborn, *J. Biol. Chem.* **254,** 4359 (1979).

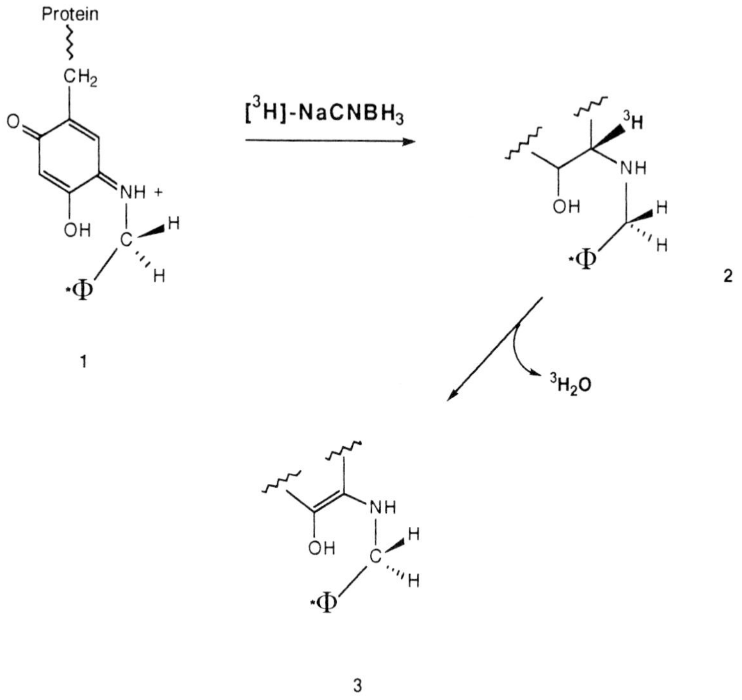

SCHEME 3. Proposed mechanism for reductive trapping of [^{14}C]amine to topa cofacto.. Substrate forms a Schiff's base with topa quinone (1), which undergoes reduction by $NaCNBH_3$. Although the use of tritiated reductant is expected to label the enzyme initially, the intermediate (2) is unstable, undergoing tautomerization to (3), with loss of tritium to solvent.[23]

enolization, resulting in the loss of tritium to the solvent and formation of the reduced aminophenol form of the cofactor (3).

Structure–Function Studies of Substrate Oxidation by BSAO: Evidence Supporting a Carbanionic Transition State

Materials

The following ring-substituted benzylamines were studied: *para*-acetyl, -bromo, -*N*,*N*-dimethylamino, -fluoro, -hydroxy, -isopropyl, -methoxy, -methyl, and -trifluoromethylbenzylamines. All substituted substrates (C-1 dideuterated and protonated) were synthesized by reduction of the corresponding benzonitriles either by $LiAlH_4$ or $LiAlD_4$ or by specific multistep syntheses.[20] Substrates were analyzed by ^{1}H-NMR, mass spectrometry, and elemental analysis for isotopic purity. Glucose oxidase, catalase, and glucose were obtained from commercial sources and were of reagent grade purity or better.

Method

Pre-Steady-State Kinetics: Single Wavelength Stopped-Flow Spectrometry. Stopped-flow kinetic experiments were carried out under conditions of substrate in excess over enzyme concentration. Enzyme activity was normalized to a single value (0.36 units/mg). The stopped-flow apparatus was prepared for anaerobiosis by overnight incubation with a solution of 100 m*M* glucose, 50 units/ml glucose oxidase, and 50 units/ml catalase in 100 m*M* potassium phosphate buffer, pH 7.2 (anaerobic buffer). Anaerobic buffer, containing the same components as earlier and sparged with argon, was used to flush the syringes and chambers immediately prior to use.

The anaerobic enzyme was prepared in a tonometer containing 50–60 μM enzyme in the anaerobic buffer described earlier. After introducing the enzyme into the tonometer, the entire system was placed under vacuum and repetitively flushed with argon. Care must be taken when placing the tonometer under vacuum to minimize frothing of the enzyme solution. Substrate solutions were prepared for anaerobiosis by bubbling high-purity argon through 8 ml of the substrate/phosphate/glucose buffer in 10-ml disposable syringes for 15 min, followed by the addition of 50 units/ml each of glucose oxidase and catalase. To minimize the introduction of small bubbles into the stopped-flow syringes, the substrate solutions were decavitated by gentle tapping of the Parafilm-covered syringes immediately prior to their placement on the apparatus. It is important to check the stability of enzyme activity to this treatment which, in the case of BSAO, remained unchanged on completion of the experiments.

Anaerobic stopped-flow kinetics were carried out at 25° in 100 m*M* potassium phosphate buffer, pH 7.2, on a modified Aminco-Morrow stopped-flow system and photomultiplier transducer (Kinetic Instruments Inc., Ann Arbor, MI). Rapid data acquisition and subsequent analyses were determined using an IBM-PC compatible set of hardware and software subsystems. Raw voltage proportional to percent transmission was offset, expanded as necessary, and digitalized via a 12-bit IBM-PC analog-to-digital converter with an on-board crystal-controlled clock to generate precise sampling intervals. Data storage and analysis were performed by a program which uses a nonlinear least-squares analysis via a gradient expansion reiterative algorithm to a large number of integral differential equations. The anaerobic reduction of serum amine oxidase by excess substrates was monitored by following the decrease in absorbance at 480 nm as a function of time. A total of 1024 data points were collected and stored. First-order rate constants were calculated as described by Ramsey *et al.*[27] using a program from Anarac Associates. Four determinations were performed and averaged to obtain the final rate parameters and associated standard deviations.

Analysis of the slopes and intercepts from double reciprocal plots for all substrates were linear with the exception of *p*-acetylbenzylamine. Nonlinear plots of this nature imply that the back rate for the formation of the ES from the EP complex is finite.[28]

$$\mathrm{E} + \mathrm{S} \underset{}{\overset{K_d}{\rightleftharpoons}} \mathrm{ES} \underset{k_4}{\overset{k_3}{\rightleftharpoons}} \mathrm{EP}. \tag{1}$$

The magnitude of this process can be estimated from the y intercept of plots of k_{obs} vs [S].[14] Once estimates were obtained for the back rate, subsequent plots of $1/(k_{obs} - k_4)$ vs 1/[S] were linear and provided values for k_3, the rate constant for substrate oxidation; K_d, the dissociation constant for substrate binding; and the isotope effects on these parameters.

Method

Steady-State Kinetics. Initial rate studies for both protonated and dideuterated substrates were carried out at 25° in 100 m*M* sodium phosphate buffer, pH 7.2. Substrate solutions were prepared immediately prior to use. The rates of the reactions were measured either by monitoring the oxygen consumption of air-saturated solutions (Method 1) or by monitoring the rate of aldehyde product formed at 250 nm (Method 2). Velocities of both protonated and dideuterated substrates were obtained on the same day. An average of four to six determinations of the rate parameters were performed yielding a single value with a known standard deviation. All data were fit to the expression

$$v/\mathrm{E}_t = V_{\max}[\mathrm{S}]/K_m + [\mathrm{S}], \tag{2}$$

using the FORTRAN program of Cleland.[29] There are now several versions of the Cleland program available commercially which are appropriate for the data analysis (e.g., Enzfitter, BioSoft, Inc., Elsevier, Amsterdam, Netherlands). Microscopic rate constants, dissociation constants, and propagated errors were calculated as described by Palcic and Klinman.[30] Calculation

[27] R. R. Ramsey, M. Husain, and D. J. Steenkamp, *Biochem. J.* **241,** 883 (1986).
[28] S. Strickland, G. Palmer, and V. Massey, *J. Biol. Chem.* **250,** 4048 (1975).
[29] W. W. Cleland, this series, Vol. 63, p. 103.
[30] M. M. Palcic and J. P. Klinman, *Biochemistry* **22,** 5957 (1983).

of the microscopic rate constants for each of the substrates is dependent on the use of an estimation for the intrinsic isotope effect on benzylamine oxidation.[30] Confirmation that the intrinsic isotope effect does not change appreciably with changes in substrate *para*-substituent was obtained from direct measurement of the isotope effect for *p*-trifluoromethyl- and *p*-acetylbenzylamines.[20]

Multiple Regression Analysis. The use of structure–function studies in enzymology is complicated by the fact that there are interactions of side chains in the active site of the enzyme which affect not only the substrate dissociation constant, but also the well-defined spatial orientation between substrate and cofactor. This complication can introduce scatter in a Hammett plot, rendering it uninterpretable. Thus, a clear-cut separation of hydrophobic, steric, and electronic effects is necessary before a transition state structure can be determined. In order to distinguish these effects through regression analysis, substrates must be chosen which minimize parameter collinearity among the substituents investigated.

Despite the fact that different parameters describe different physical properties, if two or more parameters exhibit a high degree of collinearity, their individual expression on alterations in activity will be masked due to statistically indistinguishable results. The judicious selection of substituents to minimize collinearity as much as possible avoids this statistical complication. Multiple regression analyses were then performed to determine which of these parameters exerted the greatest effect on each of the calculated constants. According to the equation below, the magnitude of the coefficients A, B, and C will indicate the extent to which observed rates and dissociation constants are dependent on electronic (σ_p), hydrophobic (π), and steric (E_s or V_w) factors;

$$\log(k_3, K_d) = \sigma_p A + \pi B + E_s(V_w)C + D, \tag{3}$$

respectively. The hydrophobic binding parameter, π, is a measure of the possible hydrophobic binding energy of a given compound, comparable to the σ parameter. It has been defined and is based on partition coefficients between octanol and water.[31]

It is important to note that the correct choice of steric parameter is not necessarily straightforward and depends on the nature of the substrate-binding pocket. The Taft steric parameter, E_s, correlates the van der Waals radius of the first atom of the substituent to the overall bulk of that substituent. This implies that the branched substituents can alter their conformations so that the majority of the substituent will not interact with the active

[31] C. Hansch and A. Leo, in "Substituent Constants for Correlation Analysis in Chemistry and Biochemistry." Wiley, New York, 1979.

site. Bondi's treatment for calculated van der Waals radii, V_w, correlates total steric bulk of the substituent within a more constrained binding pocket.[32] Thus, the effects of both parameters should be examined.

Multiple regression analysis is a method utilized for relating two or more independent variables to a dependent variable. This regression analysis approach assumes that the effects of π, σ, and E_s (V_w) increase linearly. The analyses were performed on variations of Eq. (3) with weighted values for log k_3 (log K_d) using the SAS general linear models procedure (SAS Institute Inc., Cary, NC). The experimental data points are values of $k \pm$ error, but the correlation function is linear in log k. Therefore, the weighting factors were derived from the limiting values of log ($k \pm$ error). Correlations were determined for one and two parameter fits to Eq. (3); with only nine experimental observations the full three parameter correlation is not valid.

For the C–H bond cleavage step in the oxidation of substrate by bovine serum amine oxidase, the data correlated with σ_p between the 95 and 99% level. The best two parameter correlations for log k_3 with σ_p and π produced a statistically significant improvement.[20] This result indicated that hydrophobicity inhibited, rather than facilitated, catalysis. Once this adjustment was made, the plot of log $k_3 + \pi$ vs σ_p yielded a ρ value of 1.5 ± 0.3, indicative of a transition state with considerable negative charge at C-1 of the substrate.[20]

Anaerobic Rapid-Scanning Stopped-Flow (RSSF): Evidence for a Product Schiff Base Complex

Method

RSSF experiments were performed on a Durrum D-110 rapid mixing stopped-flow spectrophotometer modified with a J-Y polychromator and a Princeton Applied Research (PAR) 1412 photo diode array detector. Enzyme, substrate solutions, and the instrument were prepared for anaerobiosis as described earlier for single wavelength stopped-flow experiments. It should be noted that K_m values obtained in the steady state do not give an indication of the actual pre-steady-state K_m values. Therefore, for the substrates included in this study, it was assumed that concentrations of 5 mM and above (which is $\geq$200 times the enzyme concentration) were saturating conditions and, hence, that comparison of the rapid scanning stopped-flow and single wavelength stopped-flow data was valid. The detector was interfaced with a PAR OMA III multichannel analyzer equipped

[32] A. Bondi, *J. Phys. Chem.* **68,** 441 (1964).

with a 1463 detector controller card. The hardware configuration for this system has been previously described.[33]

Data acquisition consisted of collection of a 100% transmission reference spectrum (defined as the light transmitted through the buffer solution) and the diode array dark current. Utilizing these spectra, rapid-scanning data were initially collected as transmission spectra, converted to absorbance, and stored directly onto a floppy disk. The computer software allows 24 spectra to be collected at various intervals. The experiments utilized 500 pixels on the photo diode array for the repetitive scan time of 8.528 msec/scan with a wavelength resolution of approximately 1 nm. The software of the PAR OMA III multichannel analyzer allows for arithmetic manipulation of the data to yield single wavelength time courses and difference spectra.

For the spectra reported, the collection of the first scan relative to cessation of flow is noted in the figure legend. A representative difference spectrum derived from data acquisition is shown in Fig. 1. The first five scans were collected every 8.528 msec. Subsequent illustrated scans were collected at the following intervals: (7) 213.3 msec; (10) 469.2 msec; (12) 810.4 msec; (15) 1578.1 msec; (20) 3284.1 msec; and (24) 4990.1 msec.

Results

Single wavelength pre-steady-state measurements of the oxidation of *p*-hydroxy and *p*-methoxybenzylamine by bovine serum amine oxidase had shown biphasic traces at 480 nm, indicating the formation of an intermediate with greater absorbance than topa quinone. This type of behavior had been previously noted by Olsson *et al.*[34] in the reaction of porcine serum amine oxidase with *p*-methoxybenzylamine, but analysis of their data to obtain rate parameters for the observed processes proved unsuccessful. To aid in the identification of these spectral intermediates, anaerobic RSSF analysis was used to obtain absorbance spectra from 300 to 540 nm during the course of enzyme reduction by benzylamine and by *p*-hydroxy- and *p*-methoxybenzylamines. As a result of these RSSF analyses, subsequent single wavelength stopped-flow studies focused on the relaxation times at 440 and 310 nm to obtain the actual rates of these processes.[35]

Reduction of BSAD by benzylamine under anaerobic RSSF conditions leads to a total of three observable relaxations in the UV/VIS range from 300 to 540 nm. These relaxations reflect the formation of an enzyme–

[33] S. C. Koerber, A. K. H. MacGibbon, H. Dietrich, M. Zepperzauer, and M. F. Dunn, *Biochemistry* **22,** 3424 (1983).

[34] B. Olsson, J. Olsson, and G. Pettersson, *Eur. J. Biochem.* **71,** 375 (1976).

[35] C. Hartmann, P. Brzovic, and J. P. Klinman, *Biochemistry* **32,** 2234 (1993).

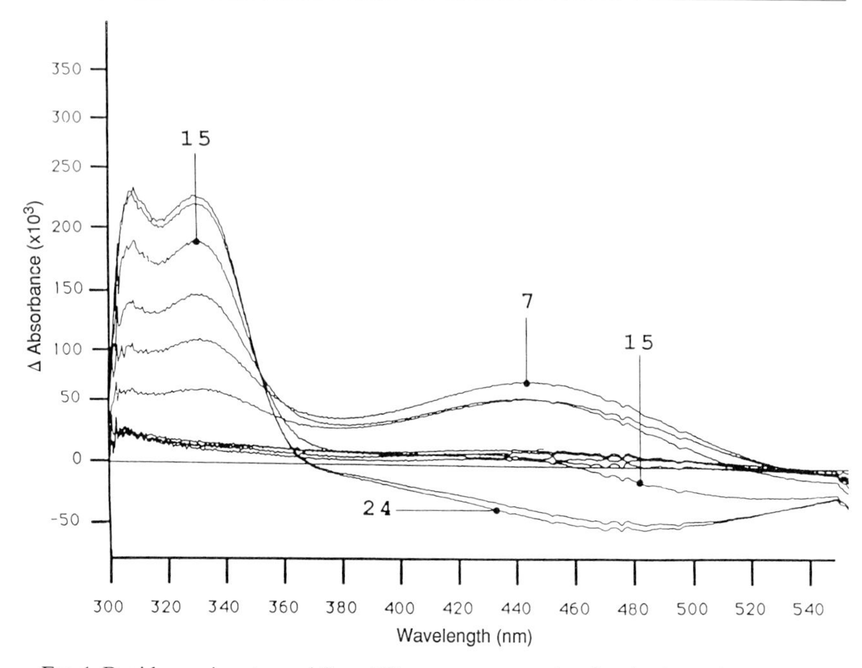

FIG. 1. Rapid scanning stopped-flow difference spectrum showing the formation and decay of the proposed quinonoid intermediate (440 nm) and the formation of the species at 310 and 330 nm, attributed to reduced cofactor and product aldehyde, respectively. The initiation of scanning occurred 7 msec after flow had stopped. See Methods for the timing sequence. Conditions after mixing were 5 m*M* *p*-hydroxy-[1,1-$^2H_2$]benzylamine and 36 μM BSAO in the glucose, glucose oxidase, and catalase cocktail at 25°. Difference spectra are computed as $(\text{scan})_t - (\text{scan})_0$ from the original spectrum.

substrate complex (340 nm) and reduction of enzyme bound cofactor (310 and 480 nm).[30,35] Reduction of enzyme by *p*-hydroxybenzylamine, *p*-methoxybenzylamine, or *p*-(*N*,*N*-dimethylamino)benzylamine introduces additional transient relaxations centered at 440, 425, and 460 nm, respectively (*p*-hydroxybenzylamine; Fig. 1). Product formation was also observed for *p*-hydroxybenzylamine at 330 nm.[36] These transient intermediates exhibit spectral properties similar to that of quinonoid intermediates of the enzyme–product complex of tryptophan synthase[37] and in model studies with *para*-substituted benzaldehydes and substituted anilines.[38]

Substitution by deuterium at the benzylic position of the substrate resulted in the reduced accumulation of the spectrally modified intermediate

[36] R. H. Suva and R. H. Abeles, *Biochemistry* **17,** 3538 (1978).
[37] W. F. Drewe and M. F. Dunn, *Biochemistry* **24,** 3977 (1985).
[38] J. Császár, J. Balog, and A. Makáry, *Acta Phys. Chem.* **24,** 373 (1978).

at 440 nm and a decrease in the rate of cofactor reduction at 310 and 480 nm. This result implied that the formation of this intermediate is preceded by deprotonation of the α-carbon of the amine substrate as originally proposed by Olsson *et al.*[34]

The appearance of the intense absorbance at 440 nm with *p*-hydroxybenzylamine as the substrate (Fig. 1) can be rationalized from the structure of the enzyme–product Schiff base complex (Scheme IV). As illustrated, placement of a hydroxyl group in the *para*-position of substrate leads to a product Schiff base complex (1) capable of tautomerization to a quinone methide species (2). Using RSSF, a shorter lived and less intensely absorbing intermediate was observed at $\approx$425 nm in the reaction of *p*-methoxybenzylamine with bovine serum amine oxidase.[35] Oxidation of *p*-*N*,*N*-(dimethylamino)benzylamine by bovine serum amine oxidase produced a similar intermediate observed at 460 nm in a anaerobic single turnover experiment.[35]

Retention of Ammonia on BSAO: Evidence for the Aminophenol Form of Reduced Cofactor

Two types of experiments that definitively establish the existence of the aminophenol form of reduced topa as an intermediate have been re-

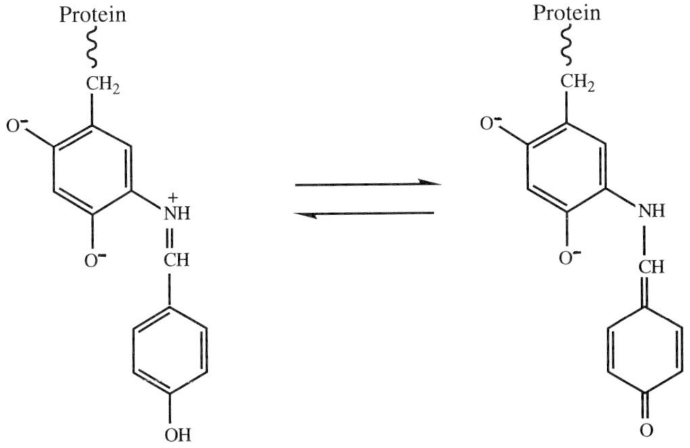

SCHEME 4. Proposed mechanism to explain spectral intermediates formed from benzylamines with electron-releasing groups. As shown for *p*-hydroxybenzylamine, the product Schiff base (1) can tautomerize to a quinone methide (2). From Hartmann *et al.*[35]

ported.[39,40] Although the work is not our own, the methods are briefly described here because the experimental approaches may be generally applicable to mechanistic studies of amine oxidases, amine dehydrogenases, and related enzymes.

Freeze–Quench Kinetics of Ammonia Release in Porcine Serum Amine Oxidase

Method

The highly purified enzyme was extensively dialyzed against 0.1 *M* sodium phosphate buffer at pH 9.0 to remove all traces of ammonia.[39] The dialyzed enzyme (approximately 10 mg/ml) was transferred to a tube fitted with a serum cap and saturated with O_2 by directing a stream of the pure gas on the surface of the solution for 15 min while gently shaking. A solution of benzylamine (10 m*M*) was prepared in the same buffer, the pH readjusted to 9.0, and then saturated with O_2 by bubbling the gas through the solution for 10 min. The syringes of a standard rapid-quench apparatus were filled with the oxygen-saturated enzyme and benzylamine solutions and equilibrated at 20° for 20 min. At various reaction times ranging from 10 msec to 20 sec, samples were extruded into weighed vials containing 0.19 ml of 1.0 *M* HCl. These quenching conditions irreversibly inactivate the enzyme. The vials were then reweighed to compensate for any small variations in the volume of the extruded samples. Subsequently, the quenched samples were centrifuged (300*g*) for 20 min, the supernatant was removed and adjusted to pH 6.5 by the addition of 0.167 ml of 2.0 *M* $KHCO_3$, and then recentrifuged as described earlier. Six 0.1-ml samples of the supernatant were taken and analyzed for ammonia as previously described.[41] The rate of ammonia release was oxygen dependent and equal to that of the reoxidation of reduced topa and H_2O_2 production by the enzyme, thereby providing unambiguous evidence for the aminotransferase mechanism. The applicability of this approach is dependent on the relative rate constants for the reduction of topa by substrates and its reoxidation by O_2.

Analysis of Product Formation under Anaerobic Conditions for BSAO

Method

The method, which is described in detail in the literature,[40] involves the analysis for ammonia and benzaldehyde production by radiochemical

[39] F. X. Rius, P. F. Knowles, and G. Pettersson, *Biochem. J.* **220,** 767 (1984).
[40] S. M. Janes and J. P. Klinman, *Biochemistry* **30,** 4599 (1991).
[41] V. F. Kalb, T. J. Donohue, M. G. Corrigna, and R. W. Beruldur, *Anal. Biochem.* **90,** 47 (1978).

methods under strictly anaerobic conditions. Bovine serum amine oxidase in 10 m*M* potassium phosphate buffer, pH 6.5, containing 100 m*M* β-D-glucose was made anaerobic by placing the solution in a septum-stoppered vessel, freezing it in liquid nitrogen, and then subjecting the system to repeated flush/purge cycles with a house vacuum and 99.999% Ar gas. After the solutions were thawed, glucose oxidase and catalase were added to achieve a final concentration of 125 units/ml of each enzyme. The substrate (benzylamine) solution was made anaerobic in a similar fashion. Both enzyme and substrate solutions were allowed to stand for 20 min at room temperature to permit the glucose oxidase and catalase to remove all remaining oxygen from the reaction vessels. Ammonia was determined using a coupled assay with α-[1-^{14}C]ketoglutarate and glutamate dehydrogenase.[40] Benzaldehyde was determined using ^{14}C-labeled benzylamine after separating the labeled benzaldehyde from unreacted benzylamine by HPLC. Using these procedures, Janes and Klinman[40] were able to show unequivocally that the moles of benzaldehyde released anaerobically correlated linearly with the number of moles of bovine serum amine oxidase active sites and that no ammonia was released under the same conditions, consistent with an aminotransferase reaction.

Detection of Topa Semiquinone during the Oxidative Half Reaction

EPR Characterization of the Topa Semiquinone Intermediate

Method

Room Temperature EPR. EPR spectra were obtained in 0.3 × 6 × 100-mm glass microslides (Vitro Dynamics) on a Bruker 220D SRC instrument immediately after samples were prepared and sealed in a nitrogen atmosphere glove box. Typical parameters for observing the copper signal and the semiquinone simultaneously were: frequency, 9.8 GHz; power, 20 mW; and modulation amplitude, 20 G. Enzymes used for this study were purified as previously described or were obtained from collaborators.[15] Enzyme concentrations varied from 0.3 to 0.8 m*M* in 50 m*M* barbitol, pH 8.0 (*Arthrobacter* P1 amine oxidase, APAO; BSAO; and pig plasma amine oxidase, PPAO), 50 m*M* piperazine-*N*,*N*′-bis(2-ethane sulfonic acid) (PIPES), pH 7.0 (pea seedling amine oxidase, PSAO), or 100 m*M* phosphate buffer, pH 7.2 (pig kidney diamine oxidase, PKAO). Substrate concentrations were either 1–3 m*M* benzylamine for APAO, BSAO, and PPAO or 1–3 m*M* cadaverine for PSAO and PKAO. Data were collected and manipulated using EPRware software (Scientific Software Services) on an IBM personal computer. EPR-detectable copper was measured by double integration of the EPR signal using standard solutions of Cu(EDTA) in 0.1 *M* HCl and $Cu(ClO_4)_2$ in 2 *M* $NaClO_4$. Baseline corrections and integrations were made using Spectracalc (Version 2.1, Galactic Industries, Inc.). Results of these EPR experiments are shown in Fig. 2, where curves A are in the absence of substrate and curves B are in the presence of substrate. The latter curves show the appearance of an organic radical at g ~ 2 and a decrease in the Cu signal (see below). The topa semiquinone of lentil

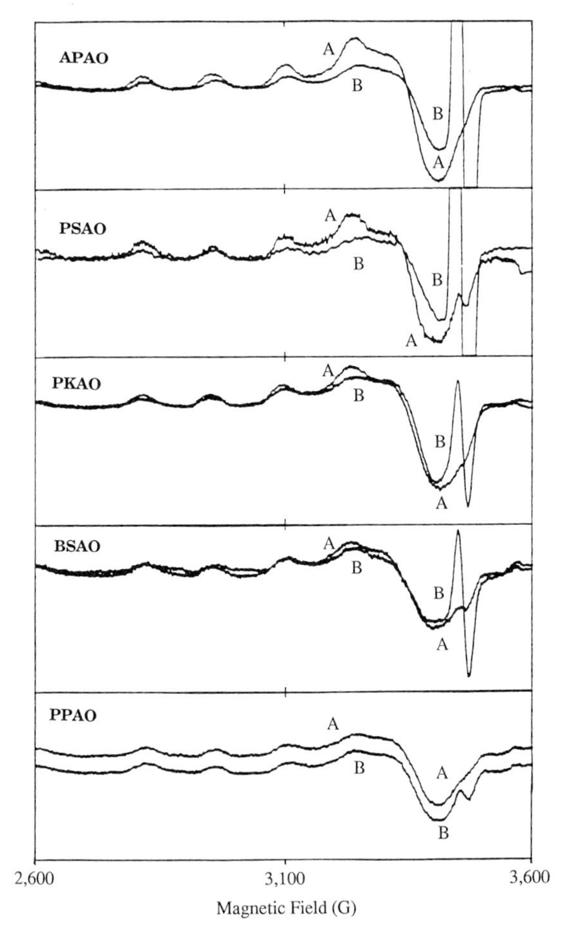

FIG. 2. Room temperature EPR spectra of amine oxidases from *Arthrobacter* P1 (APAO), pea seedling (PSAO), porcine kidney (PKAO), bovine serum (BSAO), and porcine plasma (PPAO) before (A) and after (B) the anaerobic addition of the substrate. Conditions are as described under EPR Methods.

seedling amine oxidase was also detected by EPR and compared to 6-hydroxydopamine, under essentially identical conditions as those described earlier.[42]

Quantitation of the Cu(II) EPR signals of PSAO and APAO shown in Fig. 2 was performed by two different methods. Method (i): The area of the $M_I = -3/2$ hyperfine line in a first derivative Cu(II) EPR spectrum is proportional to the total intensity.[43] Comparison of the peak areas for the resting and substrate-reduced enzymes, obtained under identical conditions, indicated that approximately 40% of the Cu(II) was reduced. Method (ii): After subtraction of the contributing sharp $g \sim 2$ signal from the EPR absorption envelope, integration yielded the Cu(II) intensity. Using the resting enzyme as the standard, the results were in good agreement with those obtained from Method (i) just described. The intensity of the EPR signal of the semiquinone radical was also measured at room temperature, but under different conditions. Spectrometer conditions were power, 6.3 mW; modulation amplitude, 1 G; and frequency, 9.8 GHz. Solutions (0.1 and 0.5 m*M*) of PROXYL (a water-soluble nitroxide spin label) were used as a standard.

Circular Dichroism and UV/VIS Spectroscopic Characterization of the Topa Semiquinone Intermediate

Method

Circular Dichroism (CD). Copper reduction by substrates was confirmed by circular dichroism spectroscopy. The Cu(II) ligand-field transitions of PSAO and APAO can be observed in CD spectra from 600 to 800 nm[2]. Spectra were obtained on a Jasco Model 40 circular dichroism spectrometer (1-cm path length cell) that had been retrofitted by Olis Instruments, Inc. A sample containing 0.2–0.3 m*M* APAO and a fivefold excess of benzylamine in 700 μl of 0.1 *M* potassium phosphate buffer, pH 7.2, was prepared in an oxygen-free atmosphere (N_2 glove box). Spectra were acquired at room temperature after calibration against 0.06% camphosulfonic acid (Fig. 3). The addition of substrate under anaerobic conditions decreased the intensity of the APAO Cu(II) CD bands approximately 35% (Fig. 3), which was consistent with the EPR data.[15]

[42] J. Z. Pederson, S. El-Sherbini, A. Finazzi-Agró, and G. Rotilio, *Biochemistry* **31,** 8 (1992).

[43] T. Vänngård, *in* "Magnetic Resonance in Biological Systems" (A. Ehrenberg, B. G. Malmström, and T. Vänngård, eds.), p. 213. Pergamon, Oxford, 1984.

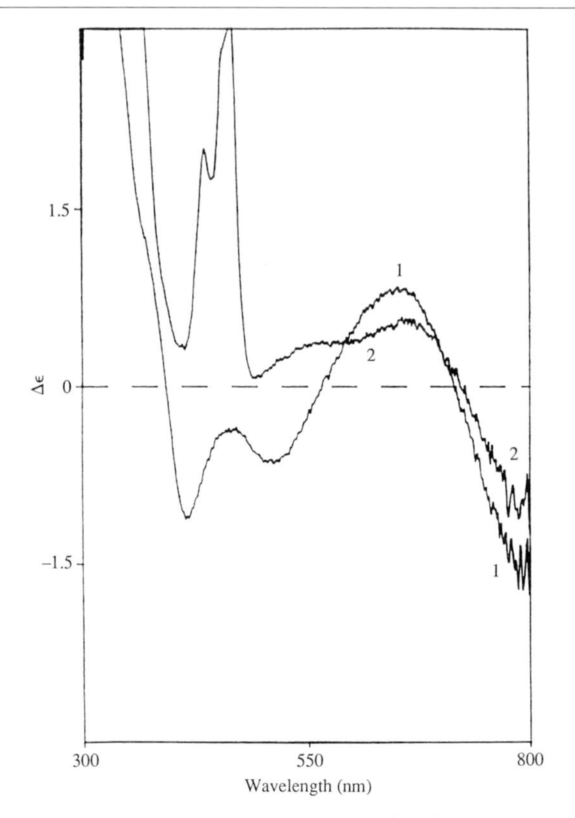

FIG. 3. Circular dichroism spectra ($\Delta\varepsilon = \varepsilon_L - \varepsilon_R$, $M^{-1}cm^{-1}$) of 15 μM APAO (1) and of 15 μM APAO after the anaerobic addition of 60 μM benzylamine (2) in 0.1 M phosphate buffer, pH 7.2, 295 K.

Method

UV/VIS Spectroscopic Characterization of the Topa Semiquinone Intermediate. UV-visible spectra of anaerobic samples of *Arthrobacter* P1 amine oxidase (+/− cyanide) were acquired at various temperatures using a Hewlett-Packard 8452A spectrophotometer (Fig. 4). Sample preparation was as described earlier for the circular dichroism experiments. Temperature at acquisition and concentrations of enzyme, substrate, and cyanide are given in the legend to Fig. 4. Anaerobic addition of the substrate to enzyme produced the characteristic semiquinone absorption bands.[13,14] Cooling the sample caused bleaching of the absorption spectrum and generated a spectrum that represented a mixture of the oxidized and fully reduced quinones. This is a reversible process since warming of the sample to 22° caused the

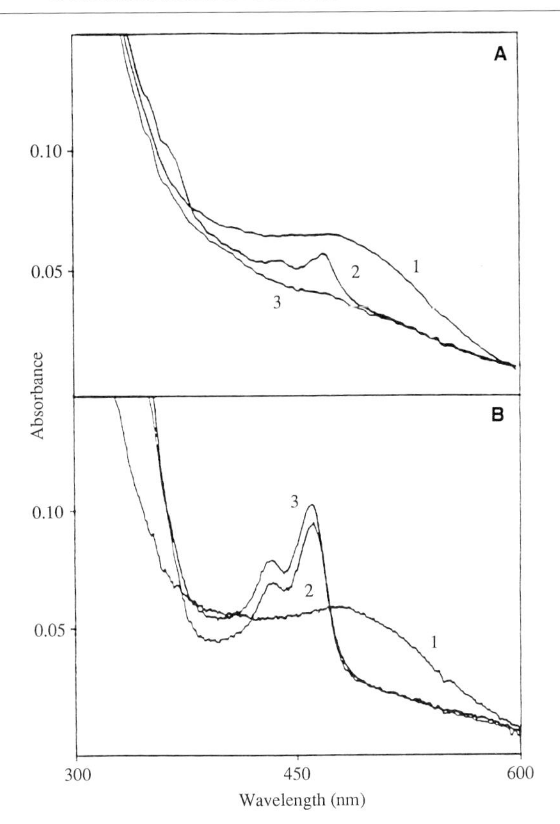

FIG. 4. Visible spectra of *Arthrobacter* P1 amine oxidase, indicating the effect of temperature on the topa semiquinone radical (A) and on the cyanide-stabilized radical (B). Conditions during measurements were: (1) 15 μM enzyme, 10 mM phosphate in methanol/water (1:4), pH 7.5, 295 K; (2) 15 μM enzyme, 60 μM benzylamine, 295 K; and (3) 15 μM enzyme, 60 μM benzylamine, 258 K. [NaCN], 5 mM for spectra (A3) and (B3).

reappearance of the semiquinone bands.[15] Cyanide addition to the samples prevents bleaching of the semiquinone spectrum on cooling (Fig. 4B), possibly by stabilizing the Cu(I) and preventing back electron transfer.

Temperature-Jump Detection of Intramolecular Electron Transfer in PSAO

Method

Pea seedling amine oxidase was purified to 95% homogeneity by a modification of the method of Kluetz *et al.*[44] The instrument was a Dialog

[44] M. D. Kluetz, K. Adamsons, and J. E. Flynn, Jr., *Prep. Biochem.* **10**, 615 (1980).

dual-heating temperature-jump apparatus with a heating time of 1 μsec. The substrate-reduced sample was generated anaerobically by the addition of 1,5-diaminopentane (8 equiv) or 4-(dimethylaminomethyl)benzylamine (1.5 equiv) to enzyme (58 μ*M*) in 0.1 *M* potassium phosphate buffer, pH 7.2. Anaerobiosis was maintained by flushing the sample with O_2-free argon for at least 2 hr. Spectra were acquired using a Hewlett-Packard 8452A spectrophotometer. A temperature jump of 1.1° at 296 K in a 5-mm cell was effected as described by Pasternack *et al.*[45] The change in light intensity at 465 nm (I_o = 5 V) was analyzed using an On-Line Systems program. ΔI_o is the transmittance intensity, and a decrease in intensity reflects an increase in absorbance. The relaxation curves were fitted to the sum of three first-order exponentials. Four temperature jumps were averaged for each kinetic trace and two different enzyme preparations were used yielding similar results. As a control, the temperature jump was repeated on each sample after oxidation in air: no rapid relaxation processes were observed.

As shown by Turowski *et al.*,[16] electron transfer between Cu(II) and the substrate reduced form of topa quinone produces Cu(I) and topa semiquinone at a rate of 20,000 sec^{-1}. Although this electron transfer is not stoichiometric with enzyme active sites, it indicates a rapid, kinetically competent electron transfer process.

Acknowledgment

We acknowledge Joseph Bell, Department of Biochemistry, School of Medicine, Temple University, Philadelphia, PA, for his assistance in the preparation of this manuscript.

[45] R. F. Pasternack, K. Kustin, L. A. Hughes, and E. Gibbs, *J. Am. Chem. Soc.* **91,** 4401 (1969).

[7] Mass Spectrometric Studies of the Primary Sequence and Structure of Bovine Liver and Serum Amine Oxidase

By GREGORY W. ADAMS, PETRA MAYER, KATALIN F. MEDZIHRADSZKY, and ALMA L. BURLINGAME

Introduction

Mass spectrometry has played an important role in the understanding and determination of the primary structure of quinoproteins. This chapter discusses primarily the use of mass spectrometry in the determination of the structure of the cofactor in bovine serum amine oxidase (BSAO) and the elucidation of the primary sequence for native BSAO. Amino acid sequences obtained from BSAO-derived peptides using tandem mass spectrometry and high energy collision-induced dissociation, together with Edman sequencing, have established 77% of the protein's sequence as deduced from the cDNA sequence recently obtained for BSAO.[1]

Experimental

Bovine serum amine oxidase was prepared from bovine plasma and inactivated by the addition of phenylhydrazine (D. Mu, UC Berkeley). An aliquot of BSAO was subjected to further purification by SDS/PAGE followed by electroelution and acetone precipitation. Tryptic digestion of the gel-isolated BSAO was carried out by incubating 0.5–2 nmol of protein in 75 m*M* NH_4HCO_3/0.02 m*M* $CaCl_2$/2.0 *M* urea (pH 8.2) at 37° for 24 hr, at an enzyme/substrate (w/w) ratio of 1:25. The BSAO protein was not reduced and carboxymethylated prior to digestion. Separation of the resulting mixture of peptides was achieved by C_{18} microbore reverse-phase HPLC. Solvent A was H_2O/0.01% TFA and solvent B was 30% H_2O/70% acetonitrile/0.01% TFA. Gradient elution involved the linear addition of solvent B to 10% at 10 min, 70% at 100 min, and 100% at 120 min. In addition, a chymotryptic digest was analyzed using the same enzyme/substrate ratio and chromatographic conditions.

Molecular weights of the peptides in each HPLC fraction were determined by positive ion liquid secondary ion mass spectrometry (LSIMS) using a Kratos MS50 double-focusing mass spectrometer, operating at a mass resolution of 1500, equipped with a high field magnet and a cesium ion source.[2] A Cs^+ primary ion beam of 10 keV was used with a 6-kV secondary ion accelerating voltage. One-fifth of each HPLC fraction was dissolved in a matrix of glycerol/thioglycerol (1:1) (v/v) with 0.1% TFA before spectra were acquired. High energy CID mass spectra were obtained using a Kratos Analytical Instruments Concept IIHH four sector high-performance tandem mass spectrometer equipped with a 4% electrooptical multichannel array detector.[3] In certain cases, sequence information was obtained using an Applied Biosystems 470A gas-phase protein sequencer with an Applied Biosystems 120A phenylthiohydantoin analyzer.[4]

[1] D. Mu, K. F. Medzihradszky, G. W. Adams, P. Mayer, W. M. Hines, A. L. Burlingame, A. J. Smith, D. Cai, and J. P. Klinman, *J. Biol. Chem.* **269,** 9926 (1994).

[2] A. M. Falick, G. H. Wang, and F. C. Walls, *Anal. Chem.* **58,** 1308 (1986).

[3] F. C. Walls, M. A. Baldwin, A. M. Falick, B. W. Gibson, S. Kaur, D. A. Maltby, B. L. Gillece-Castro, K. F. Medzihradszky, and S. Evans, *in* "Biological Mass Spectrometry" (A. L. Burlingame and J. A. McCloskey, eds.), p. 197. Elsevier, Amsterdam, 1990.

[4] M. W. Hunkapiller, R. M. Hewick, W. J. Dreyer, and L. E. Hood, this series, Vol. 91, p. 399.

Proteolytic digests derived from BSAO were also analyzed by microbore HPLC/electrospray ionization mass spectrometry using water/acetonitrile/TFA as the mobile phase. A dual syringe pump (Carlo Erba) was used to deliver mobile phase at a flow rate of 50 μl/min. HPLC separations were performed on an Aquapore 300 C_{18} microbore column, 1.0 mm i.d. $\times$ 100 mm (Applied Biosystems). Postcolumn addition of 2-methoxyethanol/isopropanol (1 : 1) to improve the mass spectrometry signal[5] was delivered by a separate syringe pump (Isco) at a flow rate of 40 μl/min to a 3.1-μl dead volume PEEK mixing tee (Upchurch Scientific) positioned after the UV detector. After the mixing tee, a zero dead volume tee was incorporated to split the column effluent 1 : 20 so that the flow rate into the mass spectrometer was 3–5 μl/min, whereas the remaining sample was manually collected for subsequent analysis. The microbore HPLC system was interfaced to a VG Biotech/Fisons BIOQ triple quadrupole mass spectrometer equipped with an electrospray source, by a 1-meter length of fused silica capillary tubing. Typical operating voltages were: probe tip, 4200 V; counter electrode, 550 V; and sampling orifice, 40–50 V. The source temperature was maintained at 60°. The mass spectrometer was scanned in noncontinuum mode over a range of m/z 350–2000 at 5 sec/scan.

Tandem Mass Spectrometry in Biological Structural Analysis

Developments in mass spectrometric soft ionization techniques such as fast atom (FABMS) bombardment[6,7] or ion (LSIMS) bombardment[8] have led to widespread use of mass spectrometry for biological structural analysis. Primary structural information of peptides may be rapidly obtained after ionization by LSIMS and collision-induced dissociation using high performance tandem mass spectrometry.[9,10]

In tandem mass spectrometry, two consecutive stages of mass analysis are employed. The initial stage permits selection and separation of the $(M+H)^+$ ion of the peptide of interest from other ions produced in the ioniza-

[5] P. R. Griffen, J. A. Coffman, L. E. Hood, and J. R. Yates, *Int. J. Mass Spectrom. Ion Processes* **111,** 131 (1991).

[6] M. Barber, R. S. Bordoli, R. D. Sedwick, and A. N. Tyler, *Nature* (*London*) **293,** 270 (1981).

[7] M. Barber, R. S. Bordoli, G. J. Elliott, R. D. Sedwick, and A. N. Tyler, *Nature* (*London*) **54,** 645 (1982).

[8] W. Aberth, K. M. Straub, and A. L. Burlingame, *Anal. Chem.* **54,** 2029 (1982).

[9] K. Biemann, this series, Vol. 193, p. 455.

[10] S. Kaur, K. F. Medzihradszky, Z. Yu, M. A. Baldwin, B. L. Gillece-Castro, F. C. Walls, B. W. Gibson, and A. L. Burlingame, *in* "Biological Mass Spectrometry" (A. L. Burlingame and J. A. McCloskey, eds.), p. 285. Elsevier, Amsterdam, 1990.

tion source. This low internal energy ion is then transmitted into a collision cell containing inert gas in the field free region between the two analyzers, where collisional deposition of additional internal energy induces the relatively stable molecular ion to undergo extensive dissociation reactions.

The fragmentation process produces different types of product ions characteristic of the structure of the peptide. The fragment ions are then transmitted into the second analyzer where they are separated and detected. The recording of the relative abundances of these product ions as a function of nominal mass constitutes a collision-induced dissociation (CID) mass spectrum. Tandem double focusing (or four sector) mass spectrometers featuring unit mass resolution allow selection of only the ^{12}C component of the monoisotopic peptide ion's stable isotope profile, and fragment ions characteristic of this particular monoisotopic precursor ion will be generated and recorded by the second mass spectrometer forming a high energy CID mass spectrum.

To elucidate the sequence of a peptide by interpretation of high energy CID data, the most important bond fragmentation occurs along the peptide backbone, -NH-CH(R)-CO-, where R represents the side chain of the amino acid. Bond cleavage may occur at three possible bond positions along the peptide backbone as illustrated in Scheme 1. When the positive charge is retained on the N terminus, the ions are designated $\mathbf{a}_i$, $\mathbf{b}_i$, and $\mathbf{c}_i$ and when the corresponding ion forms with charge retention at the C terminus the ions are designated $\mathbf{x}_i$, $\mathbf{y}_i$, and $\mathbf{z}_i$.[11] Cleavage of the peptide bond itself results in formation of $\mathbf{b}_i$ and $\mathbf{y}_{n-i}$ ions; either or both of these ions will form depending on the relative basicities of the N- or C-terminal residues. When cleavage of the same bond occurs in consecutive peptide linkages, a series of ions of increasing or decreasing mass is generated and the mass difference between consecutive pairs reveals the identities of the respective amino acids and thus the sequence of the peptide can be elucidated.[12,13]

Furthermore, double focusing tandem MS instruments allow for high (kilovolt) collision energies, necessary for deposition of sufficient internal energy for observation of side chain-specific fragmentation processes.[14] When basic amino acids are present in the peptide, cleavage of the β,γ-bond of the side chain $(\mathbf{R}_i)$ of the corresponding $\mathbf{z}_{i+1}$ ion results in the formation of a $\mathbf{w}_i$ ion; if cleavage of the α,β-bond and an additional hydrogen

[11] K. Biemann, *Biomed. Environ. Mass Spectrom.* **16,** 99 (1988).

[12] K. Biemann, *in* "Biological Mass Spectrometry" (A. L. Burlingame and J. A. McCloskey, eds.), p. 179. Elsevier, Amsterdam, 1990.

[13] K. F. Medzihradszky and A. L. Burlingame, Methods: *Companion Methods Enzymol.* **6,** 284 (1994).

[14] R. S. Johnson, S. A. Martin, and K. Biemann, *Int. J. Mass Spectrom. Ion Processes* **86,** 137 (1988).

PEPTIDE BACKBONE FRAGMENTATIONS

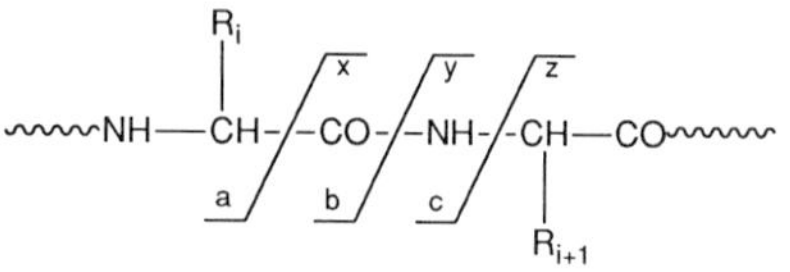

$$\sim\sim NH - \underset{\displaystyle R_i}{CH} - / - CO - / - NH - / - \underset{\displaystyle R_{i+1}}{CH} - CO \sim\sim$$

(cleavage sites: x, y, z above; a, b, c below)

Numerical Relationship between ions:

$$y_i + b_{n-i} = MH^+ + 1$$

$$z_i + 17 = y_i \qquad a_i + 28 = b_i$$

$$y_i + 26 = x_i \qquad b_i + 17 = c_i$$

SIDE CHAIN FRAGMENTATIONS

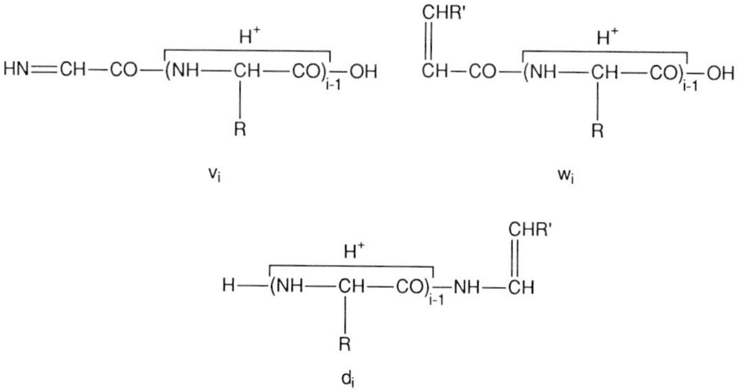

$$HN{=}CH{-}CO{-}(NH{-}\underset{\displaystyle R}{CH}{-}CO)_{i-1}{-}OH \cdot H^+$$

v_i

$$\overset{\displaystyle CHR'}{\|}\atop{CH}{-}CO{-}(NH{-}\underset{\displaystyle R}{CH}{-}CO)_{i-1}{-}OH \cdot H^+$$

w_i

$$H{-}(NH{-}\underset{\displaystyle R}{CH}{-}CO)_{i-1}{-}NH{-}CH{=}CHR' \cdot H^+$$

d_i

SCHEME 1. Nomenclature for peptide backbone fragment ions and the numerical relationship among the related ions.

occurs from a $\mathbf{y_i}$ ion, a $\mathbf{v_i}$ ion is produced (Scheme 1). Similarly, cleavage of the β,γ-bond of the C-terminal amino acid of a $\mathbf{a_{i+1}}$ ion results in the formation of a $\mathbf{d_i}$ ion (Scheme 1). Such fragmentation reactions enable differentiation between amino acid residues with the same nominal mass but isomeric structures (e.g., Leu, Ile). The isobaric amino acids, Gln and Lys, cannot be distinguished by tandem mass spectrometry; however, distinction can be made following acetylation, which only takes place on the ε-amino group of lysine.

Mass Spectrometric Determination of the Cofactor in BSAO

As described by Janes *et al.*,[15] BSAO contains a new amino acid as the cofactor: 2,4,5-trihydroxyphenylalanine quinone, subsequently known as

[15] S. M. Janes, D. Mu, D. Wemmer, A. J. Smith, S. Kaur, D. A. Maltby, A. L. Burlingame, and J. P. Klinman, *Science* **248,** 981 (1990).

topa quinone (cf. [2], this volume). Tandem mass spectrometry played a major role in the elucidation of this new prosthetic group. The high energy CID mass spectrum of a phenylhydrazine-derivatized active site pentapeptide, molecular ion MH^+ m/z 807.5, is shown in Fig. 1. The spectrum exhibits a contiguous series of $\mathbf{y_i}$ (i.e., C-terminal) and $\mathbf{b_i}$ (i.e., N-terminal) ions for the peptide backbone, assuming a molecular weight of 283 for X, ruling out possible covalent modification of residues other than X in the modified peptide. Furthermore, analysis of a ^{15}N-phenylhydrazine-derivatized peptide revealed an increase by one mass unit for the molecular ion and all peptide fragments containing the unknown amino acid X.

The low mass region (below mass 300) contains several types of informative fragment ions, including immonium and internal dipeptide ions. Upon collision-induced dissociation an amino acid produces an immonium ion (general structure $[H_2N{=}CH\text{-}R]^+$, where R is the amino acid side chain).[16,17] Such ions and their related fragment ions often yield characteristic information about the structure of the side chain of a particular amino acid. The intense peak at m/z 256 in Fig. 1 corresponds to the immonium ion derived from X. The detection of an intense immonium ion, together with the fact that Edman sequencing yielded sequence data past X, established the presence of a peptide backbone structure in X. Furthermore, several ions appear related to the immonium ion: m/z 239 corresponds to loss of NH_3 and m/z 227 corresponds to loss of NH_2CH.

Side chain fragmentation also occurs from the molecular ion, and ions can often be observed that correspond to loss of the side chain (cf. ion mass m/z 750.3 corresponds to loss of C_4H_9, characteristic of fragmentation at the β-bond of leucine).[14] Analogous fragmentations were used to distinguish between the two possible structures initially proposed for the unknown amino acid X (Scheme 2) (cf. [2], this volume). Side chain fragmentation of structure A would lead to loss of mass 227 Da for fragmentation between the β carbon and the peptide backbone or loss of mass 213 Da for fragmentation between the β carbon and oxygen. In no instance was a loss of this mass observed. In contrast, an intense peak at m/z 580 was observed in the spectra of both the ^{14}N- and ^{15}N-phenylhydrazone-derivatized peptides, which corresponds to a loss of mass 227 (or 228) from the molecular ion. This loss can be contributed to cleavage between the β carbon and the peptide backbone of structure B. Mass m/z 580 is isobaric with the $\mathbf{y_3}$ ion in the ^{14}N spectrum; however, the spectrum for the ^{15}N peptide indicated losses at both m/z 580 and 581, corresponding to the side chain fragmentation and the $\mathbf{y_3}$ ion, respectively (data not shown).[15]

[16] R. S. Johnson and K. Biemann, *Biomed. Environ. Mass Spectrom.* **18,** 945 (1989).

[17] A. M. Falick, W. M. Hines, K. F. Medzihradszky, M. A. Baldwin, and B. W. Gibson, *J. Am. Soc. Mass Spectrom.* **4,** 882 (1993).

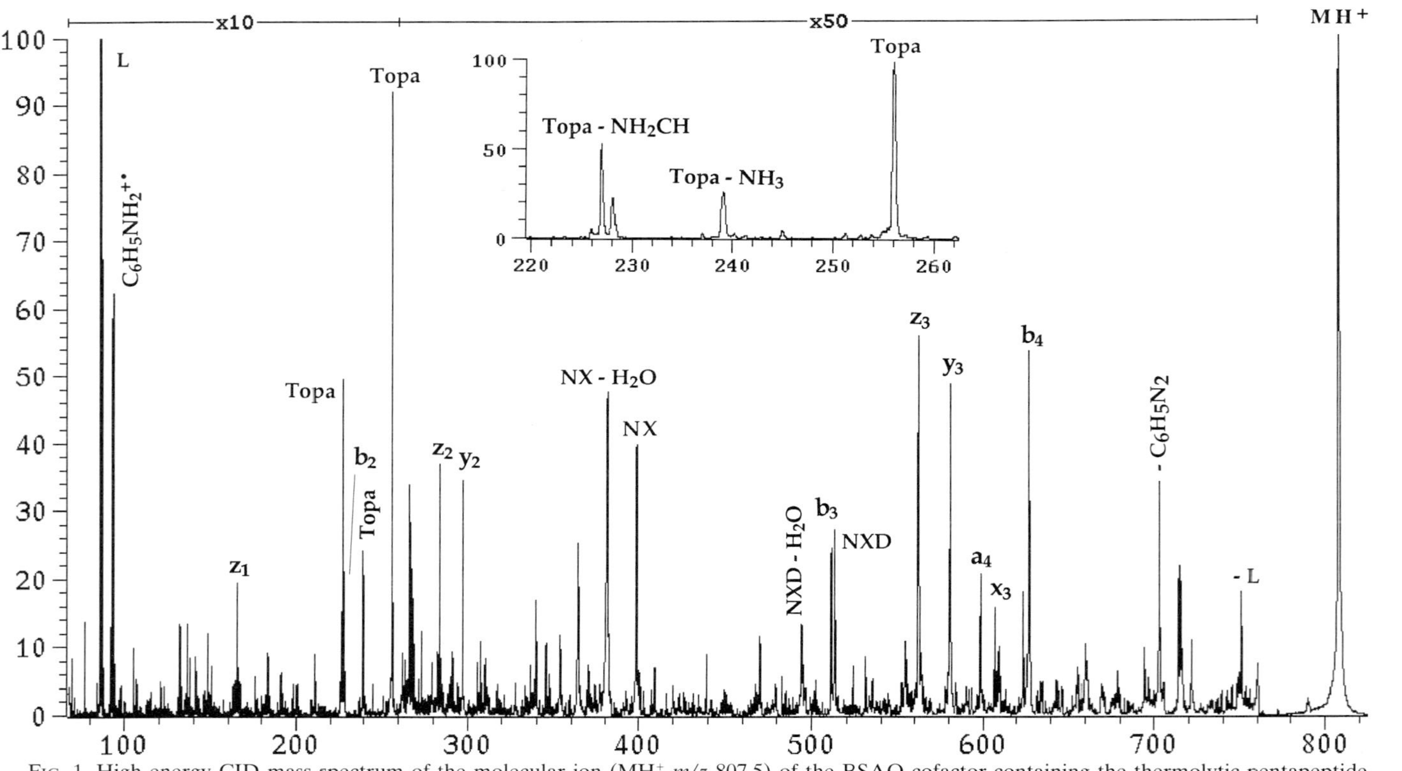

FIG. 1. High energy CID mass spectrum of the molecular ion (MH^+ m/z 807.5) of the BSAO cofactor containing the thermolytic pentapeptide. The elucidated sequence is Leu-Asn-X-Asp-Tyr, where X is topa. Significant ions for the structure are labeled and are shown in the sequence above the spectrum. Immonium ions corresponding to individual amino acid residues are denoted by the single letter code. Acylium ions are denoted by the single letter code of the amino acids in each ion (cf. the ion m/z 398 corresponds to the ion NX). Peptide backbone cleavage ions formed from fragmentation indicative of charge retention at the C terminus are denoted by x, y, and z and at the N terminus by a and b (see Scheme 1). The expanded region (m/z 220–260) illustrates the topa immonium ion at m/z 256 and corresponding losses of NH_3 (m/z 239) and NH_2CH (m/z 227) from the immonium ion.

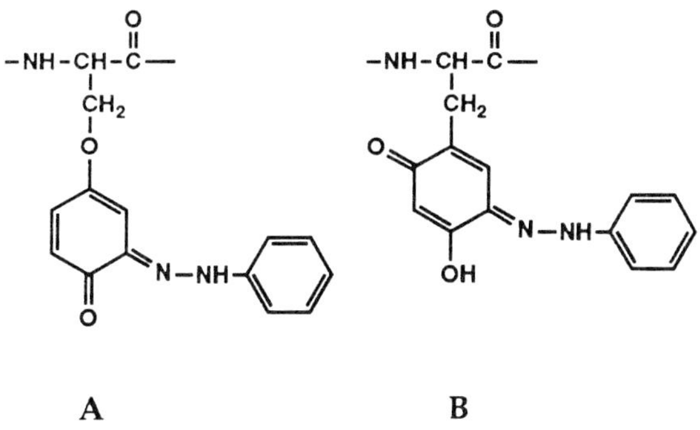

SCHEME 2. Possible cofactor structures proposed for the unknown residue X in the pentapeptide, molecular ion $MH^+ = 807.5$. (A) The phenylhydrazone derivative of a catechol linked via an ether bond to a Ser residue. (B) The phenylhydrazone derivative of 6-hydroxydopa.

The mass spectral data, in addition to elemental composition from accurate mass measurement, permitted tentative assignment of the active site cofactor in BSAO to structure B (Scheme 2). Janes *et al.*[15] further characterized the cofactor structure using UV/VIS spectroscopy and ^1H NMR of the phenylhydrazine-derivatized pentapeptide and a synthetic analog of the topa-containing peptide (cf. [2], this volume). The direct structural analysis and the additional comparison of chemical and physical properties led to unambiguous determination of the structure of the cofactor as topa quinone.

Mass Spectrometric Identification of a Second Unknown Amino Acid in an Active Site, Cofactor-Containing Peptide

The active site, cofactor-containing peptide has also been isolated from a tryptic digest. An initial analysis of this peptide by Edman degradation revealed a peptide 25 amino acids in length:

```
1                                               25
S V S T M L N J D Y V J D M V F Y P N G A I E V K
```

SCHEME 3. Amino acid sequence of the BSAO cofactor containing tryptic peptide obtained by Edman degradation. J refers to an unidentified amino acid.

In two cycles, namely the 8th and 12th cycles, it was not possible to determine the amino acid unambiguously. As described earlier, the amino acid in the 8th cycle was identified as topa quinone. To investigate the identity of the unknown amino acid at cycle 12, an amino acid analysis of the BSAO active site tryptic peptide was carried out under milder hydrolysis conditions to investigate the possible presence of tryptophan. The tryptophan analysis showed the presence of one tryptophan, implying that the 12th amino acid is tryptophan.

To characterize the amino acid at residue 12 unambiguously, an Asp-N digest of the purified BSAO active site tryptic peptide was analyzed by mass spectrometric methods. The Asp-N digest was separated using on-line HPLC/ESI/MS, and peptides were sequenced using high energy CID tandem mass spectrometry analysis. Figure 2 illustrates the UV and base peak intensity chromatograms of the Asp-N digest analyzed by LC/ESI/MS. Molecular ions detected for the four peaks are listed in Table I. The values obtained for the molecular weights of the anticipated Asp-N peptides are consistent with the presence of the phenylhydrazine-derivatized topa at residue 8 and tryptophan at residue 12.

We attempted to record LSIMS tandem MS spectra of the ions listed in Table I. Unfortunately, the ion at *m/z* 582.3 which would correspond to the sequence DYVW yielded a very weak molecular ion, and a CID spectrum was not obtained. However, the high energy CID mass spectra of the ion MH^+ *m/z* 1034.5, as well as CID spectra for the ions *m/z* 1482.7 and *m/z* 1597.7, were obtained. The sequences of these three peptides confirm the complete sequence of the BSAO active site tryptic peptide as shown

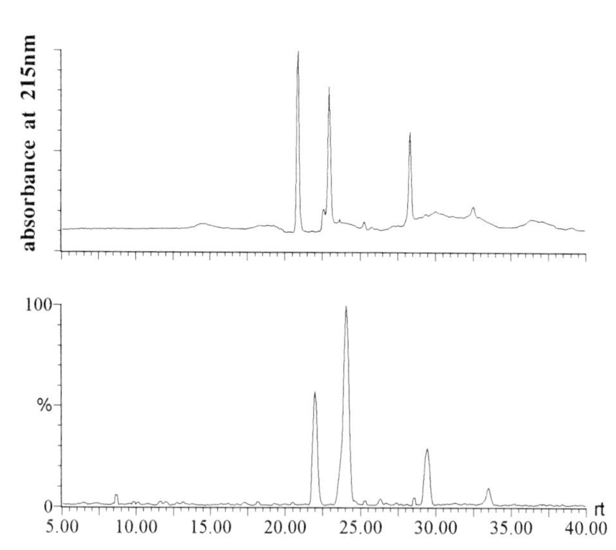

FIG. 2. UV absorbance (215 nm) (upper trace) and base peak intensity (lower trace) chromatogram of the BSAO active site peptide Asp-N digest. The digest was analyzed in LC/ESI/MS mode, using water/acetonitrile/TFA as the mobile phase for the chromatographic separation. Molecular ions detected for the four peaks are listed in Table I.

TABLE I
MOLECULAR IONS OBTAINED FROM LC/ESI/MS ANALYSIS OF AN Asp-N DIGEST OF THE BSAO ACTIVE SITE TRYPTIC PEPTIDE

Retention time[a]	MH^+ (Obs.)[b]	MH^+ (Calc.)	MH^+ (Obs.)[c]	Fragment
20.9	583.0	582.64	Not detected	9–12
23.0	1484.0	1483.73	1482.7	13–25
28.4	1035.0	1035.17	1034.5	1–8
32.5	1598.5	1598.78	1597.7	1–12

[a] Retention times refer to peaks in the lower trace of Fig. 2.
[b] Observed mass is the average mass as obtained by ESI analysis.
[c] This mass is the monoisotopic mass observed by LSIMS using tandem mass spectrometry.

in Scheme 3 and characterize the unknown amino acid in the 12th position as tryptophan.

Role of Mass Spectrometry in Obtaining Amino Acid Sequences for Cloning

Traditional molecular cloning techniques[18] involve obtaining sequence data by purification of the protein of interest using SDS–PAGE, electroblotting onto a PVDF membrane, staining with Coomassie blue, excising the band, and performing Edman sequencing to obtain a peptide sequence of the N terminus of the protein from which a probe can be designed.[19,20] Obtaining a peptide sequence for cloning using N-terminal sequence methodology has several drawbacks; first and foremost, it has been estimated that greater than 80% of mammalian proteins have blocked N termini as a result of post-translational modification.[21] The absence of a free N terminus prevents straightforward use of Edman degradation for sequencing. In these cases, internal sequence information must be obtained by chemical or proteolytic fragmentation of the protein and subsequent sequence analysis of the isolated peptides. Second, the peptide sequence obtained by Edman degradation is, in general, limited to ca. 30 amino acids or less, thus limiting the amount of sequence data from which a probe can be designed.

The approach taken to clone the BSAO gene was to use polymerase chain reaction-based, mixed oligonucleotide-primed amplification of cDNA

[18] M. Grunstein and D. S. Hogness, *Proc. Natl. Acad. Sci. U.S.A.* **72,** 3961 (1975).
[19] P. Matsudaira, *J. Biol. Chem.* **262,** (1987).
[20] S. W. Yuen, A. H. Chiu, K. J. Wilson, and P. M. Yuan, *BioTechniques* **7,** 74 (1989).
[21] H. P. C. Driessen, W. W. De Jong, G. I. Tesser, and H. Bloemendal, *CRC Crit. Rev. Biochem.* **18,** 281 (1984).

(MOPAC)[22] to amplify a BSAO cDNA probe from an appropriate bovine cDNA library (cf. [8], this volume). This experiment employed two oligonucleotide probes derived from the amino acid sequence of peptides obtained from native BSAO. Thus multiple sequences of peptides derived from BSAO were required, from which a probe containing regions of coding sequence with the least ambiguity could be selected.

As sequence information is relatively easily achieved by tandem mass spectrometry,[9] the general approach this laboratory utilizes to sequence a protein involves proteolytic digestion, separation of peptides by high performance liquid chromatography (HPLC), and analysis of selected peptides by tandem mass spectrometry.[10]

Protein Sequencing Using High Energy CID Tandem Mass Spectrometry

When peptides are ionized by LSIMS, the positive charge tends to be localized on the side chains of basic amino acids such as His, Arg, and Lys instead of on the less basic amide nitrogen atoms along the peptide backbone. Trypsin is a highly specific protease, which yields peptides with a basic amino acid, either Arg or Lys, at the C terminus. The high energy CID spectra of such peptides tend to be dominated by the C-terminal $\mathbf{x}_i$, $\mathbf{y}_i$, and $\mathbf{z}_i$ ions and are in general simpler spectra to interpret.[13] Thus, trypsin is used most frequently when a proteolytic digest is to be analyzed by mass spectrometry.

The UV chromatogram of the BSAO tryptic digest is illustrated in Fig. 3; each fraction was collected and lyophilized, and its LSIMS spectrum was recorded to obtain the molecular weights of the peptide(s) in the fraction. Molecular ions were then selected on the basis of relative abundance and mass (<1900 Da) for sequence analysis by tandem mass spectrometry to obtain the corresponding high energy CID spectra of the molecular ion.

The peptide sequences were obtained by interpretation (either manually or with the aid of a computer algorithm[23]) of the respective high energy CID mass spectra. In general, these sequences were furnished from the most abundant ions exhibited in the initial LSIMS spectra. Additional sequences were obtained by Edman sequence analysis. Fractions analyzed by Edman were selected on the basis that the fraction contained only one component (i.e., only one molecular ion detected in the LSIMS spectrum)

[22] C. C. Lee, X. Wu, R. A. Gibbs, R. G. Cook, D. M. Munzy, and C. T. Caskey, *Science* **239,** 1288 (1988).

[23] W. M. Hines, A. M. Falick, A. L. Burlingame, and B. W. Gibson, *J. Am. Soc. Mass Spectrom.* **3,** 326 (1992).

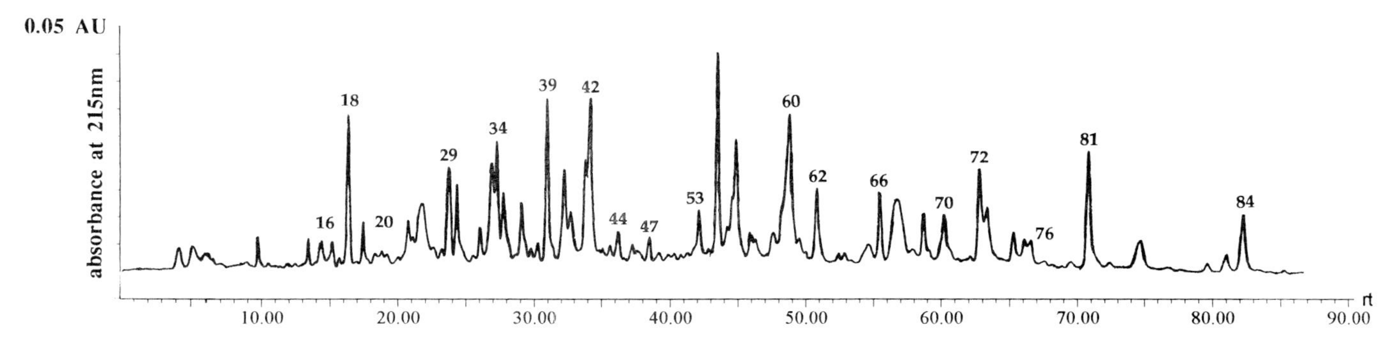

FIG. 3. UV absorbance chromatogram (at 215 nm) of a tryptic digest of BSAO. See experimental section for chromatographic conditions.

and that the mass was too large to be successfully analyzed by tandem mass spectrometry (cf. MH^+ >2000 Da).

Mass spectrometry greatly facilitated the determination of amino acid sequences for BSAO peptides. BSAO is particularly resistant to proteolytic digestion; hence enzymatic digestion of BSAO required a lengthy incubation time and denaturation of the protein by urea to promote a more effective digestion. The combination of the time required for denaturation and digestion conditions often leads to some of the peptides being N-terminally blocked due to the reaction between urea and the peptide N termini, resulting in N-carbamoyl peptide(s).[13,24] Furthermore, the chromatographic separation of a complex mixture of BSAO peptides produced by proteolytic digestion often resulted in fractions that contained more than one peptide. Unseparated peptide mixtures and peptides that are N-terminally blocked cannot be sequenced by traditional Edman techniques. Mass spectrometric methodology overcomes the limitations of Edman sequencing, including both the ability to sequence chemically modified peptides and the ability to obtain unambiguous sequence information from unseparated mixtures.

Sequence Analysis of High Energy CID Mass Spectra

Of all fractions collected in the tryptic digest, 18 CID spectra were obtained from which a complete sequence for the peptide was elucidated. The sequences are listed in Table II. Furthermore, in the course of this study, a chymotryptic digest of BSAO was also studied; from this digest an additional 13 sequences were obtained from CID mass spectra. These sequences are listed in Table III. Three examples of CID spectra of BSAO tryptic peptides are illustrated in Figs. 4–6, a further example of a chymotryptic peptide is illustrated in Fig. 7.

Figure 4 illustrates the high energy CID of a BSAO tryptic peptide, MH^+ m/z 992.6 (molecular ion not shown in the spectrum) detected in fraction 34 (Fig. 3). As this is a tryptic peptide, the C terminus is expected to be either Arg or Lys; the presence of either residue can be established by the presence of a **y**-type ion or characteristic immonium ions of each residue. The masses for $\mathbf{y_i}$ ions for Arg and Lys are 175 and 147, respectively; Arg exhibits a series of ions m/z 70, 87, 100, and 112 arising from its immonium ion at m/z 129. As Lys is isobaric with Gln, immonium ions cannot differentiate the two amino acids. In the spectrum illustrated in Fig. 4 a strong ion at m/z 175 is observed, indicating that Arg is the C-terminal

[24] D. X. Wen, B. D. Livingston, K. F. Medzihradszky, S. Kelm, A. L. Burlingame, and J. C. Paulson, *J. Biol. Chem.* **267,** 21011 (1992).

TABLE II
AMINO ACID SEQUENCES OF PEPTIDES ISOLATED FROM A TRYPTIC DIGEST OF BSAO

Entry	Fraction	MH^+ (Obs.)	MH^+ (Calc.)	Amino acid sequence
i	16[a]	616.9[b]	616.38	LQVTR
ii	18	718.4	718.37	DVTVER
iii	20	807.6	807.41	FSVQGNR
iv	22	723.6	723.36	ASFWGR
v	26	711.8	711.46	GLPLRR
vi	29	937.6	937.52	AAALAHLDR
vii	33	1002.8	1002.54	LLTMNSAPR
viii	34	753.2	753.51	RPVLLR
ix	34	992.6	992.55	YQLAITQR
x	36	857.2	857.48	YLYLASK
xi	39	2046.7	2046.21	IQTVSFAGGPMPQNSPMER[c]
xii	41	2037.6	2037.07	SQVPPGPTPPLQFHPQGPR[c]
xiii	42	1243.6	1243.63	ALDPADWTVQK
xiv	42	1900.8	1900.94	QLETEEQAAFPLGGASPR[c]
xv	44	1059.5	1059.54	HGGPLPYYR
xvi	47	900.4	900.15	(carbamoyl)YLYLASK
xvii	53	1551.4	1551.73	TLHDAFCVFEQNK
xviii	59	2596.6	2596.31	LAYEISLQEAGAVYGGNTPAAML[d]
xix	60	1844.2	1844.01	GGPYLHPVGLELLVDHK
xx	61	1781.6	1781.93	LHATGYISSAFLFGAAR[c]
xxi	62	1556.8	1556.74	EYLDIDQMIFNR
xxii	66	1925.8	1925.89	YMDSGFGMGYFATPLIR[c]
xxiii	69	2417.6	2418.79	HHSDFLSHYFGGVAQTVLVF[d]
xiv	70	3823.7[e]	3823.24	EGQDAGSCEINPLACLP[d]
xxv	72	3817.7[e]	3818.19	YYENLAQLEEQFEAGQVN[d]
xxvi	73	1987.2	1987.01	VASSLWTFSFGLGAFSGPR[c]
xxvii	76	3906.3[e]	3908.39	VDLDVGGLENWVWAEDMAFVPTAIP[d]
xxviii	81	2734.4	2735.76	EELTTVMSFLTQQLGPDLVDAAQAR[c]
xxix	84	6077.3[e]	6076.79	DLVAWVTAGFLHIPHAEDIPNTVTVGN[d]

[a] Fraction numbers refer to the HPLC chromatogram in Fig. 3.
[b] All observed masses are monoisotopic unless otherwise noted.
[c] Sequence obtained by Edman degradation.
[d] Complete tryptic peptide sequence not obtained.
[e] Observed mass is the average mass as obtained by ESI analysis; calculated mass is the average mass.

amino acid. The immonium ions detected in the low mass region of this spectrum suggest the presence of Tyr, Leu or Ile, Lys/Gln, and Arg. An initial interpretation strategy is to search for pairs of $\mathbf{b_i}$ and $\mathbf{y_{n\text{-}i}}$ ions deduced from the numerical relationship: $\mathbf{b_i} + \mathbf{y_{n\text{-}i}} = MH^+ + 1$. In this particular spectrum, identification of $\mathbf{y_i}$ ions is aided by the presence of a series of $\mathbf{y_i} - \mathbf{2}$ ions (such ions are designated $\mathbf{Y_i}$) and a series of such doublet ions are observed for $\mathbf{y_{4\text{-}7}}$. Masses corresponding to $\mathbf{y_2}$ and $\mathbf{y_3}$ are identified

TABLE III
AMINO ACID SEQUENCES OF PEPTIDES ISOLATED FROM A CHYMOTRYPTIC DIGEST OF BSAO

Entry	Fraction	MH^+ (Obs.)	MH^+ (Calc.)	Amino acid sequence
xxx	43	536.6[a]	536.28	DVRF
xxxi	43	721.6	721.42	TVQKVF
xxxii	49	831.4	831.40	GGNTPAAML
xxxiii	50	898.5	898.50	EQNKGLPL
xxxiv	51	1072.6	1072.51	THPDQSQLF
xxxv	54	1056.4	1056.57	VVGPLPQPSY
xxxvi	55	850.5	850.47	RIQTVSF
xxxvii	55	891.5	891.54	PKAAALAHL
xxxviii	59	1179.6	1179.59	EISLQEAGAVY
xxxix	60	916.6	916.57	YRRPVLL
xl	70	1220.6	1220.38	ENLAQLEEQF
xl	71	698.3	698.35	SFGLGAF
xl	73	1344.6	1344.68	KVDLDVGGLENW

[a] All observed masses are monoisotopic.

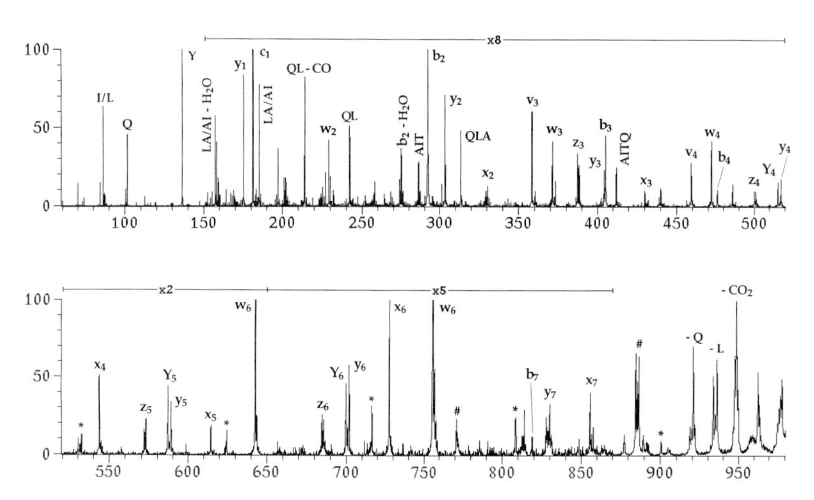

FIG. 4. High energy CID mass spectrum of a BSAO tryptic peptide, MH^+ m/z 992.6 (molecular ion not shown in the spectrum), detected in fraction 34 (Fig. 3). The elucidated sequence is Y-(Q/K)-L-A-I-T-(Q/K)-R. Fragmentations arising from cleavage of the peptide backbone are labeled as in Fig. 1. Ions formed by side chain fragmentation are denoted as v and w. Ions labeled with an asterisk refer to glycerol cluster ions, whereas ions labeled with a pound sign are thioglycerol cluster ions.[25]

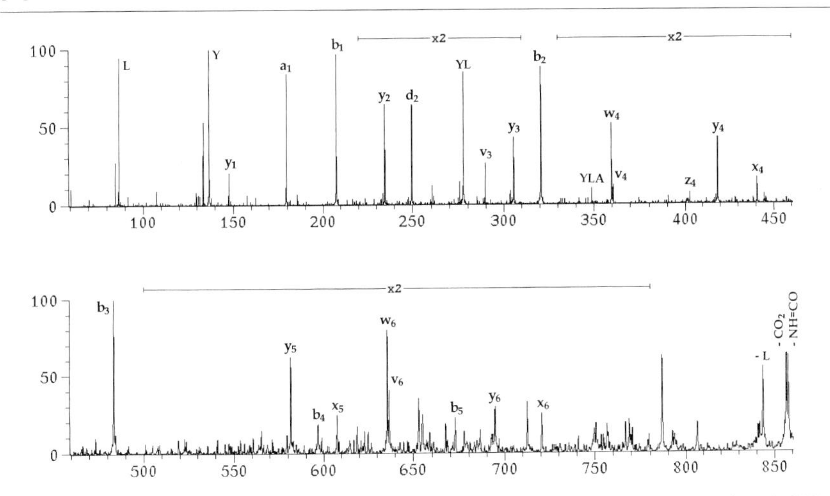

FIG. 5. High energy CID mass spectrum of a BSAO tryptic peptide, MH^+ m/z 900.4 (molecular ion not shown in the spectrum), detected in fraction 47 (Fig. 3). The elucidated sequence is N-carbamoyl-Tyr-Leu-Tyr-Leu-Ala-Ser-Lys.

from a set of $\mathbf{z_i/y_i/x_i}$ ($\mathbf{z_i}$ and $\mathbf{y_i}$ differ in mass by 17 Da, $\mathbf{y_i}$ and $\mathbf{x_i}$ differ by 26). From the mass differences among **y** ions (163/128/113/71/113/101/128/175), sequence Y-(Q/K)-(L/I)-A-(L/I)-T-(Q/K)-R can be assigned. Ions detected from side chain fragmentations exhibited in this spectrum permit differentiation between Leu and Ile. These side chain fragments, labeled $\mathbf{w_i}$ ions, are formed via cleavage of the β,γ-bond of a $\mathbf{z_i + 1}$ ion. For the amino acid Leu, this fragmentation results in an ion with a mass 54 Da higher than the preceeding $\mathbf{y_{i-1}}$ ion. Because of the different β-substitution of an Ile residue, the $\mathbf{w_i}$ ion will appear at a mass 68 Da higher than the preceeding $\mathbf{y_{i-1}}$ ion. Further inspection of the spectrum reveals an ion at m/z 471 that is identified as $\mathbf{w_4}$, establishing this residue as Ile. The abundant ion at m/z 643 corresponds to $\mathbf{w_6}$ and hence identifies this residue as Leu. The isobaric amino acids Gln and Lys cannot be distinguished by high energy CID mass spectrometry, but can by acetylation which only takes place on the ε-amino group of lysine. Also of note in this spectrum is a series of artifact peaks 92 and 108 Da apart, corresponding to glycerol and thioglycerol oligomer losses from a liquid matrix background peak which is isobaric with the chosen peptide precursor.[25]

Figure 5 illustrates the spectrum of a peptide, MH^+ m/z 900.4, detected

[25] A. M. Falick, K. F. Medzihradszky, and F. C. Walls, *Rapid. Commun. Mass Spectrom.* **4,** 318 (1990).

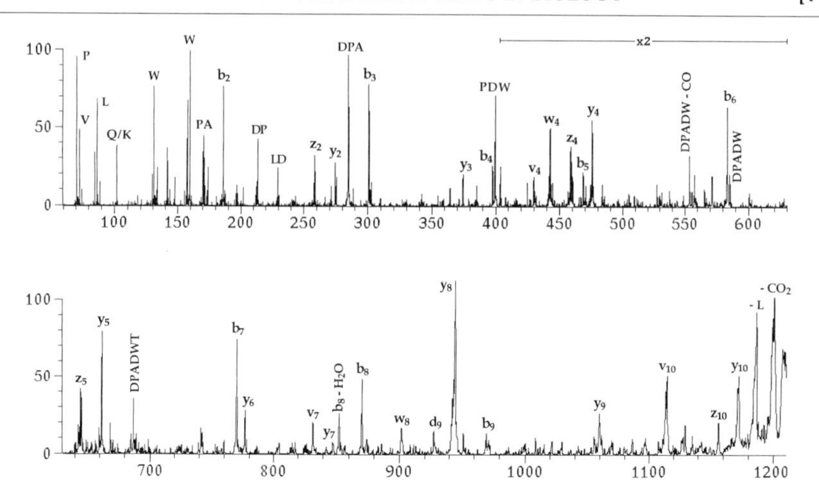

FIG. 6. High energy CID mass spectrum of a BSAO tryptic peptide, MH^+ *m/z* 1243.7 (molecular ion not shown in the spectrum), detected in fraction 42 (Fig. 3). The elucidated sequence is A-L-D-P-A-D-W-T-V-(Q/K)-K.

in fraction 47 (Fig. 3). The ion at *m/z* 147 is indicative of $\mathbf{y_1}$ for Lys, establishing the C-terminal amino acid. A series of $\mathbf{z_i/y_i/x_i}$ ions are observed up to $\mathbf{y_6}$ at *m/z* 694; the remaining mass of 206 Da does not correlate with the mass for a naturally occurring amino acid. In addition, abundant ions at *m/z* 179 and 207 are identified as $\mathbf{a_1}$ and $\mathbf{b_1}$. A residue mass combination of 206 Da has only one dipeptide combination, Cys–Cys. This combination would yield a $\mathbf{y_7}$ ion at *m/z* 707, which is not observed, nor is an immonium ion for Cys observed. As mentioned earlier, because of the digestion conditions, the possibility of N-terminal carbamoylation needs to be taken into consideration during data interpretation. The mass increase in molecular weight on carbamoylation is 43 Da. An intense ion at *m/z* 857 due to loss of 43 Da is indicative that the N terminus is carbamoylated. Subtraction of 43 Da from mass 206 corresponds to a Tyr residue (163 Da). Furthermore, during the initial LSIMS analysis of HPLC fractions, a small ion *m/z* 857 was observed in fraction 37 and corresponds to the noncarbamoylated peptide. The mass differences between ions $\mathbf{y_3}$–$\mathbf{y_4}$ and $\mathbf{y_5}$–$\mathbf{y_6}$ correspond to a Leu/Ile residue. Ions observed at *m/z* 360 and 636 correlate to $\mathbf{w_4}$ and $\mathbf{w_6}$, respectively; in both cases, the presence of these ions establishes the residue as Leu. Thus the final sequence is established as N-carbamoyl-Tyr-Leu-Tyr-Leu-Ala-Ser-Lys.

Figure 6 illustrates the spectrum of a tryptic peptide, MH^+ *m/z* 1243.7, detected in fraction 42 (Fig. 3). From the low mass region of this spectrum, immonium ions are observed that suggest the presence of Trp, Gln/Lys,

Asp, Leu/Ile, Thr, Val, and Pro. An ion at *m/z* 147 corresponds to $\mathbf{y_1}$ for Lys, hence identifying the C terminus. The initial interpretation of this spectrum is aided by the intense ion pair at *m/z* 942 and 944. The most likely explanation of this doublet is a **y** ion belonging to Pro. It is well known that Pro residues furnish more abundant $\mathbf{y_i}$ and $\mathbf{Y_i}$ ions relative to other amino acids.[9,13] A corresponding $\mathbf{b_i}$ ion is observed at *m/z* 303; using these two ions as a starting point, a complete series of $\mathbf{y_i}$ and $\mathbf{b_i}$ ions can be established, suggesting a peptide sequence: A-(L/I)-D-P-A-D-W-T-V-(Q/K)-K. The ion pair at *m/z* 1113/1114 corresponds to $\mathbf{w_{10}}$ and $\mathbf{v_{10}}$, respectively, and hence confirms the penultimate residue from the C terminus as Leu. Unambiguous determination of the postulated sequence in this example is strengthened by the presence of internal acylium ions which form by fragmentation at two positions along the peptide backbone, namely cleavage to yield a $\mathbf{y_i}$ ion and $\mathbf{b_n}$ type fragmentation further toward the C terminus.[9,13,17] Ions of this type exhibited in Figs. 4–6 are labeled by their single letter amino acid codes. In general (see Figs. 4–6), ions of this type are di- or tripeptide ions, which can fragment further by loss of H_2O or CO depending on the amino acid composition. It is known that acylium ion formation is promoted by the presence of Pro and, to a lesser extent, Gly and Asp residues, as is evident from the spectrum in Fig. 6. Abundant acylium ions as large as hexapeptides are observed (cf. ions *m/z* 584 and 685 corresponding to an internal peptide sequence DPADW and DPADWT, respectively).

The three previous examples of CID spectra are of tryptic peptides. As discussed earlier, such peptides have a basic amino acid at the C terminus and, in general, the CID spectra are dominated by C-terminal fragment ions. The CID spectrum of an ion MH^+ *m/z* 1220.4, obtained from a chymotryptic digest, is illustrated in Fig. 7. This spectrum is dominated by what appears to be a single series of ions, beginning at *m/z* 204 and ending at *m/z* 1055, that could be either $\mathbf{b_i}$ or $\mathbf{y_i}$ ions. Assuming that *m/z* 1055 is either the $\mathbf{b_i}/\mathbf{y_i}$ ion for the pentultimate amino acid in this sequence, from the equation $\mathbf{b_i} + \mathbf{y_{n\text{-}i}} = MH^+ + 1$, the complementary ion can be calculated. Subtraction of *m/z* 1055 from the molecular ion (+ 1 Da) yields a mass of 166 Da. The only possible ion with *m/z* 166 is $\mathbf{y_1}$ for Phe and hence establishes the ion series observed in the spectrum as $\mathbf{b_i}$ type ions. Furthermore, other fragment ions that are derived from $\mathbf{b_i}$ ions are observed: cf. $\mathbf{a_i}$ type ions corresponding to loss of 28 Da from the respective $\mathbf{b_i}$ ion at *m/z* 244, 556, 669, 798, 927, and 1027. The difference of 28 Da is the mass difference between $\mathbf{a_i}$ and $\mathbf{b_i}$ ions. An additional loss of H_2O from $\mathbf{a_i}$ ions is often observed from Ser, Thr, Asp, or Glu residues whereas NH_3 is eliminated from Asn, Gln, Lys, and Arg residues. Several ions of this type are observed in this spectrum: loss of H_2O from $\mathbf{b_3}$, $\mathbf{b_4}$, and $\mathbf{b_9}$ and loss of NH_3 from $\mathbf{b_5}$,

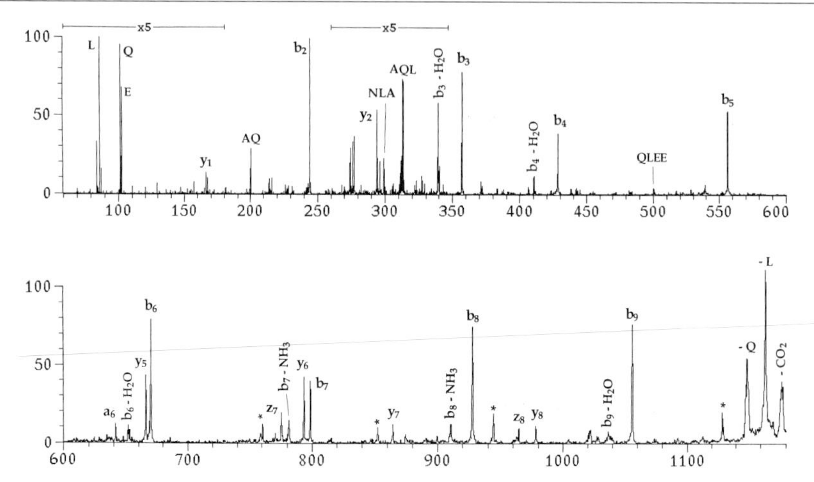

FIG. 7. High energy CID mass spectrum of a BSAO chymotryptic peptide, MH^+ m/z 1220.4. The elucidated sequence is E-N-(L/I)-A-(Q/K)-(L/I)-E-E-(Q/K)-F.

$\mathbf{b_7}$, and $\mathbf{b_8}$. Furthermore, a partial series of $\mathbf{y_i}$ type ions for $\mathbf{y_{1-2}}$ and $\mathbf{y_{5-8}}$ are observed, although the relative abundance of these ions is low. Beginning at m/z 244, which is presumably $\mathbf{b_2}$, a sequence (I/L)-A-(Q/K)-(I/L)-E-E-(Q/K)-F can be elucidated; however, the first two N-terminal residues still require identification. No apparent ion corresponding to $\mathbf{b_1}$ or $\mathbf{y_9}$ appears in the spectrum. There are five possible residue combinations that yield $\mathbf{b_2}$ at m/z 244: GW, SR, EN, and D(Q/K). This spectrum exhibits a variety of acylium ions, and closer inspection of the spectrum reveals an ion m/z 299 that corresponds to the tripeptide NLA (given the partially elucidated sequence no other tripeptide has a molecular mass of 299 Da). These data suggest that the first two N-terminal residues are Glu–Asn. From this CID spectrum the Leu/Ile residues cannot be distinguished. The sequence for this peptide as determined by CID interpretation was in agreement with the sequence determined from the cDNA sequence as E-N-L-A-Q-L-E-E-Q-F.

Molecular Cloning of BSAO

As described in another chapter of this volume (cf. [8]), a 2664-bp cDNA of BSAO containing an open reading frame of 2289 nucleotides has been cloned and sequenced.[1] The N-terminal sequence of native BSAO was identified by Edman degradation as REEGGVGSEEGVGK and is located behind a 16 residue signal peptide. The authenticity of the BSAO cDNA was confirmed by the exact matches between peptide sequences isolated from the native enzyme and the deduced protein sequence. All 28 sequences obtained by mass spectrometry/Edman degradation in Tables II and III align with regions of the cDNA-derived protein sequence and confirm 77% of the primary sequence.

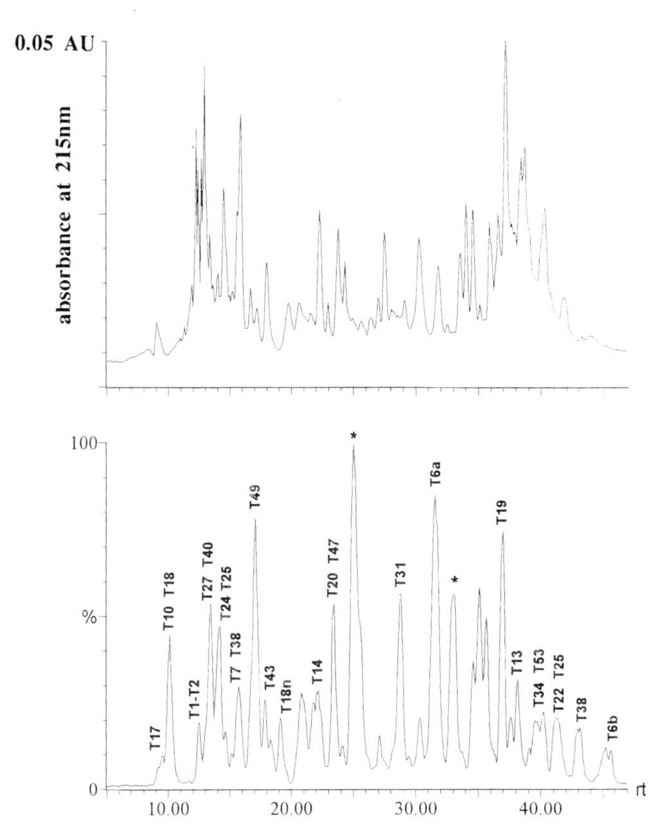

FIG. 8. HPLC/ESI/MS tryptic map of BSAO with UV monitoring at 215 nm (upper trace) and base peak intensity chromatogram as measured by the electrospray mass spectrometer (lower trace). See Experimental for details. The tryptic peptides, as determined by mass spectrometry, are numbered from the N terminus. The major peaks are labeled with the tryptic peptide detected in that peak. Table IV contains a complete list of the elution times and observed masses of each detected tryptic peptide. Two peptides resulting from uncleaved tryptic sites were observed (T_1–T_2 and T_{32}–T_{33}), and cleavage of the Arg–Pro bond of peptide T_6 results in two peptides: T_{6a} and T_{6b}. T_{18n} refers to the nonglycosylated form. Peaks labeled with an asterisk contain trypsin autolysis products.

TABLE IV
TRYPTIC PEPTIDES OBSERVED BY LC/ESI/MS ANALYSIS OF BSAO

Tryptic peptide	Amino acid	Expected mass[a]	Observed mass[b]	Retention time (min)	Sequence
1–2[c]	18–31	1390.45	1390.2	13.5	REEGGVGSEEGVGK
3	32–40	1093.25	1093.1	14.6	QBHPSLPPR
4	41–44	520.58	Not detected		BPSR
5	45–65	2355.48	Not detected		SPSDQPQTHPDQSQLFADLSR
6a[d]	66–90	2735.1	2735.2	45.2	EELTTVMSFLTQQLGPDLVDAAQAR
6b	90–106	1832.1	1832.4	32.9	PSDNBVFSVELQLPPK
7	107–115	938.08	938.3	15.7	AAALAHLDR
8	116–122	681.77	682.2	10.1	GSPPPAR
9[e]	123–153		Glycopeptide		EALAIVFFGGQPQP**N**VTELVVGPLPQPSYMR
10	154–159	718.79	719.0	10.1	DVTVER
11	160–168	1060.20	1060.7	16.7	HGGPLPYYR
12	169–174	753.97	753.9	16.9	RPVLLR
13	175–186	1557.77	1557.5	38.2	EYLDIDQMIFNR
14	187–202	1873.12	1873.1	22.1	ELPQAAGVLHHBBSYK
15	203–207	517.56			QGGQK
16	208–216	1003.21	1003.7	16.7	LLTMNSAPR
17	217–223	718.74	719.2	9.6	GVQSGDR
18n	224–235	1493.71	1493.7[f]	23.7	STWFGIYY**N**ITK
19	236–252	1845.15	1845.9	37.0	GGPYLHPVGLELLVDHK
20	253–263	1244.39	1244.9	23.5	ALDPADWTVQK
21	264–269	753.88	753.9	16.9	VFFQGR
22	270–303	3818.19	3818.0	41.2	YYENLAQLEEQFEAGQVNVVVIPDDGTGGFWSLK
23	304–322	2038.32	2038.3	20.8	SQVPPGPTPPLQFHPQGPR
24	323–329	807.89	808.2	14.1	FSVQGNR
25	330–348	1988.26	1988.7	41.3	VASSLWTFSFGLGAFSGPR
26	349–353	635.74	635.9	15.8	VFDVR
27	354–358	636.69	637.0	13.4	FQGER
28	359–383	2597.96			LAYEISLQEAGAVYGGNTPAAMLTR
29	384–400	1927.26	1927.8	39.5	YMDSGFGMGYFATPLIR

30	401–423	2746.10	2746.8	39.6	GVDBPYLATYMDWHFVVESQTPK
31	424–436	1610.79	1611.0	28.7	TLHDAFBVFEQNK
32 and 33	437–442	711.89	712.2	15.4	GLPLRR
34	443–463	2418.72	2418.7	40.2	HHSDFLSHYFGGVAQTVLVFR
35	464–488	3063.49	Not detected		SVSTMLNXDYVWDMVFYPNGAIEVK
36	489–505	1783.04	1783.6	37.6	LHATGYISSAFLFGAAR
37	506	175.21	Not detected		R
38	507–527	2333.53	2333.9	15.8	YNGQVGEHTLGPVHTHSAHYK
39	528–561	3908.39	3908.6	42.8	VDLDVGGLENWVWAEDMAFVPTAIPWSPE HQIQR
40	562–566	616.74	617.0	14.1	LQVTR
41	567–567	147.20	Not detected		K
42	568–585	1902.07	1902.8	36.8	QLETEEQAAFPLGGASPR
43	586–592	858.03	858.2	17.8	YLYLASK
44	593–596	476.51	Not detected		QSNK
45	597–601	652.74	652.9	14.1	WGHPR
46	602–604	395.44	Not detected		GYR
47	605–623	2048.35	2049.1	23.4	IQTVSFAGGPMPQNSPMER
48	624–629	723.81	723.9	19.1	AFSWGR
49	630–637	993.15	993.5	16.7	YQLAITQR
50	638	147.20	Not detected		K
51	639–672		Glycopeptide		ETEPSSSSVFNQNDPWTPTVDFSDFINNETIA GK
52	673–726	6076.79	Not detected		DLVAWVTAGFLHIPHAEDIPNTVTVGNGVG FFLRPYNFFDQEPSMDSADSIYFR
53	727–763	3997.34	3997.4	39.5	EGQDAGSBEINPLABLPQAATBAPDLPVFSH GGYPEY

[a] Expected mass calculated from average chemical weights.

[b] The mass scan starts at 350 Da: peptides having a mass below this were not observed, peptides with mass <700 Da were identified on the basis of a singly charged molecular ion.

[c] This peptide results from an uncleaved tryptic site.

[d] Peptides 6a and 6b result from cleavage of an Arg–Pro bond.

[e] This peptide contains a N-linked glycosylation site (in bold type) and was observed after cleavage with PNGase F.

[f] Nonglycosylated form.

Sequence Confirmation of Native BSAO by HPLC/Electrospray Ionization Mass Spectrometry

To further confirm the cDNA sequence obtained for BSAO, on-line microbore HPLC/electrospray ionization mass spectrometry was utilized to characterize a tryptic digest of native BSAO. Figure 8 illustrates the base peak ion (BPI) chromatogram of a tryptic digest of reduced and carboxymethylated BSAO analyzed by LC/ESI/MS (conditions are described under Experimental). The digest was also treated with peptide *N*-glycosidase F (PNGase F) to enzymatically cleave any N-linked oligosaccharides that may be present, and the PNGase F-treated tryptic digest was reanalyzed by LC/ESI/MS (chromatogram not shown).

Digestion of BSAO with trypsin yields a complex mixture of peptides, due to the presence of 38 arginine and 17 lysine residues. To map the tryptic digest by LC/ESI/MS, the theoretical masses of the tryptic peptides were calculated using the MacBioSpec program (PE Sciex, Ontario, Canada), and the LC/ESI/MS data were screened for the multiply charged ions series corresponding to the theoretical tryptic peptide mass. The peptides, as observed by mass spectrometry, are labeled in the chromatogram shown in Fig. 8 and are listed in Table IV with the expected average mass, the observed average mass calculated from the multiply charged ion series, and the retention time for each peptide. Because of the mass scan commencing at m/z 350, tryptic peptides with a molecular mass <700 were observed by a singly charged molecular ion. Forty of the possible 53 tryptic peptides were observed in the BPI chromatogram. BSAO contains three N-linked glycosylation sites, within peptides T_9 (Asn-137), T_{18} (Asn-232), and T_{51} (Asn-665); treatment of the digest with the glycosidase PNGase F cleaves all classes of N-linked oligosaccharides attached to the Asn residue and further hydrolyzes the Asn residue to Asp. LC/ESI/MS analysis of the PNGase F-treated digest will therefore yield the peptide components of any possible N-linked glycopeptides present in BSAO. Multiply charged ion series of peptides with molecular masses 1494.6, 3357.9, and 3776.9 were observed in the tryptic/PNGase F digest corresponding to nonglycosylated peptides T_{18}, T_9, and T_{51} with conversion of Asn to Asp (data not shown) of these peptides. The only peptide observed in the untreated digest was T_{18} MH^+ = 1493.7, indicating that each of the potential glycopeptide sites at Asn-137 and Asn-665 are fully occupied by oligosaccharides moieties and that Asn-232 is partially occupied. Furthermore, peptide T_{28} MH^+ = 2597.6, that was not observed in the tryptic digest, was observed in the tryptic/PNGase F digest.

Figure 9 illustrates the cDNA-derived sequence for BSAO. From the combination of analysis of a tryptic digest and a PNGase F/tryptic digest

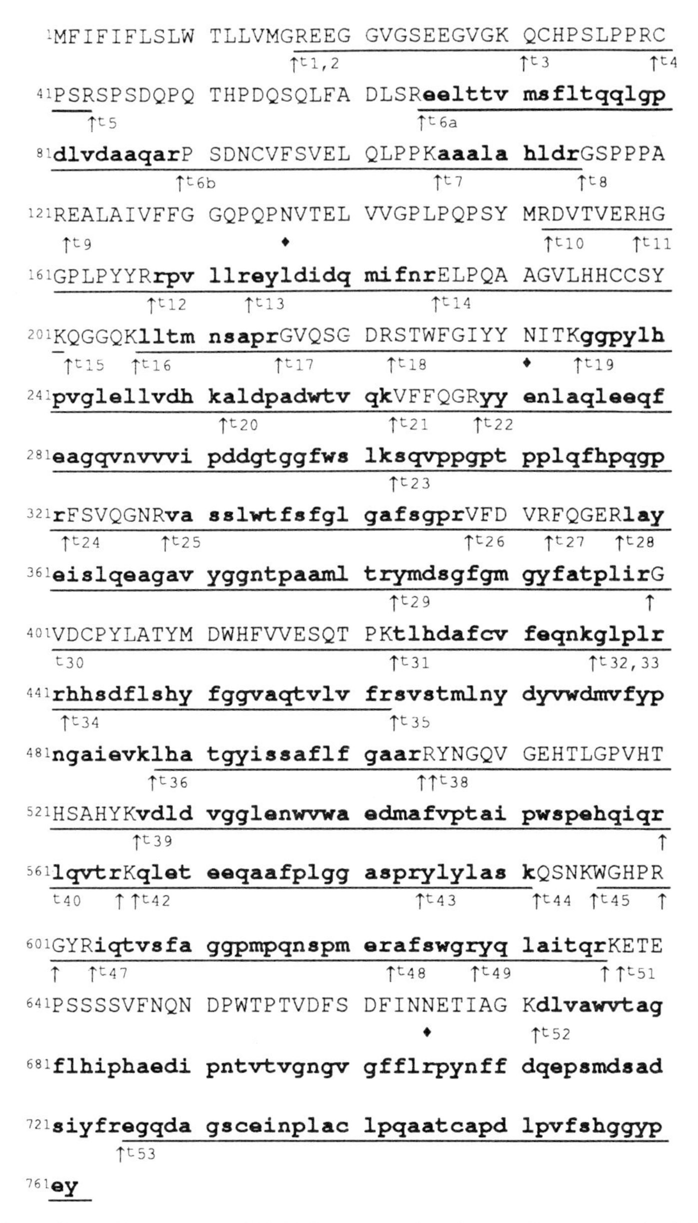

FIG. 9. Amino acid sequence of BSAO representing the tryptic map obtained by CID mass spectrometry/Edman sequencing and LC/ESI/MS analysis. Arrows beneath the sequence indicate positions of expected tryptic cleavage sites and are numbered from the N-terminal Arg residue at position 17. Asn residues labeled with a diamond indicate possible N-linked glycosylation sites. Sequences in lowercase bold type represent primary amino acid sequence data obtained from CID analysis/Edman sequencing (these sequences are listed in Table II). Underlined sequences indicate those identified by LC/ESI/MS analysis. When this digest was treated with PNGase F, deglycosylated peptides T_9 and T_{51} were observed.

by LC/ESI/MS, we accounted for 85% of the amino acid sequence. Furthermore, taking into account additional sequence information obtained by CID analysis and Edman sequencing, 96% of thc 746 amino acid sequence was determined by tryptic mapping.

Acknowledgments

Financial support was provided by National Institutes of Health Grant RR 01614 to the UCSF Mass Spectrometry Facility (A. L. Burlingame, Director), the Tobacco-Related Disease Research Program, No. 2RT0089 (to A.L.B), and the USCF Liver Center Grant NIH DK 26743. We thank Fred C. Walls for expertise in carrying out tandem mass spectrometry and Sharon Walker for obtaining preliminary mass spectrometric sequence information. We also thank Corey Schwartz and Zhonghua Yu for obtaining Edman data.

[8] Cloning of Mammalian Topa Quinone-Containing Enzymes

By DAVID MU and JUDITH P. KLINMAN

Introduction

Topa quinone-dependent enzymes appear to be ubiquitous *in vivo*, with enzymes having been purified from a variety of sources, including mammals, plants, yeast, fungi, and bacteria (cf. McIntire and Hartman[1] for a complete summary). To date, all known topa-dependent enzymes are copper-containing and capable of catalyzing the reaction of oxidative deamination. Since the earliest study reported,[2] the majority of investigations have focused on understanding the enzymatic and biochemical properties of these copper amine oxidases.[1,3] Molecular biological studies of topa quinone enzymes were not initiated until the late 1980s and early 1990s, leading to the cloning of only a few nonmammalian enzymes, including yeast amine oxidase from *Hansenula polymorpha*,[4] lentil seedling amine oxidase,[5] and bacterial amine oxidase from *Arthrobacter* P1.[6] Although an extensive back-

[1] W. S. McIntire and C. Hartman, *in* "Principles and Applications of Quinoproteins" (V. L. Davidson, ed.), p. 97. Dekker, New York, 1993.
[2] E. A. Zeller, *Proc. R. Soc. London, Ser. B.* **161,** 153 (1938).
[3] J. P. Klinman and D. Mu, *Annu. Rev. Biochem.* **63,** 299 (1994).
[4] P. G. Bruinenberg, M. Evers, H. R. Waterham, J. Kuipers, A. C. Amberg, and G. Ab, *Biochim. Biophys. Acta* **1008,** 157 (1989).
[5] A. Rossi, R. Petruzzelli, and A. Finazzi-Agró, *FEBS Lett.* **301,** 253 (1992).
[6] X. Zhang, J. H. Fuller, and W. S. McIntire, *J. Bacteriol.* **175,** 5617 (1993).

ground of mechanistic and spectroscopic data exists for copper amine oxidases from porcine and bovine plasma (cf. McIntire and Hartman[1] and Klinman and Mu[3] and references therein), efforts at cloning one of these mammalian proteins had not been successful until very recently. This chapter reports the strategy and methods for the cloning of the first mammalian topa quinone-containing protein. The availability of appropriate cDNA probes is expected to prove extremely valuable in elucidating the as yet unresolved physiologic function of the various mammalian copper amine oxidases located both intra- and extra-cellularly.

Strategy

The initial obstacle in the cloning of a plasma amine oxidase was the uncertainty of the enzyme's site of biosynthesis. Even though aorta smooth muscle[7,8] and small intestine[9] have been independently implicated to be the source of serum topa enzymes, a commercially prepared bovine liver cDNA library was chosen for trial experiments in light of the fact that most serum proteins are secreted from liver.[10] Initial screening of this cDNA library used oligonucleotide probes deduced from the bovine serum amine oxidase (BSAO) active-site peptide sequence.[11,12] Although these studies yielded several positive recombinants with small cDNA inserts (<1 kb), the translated amino acid sequences were not found to correspond to any known bovine serum amine oxidase peptide sequences.

A turning point came during a routine protein sequence data bank search using known topa quinone-containing peptide sequences. This search revealed a previously cloned protein, amiloride-binding protein from human kidney,[13] with extensive homology to pig kidney diamine oxidase as well as bovine serum amine oxidase. Subsequent demonstration of amiloride binding to several copper amine oxidases led to the redesignation of

[7] R. M. Hysmith and P. J. Boor, *Biochem. Cell. Biol.* **66,** 821 (1987).

[8] R. Lewinsohn, *J. Pharm. Pharmacol.* **33,** 569 (1981).

[9] L. D'Agostino, S. Pignata, B. Daniele, R. Ventriglia, G. Ferrari, C. Ferrari, S. Spaguolo, P. Lucchelli, and G. Mazzacca, *Biochim. Biophys. Acta* **993,** 228 (1989).

[10] C. A. Alper, *in* "Hematology" (W. J. Williams, E. Beutler, A. J. Erslev, and M. A. Lichtman, eds.), p. 1616. McGraw-Hill, New York, 1990.

[11] D. Mu, S. M. Janes, A. J. Smith, D. E. Brown, D. M. Dooley, and J. P. Klinman, *J. Biol. Chem.* **267,** 7979 (1992).

[12] S. M. Janes, M. M. Palcic, C. H. Scaman, A. J. Smith, D. E. Brown, D. M. Dooley, M. Mure, and J. P. Klinman, *Biochemistry* **31,** 12147 (1992).

[13] P. Barbry, M. Champe, O. Chassande, S. Munemitsu, G. Champigny, E. Lingueglia, P. Maes, C. Frelin, A. Tartar, A. Ullrich, and M. Lazdunski, *Proc. Natl. Acad. Sci. U.S.A.* **87,** 7347 (1990).

amiloride-binding protein as human kidney diamine oxidase.[14] This fortuitous finding provided a chance to employ a more sensitive, polymerase chain reaction[15] (PCR)-based method to approach bovine serum amine oxidase cloning. As discussed herein, this has yielded a full-length cDNA from a bovine liver cDNA library, providing an answer to the long-standing question regarding the source of the circulating serum topa quinone amine oxidases. Subsequent sequence alignments among eukaryotic topa-containing enzymes have also led to an estimation of the conserved copper-binding ligands.[14]

Cloning of a Partial BSAO cDNA by PCR

The ability of PCR to generate micrograms of a specific DNA fragment has been used to design an efficient method to yield cDNA probes for molecular cloning, termed mixed oligonucleotide primed amplification of cDNA (MOPAC).[16] By synthesizing perfectly matched cDNA probes, one can perform hybridization under very stringent conditions, thereby eliminating spurious hybridization signals. A limitation of this method compared to the conventional protocols is that some sequence information flanking the desired fragment is required. At the time the experiments described herein were performed, several bovine serum amine oxidase peptide sequences were available (Table I).[17] However, their relative positions in the enzyme were unknown. The finding that a recently cloned and sequenced human kidney amiloride-binding protein is, in actuality, a topa quinone-dependent amine oxidase[14] was triggered by the observation that peptides I and II of bovine serum amine oxidase (Table I) are highly homologous to two separate regions in human kidney amiloride-binding protein. Assuming a similar relative position for these peptides in bovine serum amine oxidase, the MOPAC technique could be used to amplify the authentic cDNA of ca. 0.7 kb flanked by the two peptides (Fig. 1).

Although a tolerance of up to a 20% base pair mismatch between the primer and the template has been shown for the MOPAC reaction,[18]

[14] D. Mu, K. F. Medzihradszky, G. W. Adams, P. Mayer, W. M. Hines, A. L. Burlingame, A. J. Smith, D. Cai, and J. P. Klinman, *J. Biol. Chem.* **269,** 9926 (1994).

[15] K. Mullis, F. Faloona, S. Scharf, R. Saiki, G. Horn, and H. Erlich, *Cold Spring Harbor Symp. Quant. Biol.* **51,** 263 (1986).

[16] C. C. Lee, X. Wu, R. A. Gibbs, R. G. Cook, D. M. Muzny, and C. T. Caskey, *Science* **239,** 1288 (1988).

[17] D. Mu, Ph.D. Thesis, University of California at Berkeley (1993).

[18] C. C. Lee and C. T. Caskey, *in* "PCR Protocols: A Guide to Methods and Applications" (M. A. Innis, D. H. Gelfand, J. J. Sninsky, and T. J. White, eds.), p. 46. Academic Press, San Diego, CA, 1990.

TABLE I
COMPARISON OF BOVINE SERUM AMINE OXIDASE (BSAO) PEPTIDE SEQUENCES TO HUMAN KIDNEY AMILORIDE-BINDING PROTEIN[a]

BSAO[b] (active site-derived peptide I)	SVSTMLNYDYV*W*[c]DMVFYPNGAIEVK
Human kidney amiloride-binding protein[d]	[453]TTSTVYNYDYIWDFIFYPNGVMEAK[477]
BSAO (peptide II)[e]	DLVAWVTAGFLHIPHAEDIPNTVTVGNGVGFFLRPYNF
Human kidney amiloride-binding protein[d]	[662]DLVAWVTVGFLHIPHSEDIPNTATPGNSVGFLLRPFNF[699]

[a] Human kidney amiloride-binding protein is now designated human kidney diamine oxidase.[14]
[b] Data from Janes *et al.*[12] Instead of topa quinone at position 8, we include its precursor, tyrosine.[11] Underlined region represents the sequence used for the synthesis of the degenerate sense MOPAC primer.
[c] This previously unidentified residue was shown to be tryptophan (cf. Mu[17] and [7], this volume).
[d] Barbry *et al.*
[e] Mu *et al.*[14] Underlined region represents the sequence used for the synthesis of the degenerate antisense MOPAC primer.

selection of amino acids with minimal degeneracy is desired. Figures 2A and 2B show the number of codons for each amino acid residue in peptides I and II. For the purpose of reducing degeneracy, the residue 8–15 of peptide I and the residue 12–19 of peptide II were selected for primer synthesis. A mixed sense primer [5′-TA(C/T)GA(C/T)TA(C/T)GT(C/G)TGGGA(C/T)ATGGT-3′] was synthesized to the amino acid residue 8–15 (Y-D-Y-V-W-D-M-V) of the active-site peptide I and a mixed anti-sense primer [5′-AT(G/A)TC(T/C)TCGGC(G/A)TG(G/T)GG(T/G/A)AT(G/A)TG-3′] was synthesized to the amino acid residue 12–19 (H-I-P-H-A-E-D-I) of the peptide II close to the C terminus. The degeneracies for the sense and antisense primers are 32- and 96-fold, respectively. This is within the range of the recommended primer degeneracy by Compton.[19] Since most serum proteins are known to originate from liver,[10] a commercially prepared bovine (female) liver cDNA library (in λgt10 vector, Clontech, Palo Alto, CA) was chosen for amplification by MOPAC. A typical MOPAC experiment consists of the following: 50% glycerol, 20 μl; 100 m*M* Tris, pH 8.3/500 m*M* KCl, 10 μl; sense primer, 120–150 pmol (5 μl); antisense primer, 120–150 pmol (5 μl); 2 μl of bovine liver cDNA library (10^6 plaque-forming units, frozen and thawed twice) mixed with 52 μl water; 4 m*M* of dNTP, 5 μl; 1–3 units of *Taq* polymerase (Perkin

[19] T. Compton, *in* "PCR Protocols: A Guide to Methods and Applications" (M. A. Innis, D. H. Gelfand, J. J. Sninsky, and T. J. White, eds.), p. 39. Academic Press, San Diego, CA, 1990.

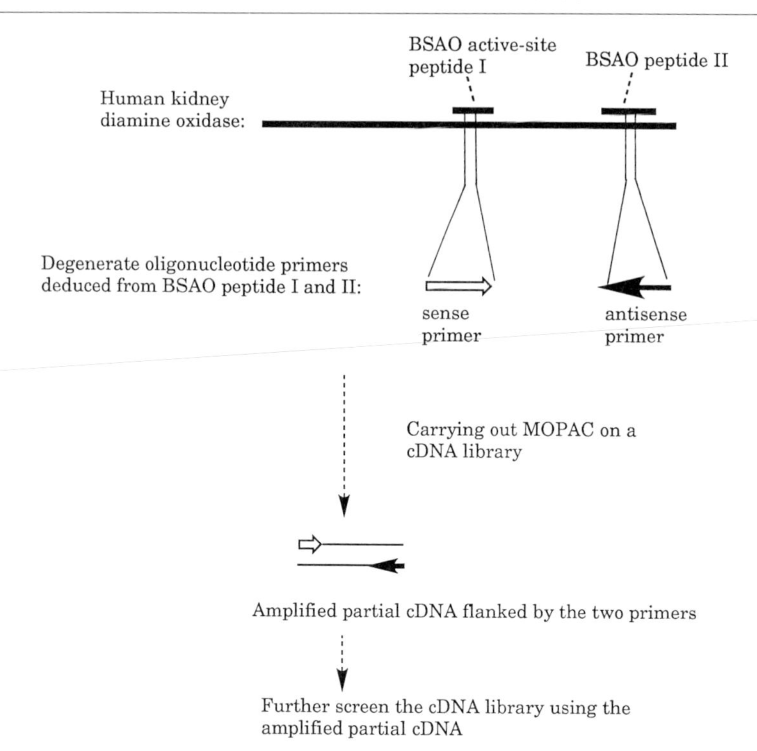

FIG. 1. Scheme for the MOPAC experiments which are based on the fact that two bovine serum amine oxidase (BSAO) peptides are highly homologous to two segments in human kidney diamine oxidase.

Elmer/Cetus, Norwalk, CT), 1 μl. Reactions are overlaid with 25–30 μl of mineral oil and amplified using a Perkin Elmer DNA thermal cycler by following this program: step (i) 94° for 5 min; step (ii) 94° for 30 sec, 44–51° (variable annealing temperatures) for 30 sec, 72° for 1 min; step (iii) 72° for 1 min. Amplification step (ii) is repeated 30 times. Following amplification, the reactions are analyzed by gel electrophoresis of 5 μl of the reaction mix on 2% agarose gels. The key to success in applying degenerate primers in PCR is to find conditions for optimal complementation between the target template and the primers. Initially, reactions are performed at a low, nonstringent annealing temperature of 44°, leading to the expected 0.7-kb product DNA (Fig. 3, lane 1). The gel in Fig. 3 shows that unknown DNA, at higher molecular weight, is amplified as well. To check the authenticity of the 0.7-kb DNA, experiments are repeated under more stringent conditions of elevated annealing temperatures (46–51°). As shown in Fig. 3 (lanes 2–5), band intensities at higher molecular weight decrease to zero through

A SVSTMLN**YDYVWDMV**FYPNGAIEVK

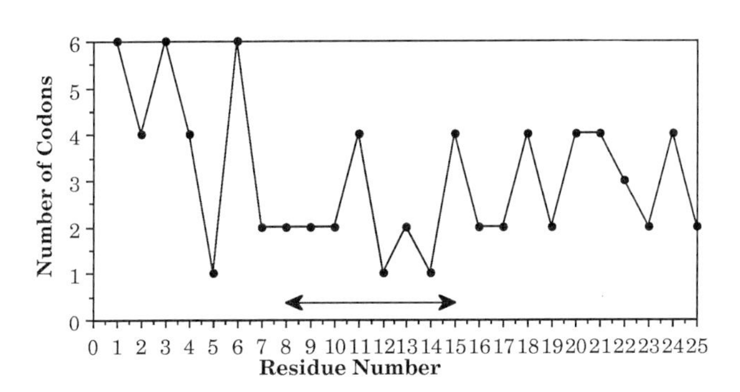

B DLVAWVTAGFL**HIPHAEDI**PNTVTVGNGVGFFLRPYNF

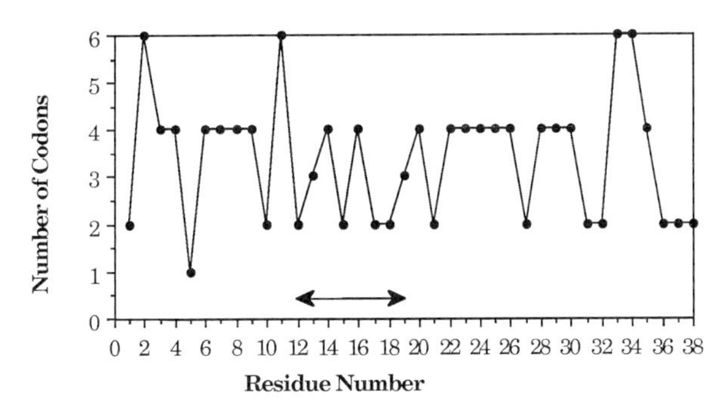

FIG. 2. Number of codons for each amino acid residue in the bovine serum amine oxidase peptide I (A) and peptide II (B). The corresponding peptide sequence is shown above the graph. The arrow covers the region selected for the synthesis of degenerate MOPAC primers.

the experiments, indicating that the 0.7-kb dominant band is likely to be the authentic cDNA fragment flanked by the two degenerated primers. Since the primers for MOPAC are incorporated at the ends of the 0.7-kb MOPAC DNA, the same primers can be used to obtain the nucleotide sequence near the ends of gel-purified MOPAC product DNA. Comparison of the translated amino acid sequence to the original peptide sequence confirmed the authenticity of the 0.7-kb DNA fragment (Fig. 4).

Isolation of Clones Containing Complete Coding Sequence

To isolate the remainder of the coding sequence for the bovine serum amine oxidase cDNA, the 0.7-kb MOPAC product is used to screen the

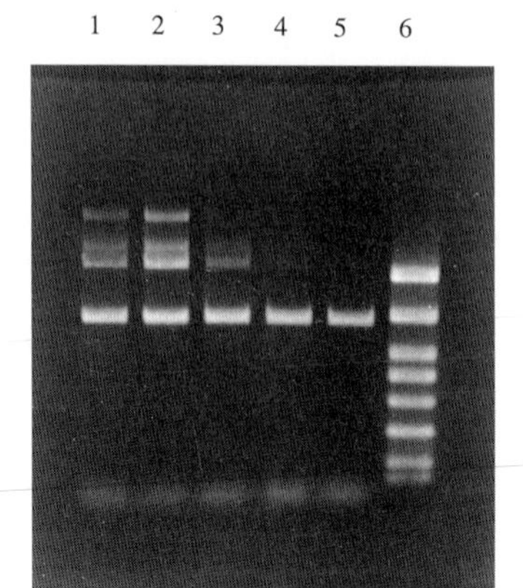

FIG. 3. Analysis of the MOPAC reaction products on bovine liver cDNA library by agarose (2%) gel electrophoresis. Lanes 1–5 are the MOPAC reactions performed at different annealing temperatures: lane 1, 44°; lane 2, 46°; lane 3, 48°; lane 4, 50°; and lane 5, 51°. Molecular weight markers were run in lane 6 with bands (from top) of 1000-, 700-, 500-, 400-, 300-, 200-, 100-, and 50-bp linear DNA (from Research Genetics). It appears that an annealing temperature of 49° represents the optimal condition. (Reproduced with permission from Mu *et al.*[14])

same bovine liver cDNA library. Preparation of the probes for screening the library is initiated by gel purifying one optimal MOPAC reaction using a Geneclean kit (Bio 101, San Diego, CA) and dissolving it in 40 μl of 1 mM Tris–HCl, pH 8.0, containing 0.1 mM EDTA. Twenty-five nanograms of this purified DNA is then labeled with [α-^{32}P]dATP using a random priming kit ("prime-a-gene," Promega, Madison, WI) to a specific activity of ca. 6×10^8 counts/min/μg.

Approximately 1.0×10^6 recombinants from the bovine liver cDNA library are plated and duplicated plaque lifts made using standard proce-

active site peptide I: SVSTMLNYDYVWDMVFYPNGA**IEVK**

→ sense primer **IEVK**LHATGYISSAFLF...

near C terminus peptide II: **DLVAWVTAG**FLHIPHAEDIPNTVTVGNGVG......

...NNETIAGK**DLVAWVTAG** ← antisense primer

FIG. 4. The translated amino acid sequence near the ends of the initially amplified DNA of 0.7 kb generated by MOPAC. The complete cDNA-deduced primary structure of bovine serum amine oxidase is published in Mu *et al.*[14]

dures.[20] Filters are prehybridized for 2–3 hr at 42° in 50% formamide, 0.15 *M* NaCl in 15 m*M* sodium citrate (6 × SSC), 0.1% sodium dodecyl sulfate (SDS), 5× Denhardt's solution,[21] 150 μg/ml denatured salmon sperm DNA, 50 m*M* Tris–HCl, pH 7.5. Then 8 × 10^5 counts/min/ml of ^{32}P-labeled probe are added and the hybridization is continued for 24–36 hr. Following hybridization the filters are washed twice for 20 min each at room temperature in 2 × SSC/0.1% SDS, washed three times each for 1 hr at 68° in 0.1 × SSC/0.1% SDS, air-dried, and autoradiographed at −70° with an intensifying screen for 3–7 days.

The initial round of library screening for BSAO cDNA resulted in the isolation of 25 recombinants. To analyze the size of the cDNA inserts and to eliminate false positive signals, each of the 25 clones was subject to two verifying PCR reactions:

i. For cDNA insert size: amplification using primers complementary to the *Eco*RI linker which flanks all library cDNA inserts

ii. To eliminate spurious hybridization signals: amplification using the two primers for MOPAC.

The former PCR reaction was performed using the cDNA insert screening amplimers (Clontech) according to the manufacturer's protocol; conditions for the latter PCR reaction and analysis of reaction products are identical to those described earlier. Thirteen of the initial 25 clones failed to yield the expected 0.7-kb amplified DNA. One of the 12 verified positive clones was found to be full-length as judged by the cDNA insert size (2.7 kb). Subsequently, the 1 phage recombinant containing this 2.7-kb cDNA was isolated using lambda-sorb (Promega) as recommended by the manufacturer, digested with *Eco*RI, and the insert subcloned into *Eco*RI-cut pGEM3Z (Promega) by following the standard exonuclease III-directed deletion method.[20] The DNA sequence of the insert (both strands) was then determined using a sequenase 2.0 dideoxy-sequencing kit (US Biochemical, Cleveland, OH). The translated amino acid sequence was matched exactly with a total of 11 peptide sequences[22] isolated from native bovine serum amine oxidase, confirming the authenticity as the cDNA of bovine serum amine oxidase.[14]

In summary, the use of degenerate PCR primers provides a powerful approach for the identification of mammalian topa quinone-containing enzymes. The fact that the two MOPAC primers, derived from bovine serum

[20] T. Maniatis, E. F. Fritsch, and J. Sambrook, "Molecular Cloning: A Laboratory Manual." Cold Spring Harbor Lab., Cold Spring Harbor, NY, 1982.

[21] D. T. Denhardt, *Biochem. Biophys. Res. Commun.* **23,** 641 (1966).

[22] G. W. Adams, P. Mayer, K. F. Medzirhadszky, W. H. Hines, and A. L. Burlingame, in preparation.

amine oxidase peptides I and II, are conserved between bovine serum amine oxidase and human kidney diamine oxidase suggests that this set of primers may be used as generic cloning primers for isolating additional mammalian topa quinone enzymes.

[9] Isolation of Active Site Peptides of Lysyl Oxidase

By HERBERT M. KAGAN and PING CAI

A variety of copper-dependent amine oxidases have been isolated from mammalian, plant, and yeast sources. These enzymes play critical roles in the metabolism of a variety of organic amines, including neurotropic monoamines as well as diamines such as putrescine and cadaverine.[1] In addition to the common requirement for a tightly bound Cu(II) cofactor at their active sites, this group of enzymes is catalytically inactivated by reagents known to form covalent adducts with active site carbonyl functions, including phenylhydrazine, semicarbazide, and others.[1,2] Indeed, several of these enzymes are known to contain trihydroxyphenylalanine (topa) quinone.[3,4] This quinone residue, which appears to derive post-translationally from a tyrosyl residue,[3] is the site of reaction with carbonyl reagents.[3,4] It is likely that this carbonyl cofactor serves as a transient electron sink operating during the oxidative deamination of the amine substrate.[5]

Lysyl oxidase shares certain features in common with this group of catalysts. Thus, this connective tissue amine oxidase contains a tightly bound Cu(II) cofactor at its active site[6,7] and is inactivated by various carbonyl

[1] U. Bachrach, *in* "Structure and Functions of Amine Oxidases" (B. Mondovi, ed.), p. 5. CRC Press, Boca Raton, FL, 1985.

[2] W. G. Bardley, *in* "Structure and Functions of Amine Oxidases" (B. Mondovi, ed.), p. 135. CRC Press, Boca Raton, FL, 1985.

[3] D. Mu, S. M. Janes, A. J. Smith, D. E. Brown, D. M. Dooley, and J. P. Klinman, *J. Biol. Chem.* **267,** 7979 (1992).

[4] D. E. Brown, M. A. McGuirl, D. M. Dooley, S. M. Janes, D. Mu, and J. P. Klinman, *J. Biol. Chem.* **266,** 4049 (1991).

[5] D. M. Dooley, M. A. McGuirl, D. E. Brown, P. N. Turowski, W. S. McIntire, and P. F. Knowles, *Nature* (*London*) **349,** 262 (1991).

[6] H. M. Kagan, *in* "Biology of Extracellular Matrix" (R. P. Mecham, ed.), Vol. 1, p. 321. Academic Press, Orlando, FL, 1986.

[7] S. N. Gacheru, P. C. Trackman, M. A. Shah, C. Y. O'Gara, P. Spacciapoli, F. T. Greenaway, and H. M. Kagan, *J. Biol. Chem.* **265,** 9022 (1990).

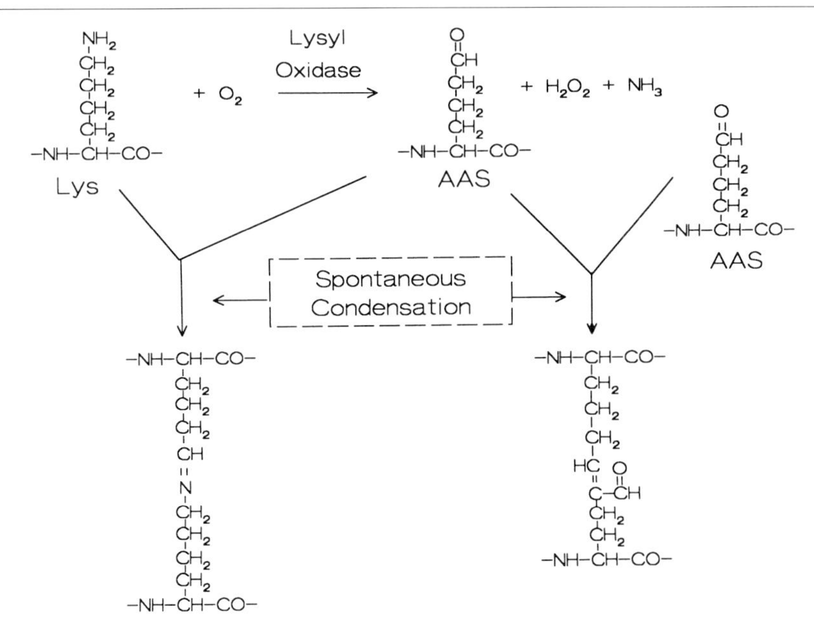

FIG. 1. Oxidation of peptidyl lysine by lysyl oxidase. The peptidyl product of peptidyl lysine oxidation, α-aminoadipic-δ-semialdehyde (AAS), can spontaneously condense to give the Schiff base (left) and aldol condensation product (right) cross-linkages, as well as other more complex products not shown.[6]

reagents.[6,8] The biological role of lysyl oxidase differentiates it from these other amine oxidases since its principal substrates *in vivo* appear to be peptidyl lysine residues contained within collagen and elastin,[6] the structural macromolecules of connective tissue. The resulting residues of peptidyl α-aminoadipic-δ-semialdehyde can undergo spontaneous condensations with unreacted ε-amino groups of neighboring lysine residues or with other peptidyl aldehyde residues to generate covalent cross-linkages within and between the individual molecules of collagen and of elastin (Fig. 1). This process underlies the formation of mechanically strong connective tissue fibers, thus providing the fibrous framework for cell growth and tissue development.

Like other members of the broader family of amine oxidases, lysyl oxidase is kinetically classified as a "ping pong" catalyst,[9] carrying out amine oxidation in two kinetically distinguishable steps:

[8] H. M. Kagan, N. A. Hewitt, L. L. Salcedo, and C. Franzblau, *Biochim. Biophys. Acta* **365,** 223 (1974).

[9] P. R. Williamson and H. M. Kagan, *J. Biol. Chem.* **261,** 9477 (1986).

$$RCH_2NH_2 + Enz_{Oxidized} \rightarrow RCH{=}O + Enz_{Reduced} \quad (1)$$
$$Enz_{Reduced} + O_2 + H_2O \rightarrow Enz_{Oxidized} + H_2O_2 + NH_3 \quad (2)$$

It is likely that the $Enz_{Reduced}$ species contains the two-electron-reduced form of the carbonyl cofactor, the oxygen-dependent reoxidation of which in the second of these two half reactions appears to be rate limiting in lysyl oxidase catalysis.[10]

Characterization of the Carbonyl Cofactor in Lysyl Oxidase

Various carbonyls have been implicated as the organic cofactor in lysyl oxidase. Early studies indicated that pyridoxal 5′-phosphate reversed the inhibition of collagen cross-linking by isoniazid and other carbonyl reagents *in vivo*[11] whereas PLP-deficient diets appeared to decrease collagen insolubilization in animal tissues.[12] Moreover, PLP stimulated lysyl oxidase activity of a preparation of chick aorta lysyl oxidase.[13] However, other studies revealed the apparent absence of PLP or PL from highly purified, active bovine aortic lysyl oxidase.[14] It was subsequently reported that the resonance Raman spectrum of an active site peptide isolated from bovine aortic lysyl oxidase which had been modified with phenylhydrazine strongly resembled that of the phenylhydrazone spectrum of bovine serum amine oxidase,[15] an enzyme which had been considered at that time to contain pyrroloquinoline quinone (PQQ) as its carbonyl cofactor,[16,17] whereas HPLC characterization of a product proteolytically released from human placental lysyl oxidase was consistent with PQQ.[18] Moreover, deprivation of PQQ from the diets of rodents decreased lysyl oxidase levels and increased the amount of soluble collagen isolable from connective tissues.[19] Although these studies implicated the presence of a PQQ-like cofactor in lysyl oxidase, the subsequent demonstration that bovine serum amine oxi-

[10] M. A. Shah, C. H. Scaman, M. M. Palcic, and H. M. Kagan, *J. Biol. Chem.* **268,** 11573 (1993).
[11] C. Levene, *J. Exp. Med.* **114,** 295 (1961).
[12] J. C. Murray, D. R. Fraser, and C. A. Levene, *Exp. Mol. Pathol.* **28,** 301 (1978).
[13] T. A. Bird and C. A. Levene, *Biochem. Biophys. Res. Commun.* **108,** 1172 (1982).
[14] P. R. Williamson, J. M. Kittler, J. W. Thanassi, and H. M. Kagan, *Biochem. J.* **235,** 597 (1986).
[15] P. R. Williamson, R. S. Moog, D. M. Dooley, and H. M. Kagan, *J. Biol. Chem.* **261,** 16302 (1986).
[16] C. L. Lobenstein-Verbeek, J. A. Jonjegan, J. Frank, and J. A. Duine, *FEBS Lett.* **170,** 305 (1984).
[17] M. Ameyama, M. Hayashi, K. Matsushita, E. Shinagawa, and O. Adachi, *Agric. Biol. Chem.* **48,** 561 (1984).
[18] R. A. van der Meer and J. A. Duine, *Biochem. J.* **239,** 789 (1986).
[19] J. Killgore, C. Smidt, L. Duich, N. Romero-Chapman, N. Tinker, K. Reiser, M. Melko, D. Hyde, and R. B. Rucker, *Science* **245,** 850 (1989).

dase contained topa quinone and not PQQ[20] has led to the reinvestigation of the nature of the enzyme cofactor in lysyl oxidase.

Unlike the gram quantities of amine oxidase which can be purified from bovine serum, the yields of purified lysyl oxidase available from typical purification procedures have been limited to 2–4 mg starting with 0.5 to 1 kg of cleaned bovine aorta. Moreover, the carbonyl content of various purified preparations of bovine aorta lysyl oxidase is commonly in the range of 0.25–0.5 mol/mol of the 32-kDa enzyme.[14] While the basis of this less than stoichiometric relationship between enzyme protein and the carbonyl cofactor remains unknown, this limitation exacerbates the difficulty of characterizing the chemical identity of the cofactor in the limiting quantities of lysyl oxidase which have been purified from mammalian connective tissues. Nevetheless, the development of methodology needed to achieve this goal has progressed and is described in this chapter.

Purification of Lysyl Oxidase

Procedures used for the purification of lysyl oxidase have routinely employed 2 to 6 *M* urea in extracting and chromatography buffers to increase the solubility, stability, and chromatographic resolution of this connective tissue enzyme.[6]

Extraction

Bovine aortas freshly obtained from 2- to 6-week-old calves are cleaned of adhering tissue and coarsely ground in the absence of added buffer. The ground tissue (700 g) is extracted twice with 1500 ml of Buffer I (0.4 *M* NaCl, 16 m*M* potassium phosphate, pH 7.8) in a Waring blender at 4°, separating the pellet by centrifugation after each extraction. The extracted pellets are combined and then extracted with Buffer II (16 m*M* potassium phosphate buffer, pH 7.8). The pellets are isolated by centrifugation, combined, and then extracted with 1500 ml of Buffer III (4 *M* urea, 16 m*M* potassium phosphate, pH 7.8). The pellet is isolated and reextracted twice more with Buffer III.

Treatment with Hydroxyapatite

Preliminary trials established that lysyl oxidase does not bind to hydroxyapatite under specific conditions whereas other contaminating proteins do. Thus, the 4 *M* urea extracts are pooled (4.5 liters) and mixed with

[20] S. M. Janes, D. Mu, D. Wemmer, A. J. Smith, S. Kaur, S. Maltby, A. L. Burlingame, and J. P. Klinman, *Science* **248,** 981 (1990).

500 g of Bio-Gel HTF hydroxyapatite previously equilibrated with 1 liter of Buffer III. The suspension is stirred for 10 min at 4° and allowed to settle for 30 min. The supernatant is decanted, clarified by centrifugation at 10,000*g* for 10 min, concentrated to 750 ml in a S1Y10 ultrafiltration cartridge (Amicon), and dialyzed against 50 liters of Buffer II. The crude enzyme is precipitated by the addition of an equal volume of 1 *M* potassium phosphate, pH 7.8, and the precipitate is collected by centrifugation.

Gel Filtration Chromatography

The potassium phosphate precipitate is dissolved in Buffer IV (16 m*M* potassium phosphate, 6 *M* urea, pH 7.8) and resolved by chromatography through a column of Sephacryl S-200 (120 × 5 cm) eluting with Buffer IV. Enzymatically active fractions are identified by a fluorometric assay for hydrogen peroxide production, using 1,5-diaminopentane as substrate.[21] Active fractions are pooled and concentrated to 25 ml by pressure filtration through an Amicon YM-10 membrane. The concentrated enzyme is dissolved in a limiting quantity (25–30 ml) of Buffer IV and is further purified by passage through Sephacryl S-100 (120 × 2.5 cm) previously equilibrated in this buffer. The enzymatically active fractions are pooled (84 ml) and analyzed by SDS–PAGE. This electrophoretogram revealed the presence of the 32-kDa enzyme as well as bands at 22 and 24 kDa. The copurification with lysyl oxidase of proteins in this range of molecular weight has been described.[22]

Ion-Exchange Chromatography

The pooled enzyme from the previous step is added to 30 ml of a thick slurry of DEAE-Trisacryl in Buffer V (2 *M* urea, 16 m*M* potassium phosphate, pH 7.8). This mixture is packed into a 10 × 2.5-cm column which is then washed with Buffer V until the OD_{280} of the effluent is <0.002. The column is further washed through with 16 m*M* potassium phosphate, pH 7.8, and then with 0.4 *M* NaCl in 16 m*M* potassium phosphate (Buffer I). This treatment largely removes the 22- and 24-kDa contaminants from the DEAE column. Lysyl oxidase is then eluted with 6 *M* urea, 0.4 *M* NaCl, 16 m*M* potassium phosphate, pH 7.8. The enzymatically active fractions are pooled, dialyzed against 4 liters of 16 m*M* potassium phosphate for 4 hr, and stored as 0.2- and 0.5-ml aliquots at −80°. Typical yields of enzyme using this protocol amount to 2 mg of pure lysyl oxidase with a specific

[21] P. C. Trackman, C. G. Zoski, and H. M. Kagan, *Anal. Biochem.* **113,** 336 (1981).

[22] A. D. Cronshaw, J. R. E. MacBeath, D. R. Shackleton, J. F. Collins, L. A. Fothergill-Gilmore, and D. J. S. Hulmes, *Matrix* **13,** 255 (1993).

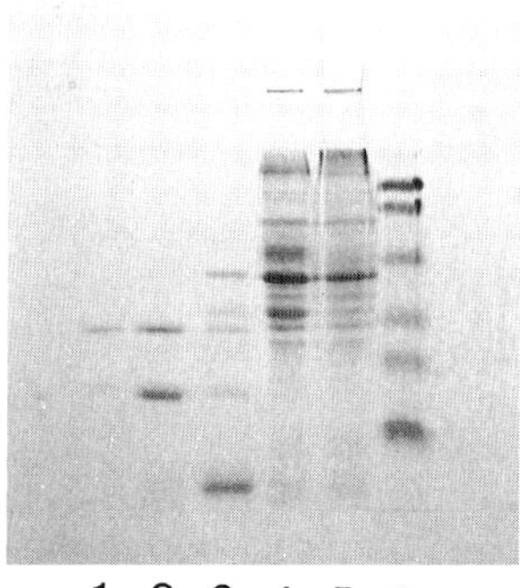

FIG. 2. SDS–PAGE electrophoresis of fractions purified from bovine aorta. Lanes: 1, final, purified lysyl oxidase; 2, Sephacryl S-100 pool; 3, Sephacryl S-200 pool; 4, precipitate of hydroxyapatite supernatant; 5, pool of 4 *M* urea extracts; and 6, molecular weight standards, top to bottom, 106,000, 80,000, 49,500, 32,500, 27,500 and 18,500.

activity of 2.5×10^6 dpm mg^{-1} assaying at 37° for 2 hr against 4.38×10^5 dpm of human recombinant tropoelastin substrate, prepared as described.[23] A SDS–PAGE electrophoretogram of various stages in the purification procedure is shown in Fig. 2.

Labeling and Isolation of Active Site Peptides

Two strategies have been adopted toward the goals of locating the carbonyl cofactor of lysyl oxidase within the sequence of the enzyme and, ultimately, of the chemical identification of the cofactor. The active site is first chemically modified by reaction of the enzyme with an isotopically labeled reagent which previous studies have indicated to inactivate the enzyme apparently by reaction with the carbonyl cofactor. Initial studies have employed [1,2-^{14}C]ethylenediamine, an irreversible inhibitor of lysyl oxidase. Vicinal diamines such as ethylenediamine inactivate the enzyme apparently by forming covalent pyrazine adducts with the carbonyl cofactor.[24] These adducts can be generated if both amino groups of the inhibitor form imine bonds with each of the carbonyl groups of an *o*-carbonyl cofactor followed by two-electron oxidation of each α-carbon of the diamine.[24] Thus, topa quinone can participate in this mechanism by reaction of its *o*-quinone tautomer with a vicinal diamine. Following chemical modification, the en-

[23] D. Bedell-Hogan, P. Trackman, W. Abrams, J. Rosenbloom, and H. Kagan, *J. Biol. Chem.* **268,** 10345 (1993).

[24] S. N. Gacheru, P. C. Trackman, S. D. Calaman, F. T. Greenaway, and H. M. Kagan, *J. Biol. Chem.* **264,** 12963 (1989).

zyme is chemically or enzymatically digested to peptides which are then resolved by electrophoresis or by chromatography, respectively.

Method 1: Chemical Cleavage/Electrophoresis

Lysyl oxidase exhibits unusual solubility properties in that the 32-kDa enzyme tends to aggregate in concentrated solutions in the absence of urea. Similar behavior has been noted with some peptide preparations derived from the enzyme. Coupled with the limitation imposed by the relatively small quantities of purified enzyme available from tissue sources, the search for and identification of the carbonyl cofactor of this catalyst have proven to be formidable tasks. To facilitate these efforts, a rapid, simple method for the identification of the peptide bearing the carbonyl cofactor was developed. The purified 32-kDa bovine aorta lysyl oxidase (320 μg, 10 nmol) is inactivated and irreversibly labeled by modification for 3 hr at 37° in 2 *M* urea, 0.05 *M* sodium borate, pH 8.2, with 1 μmol of [1,2-^{14}C]ethylenediamine (Sigma Chemical Co., specific activity 9.9 μCi μmol^{-1}). Excess reagent is removed by dialysis, and the enzyme is then cleaved by incubation with 1 μM CNBr in 70% formic acid at room temperature for 24 hr under an atmosphere of 100% nitrogen gas. After completion of this reaction, volatile materials are removed by evaporation under a stream of dry nitrogen gas and the residue is further dried *in vacuo*. The dried residue is dissolved in a SDS–PAGE sample buffer containing 2 *M* urea and the solution is directly subjected to tricine SDS–PAGE using a discontinuous gradient among 5, 10, and 16% polyacrylamide.[25] The resolved peptides are transferred by electroelution onto a polyvinylidene difluoride (PVDF) membrane (pore size, 0.45 μm; Millipore). The blot is lightly stained with Coomassie blue. This procedure reveals bands at 24, 18, 14, and 6 kDa, in addition to residual undigested protein at 32 kDa. Autoradiography of the blot showed that only the bands at 24 and 14 kDa were radioactive, as diagrammed in Fig. 3. In a parallel experiment, the 32-kDa enzyme was digested and otherwise analyzed identically but without prior modification of the active site of the protein. After electroblotting, the PVDF membrane was probed for the presence of the *o*-quinone cofactor by a colorimetric redox cycling assay.[26] The smallest positive band was 14 kDa. The area of the blot corresponding to this band was cut out and an amino acid sequence of the N-terminal region of this peptide was determined and found to match that of the C-terminal CNBr peptide predicted to be cleaved from the

[25] H. Schagger and J. Gebhard, *Anal. Biochem.* **166,** 369 (1987).

[26] M. A. Paz, R. Flückiger, A. Boak, H. M. Kagan, and P. M. Gallop, *J. Biol. Chem.* **266,** 689 (1990).

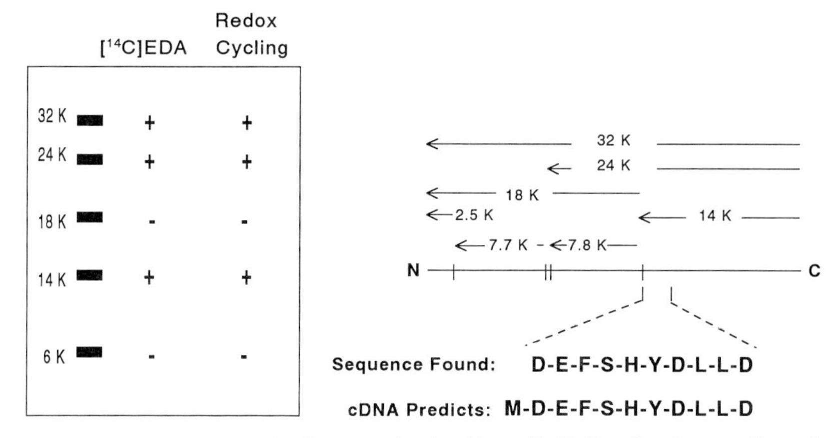

FIG. 3. CNBr cleavage of bovine aorta lysyl oxidase. (Left) Results of autoradiography for bands derived from the enzyme chemically modified with [^{14}C]ethylenediamine and results of redox cycling assay for quinones determined with a digest of enzyme that had not been chemically modified with ethylenediamine. (Right) Molecular masses (in kilodaltons) of peptide fragments expected from cleavage at individual methionines (approximate loci indicated by vertical lines) are indicated above the linear representation of the 32-kDa enzyme. The expected content and positions of methionine residues are predicted from the cDNA sequence of rat aorta lysyl oxidase.[27,28] The N-terminal sequence of the isolated 14-kDa peptide as found and predicted by the cDNA sequence is shown.

32-kDa enzyme[27,28] (Fig. 3). Moreover, it is within this region of the sequence that copper-binding consensus sequences are found, consistent with the locus of catalytically active residues within this C-terminal sequence.[27,28]

Attempts were made to isolate a smaller labeled peptide from the ^{14}C-labeled 14-kDa peptide to obtain a complete sequence surrounding the ethylenediamine-labeled residue by further digestion of this band with trypsin. However, these efforts were unsuccessful because of the limiting quantities of the 14-kDa peptide available and because of the difficulty of identifying very small peptide fragments by SDS–PAGE. An alternative method for peptide generation and isolation was devised.

Method 2: Enzymatic Degradation/HPLC

Reductive Alkylation. Preliminary experiments established that reductive alkylation of cysteine residues in lysyl oxidase prior to protease diges-

[27] P. C. Trackman, A. M. Pratt, A. Wolanski, S. S. Tang, G. D. Offner, R. T. Troxler, and H. M. Kagan, *Biochemistry* **29,** 4863 (1990).

[28] P. C. Trackman, A. M. Pratt, A. Wolanski, S. S. Tang, G. D. Offner, R. T. Troxler, and H. M. Kagan, *Biochemistry* **30,** 8282 (1991).

tion yielded the most useful peptide profiles. For this purpose, lysyl oxidase (1.5 mg) is labeled with [^{14}C]ethylenediamine as described. Preliminary trials established that the ratio of bound diamine per enzyme increased from approximately 0.25 to 0.5 mol mol^{-1} if the enzyme was reduced with a 100-fold molar excess of $NaCNBH_3$ after modification with the diamine. This result likely reflects incomplete oxidative processing of the diamine bound at the active site. The modified enzyme is then reduced in this fashion and dialyzed against 2 *M* urea, 16 m*M* potassium phosphate, pH 7.8, for 8 hr, then against distilled water for 4 hr, and then dried by lyophilization. The product is redissolved in 125 μl of 8 *M* urea, 0.4 *M* NH_4HCO_3 at pH 8.5. Dithiothreitol (1.13 μmol) is added, the mixture is incubated at 50° for 1 hr under nitrogen, cooled to room temperature, and protein sulfhydryls are alkylated by the addition of 2.5 μmol in 25 μl of iodoacetamide followed by incubation in the dark in a sealed tube for 1 hr at room temperature. The mixture is then digested with trypsin (see below) without prior removal of the excess alkylation reagents. As shown in Fig. 4, the HPLC profiles of lysyl oxidase peptides derived by trypsin digestion with (Fig. 4, bottom) or without (Fig. 4, top) prior reductive alkylation indicate that digestion of the previously alkylated enzyme yielded a more well-resolved radioactive peak.

Proteolytic Digestion. Various proteases (thermolysin, trypsin, lys C) were assessed as catalysts of the hydrolysis of the reductively alkylated enzyme. Trypsin proved to yield the clearest peptide maps among those tested. Following carboxamidomethylation of lysyl oxidase as described, the urea concentration is diluted with water to 1.2 *M* prior to addition of trypsin. A 1 : 100 (w/w) ratio of trypsin to lysyl oxidase proved to be sufficient to adequately digest lysyl oxidase without generating excess fragments derived from trypsin itself. The reductively alkylated enzyme is incubated with trypsin at 37° and the progress of the digestion is monitored by HPLC analysis of aliquots removed at selected time intervals. A total digestion time of 4–5 hr proved optimal. Digestion is terminated by acidification to pH 3 with 1% trifluoroacetic acid. The acidified mixture is sedimented for 1 min at high speed in a microcentrifuge and the clarified supernatant is immediately injected into a HPLC column.

HPLC Separation of Peptides. A Dynamax-C_8 (4.6 × 250 mm) analytical scale HPLC column is used for the separation of digested peptides. Elution gradients are generated between Buffer A (0.05% trifluoroacetic acid in water) and Buffer B [0.038–0.04% trifluoroacetic acid in acetonitrile : water (80 : 20, v/v)]. The optical densities of Buffers A and B are matched by titration of the trifluoroacetic acid content of Buffer B. Chromatography is performed at ambient temperature (25°) at a flow rate of 0.65 ml min^{-1}. Peak elution is monitored at 214 nm. Buffer A is pumped through the

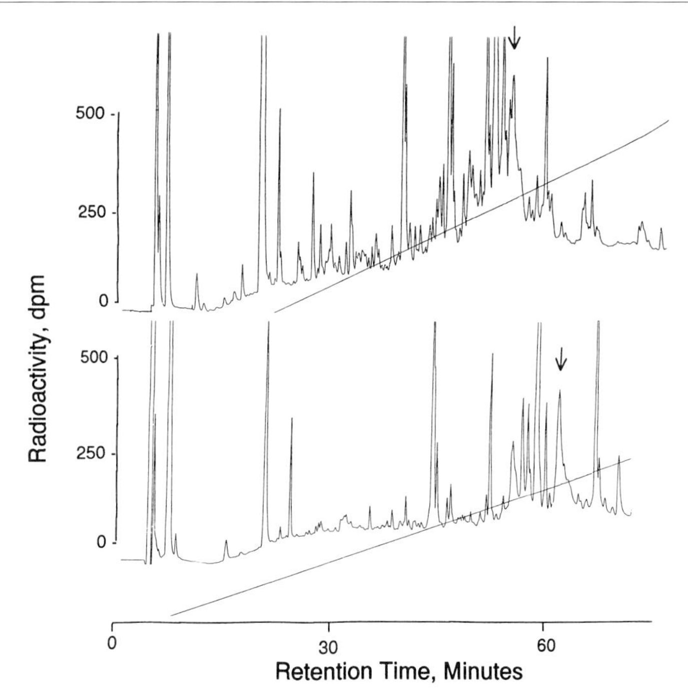

FIG. 4. HPLC peptide maps of trypsin digests of lysyl oxidase. Lysyl oxidase modified with [^{14}C]ethylenediamine was digested without (top) or with (bottom) reductive alkylation of sulhydryl groups prior to trypsin digestion. Arrows indicate positions of radioactive peaks. The linear plots represent the elution gradients applied as described in the text. Elution of peaks was monitored by recording the OD at 214 nm.

column for 3 min at a flow rate of 0.8 ml min^{-1} immediately following injection of the acidified supernatant of the trypsin digest. The gradient is then initiated at a flow rate of 0.65 ml min^{-1}. Several trials established that the best resolution was obtained with the following linear gradient: addition of Buffer B to 60% at 70 min and then to 70% at 80 min. This gradient gave the peptide profiles shown in Fig. 4. One-minute fractions are collected and radioactive peaks are identified by liquid scintillation spectrometry of aliquots of each fraction. Radioactivity was detected in a peak eluting at 62 min (55% Buffer B). Studies are in progress to scale up this procedure to isolate sufficient peptide for both sequence determination and cofactor characterization. The determination of the amino acid sequence will benefit from the availability of the full amino acid sequence of the proenzyme of lysyl oxidase predicted from lysyl oxidase cDNA.[27,28]

The methodologies described in this chapter present a workable path toward the isolation of active site peptides of lysyl oxidase. Given sufficient amounts of enzyme and its modification with a probe of sufficiently high specific radioactivity, this approach ultimately should permit the isolation and unequivocal characterization of the identity of the carbonyl cofactor in lysyl oxidase.

Acknowledgment

This research was supported by NIH Grant R37 AR 18880.

[10] Resonance Raman Spectroscopy of Quinoproteins

By DAVID M. DOOLEY and DOREEN E. BROWN

Introduction

An extremely wide range of questions and problems in the structure and function of biomolecules can be addressed by Raman spectroscopy. There are two basic reasons for the great attractiveness of Raman spectroscopy as a probe of biological systems. First, the information content of Raman spectra is high, reflecting the frequencies and intensities of molecular vibrations. Second, water is an excellent solvent for Raman spectroscopy, with practically no interference over the range of vibrational energies up to 3000 cm^{-1}. If the biological system contains an intrinsic chromophore or if a chromophore can be introduced, such that the incident laser light is absorbed by the sample, then resonance Raman spectroscopy offers additional advantages. Under these circumstances, the intensities of certain bands of the absorbing species may be enhanced by several orders of magnitude. This phenomenon is termed resonance enhancement. Because at least some of its Raman bands are resonance enhanced, the vibrational spectrum of a chromophore, in whole or in part, may be selectively obtained. If the chromophore is within, or is a component of, a binding site or an active site, then resonance Raman spectroscopy may provide functional as well as structural information. Resonance enhancement also increases the sensitivity of Raman spectroscopy dramatically. In favorable situations, only nanomoles of material are required to obtain excellent data. The rich biological applications of Raman spectroscopy have been reviewed extensively.[1–4] Excellent discussions on various practical and technical as-

[1] D. M. Dooley and D. E. Brown, *in* "Principles and Applications of Quinoproteins" (V. L. Davidson, ed.), p. 275. Dekker, New York, 1992.

pects of Raman spectroscopy are available.[5,6] This chapter concentrates on the practical requirements and procedures for obtaining resonance Raman spectra of quinoproteins. We have reviewed the results from resonance Raman studies of quinoproteins[1] and this chapter may be consulted for discussions of the relevant literature. Other related articles that provide a wealth of experimental information are those by Wang and Van Wart,[7] and by Loehr and Sanders-Loehr.[8] Theoretical aspects, Raman instrumentation, and laser sources have been covered by Wang and Van Wart,[7] whereas sampling techniques and requirements, isotope exchange, temperature control, and data collection, reduction, and interpretation are covered by Loehr and Sanders-Loehr.[8] Since the techniques and approaches discussed in these earlier articles may be directly applied to resonance Raman studies of quinoproteins, this material will not be repeated here.

Preparation of Samples for Resonance Raman Spectroscopy

Derivatization of Quinoproteins

The 2,4-dinitrophenylhydrazone (DNP) and the phenylhydrazone (PHZ) adducts of amine oxidases may be prepared by adding a 10% molar excess (over quinone) of the appropriate hydrazine to the protein in 0.1 *M* potassium phosphate buffer, pH 7.2. The hydrazines are first dissolved in a small amount of absolute ethanol (<1 ml). The amine oxidase–phenylhydrazone adducts can be used immediately, whereas the DNP derivatives should be incubated at room temperature for 3 hr before any Raman experiments. Reactions between the amine oxidases and phenylhydrazine or DNP are readily monitored by absorption spectroscopy; a typical spectrum is shown in Fig. 1. Extinction coefficients have been determined in

[2] T. G. Spiro, ed., "Biological Applications of Raman Spectroscopy," Vols. 1–3. Wiley, New York, 1987.

[3] P. R. Carey, "Biochemical Applications of Raman and Resonance Raman Spectroscopy." Academic Press, New York, 1982.

[4] P. R. Carey, *in* "Modern Physical Methods in Biochemistry Part B" (A. Neuberger and L. L. M. Van Deenen, eds.), p. 27. Elsevier, Amsterdam, 1988.

[5] D. P. Strommen and K. Nakamoto, "Laboratory Raman Spectroscopy." Wiley, New York, 1984.

[6] D. J. Gardiner and P. R. Graves, eds., "Practical Raman Spectroscopy." Springer-Verlag, Berlin, 1989.

[7] Y. Wang and H. E. Van Wart, this series, Vol. 226, p. 319.

[8] T. M. Loehr and J. Sanders-Loehr, this series, Vol. 226, p. 431.

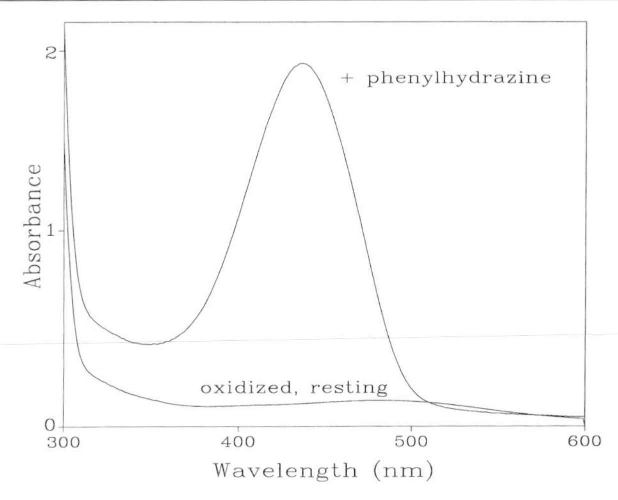

FIG. 1. Absorption spectra of the phenylhydrazone of *Arthrobacter* P1 methylamine oxidase and oxidized methylamine oxidase at pH 7.0.

some cases[9,10]: bovine plasma amine oxidase, (DNP)ε_{470} = 44,000 M^{-1}cm^{-1}, (PHZ)ε_{447} = 32,400 M^{-1}cm^{-1}; methylamine oxidase from *Arthrobacter* P1, (PHZ)ε_{444} = 36,400 M^{-1}cm^{-1}. DNP does not react with methylamine oxidase.

p-Nitrophenylhydrazone (NPH) derivatives of amine oxidases can be prepared by adding a 10-fold excess (over quinone) of NPH, dissolved in a small amount of 95% ethanol (<1 ml), to the protein in 0.1 *M* potassium phosphate buffer, pH 7.2. Samples are incubated at 5° overnight and then dialyzed to remove unreacted NPH. Alternatively, a 0.1 *M* solution of NPH is prepared in 1 ml of water and 15 μl of concentrated HCl. The mixture is sonicated for 1–2 min and filtered. A sufficient volume of the filtered solution to give a 20-fold excess (over quinone) of NPH is then added to the protein in an appropriate buffer, e.g., 0.05–0.1 *M* phosphate, pH 7.0–7.2. Reactions between amine oxidases and NPH are monitored by absorption spectroscopy; λ_{max} of amine oxidase–nitrophenylhydrazone derivatives is 450–470 nm, with, for example, ε_{466} = 44,400 M^{-1} cm^{-1} for *Arthrobacter* P1 methylamine oxidase.[11] A reaction is judged complete when no further increases in the absorbance at λ_{max} are observed. Depending on the amine oxidase, the derivatization is complete in 1.5 hr (at room temperature) or

[9] L. Morpurgo, O. Befani, S. Sabatini, B. Mondovì, M. Artico, F. Corelli, S. Massa, G. Stefancich, and L. Avigliano, *Biochem. J.* **256,** 565 (1988).

[10] S. M. Janes and J. P. Klinman, *Biochemistry* **30,** 4599 (1991).

[11] W. S. McIntire, unpublished observations (1989).

overnight at 5°. After the reaction is complete, excess NPH must be removed by dialysis or gel filtration.

The 2-hydrazinopyridine (2-HP) derivatives of amine oxidases are prepared by adding a 10-fold excess (over quinone) of 2-HP, dissolved in 1 ml of 0.1 *M* potassium phosphate buffer, pH 7.2, to the protein in the same buffer. 2-Hydrazinopyridine should be recrystallized from methanol prior to use. The initial product having λ_{max} at 415 nm spontaneously converts at room temperature to the final product, which absorbs at 520 nm. The rate of conversion depends on the identity of the amine oxidase (several minutes to hours for most enzymes), but the final products have identical visible absorption spectra in every case. The 2-HP derivative of porcine plasma amine oxidase is converted to the 520-nm form by warming the sample for 2 hr at 60°. In contrast, the bovine plasma amine oxidase and methylamine oxidase derivatives require only 10 min at 60°. Incubation for 1 hr at room temperature is sufficient to produce the 520-nm forms of the 2-HP derivatives of pea seedling and porcine kidney amine oxidases.

The phenylhydrazine derivative of methylamine dehydrogenase from bacterium W3A1 is prepared as described below.[12] We expect this procedure to work well, with at most minor changes, for other methylamine dehydrogenases. Small amounts of the reduced enzyme in samples of oxidized methylamine dehydrogenase are first reoxidized by incubation with 0.05 *M* potassium ferricyanide [$K_3Fe(CN)_6$] overnight at 4°. Oxidized methylamine dehydrogenase (12.8 mg/ml) is then reacted with 1 m*M* phenylhydrazine (dissolved in buffer) at pH 7.0 and 23°; under these conditions the reaction is immediate. Since trace amounts of phenylhydrazine and/or ferricyanide increase the photosensitivity of methylamine dehydrogenase, these are removed from samples by chromatography on a Pharmacia Superose column 12HR 10/30 (1 × 30 cm) using 0.05 *M* potassium phosphate/0.15 *M* KCl, pH 7.2, as the eluant. The 2-HP derivative of methylamine oxidase may be prepared using the same procedures described for amine oxidases. Representative absorption spectra of W3A1 methylamine dehydrogenase and some derivatives are shown in Fig. 2.

Preparation of Samples

Derivatized samples were concentrated using a Micro-ProDiCon (Bio-Molecular Dynamics, Beaverton, OR) or Centricon-10 and -30 centrifuge concentrators (Amicon, Beverly, MA). The primary criterion for a suitable concentration is the absorbance of the sample at the laser excitation wavelength. Our experience has been that a very good to excellent signal-to-noise

[12] W. S. McIntire, J. L. Bates, D. E. Brown, and D. M. Dooley, *Biochemistry* **30,** 125 (1991).

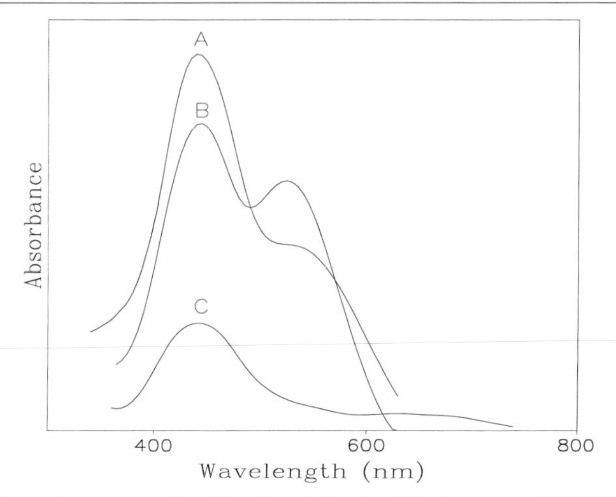

FIG. 2. Absorption spectra of oxidized methylamine dehydrogenase (MADH) from bacterium W3A1 and certain derivatives. (A) MADH-2-hydrazinopyridine derivative, (B) MADH phenylhydrazone at pH 7.0, and (C) oxidized MADH. These data were originally published in McIntire *et al.*[12]

ratio could be obtained in reasonable times if the quinoprotein samples were concentrated to give an absorbance of 2–15 at λ_{max} in a 1-cm path length cell. It is critical for the samples to be optically clear, i.e., without any turbidity detectable by eye. If necessary, samples should be centrifuged (2–3 min at 9000 rpm in a microcentrifuge) until there is no detectable turbidity.

Data Collection and Reduction

As noted earlier, instrumentation for resonance Raman spectroscopy, including spectrometers, lasers, and detectors, has already been described and evaluated in detail. All of the standard configurations will enable one to collect Raman or resonance Raman data on quinoproteins. To benefit from resonance enhancement, it is necessary that the laser line be absorbed by the sample. Given the absorption spectra of the native or derivatized proteins, an Ar^+ laser is the best general purpose source as it provides lines with adequate power at 457.9, 488, and 514.5 nm. Red excitation may be provided by Kr^+ or dye lasers, violet excitation is available from Kr^+ lasers, and ultraviolet lines may be provided by either Ar^+ or Kr^+ lasers. To prevent or minimize laser-induced photochemical reactions, or denaturation of the proteins, the power at the sample should be 50 mW or less. With the exception of the methylamine dehydrogenase–ammonia complex, the quinoprotein samples we have examined are sufficiently stable so that data may be obtained on small amounts (10–30 μl) of sample contained in glass capillary tubes. Less concentrated samples, which generally required longer acquisition times, were cooled by blowing cold nitrogen gas over the sample capillary and holder in the sample compartment of the Raman spectrometer. It is vital to ensure that no laser-induced changes in the sample occur during data collection. There are several ways in which this can be checked, including: (1) measuring the specific activity of a native enzyme (e.g., underivatized methylamine dehydrogenase) before and after the resonance Raman experiment; (2) measuring a key spectroscopic property, e.g., the absorption spectrum, before and after the experiment; and (3) using the reproducibility of the spectra themselves to monitor the integrity of the sample. We have found the final approach to be generally useful and straightforward, especially since it is compatible with common data collection strategies, as explained below.

Both native and derivatized quinoproteins typically display resonance-enhanced vibrational bands over the 200- to 1700-cm^{-1} range. Instead of collecting the data over this range as a single block, it is preferable to divide the region into two or three sections. With CCD or diode-array detectors this is often necessary because these devices, in conjunction with the typical

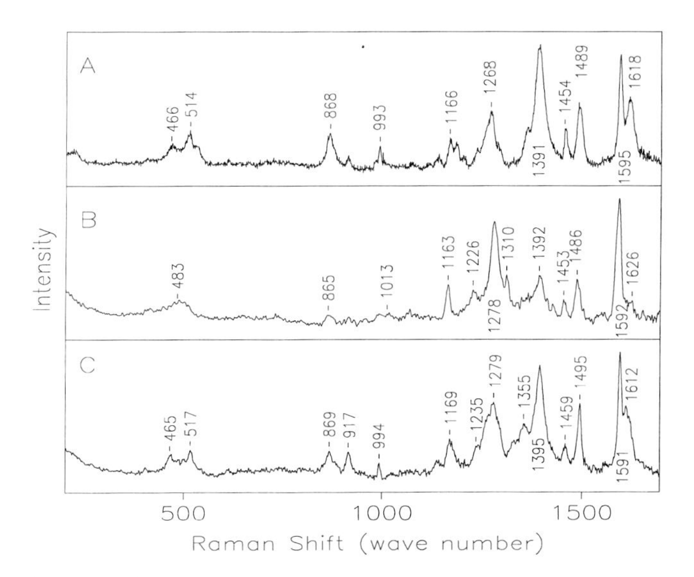

FIG. 3. Resonance Raman spectra of amine oxidase phenylhydrazones. (A) *Arthrobacter* P1 methylamine oxidase, (B) pea seedling diamine oxidase, and (C) bovine plasma amine oxidase. The excitation wavelength was 457.9 nm and the laser power was 20–40 mW. These experiments were originally reported in Brown *et al.*[13]

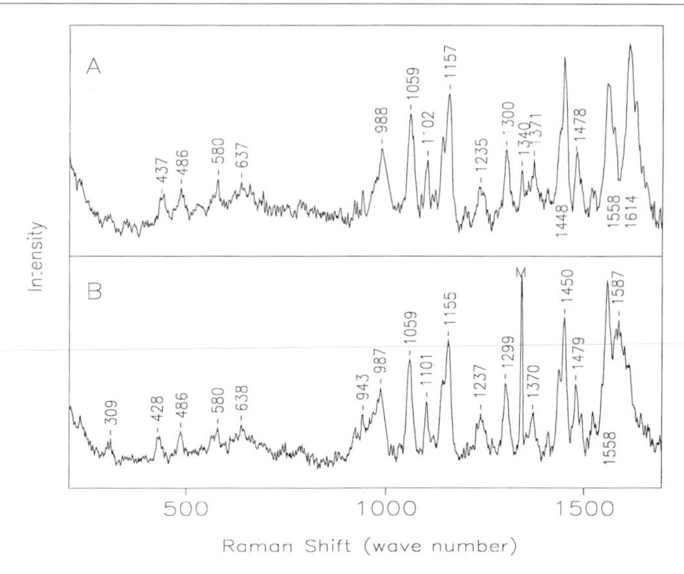

FIG. 4. Resonance Raman spectra of underivatized methylamine dehydrogenase (MADH) from bacterium W3A1. (A) Native oxidized MADH, and (B) native oxidized MADH following exchange with ^{18}O. The band marked M is a wavelength marker. The excitation wavelength was 457.9 nm. Data were originally reported in McIntire *et al.*[12]

gratings used in Raman spectrometers, will not cover such a large spectral region at the desired resolution, generally 5–10 cm^{-1}. Most of the published data have been obtained with a photomultiplier tube detector in blocks, usually 200–1000 cm^{-1} and 1000–1700 cm^{-1}. Our basic approach is to collect several scans (3–10) in each region, check them for consistency, and then add them. Any laser-induced changes are readily apparent as changes in the vibrational frequencies or signal-to-noise ratio from one scan to another. Scans that are inconsistent with the initial scan are discarded. Depending on the intensities of the resonance Raman bands, 2–10 spectra are collected and added. After a data set is collected in each region with comparable signal-to-noise ratios, the two data sets are spliced together to form a single spectrum over the entire range of interest. There are several commercially available software packages that permit the data to be processed in this manner. Spectra that are representative of the signal-to-noise ratio that is attainable are shown in Fig. 3[13] for amine oxidase derivatives and in Figs. 4 and 5 for methylamine dehydrogenase samples. The spectra shown in

[13] D. E. Brown, M. A. McGuirl, D. M. Dooley, S. M. Janes, D. Mu, and J. P. Klinman, *J. Biol. Chem.* **266,** 4049 (1991).

Fig. 3 are generally similar (particularly A and C), but significant differences are apparent (compare B to A or C), especially in the relative intensities of certain bands. The similarities and differences among the spectra in Fig. 3 are consistent with the presence of an identical chromophore in all three enzymes, but with differences in the microscopic environment of the chromophore. This conclusion is confirmed by the fact that the resonance Raman spectra of the corresponding active-site peptide derivatives (with the same microenvironment in solution) are practically identical. An even more dramatic example of environmental effects on resonance Raman is evident in Fig. 5: the spectrum of intact MADH differs substantially from that of the isolated β subunit or the cofactor-containing peptide. This was attributed to the fact that the TTQ cofactor in MADH is bound at the $\alpha\beta$ subunit interface. Figure 4 illustrates the use of isotopic substitution ($^{18}O/^{16}O$, in this case) to identify bands in the resonance Raman spectrum with the vibrational motions of particular nuclei. Note that only two bands shift significantly following ^{18}O exchange into the carbonyl groups of the TTQ cofactor: 437 cm^{-1} $\rightarrow$ 428 cm^{-1} and 1614 cm^{-1} $\rightarrow$ 1587 cm^{-1}. Hence these bands must reflect modes involving substantial carbonyl deformation. The high-frequency bands may be plausibly assigned to C=O stretching vibrations.

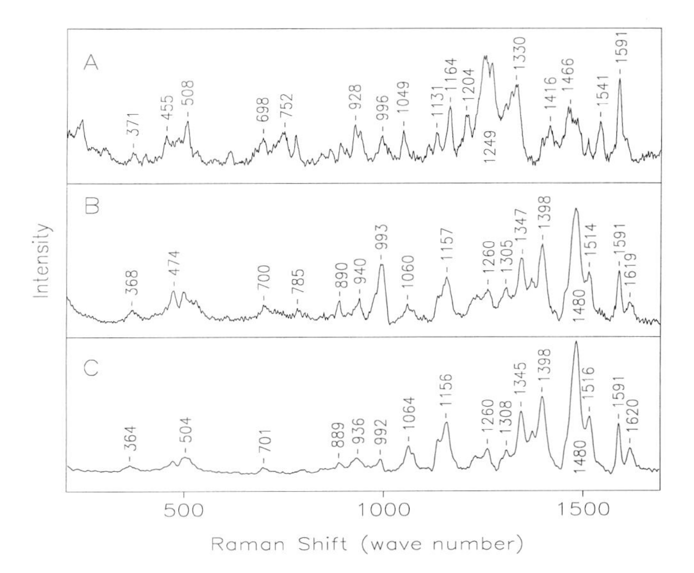

FIG. 5. Resonance Raman spectra of methylamine dehydrogenase (MADH) phenylhydrazones. (A) Oxidized MADH, (B) isolated β subunit of MADH, and (C) isolated TTQ-containing peptide of MADH. The excitation wavelength was 457.9 nm. These experiments were originally reported in McIntire *et al.*[12]

Baselines in resonance Raman spectra are frequently sloped or nonlinear because of the sample fluorescence or scattering from the sample holder. In such cases a linear or nonlinear baseline can be estimated from the data and digitally subtracted. Some caution is required because baseline subtraction can shift the apparent vibrational frequencies and alter band shapes. Good practices include carefully comparing frequencies and intensities of the raw and baseline-subtracted data and measuring the effects of subtracting different baselines on the data. Any variations in frequency or integrated intensity should be included when uncertainties in these quantities are cited.

[11] Redox-Cycling Detection of Dialyzable Pyrroloquinoline Quinone and Quinoproteins

By RUDOLF FLÜCKIGER, MERCEDES A. PAZ, and PAUL M. GALLOP

Introduction

Quinone and/or hydroquinone moieties in quinoproteins are either noncovalently or covalently bound to the protein. The noncovalently attached cofactor occurring in various bacterial alcohol dehydrogenases represents the tricarboxylated pyrroloquinoline quinone (PQQ).[1] The covalently attached quinone cofactors of known structure are dopaquinone,[2] 6-hydroxydopaquinone,[3] and tryptophan tryptophylquinone.[4]

The covalently attached quinone cofactors arise by hydroxylation of side chains of aromatic amino acid residues: hydroxylation of tyrosyl residues gives rise to 3,4-dihydroxyphenylalanyl residues which can be oxidized to the *o*-quinone and they occur in the vitelline proteins. 6-Hydroxylation of specific dopa residues in amine oxidases gives rise to 6-hydroxydopa residues that can generate an *o*- or *p*-quinone.[3] Hydroxylation of a tryptophanyl residue in methylamine dehydrogenase initiates the formation of the cross-linked redox-cofactor tryptophan tryptophylquinone.[4] Dopa-containing residues can also arise from covalent attachment of free dopaqui-

[1] S. A. Salisbury, H. S. Forrest, W. B. T. Cruse, and O. Kennard, *Nature (London)* **280,** 843 (1979).

[2] J. H. Waite and A. C. Rice-Ficht, *Biochemistry* **28,** 6104 (1989).

[3] S. M. Janes, D. Mu, D. Wemmer, A. J. Smith, S. Kaur, D. Maltby, A. L. Burlingame, and P. J. Klinman, *Science* **248,** 981 (1990).

[4] W. S. McIntire, D. E. Wemmer, A. Christoserdov, and M. E. Lindstrom, *Science* **252,** 817 (1991).

none to reactive thiol groups in proteins. In this process, which may be termed thioquinolation, dopa becomes linked through a thioether bond at its two or three ring position to the sulfur of cysteine.[5]

The redox-cycling methodology described in this chapter is useful for the detection of covalently bound quinone cofactors as well as for the detection of dialyzable PQQ released from quinoproteins or isolated from tissues and fluids. Other methods for the detection and isolation of protein-bound dopa have been described.[6]

Principle of Redox-Cycling Detection of Quinones

Quinones, at an alkaline pH, can oxidize glycine in an oxidative amine degradation reaction. The hydroquinones formed react with molecular oxygen, causing superoxide anion formation which is monitored by the reduction of nitroblue tetrazolium (NBT) to the mono- or diformazan.[7] Sterically hindered valine is not oxidized at a significant rate in this reaction:

$$\begin{bmatrix} NH_2{-}CH_2{-}COO^- \\ NH{=}CH{-}COO^- \end{bmatrix} \begin{bmatrix} PQQ({-}OH)_2 \\ PQQ({=}O)_2 \end{bmatrix} \begin{bmatrix} 2\,O_2^- + 2H^+ \\ 2\,O_2 \end{bmatrix} \begin{bmatrix} HT{-}TH \text{ (formazan)} \\ T^+{-}T^+ \text{ (NBT)} \end{bmatrix}$$

The efficiency of this quinoid-catalyzed redox cycling depends on the nature of the quinone and the tendency of the reduced quinoid intermediates to form redox-inactive melanin-like polymers. Such polymerization is limited in quinoprotein-catalyzed redox cycling. Among the free quinones, PQQ is most efficient in catalyzing this redox cycling because mutual charge–repulsion prevents polymerization. Under the conditions of the NBT/glycinate assay, PQQ undergoes an estimated 2000 redox cycles/hr, which renders the assay highly sensitive and allows picomoles of PQQ to be detected. The specificity of the assay is increased by the addition of borate which renders ascorbate, an enediol anion and common constituent of biological samples, unreactive by complex formation.[8] Redox cycling of PQQ and many other quinones is only marginally affected by the concentration of borate used in the assay (Fig. 1).

Factors Affecting Formazan Production

The color yield in the NBT–glycinate redox-cycling assay depends on pH and the formazan solubilizer. With albumin, the purple monoformazan

[5] S. Ito, T. Kato, and K. Fujita, *Biochem. Pharmacol.* **37,** 1707 (1988).

[6] J. H. Waite and C. V. Benedict, this series, Vol. 107, p. 397.

[7] R. Flückiger, T. Woodtli, and P. M. Gallop, *Biochem. Biophys. Res. Commun.* **153,** 353 (1988).

[8] M. A. Paz, R. Flückiger, and P. M. Gallop, *FEBS Lett.* **264,** 283 (1990).

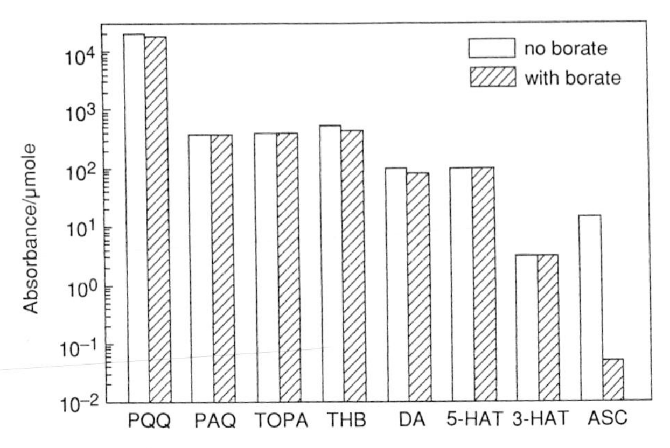

FIG. 1. Formazan production by quinones and ascorbate with and without borate in the assay. PQQ, pyrroloquinoline quinone; PAQ, phenanthrenequinone; TOPA, trihydroxyphenylalanine; THB, trihydroxybenzene; DA, dopamine; 3/5-HAT, hydroxyanthranilic acid; ASC, ascorbate. Absorbance at 530 nm was measured after a 1-hr reaction at 25°; the values are selected from Table I.[15]

(ε_{525}23,400) is formed, and in the presence of Triton the blue diformazan is formed (ε_{605}40,200) (extinction coefficients from[9]). Formazan production is totally suppressed by high concentrations (20 U) of superoxide dismutase, indicating that superoxide anion is generated during redox cycling.

Inhibition of PQQ-Catalyzed Redox Cycling

The redox cycling of PQQ is strongly inhibited by certain di- and trivalent metal cations. Strong inhibition (IC_{50} <10 μM) is observed with cations with ionic radii of 0.8–0.93 Å (see Table I), and this inhibition is competitively reversed by the chelator Tiron (4,5-dihydroxy-1,3-benzene disodium sulfonate). With certain biological samples, the sensitivity of the redox-cycling assay toward metal cations renders it necessary to remove divalent cations by cation exchangers.

Several aromatic cations such as phenazine methosulfate[10] and iodonium compounds[11] also specifically and strongly inhibit PQQ-catalyzed redox cycling. This inhibition is reversed in part by Tiron and thyroxine (T_4).[10]

[9] M. J. Eadie, J. H. Tyrer, J. R. Kukums, and W. D. Hooper, *Histochemie* **21,** 170 (1970).

[10] R. Flückiger, M. A. Paz, J. Mah, A. Bishop, and P. M. Gallop, *Biochem. Biophys. Res. Commun.* **196,** 61 (1993).

[11] P. M. Gallop, M. A. Paz, R. Flückiger, P. J. Stang, V. V. Zhdankin, and R. Tykwinski, *J. Am. Chem. Soc.* **115,** 11702 (1993).

TABLE I
IC_{50} OF SELECTED INHIBITORS OF PQQ-CATALYZED REDOX CYCLING[a]

Inhibitor	IC_{50}
Indium(III)	0.5 μM
Manganese(II)	3 μM
Phenazine methosulfate	67 nM
Diphenyleneiodonium	3 μM

[a] From Flückiger *et al.*[10]

Aromatic cations interact with reduced PQQ through charge-transfer and π–π interactions.[12]

The inhibitors of PQQ-catalyzed redox cycling recognize basic structural features of PQQ; the metal cations occupy the metal binding site derived from the Schiff base, quinoline nitrogen, and the 7-carboxylate group. Aromatic cations recognize the extended anionic aromatic ring system of PQQ redox species. The specificity of these inhibitors for PQQ is illustrated in Fig. 2.

The sensitivity toward these inhibitors was instrumental in establishing the occurrence of free PQQ in animal tissues and fluids.[10] The inhibition pattern with PQQ and with the putative PQQ isolated from red cells by HPLC is identical. The redox activity of a fungal isolate, containing *o*- and *p*-quinones derived from veratryl alcohol,[13] which are not related to PQQ, is not inhibited by these agents (Fig. 3).

Strategies for the Identification of PQQ and Quinoproteins in Fluids and Tissues

Freshly removed tissues are homogenized and centrifuged, and the supernatant is dialyzed against cold water. Biological fluids such as raw skim milk, plasma, and serum are also dialyzed. Both the protein-containing dialysate and the protein-free dialyzable material are collected and concentrated. Precipitation of protein by TCA or perchloric acid is not useful because the dialyzable redox-cycling activity from putative PQQ is lost. The crude nondialyzable material is examined for the presence of quinoproteins after SDS–PAGE separation and electroblotting by redox-cycling staining.

[12] T. Ishida, M. Doi, K. Tomita, H. Hayashi, M. Inoue, and T. Urakami, *J. Am. Chem. Soc.* **111**(17), 6822 (1989).

[13] H. W. H. Schmidt, S. D. Haemmerli, H. E. Schoemaker, and M. S. A. Leisola, *Biochemistry* **28,** 1776 (1989).

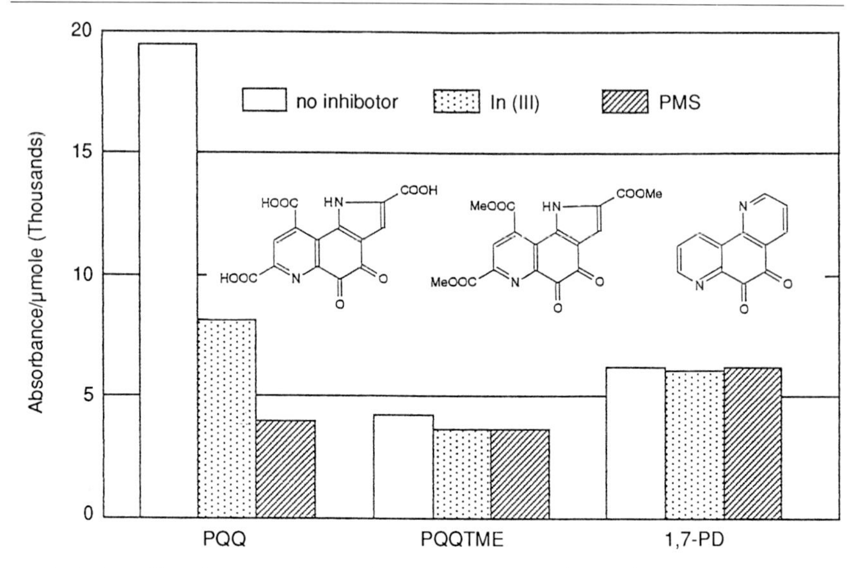

FIG. 2. Inhibition of redox cycling by the PQQ-specific inhibitors indium(III) and phenazine methosulfate (PMS). Absorbance was read at 605 nm after a 20-min incubation at 37°. The inhibitors $InCl_3$ (0.6 μM) and PMS (0.2 μM) do not affect formazan production by the trimethylester of PQQ (PQQTME, 90 nM) or 1,7-phenanthrolinedione (1,7-PD, 1.6 μM). Color production by the quinones (PQQ, 30 nM) in the absence of inhibitors was comparable.

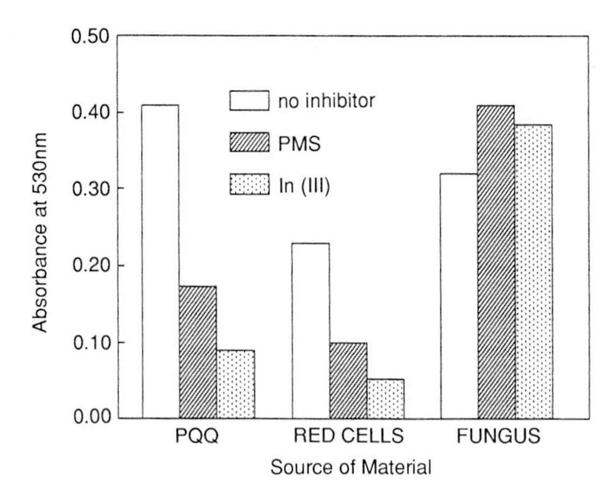

FIG. 3. Inhibition of PQQ-catalyzed redox-cycling activity and of HPLC-purified redox-active material from red cells and from the fungus *Phenerochaete chrysosporium*. Red cells contain PQQ, whereas the fungal quinones are derived from veratryl alcohol. Values for red cells are taken from Flückiger *et al.*[10]

Partial purification to obtain fractions enriched in quinoproteins is generally necessary.

To detect dialyzable quinones in the protein-free material, the NBT/glycinate redox-cycling assay is run with borate to suppress reactivity originating from ascorbate. A NBT/valinate control is also run to ascertain that the reactivity does not originate from direct reduction of NBT by reducing compounds that are not related to PQQ.

To further characterize putative PQQ, the protein-free material is separated by HPLC. For HPLC, a semipreparative SUPELCOSIL LC-18 column has been found to be suitable since 1-ml vol can be injected. The column is eluted with 0.1 *M* ammonium formate, pH 3.3, with a linear gradient to 100% acetonitrile. The elution from the column is followed by the NBT/glycinate redox-cycling assay. The putative PQQ elutes in the exact position of authentic PQQ. Some redox-cycling activity which elutes early from the column is always present and may be a PQQ–amino acid adduct. The HPLC column gets contaminated rapidly with hydrophobic material which causes retention of PQQ. Extensive washing of the column with acetonitrile after each run is necessary to elute this hydrophobic material. The putative PQQ, which elutes in the position of authentic PQQ, is further characterized by its specific response to inhibitors and activators as described.[10] For HPLC elution profiles of putative PQQ, see Flückiger *et al.*[10,14]

NBT/Glycinate Assay

Materials

- Nitroblue tetrazolium chloride monohydrate
- Glycine (BDH, AnalaR)
- Valine
- Bovine serum albumin (BSA)
- Triton X-100
- PQQ
- Chelex-100 (100–200 mesh)
- Phenazine methosulfate (PMS)
- Indium chloride
- Manganous sulfate
- L-thyroxine
- Tiron (4,5-dihydroxy-1,3-benzenedisulfonic acid, disodium salt)

[14] R. Flückiger, M. A. Paz, P. R. Bergethon, E. Henson, and P. M. Gallop, *in* "Principles and Applications of Quinoproteins" (V. L. Davidson, ed.), p. 331. Dekker, New York, 1992.

Stock Solutions

2 *M* potassium glycinate, pH 10 (150 g glycine/liter, glycine goes into solution when pH is adjusted with 2 *M* KOH)
0.5 *M* potassium valinate, pH 10 (59 g valine/liter)
0.1 *M* borate/glycinate buffer, pH 10 (38 g sodium tetraborate decahydrate/liter in glycinate buffer)
0.1 *M* borate/valinate buffer, pH 10 (38 g sodium borate as above in valinate buffer)
BSA (approximately 50 mg/ml in 10 m*M* potassium phosphate, pH 7): The albumin solution is reduced with a spatula tip of sodium borohydride at 4° overnight. Excess borohydride is destroyed by acidification with glacial acetic acid to pH 4. The albumin solution is dialyzed overnight against distilled water, and 1-ml aliquots are kept frozen in microcentrifuge tubes.
Triton: 0.1% in water
PQQ stock: 1 m*M* (3.3 mg/10 ml 0.05 *M* potassium phosphate, pH 7). This stock solution keeps for months at 4° when protected from light.
Indium chloride: 1 m*M* indium chloride (2.2 mg/10 ml water)

Reagents Prepared Daily

NBT/glycinate reagent: 0.24 m*M* NBT (2 mg/10 ml) in potassium glycinate. Prepare shortly before the assay and keep protected from light on ice.
NBT/valinate reagent: as described for potassium valinate
PQQ standard: 1:1000 dilution of stock in water
BSA: 1:7 dilution of stock in water
Phenazine methosulfate (1 m*M*; 3.1 mg in 10 ml water)
Manganous sulfate (1 m*M*; 1.69 mg $MnSO_4 \cdot H_2O$ in 10 ml water)
L-Thyroxine (1 m*M*; 8.89 mg L-thyroxine, sodium salt pentahydrate in 10 ml water). To help solubilize the T_4 a drop of 2 *M* KOH is added before completing the volume to 10 ml.
Tiron (1 m*M*; 3.3 mg in 10 ml water)

Procedure

For Chelex pretreatment, add one spatula tip of Chelex-100 to ca. 1 ml of sample, vortex well, spin resin down, and use supernatant in assay.

For the PQQ standard curve, dispense 0, 10, 15, 20, and 30 μl of PQQ standard into test tubes and adjust to a final volume of 300 μl with water.

Add 100 μl borate/glycinate or borate/valinate reagent and 100 μl BSA or Triton, followed by 1 ml NBT/glycinate (or NBT/valinate) reagent. Incubate in waterbath for 1 hr at 25°. Read absorbance at 530 nm with albumin or at 605 nm with Triton. Calculate PQQ concentrations from the standard curve. One picomole of PQQ yields an OD of approximately 0.02.

If the reaction is performed in 96-well microtiter plates, use 60 μl sample/standard/inhibitor, 20 μl borate, 20 μl BSA, and 200 μl NBT/glycinate reagent and read absorbance with a microplate reader. At 37° and a 20-min reaction time, absorbance readings are comparable to 1-hr readings at 25°.

Inhibitors are used at approximately IC_{50} values. Thyroxine and Tiron enhance the redox-cycling activity of PQQ by about 40% with 100 nmol in the reaction. Tiron is generally used with the metal inhibitors and thyroxin with PMS and diphenyleneiodonium. Tiron is more potent at reversing the manganese inhibition than that of indium. Controls with inhibitors, thyroxine, and Tiron are run as these compounds alone cause some formazan production. Thyroxine cannot be used with Triton replacing albumin because turbidity develops. A close correspondence of the response to inhibitors/activators exists for PQQ isolated from biological sources and authentic PQQ.[10]

Quinoprotein Detection on PAGE Electroblots

Reagents

Blotting buffer: 25 m*M* Tris, 192 m*M* glycine, pH 8.3, containing 20% methanol

NBT/glycinate reagent as described earlier

Ponceau S staining solution: 0.1% Ponceau S in 5% acetic acid

Borate wash buffer: 0.1 *M* sodium borate, pH 10

Procedure

Partially purified protein preparations are separated by SDS–polyacrylamide gel electrophoresis. For staining, the proteins are electroblotted onto a nitrocellulose membrane. Electrophoretic transfer is performed at 100 V for 1 hr at 10°. Quinoproteins are stained by immersing the nitrocellulose membrane in NBT/glycinate reagent for 45 min in the dark. This results in a blue-purple stain of quinoproteins without staining of other proteins (Fig. 4). The nitrocellulose membrane is washed with borate buffer and counterstained for protein with Ponceau S. For storage, the nitrocellulose membrane is washed with the borate wash buffer and kept at 4°.

Comment

Use of prepurified protein preparations is recommended because some quinoproteins stain rather faintly. The staining properties of the quinopro-

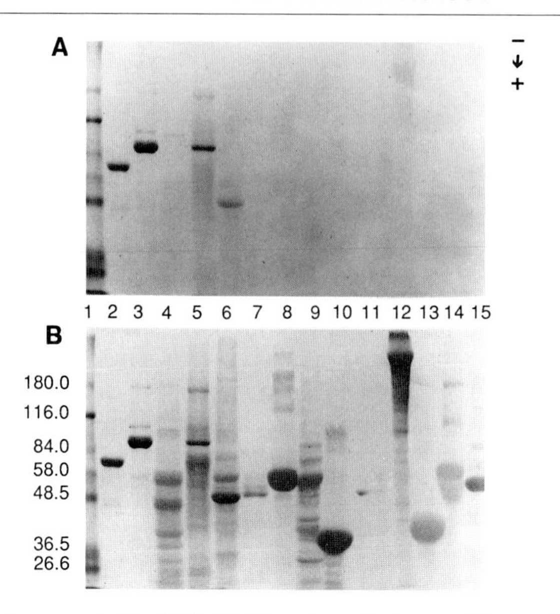

FIG. 4. Proteins separated by SDS–PAGE under reducing conditions and electroblotted onto nitrocellulose. (A) Staining with NBT/glycinate, and (B) counterstained with Ponceau S. Lane 1, Coomassie blue prestained molecular weight standard; lanes 2–4, established quinoproteins (2, methylamine oxidase from *Arthrobacter* P1; 3, bovine serum amine oxidase; 4, pig kidney diamine oxidase); lane 5, dopamine-β-hydroxylase (commercial preparation) containing an unidentified 97-kDa quinoprotein; lane 6, rat parotid gland homogenate containing an unidentified quinonoid component chromatographing with α-amylase; lane 7, salivary human α-amylase; and lanes 8–15, negative controls (8, bovine serum albumin; 9, isocitric dehydrogenase; 10, alcohol dehydrogenase; 11, glucose-6-phosphate 1-dehydrogenase; 12, hemocyanine from keyhole; 13, ovalbumin; 14, fetuin; 15, carnitine acetyltransferase).

teins tested varied greatly: fast and intense color development was observed with serum amine oxidase, methylamine dehydrogenase, methylamine oxidase, the vitellines, and a model quinoprotein. This model quinoprotein was prepared by reacting bovine serum albumin with hematoxylin followed by alcohol precipitation and extensive washing. Hematoxylin is a diquinoid compound that covalently attaches to arginine residues.[15] Intense staining is found at the molecular weight site for serum albumin and for higher molecular weight aggregates.

Prolonging the staining of electroblotted quinoproteins to over 45 min did not increase the intensity of the stain with the less reactive quinoproteins diamine oxidase or lysyl oxidase. If faint staining is encountered and purified protein preparations are available, these can also be tested in solution by

[15] P. M. Gallop, M. A. Paz, and R. Flückiger, *Chemtracts Biochem. Mol. Biol.* **1,** 357 (1990).

allowing the reaction to proceed overnight at 4° in order to increase sensitivity.

SDS–PAGE and subsequent redox-cycling staining are useful for the specific detection of quinoproteins where the quinoid component is covalently attached to the protein. However, it does not yield information about the structure of the quinone present.

In the past, several proteins were wrongly identified as quinoproteins with a method based on the formation of a hydrazine adduct.[16] The absence of staining with the NBT/glycinate reagent showed that the putative identification of covalently bound PQQ in dopamine-β-hydroxylase (DBH) and soybean lipoxygenase-1[17,18] was incorrect. It also revealed the presence of a quinoid 97-kDa contaminant in some DBH preparations. The application of this methodology has established that bacterial and mammalian apoferritin are quinoproteins.[19]

It is to be expected that additional quinoproteins will be detected as the redox-cycling technique is further applied.

Acknowledgments

This work was supported in part by Grant NIA 07723 and a National Dairy Promotion and Research Board Grant administered by the National Dairy Council. The authors thank Drs. V. Apekin and J. Mah for their valuable contributions, and Dr. T. Choinowski, Swiss Federal Institute of Technology Zürich, for providing the fungal cultivation media.

[16] R. A. van der Meer, J. A. Jongejan, and J. A. Duine, *FEBS Lett.* **221**(2), 299 (1987).
[17] M. A. Paz, R. Flückiger, A. Boak, H. M. Kagan, and P. M. Gallop, *J. Biol. Chem.* **266,** 689 (1991).
[18] G. A. Veldink, H. Boelens, M. Maccarone, F. van der Lecq, J. F. G. Vliegenthart, M. A. Paz, R. Flückiger, and P. M. Gallop, *FEBS Lett.* **270,** 135 (1990).
[19] F. K. Al-Massad, F. H. A. Kadir, and G. R. Moore, *Biochem. J.* **283,** 177 (1992).

[12] Tryptophan Tryptophylquinone in Bacterial Amine Dehydrogenases

By WILLIAM S. MCINTIRE

Introduction

Bacterial oxidations of alkylamines are carried out by several different types of enzymes. One type contains flavin adenine dinucleotide (FAD),

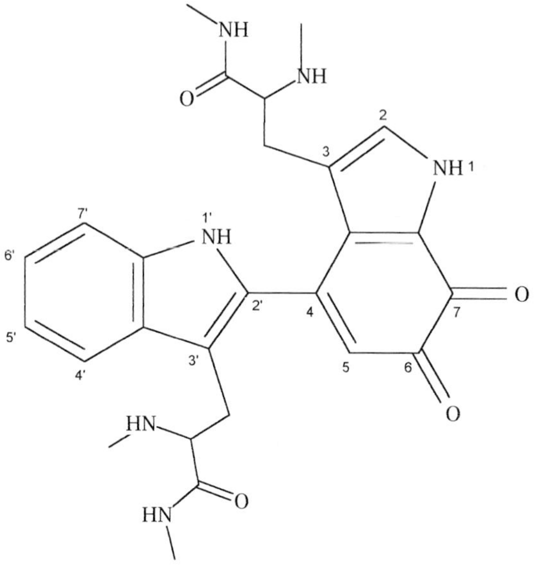

FIG. 1. Structure of 2′,4-bitryptophan-6,7-dione (tryptophan tryptophylquinone, TTQ), the redox cofactor of bacterial methylamine dehydrogenase.

and all in this group are oxidases.[1] Trimethyl- and dimethylamine are oxidized via flavin mononucleotide (FMN)/Fe_4S_4-containing dehydrogenases in numerous methylotrophs.[2] Another group is composed of copper/topa quinone-containing amine oxidases, found in gram-positive and gram-negative bacteria.[1,3] A number of monoalkylamine dehydrogenases have been identified that contain 2,4′-bitryptophan-6′,7′-dione(tryptophan tryptophylquinone, TTQ) (Fig. 1) as the redox-active cofactor. TTQ results from the oxidation/hydroxylation of a tryptophyl residue to an *o*-quinone, before or after it cross-links to a second, otherwise unmodified tryptophyl residue.[4,5] The reaction carried out by TTQ-containing amine dehydrogenases is

$$R{-}CH_2NH_3^+ + H_2O + \text{acceptor} \rightarrow R{-}CHO + NH_4^+ + \text{reduced acceptor}.$$

[1] W. S. McIntire and C. Hartmann, *in* "Principles and Applications of Quinoproteins" (V. L. Davidson, ed.), Chapter 6, p. 97. Dekker, New York, 1993.

[2] W. S. McIntire, this series, Vol. 188, p. 250.

[3] R. A. Cooper, P. F. Knowles, D. E. Brown, M. A. McGuirl, and D. M. Dooley, *Biochem. J.* **288,** 337 (1992).

[4] A. Y. Chistoserdov, C. Boyd, F. S. Mathews, and M. E. Lidstrom, *Biochem. Biophys. Res. Commun.* **184,** 1226 (1992).

[5] W. S. McIntire, D. E. Wemmer, A. Chistoserdov, and M. E. Lidstrom, *Science* **252,** 817 (1991).

For routine assays, the acceptor can be any one of the following dyes: phenazine methosulfate, phenazine ethosulfate (PES), or Wurster's blue (the radical of *N,N,N′,N′*-tetramethyl-*p*-phenylenediamine).[6,7]

Methylamine dehydrogenase (MADH) is a periplasmic enzyme that passes electrons, extracted from methylamine, into the electron transport chain in the inner membrane of the bacteria for the production of ATP. The natural electron acceptor for most TTQ-containing enzymes is the copper protein amicyanin, although cytochrome c_H can also act as the electron acceptor for MADH from *Methylobacterium extorquens* strain AM1 (former known as *Pseudomonas* sp. AM1).[8] The author has been unable to detect amicyanin in bacterium W3A1 grown on methylamine. Several *c*-type cytochromes have been purified from this organism, and preliminary work, by the author, indicates that two of these can accept electrons from reduced MADH. Based on its properties, an aromatic amine dehydrogenase[9] likely contains TTQ as the essential redox cofactor.

Preliminary Identification of an Enzyme Containing TTQ

All of the monoamine dehydrogenases have an $\alpha_2\beta_2$ structure, with the exception of methylamine dehydrogenase from *Methylobacillus flagellatum*. It has an $\alpha\beta$ structure.[7,10] The large α subunits have molecular weights in the range from 40,000 to 48,000, whereas the small, cofactor-containing β subunits have molecular weights in the range of 8000–16,000, depending on the source of the enzyme. In addition to the cross-link provided by TTQ, the small subunit contains a large number of disulfide bonds: six in MADH from *M. extorquens*, *Paracoccus denitrificans*, and *Thiobacillus versutus*,[4] and seven in MADH from bacterium W3A1 (F. S. Mathews, personal communication). The large subunit does not contain any cysteinyl groups. The molecular weight of the native enzyme can be estimated by traditional methods (e.g., size exclusion chromatography), whereas the subunit molecular weights can be estimated by any number of polyacrylamide gel electrophoresis methods or by size exclusion chromatography under denaturing conditions.[11,12] By comparison of native and denatured protein molecular weights, the subunit composition can be deduced.

[6] W. S. McIntire, *J. Biol. Chem.* **262,** 11012 (1987).

[7] V. L. Davidson, this series, Vol. 188, p. 241.

[8] P. J. Large, *FEMS Microbiol. Rev.* **87,** 235 (1990).

[9] M. Nozaki, this series, Vol. 142, p. 650.

[10] M. Y. Kiriukhin, A. Y. Chistoserdov, and Y. D. Tsygankov, this series, Vol. 188, p. 247.

[11] A. J. Rowe, *Tech. Life Sci.* [*Sect.*]: *Biochem.* **B105a,** 1 (1978); T. J. Mantle, *ibid.* **B105b,** 1 (1978); J. O. Thomas, *ibid.* **B106,** 1 (1978).

[12] R. Scopes, "Protein Purification." Springer-Verlag, New York, 1982.

All enzymes known to contain TTQ are monoalkylamine dehydrogenases. Thus, any enzyme suspected of having this cofactor should be tested for this dehydrogenase activity. The following assay, used for MADH from bacterium W3A1, should work well for all TTQ-containing amine dehydrogenases.

Reagents

Potassium phosphate buffer, 50 m*M*, pH 7.5
Phenazine ethosulfate, 20 m*M* in H_2O
Disodium 2,6-dichlorophenolindophenol (DCIP), 0.05% (w/v in H_2O)
Alkylamine hydrochloride, 0.1 *M* in H_2O

Assay Procedure

Mix 2.46 ml of buffer, 0.2 ml DCIP, 0.3 ml PES, and 40 μl of the amine hydrochloride in a 3-ml cuvette (1-cm light path) in the cell holder of a visible spectrophotometer set to record the absorbance change at 600 nm. A blank rate is measured, before adding 2–50 μl of enzyme. The rate of change at 600 nm, caused by the reduction of DCIP by enzyme-reduced PES, is recorded. The ε_{600} value for DCIP is 21.6 mM^{-1} cm^{-1}, at pH 7.5. One unit of activity is defined as the amount of dehydrogenase required to reduce 1 μmol of DCIP per minute at 30°. A number of precautions have been reported regarding this assay.[2]

It is possible that this cofactor could be involved in the oxidation of substances other than amines. A related cofactor, pyrroloquinoline quinone, is required for oxidation of alcohols and aldehydes by a number of dehydrogenases.[13]

Spectral and Oxidation–Reduction Properties

An obvious feature of the amine dehydrogenases is the color given to them by TTQ. The color is yellow in dilute solutions and is blackish-green in concentrated solutions. The UV-visible spectrum of oxidized MADH from bacterium W3A1 is shown in Fig. 2. This spectrum is altered by the presence of monovalent cations (W. S. McIntire, personal observation). The enzyme was dissolved in the nonreacting buffer, Bis Tris Propane {1,3-

[13] C. Anthony, *Int. J. Biochem.* **24,** 29 (1992).

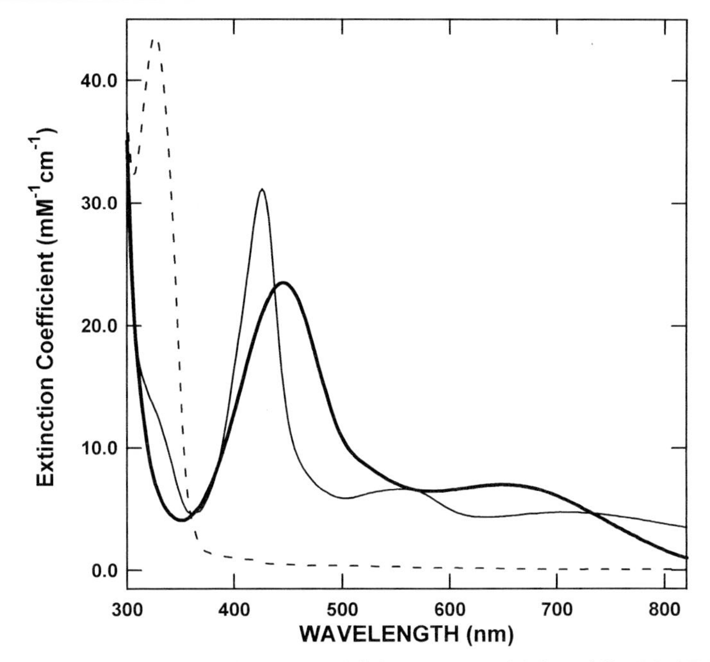

FIG. 2. The thick solid line is the UV-visible spectrum of fully oxidized MADH from bacterium W3A1 in 10 m*M* Bis Tris Propane, pH 7.43. The thin solid line is the calculated spectrum of the two-electron-reduced form of the enzyme, when one electron resides with each of the two TTQ groups. The dashed line is the spectrum of the four-electron-reduced enzyme (each TTQ moiety is two electron reduced). The spectrum of the two-electron-reduced enzyme was obtained by a factor analysis[13a] of data obtained during a sodium dithionite titration of MADH. The factor analysis program was written by V. Kuusk.

bis[tris(hydroxymethyl)methylamino]propane}, for the spectra shown in Fig. 2. Being an *o*-quinone, TTQ can exist in a one or two electron reduced state. It has been shown that during an anaerobic titration of MADH with substrate or reductant, sodium dithionite, one observes the one and two electron reduced state of the enzyme in UV-visible spectra.[14,15] Figure 2 displays the spectral changes that occur during a dithionite titration of

[13a] P. J. Gemperline, *J. Chemom.* **3,** 549 (1989); J. C. Hamilton and P. J. Gemperline, *ibid.* **4,** 1 (1990).

[14] W. C. Kenney and W. S. McIntire, *Biochemistry* **22,** 3858 (1983).

[15] W. C. Kenney and W. S. McIntire, *in* "Microbial Growth on C_1 Compounds" (R. L. Crawford and R. S. Hanson, eds.) p. 165. Am. Soc. Microbiol., Washington, DC, 1984.

MADH from bacterium W3A1. The spectra for these two redox state are diagnostic for the presence of TTQ.

TTQ, like other *o*- or *p*-quinones, can react with carbonyl reagents, e.g., semicarbazide, substituted phenylhydrazines, hydroxylamine, or hydrazine.[14,16] For any of these reagents, very distinct, well-defined UV-visible spectral changes should occur, indicating a specific reaction. These reagents may be so specific that they will react stoichiometrically,[14] thus providing a method for quantitating the cofactor content of a particular enzyme. The reaction with any of these reagents can be used as a preliminary diagnostic tool for the presence of TTQ, pyrroloquinoline quinone, topa quinone, and yet to be discovered quinone cofactors.

The unmodified enzyme-bound cofactor and the protein-bound and protein-free phenylhydrazine-derivatized cofactors provide excellent resonance Raman spectra. This is due to the stability of the cofactor when irradiated with laser light and the lack of cofactor fluorescence.[17,18] This technique has confirmed the existence of several important functional groups of the cofactor (i.e., the reactive carbonyl). Significant differences exist among the resonance Raman spectra of derivatized pyrroloquinoline quinone, topa quinone, and TTQ.[19] Thus, resonance Raman spectroscopy offers a valuable method for distinguishing the known classes of quinoproteins and it can be used in a screening regiment for an enzyme suspected of containing a quinonoid cofactor.

Caveats

Enzymes may exist that contain TTQ that will not display the properties mentioned in the previous sections. TTQ-containing amine dehydrogenases may exist that have different subunit molecular weights and quaternary structures. An enzyme other than an amine dehydrogenase may require TTQ. As is the case for other redox cofactor (e.g., riboflavin, topa quinone,[1] pyrroloquinoline quinone[20]), a TTQ-containing quinoprotein could contain a redox-active metal cofactor, and reduced TTQ could transfer electrons to a metal center, followed by a transfer of the electrons to oxygen. An enzyme-bound TTQ moiety could transfer electron directly to oxygen, as occurs with enzyme-free quinones in solution. These type enzymes would

[16] W. S. McIntire and J. T. Stults, *Biochem. Biophys. Res. Commun.* **141,** 562 (1986).

[17] W. S. McIntire, J. L. Bates, D. E. Brown, and D. M. Dooley, *Biochemistry* **30,** 125 (1991).

[18] G. Backes, V. L. Davidson, F. Huitema, J. A. Duine, and J. Sanders-Loehr, *Biochemistry* **30,** 9201 (1991).

[19] D. M. Dooley and D. E. Brown, *in* "Principles and Applications of Quinoproteins" (V. L. Davidson, ed.), Chapter 11, p. 275. Dekker, New York.

[20] J. A. Duine, *Eur. J. Biochem.* **200,** 271 (1991).

be oxidases. The environment of the enzyme surrounding TTQ might dramatically alter the spectral properties of the cofactor or may not be conducive to stabilizing the radical form of TTQ. In such cases, one could compare the spectra of the denatured enzyme with the spectrum of denatured MADH.[14,21]

With regard to carbonyl reagents, for various reasons (e.g., electrostatic, steric), the reagents may not react with native enzyme. The reagents may react with the cofactor once the enzyme is denatured; however, one runs the risk of the quinone reacting with the denaturant and/or unshielded amino groups and sulfhydryl groups of the denatured protein (vide infra). Other nonquinonoid carbonyl-containing cofactors (e.g., pyridoxal phosphate) will react with the carbonyl reagents, and FMN or FAD bound to enzymes may react with phenylhydrazine as do monoamine oxidase and trimethylamine dehydrogenase.[22]

Early work aimed at deciphering the structure of the cofactor bound in MADH resorted to indirect chemical methods (e.g., attempts to release the cofactor by relatively simple chemical treatments) and comparison of properties to those of known substances (e.g., pyridoxal phosphate, pyrroloquinoline quinone). This approach uniformly provided incorrect identifications in enzymes containing pyrroloquinoline quinone (promoted to contain pyridoxal phosphate), enzymes harboring topa quinone (said to contain pyridoxal phosphate or pyrroloquinoline quinone), and those with TTQ ("proved" to contain pyrroloquinoline quinone).[23] While such approaches are good as an initial screen, they cannot and should not substitute for a direct structural analysis, after the cofactor is stabilized and carefully extracted, intact, from the proteinous matrix.

This brings up another important point: How to stabilize an apparently unknown cofactor without altering its basic structure? Obviously, this is an after the fact question. Once the true structure is known, this question can be answered directly. If there is doubt about the structural integrity of a derivatized form of the cofactor, another unrelated stabilization method should be used.

Isolation and Structural Analysis of TTQ from MADH

The first steps in isolation of TTQ from an enzyme require stabilization of the cofactor and amino acid side chains, denaturation of the enzyme, and separation of the cofactor-containing subunits from other subunits and

[21] S. Itoh, M. Ogino, M. Komatsu, and Y. Oshiro, *J. Am. Chem. Soc.* **114,** 7294 (1992).

[22] J. Nagy, W. C. Kenney, and T. P. Singer, *J. Biol. Chem.* **254,** 2684 (1979).

[23] W. S. McIntire, *Essays Biochem.* **25,** 119 (1992).

contaminating proteins. Because of the *o*-quinone nature of TTQ, it will have a propensity to react with amino groups, as does pyrroloquinoline quinone. The latter cofactor reacts with amino acids to form redox-inactive oxazoles.[24] Pyrroloquinoline quinone also reacts with the protein denaturant urea,[24] and possibly guanidine,[25] to form (semi)stable adducts. We chose to use the carbonyl reagent semicarbazide to convert TTQ bound in MADH from bacterium W3A1 to a stable redox-inactive species. The cofactor in native MADH from bacterium W3A1 also reacts with phenylhydrazine derivatives;[17] however, the phenylhydrazine adduct is not stable.

Since MADH contains six to seven disulfide bonds, it is advisable to reduce these, after the redox property of TTQ has been arrested, and to alkylate all resulting sulfhydryl groups. Otherwise a great number of randomly cross-linked sulfhydryl peptides could form, due to disulfide exchange. The following procedure is used for MADH from bacterium W3A1.[5,16,17]

Step 1: Stabilizing the Cofactor and Sulfhydryl Groups

MADH was purified by a published procedure from bacterium W3A1 grown on methylamine hydrochloride.[6,14] The culture of bacterium W3A1 was obtained from the National Collection of Marine and Industrial Bacteria, Aberdeen, Scotland (NCIMB No. 11348). The culture was purified as described earlier.[2]

To a solution of pure enzyme (1.09 g, 8.35 μmol of enzyme or 16.7 μmol of cofactor) dissolved in 6.5 ml of 1.44 *M* Tris–HCl, pH 8.75, 0.7 ml of 1 *M* reagent grade semicarbazide hydrochloride (700 μmol) was added. The dark green solution immediately turned deep orange-red. To this solution, 4.8 g of solid guanidine hydrochloride (ultra pure grade, Schwarz/Mann Biotech, Cleveland, OH) was slowly added with stirring. The volume increased to 10 ml, and the concentration of guanidine hydrochloride was 5 *M*. The vessel containing the denatured enzyme was fitted with a rubber septum, and the system was flushed for 10 min with purified argon through inlet and outlet syringe needles. Two hundred microliters of reagent grade mercaptoethanol was added (2.84 mmol; 16.7 μmol $\times$ 14 $=$ 0.23 mmol of sulfhydryl groups from the small subunit present; total SH groups $=$ 3.07 mmol). After 2 hr under argon, 3 g (14.4 mmol) of solid sodium iodoacetate (approximately 80%; Sigma Chemical Co., St. Louis, No. I-4511) was slowly added to the solution and incubated at room temperature for 2 hr.

[24] E. J. Rodriquez and T. C. Bruice, *J. Am. Chem. Soc.* **111,** 7947 (1989).
[25] M. Mure, K. Nii, S. Itoh, and Y. Oshiro, *Bull. Chem. Soc. Jpn.* **63,** 417 (1990).

Step 2: Isolation of the Small Cofactor-Containing Subunit

The solution from step 1 was brought to 60 ml in 200 m*M* potassium phosphate, pH 7.0, and 2.5 *M* guanidine hydrochloride. The final concentration of guanidine hydrochloride was 2.5 *M*. The solution was split into three equal portions, and each was separately chromatographed with 200 m*M* potassium phosphate and 2.5 *M* guanidine hydrochloride, pH 7.0, on a 5 × 55-cm column containing Bio-Gel A-0.5m (200–400 mesh; Bio-Rad Laboratories, Hercules, CA). The flow rate was 25–30 min/tube, and 9.3-ml fractions were collected. Tubes 54–63 contained the large subunit, whereas tubes 74–83, containing the small subunit, were pooled.

The combined fractions of the small subunit were brought to 10% acetic acid (v/v) and were very slowly applied to a 1.5 × 12-cm Florisil (60–100 mesh; Fisher Scientific, Co., Fair Lawn, NJ) column equilibrated with 5% (v/v) acetic acid. (The Florisil was washed several times with 5% acetic acid in a flask, before pouring the column.) Some colored material did not bind. This was reapplied to a 0.9 × 5-cm column of Florisil. No colored material washed off in this step. Both columns were washed with 5% acetic acid, and no colored material leached off. After all the guanidine hydrochloride was washed away, the columns were washed with water to ~pH 6.0, and then most of the colored material was eluted with 20% (v/v) reagent grade pyridine. The pyridine eluant was lyophilized. The column was allowed to stand overnight in 20% pyridine and was reeluted with more of this eluant. A fair amount of color material eluted, which was combined with the lyophilized material, and this was relyophilized. The lyophilized material was treated with 1.5 ml of 20% pyridine to wash all the colored material into the bottom of the lyophilization vessel. The mixture was centrifuged to give a liquid phase and a fair size pellet consisting of remnant Florisil. The vessel and pellet were washed three more times, in the same manner, each time with 1 ml of 20% pyridine. The combined liquid was relyophilized to give a yellow powder.

Step 3: Pronase Treatment

The solid from step 2 was dissolved in 10.5 ml of 50 m*M* MOPS (3-[*N*-morpholino]propanesulfonic acid) plus 10 m*M* $CaCl_2$, pH 7.0. To this solution, 0.75 ml of 10 mg/ml Pronase CB protease from *Streptomyces griseus* was added (Grade A, No. 537011, Lot No. 023054, 120,000 proteolytic units/g from CalBiochem, Los Angeles, CA). The reaction was monitored by high pressure liquid chromatography (HPLC) on a TSK-DEAE PW column, 0.75 × 7.5 cm (Phenomenex, Torrance, CA), using a 30-min gradient of 20 to 300 m*M* ammonium phosphate, pH 7.0 ("Baker Analyzed" reagent grade, J. T. Baker, Phillipsburg, NJ) and 410 nm detection. The ammonium phosphate buffer was used so the counter cation for groups on peptides would be NH_4^+. (Potassium or sodium ions cause problems in mass spectral analyses, vide infra.) The mixture was incubated from 4.5 hr at 37°. The reaction was stopped by freezing.

The mixture was thawed and adjusted to pH 2.0 with H_3PO_4. Two milliliters of the solution was chromatographed on a PepRPC HR 5/5 reverse-phase column (Pharmacia LKB Biotechnology, Alameda, CA), with a gradient of water → 40% (v/v) acetonitrile in 20 min. [Both solvents were HPLC grade and both contained 0.1% (v/v) trifluoroacetic acid.] This procedure was repeated until all the sample was processed. The runs were monitored at 214 and 410 nm. Major cofactor-containing peptide fractions from each run, eluting from 14.5 to 17.5 min, were pooled (fraction F1). Minor cofactor-containing-peptide fractions (12–14.5 and 17.5–23 min) were pooled as one (fraction F2). Most cofactor-free peptides eluted before 12 min. The two pooled fractions were lyophilized.

Step 4: Leucine Aminopeptidase Treatment

Fractions F1 and F2 were dissolved in 10 and 1.5 ml of 20 m*M* potassium phosphate, pH 7.0, respectively. To F1, 750 μl of a 7.6-mg/ml solution of leucine amino peptidase (Sigma Chemical Co., No. L-0632, Lot No. 56F-0321, 13.6 units/mg) was added. Only 100 μl of the enzyme solution was added to F2. The digests were carried out at 37° and monitored by HPLC on the TSK-DEAE PW column as in step 3. The proteolysis was carried out for 26 hr. The fractions were processed by reverse-phase chromatography as described in step 3. Major cofactor-containing peaks, eluted from 13.7 to 17.5 min for the F1 digest, were combined (fraction Fa). F1 cofactor-containing peptide side fractions (12.6–13.7 and 17.5–21 min) were combined with cofactor-containing peptides derived from the F2 digest (12.1–20 min) (fraction Fb). Fractions Fa and Fb were lyophilized separately.

Step 5: Carboxypeptidase-Y Treatment

Fractions Fa and Fb were dissolved in 10 and 1.5 ml of 50 m*M* pyridinium acetate, pH 5.5, respectively. Five milligrams of carboxypeptidase-Y from Baker's yeast (Sigma Chemical Co., No. C-3888, Lot No. 42F-8150, 100 unit/mg) was dissolved in 1 ml of the pyridinium acetate buffer. Fraction Fa was treated with 0.75 ml of the enzyme solution, while Fb received 75 μl of the solution. The digests, which were carried out at 37° for 6 hr, were monitored by HPLC with the TSK-DEAE PW column as in step 3. The resulting digests were first chromatographed by consecutive 2- to 2.5-ml injections onto the TSK column (see step 3). The major cofactor-containing peptide eluted at 20 min. Next, the Fa and Fb aliquots containing the main peak were acidified to pH 2.0 with H_3PO_4 and chromatographed on the PepRPC column as in step 3. The main cofactor-containing peptide eluted at 13.9 min, and the fractions for Fa and Fb were combined and lyophilized. Chromatography on the TSK-DEAE column was repeated for the final step in purification. The sample of the pure cofactor-containing peptide was acidified to pH 2.0 with HCl, and then applied to a 0.5×4-cm column containing octadecylsilyl-derivatized silica gel (40-μm particle size, "bulk packing for flash chromatography," J. T. Baker, Inc.). The silica gel was washed extensively with a dilute solution of HCl, pH 2.0, and eluted with a 50:50 (v/v) mixture of this solution and methanol. This step removes all the salts and buffer from the sample. The eluant was diluted with water, lyophilized, and stored at $-70°$.

Analysis of the Pure Semicarbazide-Derivatized Bispeptidyl Cofactor

Amino Acid Composition and Sequence Analysis[16]

Amino acid analysis indicated that the major peptide obtained after pronase treatment had the following composition: Ala, carboxymethyl-Cys, Phe, Pro, Val_2, Ser_3. A careful amino acid analysis indicated that there was 4.09 μmol of the material. As we started with 16.7 μmol of the small subunit, the recovery of cofactor in this one peptide was 25%.

Amino acid sequence analysis provided the following results: cycle 1, Ser and Val; cycle 2, Pro; cycle 3, Ser and carboxymethyl-Cys; cycle 4, Ser and Phe; cycle 5, Ser(?); cycle 6, Val; cycle 7, Ala. It is clear from these results that the cofactor is attached at two sites to the polypeptide backbone of the small MADH subunit, thus confirming an earlier observation by Tobari's group for MADH from *M. extorquens* AM1.[26] By comparison of our sequence data to the sequence of the small subunit for MADH from *M. extorquens* AM1,[26] we deduced the sequence of the two peptides attached to the cofactor for the material from bacteria W3A1. The portions left after leucine amino peptidase and carboxypeptidase-Y treatment are shown in Fig. 3.

Mass Spectral Analysis

Sample preparation and fast atom bombardment mass spectral analyses are described in other publications.[5,16] At the time these experiments were done, we thought that the cofactor might be pyrroloquinoline quinone or a pyrroloquinoline quinone analog. As a control, fast atom bombardment

[26] Y. Ishii, T. Hase, Y. Fukumori, H. Matsubari, and J. Tobari, *J. Biochem.* (*Tokyo*) **93,** 107 (1983).

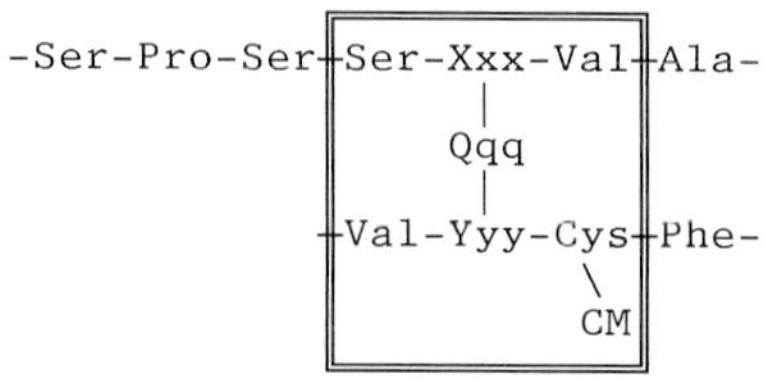

```
-Ser-Pro-Ser-|Ser-Xxx-Val|-Ala-
             |     |     |
             |    Qqq    |
             |     |     |
            -|Val-Yyy-Cys|-Phe-
             |          \|
             |         CM|
```

FIG. 3. The semicarbazide-derivatized bispeptidyl cofactor from bacterium W3A1 MADH obtained after treatment with pronase. Xxx and Yxx are the amino acyl groups bound to the semicarbazide-derivatized quinonoid cofactor, Qqq. CM represents the carboxymethyl group. The boxed portion is the bispeptidyl cofactor remaining after further treatment with leucine amino peptidase and carboxypeptidase-Y.

mass spectral (FAB-MS) data were collected for pyrroloquinoline quinone (Fluka Chem. Corp., Ronkonkoma, NY) and its semicarbazide derivative. Mass spectral analyses of pyrroloquinoline quinone provide a molecular weight of 332, two mass units higher than expected for this *o*-quinone. Pyrroloquinoline quinone picks up two hydrogen atoms from the matrix material when bombarded with 6 keV Xe atoms to convert it to the dihydroquinone. In contrast, semicarbazide-derivatized pyrroloquinoline quinone gave the correct $(M-H)^+$ mass of 388 early in an analysis, but as the analysis progressed, more of a $(M-H)^+$ peak with a mass of 390 appeared. Even the semicarbazide derivative is capable of being converted to a reduced dihydro form. For the semicarbazide derivative of pyrroloquinoline quinone, daughter ion spectra clearly indicated sequential loss of carboxyl groups. Very little fragmentation of the tricyclic pyrroloquinoline quinone ring system was detected.

A sample of a pure bispeptidyl cofactor obtained after pronase treatment provided a $(M-H)^+$ ion mass of 1429 early in the run, with progressively more of a 1431 peak appearing late in the analysis. This observation provides evidence for the quinone nature of the cofactor from MADH. The final bispeptidyl cofactor (step 5) provided a $(M-H)^+$ parent ion mass of 940. Both positive and negative ion daughter spectra displayed considerable fragmentation, some of which could be interpreted in terms of the amino acyl groups present, but there was no evidence for carboxyl groups as part of the cofactor. It was not possible to interpret the masses of all major fragments in terms of structural elements. The mass spectra showed considerable background ion intensity derived from the matrix, and spectra were complicated by adduct ions formed by the matrix material and contaminating sodium and potassium ions in the cofactor sample. Although there was substantial fragmentation of the known amino acyl portion, very little useful fragmentation of the cofactor had occurred. Peak matching FAB-MS offered the exact mass of 940.3262. This puts a very severe restriction on the elemental composition of the semicarbazide-derivatized bispeptidyl cofactor. Empirical formulae consistent with the exact mass measurement are: $C_{33}H_{58}N_{13}O_{15}S_2$, $C_{38}H_{58}N_{11}O_{13}S_2$, and $C_{41}H_{53}N_{11}O_{13}S$.

^1H and ^{13}C Nuclear Magnetic Resonance (NMR) Analyses

Large amounts of protein, as a source of an unidentified coenzyme, may not always be available for structural analysis by NMR. It is hoped that the development of more sensitive instrumentation will minimize this restriction. In the case where little material is available, isotopic enrichment of the enzyme can be considered for ^{13}C or ^{15}N NMR analyses. Although it is very expensive to grow a bacterium on enriched compounds, in some cases it may be the only method for definitive structural analysis by NMR. Even a small amount of highly enriched material is very helpful. Obviously, this would be impossible for an enzyme from a higher organism, unless the "native and active" protein can be overexpressed in a bacterial system from its cloned structural gene(s).[27]

It is fortunate that gram quantities of MADH can be purified from bacterium W3A1. (A typical purification provides 400–700 mg of enzyme.) Although we only had 4.09 μmol (3.84 mg) of semicarbazide-derivatized bispeptidyl cofactor, we were still able to obtain high quality NMR spectra. Complete details of the NMR experiments can be found in McIntire *et al.*[5] The material (step 5) was dissolved in dimethyl sulfoxide-d_6 (DMSO-d_6) or D_2O. One-dimensional ^1H NMR analysis of the material indicated that two conformations for this material were slowly interconverting. While this complicated the spectra somewhat, the NMR data were still interpretable.

Double quantum filtered correlated spectroscopy (^1H DQF COSY) allowed the identification of resonances associated with the two Val, the carboxylmethyl-Cys, and the Ser, and the identification of the α and β protons of the amino acyl residues connected to the cofactor. A set of five aromatic resonances conforming to the A(X)MP + A coupling pattern was also seen.[28] This pattern is the same as that seen for tryptophan, thus it was concluded that a tryptophyl residue was part of the bispeptidyl cofactor system. Since Trp was not detected during amino acid sequence analysis, it was concluded that Trp was attached to the cofactor in some fashion.

[27] A. Bax, *Annu. Rev. Biochem.* **58,** 223 (1989); D. C. Muchmore, L. P. McIntosh, C. B. Russell, D. E. Anderson, and F. W. Dahlquist, this series, Vol. 177, p. 44; D. W. Hibler, L. Harpold, M. Dell'Acqua, T. Pourmotabbed, J. A. Gerlt, J. A. Wilde, and P. H. Bolton, *ibid.*, p. 74.

[28] K. Wüthrich, "NMR of Proteins and Nucleic Acids." Wiley (Interscience), New York, 1986.

One-dimensional NOE (nuclear Overhauser effect) and NOESY (two-dimensional nuclear Overhauser and exchange spectrometry) spectra offer information regarding the spatial disposition of protons relative to one another. Unfortunately, this only confirmed what was already known about the structure.

Finally, natural abundance ^{13}C NMR spectra of the sample dissolved in DMSO-d_6 were recorded. A distortionless enhancement polarization transfer sequence was used to identify all hydrogen-bearing carbons, and a proton-decoupled ^{13}C spectrum provided the chemical shifts of all carbon centers. The carbon count was consistent with only one of the predicted empirical formuli for the semicarbazide-derivatized bispeptidyl cofactor, i.e., $C_{41}H_{53}N_{11}O_{13}S$. When the contributions of all confirmed atoms (i.e., those from carboxymethyl-Cys, Ser, Val, the α and β portion of the amino acyl groups attached to the cofactor, and the semicarbazone moiety) were eliminated, the remaining formula is $C_{16}H_8N_2O_2$. To summarize, the ^1H NMR experiments indicated the presence of a benzenoid ring with four adjacent protons, as observed for Trp, two isolated, unsaturated C–H groups, and two heterocyclic ring N–H groups in the cofactor. On the other hand, the ^{13}C NMR data revealed that the cofactor contains two carbonyl groups (evidence for the quinonoid nature of the cofactor) and eight unsaturated carbon centers. Although we had all the necessary information to deduce the structure of the cofactor and were certain that the cofactor was "attached" to a tryptophyl residue, we did not consider that the cofactor could be constructive from two cross-linked tryptophyl residues until the DNA-derived amino acid sequence was in hand.

Cloning and Sequencing of the Structural Gene of the Small Subunit of MADH for *M. extorquens* AM1

A definitive way to confirm an amino acid sequence and to establish the identity of the precursor to a modified amino acyl residue is to sequence the structural gene for the protein in question. The amino acid sequence for the small subunit of MADH from *M. extorquens* AM1 had been determined by chemical methods in 1983.[26] All amino acid residues in the sequence were tentatively identified, except for those at positions 55 and 106. It was concluded that these residues were covalently bound to the quinonoid cofactor. In 1990 the structural gene of the small subunit for MADH from *M. extorquens* AM1 was cloned and sequenced by Chistoserdov *et al.*[4] A comparison of the translated amino acid sequence with the earlier protein sequence revealed that Asn_{17} is in fact Asp_{17} and that positions 55 and 106 are occupied by Trp residues. Chistoserdov *et al.*[4] surmised that the two Trp residues were the amino acyl linkers to the cofactor. As mentioned in the previous section, the presence of one tryptophyl residue was expected, but the presence of two was quite surprising. With this essential information, the true nature of the cofactor immediately was deduced from the amino acid sequence, NMR, and mass spectral data. It was clear that the two Trp residues were not just the links to the cofactor, but constituted the cofactor itself. ^{1}H NOESY NMR data were the most important for determining the correct structure of the cofactor (Fig. 1).[5] This structure (in the semicarbazone form) has the correct mass and empirical formula and all of the structural elements predicted by the mass spectral and NMR data. Most interesting are its quinonoid nature and the unique cross-linking of the indole rings.

Confirmation of the Structure of the Quinone Cofactor of MADH

While quite certain that we have arrived at the true structure of the cofactor, it was possible that an isomer of this structure was the correct one. The first confirmation of the structure in Fig. 1 came from X-ray crystallographic electron density patterns at the active sites of MADH from *P. denitrificans* and *T. versutus*.[29] The skeleton of the structure displayed in Fig. 1 easily conformed to the electron density at the suspected cofactor site. This structure is also accommodated in the X-ray map of MADH from bacterium W3A1 (F. S. Mathews, personal communication).

The synthesis of the cofactor provided final, persuasive conformation of the proposed structure.[21] NOE, UV-visible, resonance Raman, and X-ray analyses were absolutely consistent with the structure shown in Fig. 1. The ε_{440} for TTQ in MADH from bacterium W3A1 is 11,760 M^{-1}cm^{-1} (W. S. McIntire, unpublished), ε_{407} for the synthetic cofactor in acetonitrile is 10,700 M^{-1}cm^{-1},[21] and ε_{410} for the cofactor in the small MADH subunit when dissociated from the large subunit is 11,000 M^{-1}cm^{-1}.[14] The same general spectral features were observed for the last two cases. The redox potential for the synthetic material is 56 mV (NHE, pH 7.4), whereas it is 96 mV (pH 7.5)[30] and 100 mV (pH 7.5)[31] for TTQ bound to MADH from bacterium W3A1 and *P. denitrificans*, respectively.

Final Comments

Without a great deal of luck, no one method will reveal the unambiguous structure for an unknown substance found in nature. In the example present

[29] L. Chen, F. S. Mathews, V. L. Davidson, E. G. Huizinga, F. M. D. Vellieux, J. A. Duine, and W. G. J. Hol, *FEBS Lett.* **287,** 163 (1991).

[30] A. L. Burrows, H. A. O. Hill, T. A. Leese, W. S. McIntire, H. Nakayama, and G. S. Sanghera, *Eur. J. Biochem.* **199,** 73 (1991).

[31] M. Husain, V. L. Davidson, K. A. Gray, and D. B. Knaff, *Biochemistry* **26,** 4139 (1987).

in this chapter, some aspects of the basic properties of the MADH cofactor were known, allowing numerous biochemists to deduce that it was a quinone-type cofactor (e.g., its reaction with carbonyl reagents and its redox properties). However, these properties are also features of known redox cofactors, i.e., riboflavin. By using a variety of methods, each providing exact structural information, it was possible to deduce the true structure for TTQ. Perhaps we would have arrived at the structure without the results from gene sequencing, but this information greatly accelerated the process, showing that sophisticated, expensive, and analytical instrumentation are not always sufficient.

Acknowledgments

The research by the author was supported by a Department of Veterans Affairs Merit Review Grant, Program Project Grant HL-16251 from The Heart, Lung, and Blood Institute of The National Institutes of Health, National Science Foundation Grant MCB-9206952, and a Research and Education Allocation Grant from the University of California, San Francisco, California. The author thanks Vladislov Kuusk for writing the factor analysis program for UV-visible spectral data.

[13] Model Studies of Cofactor Tryptophan Tryptophylquinone

By SHINOBU ITOH and YOSHIKI OHSHIRO

Introduction

Tryptophan tryptophylquinone (TTQ) is a novel amino acid-derived cofactor that was found in bacterial methylamine dehydrogenases (EC 1.4.99.3, MADH) in 1991.[1,2,2a] It has a unique heterocyclic *o*-quinone structure of 6,7-indolequinone with a 2-indolyl group at the 4-position. This cofactor is the redox center of MADH that catalyzes the oxidation of methylamine to formaldehyde and ammonia and donates two electrons to

[1] W. S. McIntire, D. E. Wemmer, A. Chistoserdov, and M. E. Lidstrom, *Science* **252,** 817 (1991).
[2] L. Chen, F. S. Mathews, V. L. Davidson, E. G. Huizinga, F. M. D. Vellieux, J. A. Duine, and W. G. J. Hol, *FEBS Lett.* **287,** 163 (1991).
[2a] L. Chen, F. S. Mathews, V. L. Davidson, E. G. Huizinga, F. M. D. Vellieux, and W. G. J. Hol, *Proteins: Struct., Funct., Genet.* **14,** 288 (1992).

a blue copper protein, amicyanin.[3] For the enzymatic amine oxidation reaction, a transamination mechanism has been proposed[1,4,5] that has been confirmed by an EPR study showing that the substrate nitrogen atom is incorporated into the substrate-reduced TTQ.[6] However, much yet remains to be done to fully understand the chemistry of TTQ at a molecular level.

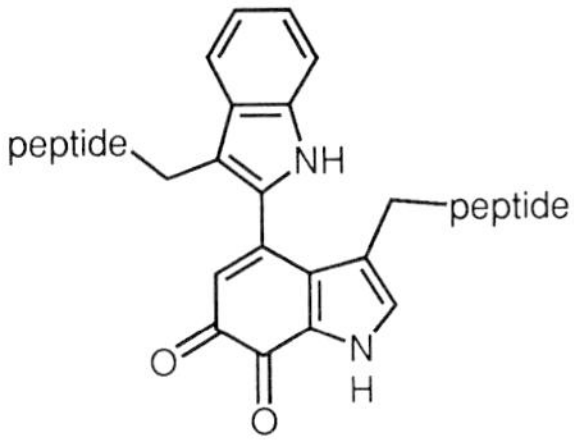

TTQ

TTQ is derived from two tryptophan residues of the MADH subunit peptide chain[7]; thus it is tightly associated in the enzyme matrix through an amide linkage. This makes it difficult to isolate the cofactor intact and to perform mechanistic studies of the redox reaction. This chapter demonstrates model studies of cofactor TTQ in order to obtain further information on the structure and reactivity of the active site cofactor of MADH.[8] The biosynthetic pathway of TTQ has also been investigated from the viewpoint of organic chemistry.

Synthesis of the TTQ Model Compound

Model compound **1**, which has exactly the same ring skeleton of cofactor TTQ except that the two peptide chains at the 3- and 3′-positions are

[3] L. Chen, R. Durley, B. J. Poliks, K. Hamada, Z. Chen, F. S. Mathews, V. L. Davidson, Y. Satow, E. Huizinga, F. M. D. Vellieux, and W. G. J. Hol, *Biochemistry* **31,** 4959 (1992); L. Chen, F. S. Mathews, V. L. Davidson, M. Tegoni, C. Rivetti, and G. L. Rossi, *Protein Sci.* **2,** 147 (1993).

[4] G. Backes, V. L. Davidson, F. Huitema, J. A. Duine, and J. Sanders-Loehr, *Biochemistry* **30,** 9201 (1991).

[5] V. L. Davidson, L. H. Jones, and M. E. Graichen, *Biochemistry* **31,** 3385 (1992); V. L. Davidson and L. H. Jones, *Biochim. Biophys. Acta* **1121,** 104 (1992); H. B. Brooks, L. H. Jones, and V. L. Davidson, *Biochemistry* **32,** 2725 (1993).

[6] K. Warncke, H. B. Brooks, G. T. Babcock, V. L. Davidson, and J. McCracken, *J. Am. Chem. Soc.* **115,** 6464 (1993).

[7] A. Y. Chistoserdov, Y. D. Tsygankov, and M. E. Lidstrom, *Biochem. Biophys. Res. Commun.* **172,** 211 (1990).

[8] Preliminary results have been published in S. Itoh, M. Ogino, M. Komatsu, and Y. Ohshiro, *J. Am. Chem. Soc.* **114,** 7294 (1992).

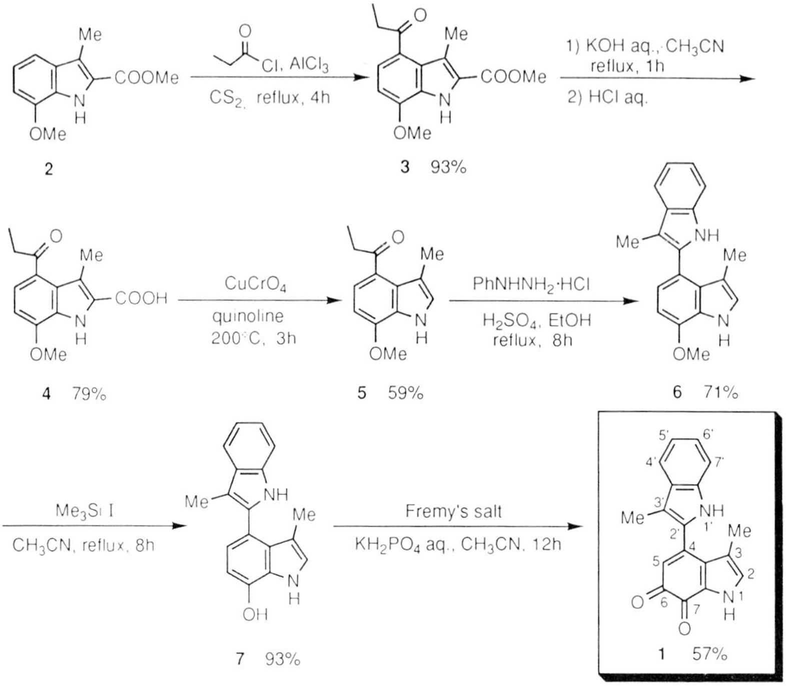

SCHEME 1

displaced by simple methyl groups, was designed and synthesized as follows (Scheme 1). Friedel-Crafts acylation on indole derivative **2** (2.4 mmol)[9] with propionyl chloride (7.5 mmol) by the standard method using $AlCl_3$ (4.8 mmol) as a catalyst in refluxing CS_2 (40 ml) for 4 hr gave the 4-acylated derivative **3** in 93%. The position of the acylation in **3** was confirmed by ^{1}H NMR; a large NOE (nuclear Overhauser effect) (18%) was detected on H(6) with irradiation at 7-OMe. When the Friedel-Crafts acylation was carried out on 7-methoxyskatole, a propionyl group was introduced predominantly at the 2-position. Indole derivative **3** was then converted into **5** by ester hydrolysis [KOH in CH_3CN–H_2O (1 : 1), refluxed for 1 hr 79%] followed by thermal decarboxylation using $CuCrO_4$ in quinoline at 200° for 3 hr (71%). When the thermal decarboxylation was carried out at a higher temperature (240°) in glycerol, deacylation occurred together with the decarboxylation to afford 7-methoxyskatole. The second indole ring

[9] The starting material **2** was prepared by the reported procedure; K. G. Blaikie and W. H. Perkin, *J. Chem. Soc.* **125,** 269 (1924).

was then constructed by the Fischer indolization reaction on **5** (1.7 mmol) with phenylhydrazine hydrochloride (2.6 mmol) in refluxing ethanol (20 ml) containing H_2SO_4 (1 ml) for 8 hr (71%). Deprotection of the methoxy group of **6** (0.78 mmol) by trimethylsilyl iodide (8.4 mmol) in refluxing CH_3CN (10 ml) gave the 7-hydroxy derivative **7** (93% yield), which was finally converted to the expected quinone **1** by oxidation with Fremy's salt (4 equiv) in CH_3CN–KH_2PO_4 aqueous solution in a 57% yield. Each product was purified by flash column chromatography on SiO_2.

Compound **1** was readily reduced to the quinol derivative (**1H₂**) by treatment with methylhydrazine (10 equiv) in CH_3CN under anaerobic conditions (Ar). Removal of the solvent under reduced pressure gave a yellow residue from which **1H₂** was precipitated by adding *n*-hexane in 73% [Eq. (1)].

$$\mathbf{1} \xrightarrow[\text{under Ar}]{\text{MeNHNH}_2\text{ / MeCN}} \mathbf{1H_2} \qquad (1)$$

Physicochemical Properties of Model Compound 1

Some spectral characteristics of the model compound are worth noting (Table I). In the mass spectrum, a strong peak of M + 2, which is a

TABLE I
ANALYTICAL DATA OF COMPOUND **1**

mp	>300°
MS (EI) *m/z*	290 (M^+)
	292 (M^+ + 2, characteristic peak for *o*-quinone compounds)
IR (ν, cm^{-1})[b]	3412 (NH), 1628 (C=O)
^{1}H-NMR[a] (δ, ppm)	1.53 (3 H, s, 3-Me), 2.31 (3 H, s, 3′-Me), 5.86 (1 H, s, 5-H), 7.04 (1 H, t, J = 7.9 Hz, 5′-H), 7.15 (1 H, s, 2-H), 7.15 (1 H, t, J = 7.9 Hz, 6′-H), 7.35 (1 H, d, J = 7.9 Hz, 7′-H), 7.56 (1 H, d, J = 7.9 Hz, 4′-H), 11.24 (1 H, br s, 1′-H), 12.74 (1 H, br s, 1-H)
^{13}C-NMR[a] (δ, ppm)	9.4 (-Me), 10.7 (-Me), 110.0, 111.6, 119.1, 119.3, 121.4, 122.8, 123.6, 127.4, 128.4, 128.7, 130.3, 130.7, 136.4, 144.2 (14 sp^2 carbons), 167.3 (d, $^2J_{CH}$ = 4.9 Hz, 6-C), 183.1 (s, 7-C)

[a] In DMSO-d_6.
[b] KBr disk.

characteristic one for *o*-quinone compounds,[10] is seen at 292 together with the parent peak at 290. A strong IR absorption of carbonyl stretching of the quinone function appears at 1628 cm^{-1}, which is lower by about 50 cm^{-1} than that of ordinary *o*-quinones.[11] This result may indicate that the conjugation between the *o*-quinone group and the indole ring and C(4)–C(5) double bond is relatively strong. In the ^{13}C NMR, quinone carbonyl carbons C(6) and C(7) appear at 167.3 and 183.1 ppm, respectively. The chemical shift of C(6) is characteristically higher than that of normal quinones, also indicating a relatively higher degree of conjugation, particularly at the C(6) carbonyl group. The ^{1}H-NMR spectra in dimethy sulfoxide (DMSO) fitted the structure as well. The signal pattern of the aromatic protons of the second indole ring and the vinyl proton at the 5-position is similar to that of the hydrazone derivative of the cofactor-bearing peptide.[1] It is interesting to note that the chemical shift of 3-Me is relatively higher than that of 3′-Me. This could be attributed to the molecular geometry of the model compound as discussed next.

X-ray crystallographic investigation of MADHs from *Paracoccus denitrificans* and *Thiobacillus versutus* indicated that the dihedral angle of the two indole rings of TTQ is about 42°.[2] Molecular orbital calculations by the AM1 method indicated that the dihedral angle of the two indole rings, defined by C(5)–C(4)–C(2′)–C(3′), in the optimized structure of compound **1** is 46.9° and the distances between the protons of 3′-Me and those of 3-Me and 5-H are about 4.7 and 2.7 Å, respectively (Fig. 1A).[12] This molecular geometry was also suggested by the observed NOE correlation for **1** (Fig. 1B). A NOE of approximately 4% was detected between the two methyl groups, but this value is relatively small compared to that between 3′-Me and 5-H (20%). The upfield shift of the methyl group at the 3-position (vide ante) is due to the ring current effect by the second indole moiety which exists just below 3-Me. In the case of molecular mechanic calculations on compound **1**, minimum steric energy was obtained when the dihedral angle of the two indole rings was about 55°, although the difference in the steric energy is within 1 kcal/mol at a dihedral angle from 40 to 120° (Fig. 2).[13] It is interesting to note that the molecular

[10] K.-P. Zeller, *in* "The Chemistry of the Quinonoid Compounds. Part 1" (S. Patai, ed.), p. 231. Wiley, London, 1974.

[11] St. Berger and A. Rieker, *in* "The Chemistry of the Quinonoid Compounds. Part 1" (S. Patai, ed.), p. 193. Wiley, London, 1974.

[12] Molecular orbital calculations were performed with the MOPAC program (version 6.1, AM1 method) using a CAChe WorkSystem (SONY Tektronix); M. J. S. Dewar, E. G. Zoebisch, E. F. Healy, and J. J. P. Stewart, *J. Am. Chem. Soc.* **107,** 3902 (1985).

[13] Molecular mechanics calculations were performed using a CAChe WorkSystem (SONY Tektronix).

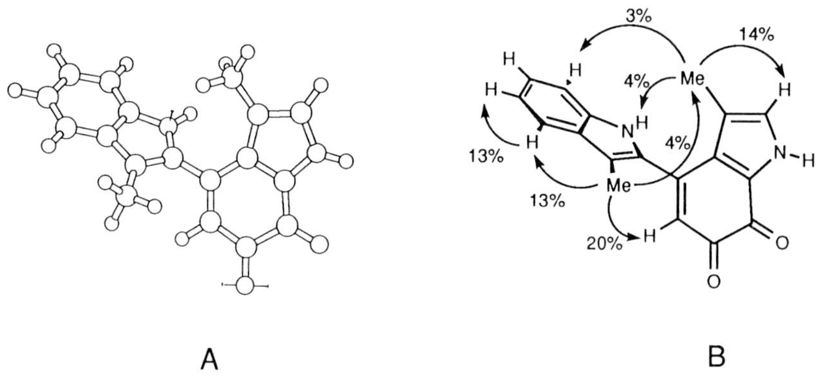

FIG. 1. (A) Fully optimized structure of **1** by AM1. The dihedral angle of the two indole rings defined by C(5)–C(4)–C(2′)–C(3′) is 46.9°. (B) Selected NOE correlations for **1**.

geometry of TTQ in the enzyme active site is almost adjusted to that having the minimum steric energy of the molecule despite the existing steric restriction through space and/or through peptide bonding around the enzyme active site.

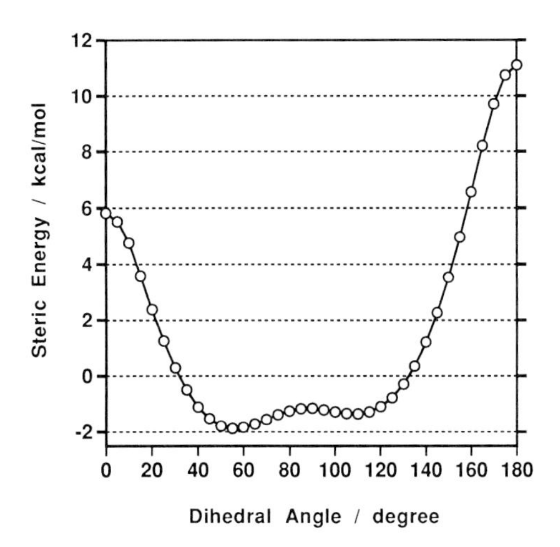

FIG. 2. Plot of steric energy of **1** vs the dihedral angle of the two indole rings defined by C(5)–C(4)–C(2′)–C(3′).

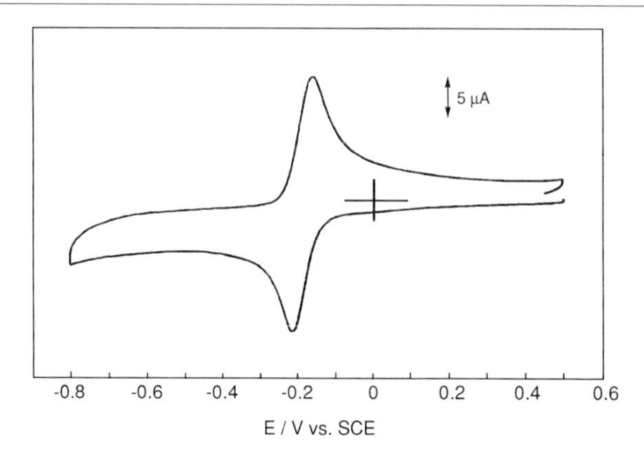

FIG. 3. Cyclic voltammogram of **1** in a 0.1 *M* phosphate buffer solution containing 30% CH_3CN (pH 7.4) at a scan rate of 100 mV/sec.

Figure 3 shows a cyclic voltammogram of compound **1** obtained in a 0.1 *M* phosphate buffer solution containing 30% CH_3CN (pH 7.4) at a scan rate of 100 mV/sec. Compound **1** gave a pair of cathodic and anodic waves with a midpoint potential of −188 mV vs SCE. This value is comparable to that of native MADH from bacterium W3A1 ($E_{1/2} = -148$ mV vs SCE at pH 7.5)[14] and is almost the same as that of another type of heterocyclic *o*-quinone cofactor, PQQ (pyrroloquinolinequinone, $E_{1/2} = -187$ mV vs SCE at pH 7.2).[15]

Compound **1** shows a strong absorption at 407 nm ($\varepsilon = 1.07 \times 10^4$ M^{-1} cm^{-1}) with a broad shoulder at 500–650 nm in CH_3CN (Fig. 4); thus the color of the solution is reddish-brown. Reduction of **1** to **1H$_2$** by methylhydrazine caused the complete disappearance of the absorptions but gave a new one at 306 nm ($\varepsilon = 1.62 \times 10^4$ $M^{-1}cm^{-1}$). The spectra of **1** and **1H$_2$** resemble those of MADH from *P. denitrificans* in shape ($MADH_{ox}$; 440 nm; $MADH_{red}$; 326 nm).[16] In the resonance Raman spectrum, there is also a good spectral similarity between model compound **1** and the TTQ cofactor in the enzyme (Fig. 5).[4,17]

[14] A. L. Burrows, H. A. O. Hill, T. A. Leese, W. S. McIntire, H. Nakayama, and G. S. Sanghera, *Eur. J. Biochem.* **199,** 73 (1991).

[15] K. Kano, K. Mori, B. Uno, T. Kubota, T. Ikeda, and M. Senda, *J. Electrochem.* **299,** 193 (1990).

[16] M. Husain and V. L. Davidson, *Biochemistry* **26,** 4139 (1987).

[17] W. S. McIntire, J. L. Bates, D. E. Brown, and D. M. Dooley, *Biochemistry* **30,** 125 (1991).

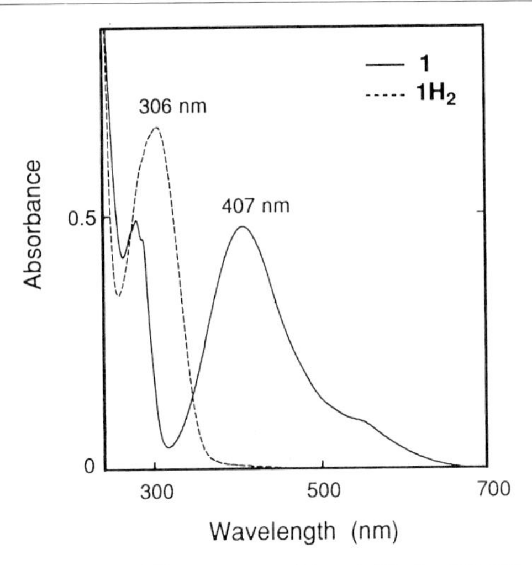

FIG. 4. UV-visible spectra of **1** and **$1H_2$** in CH_3CN.

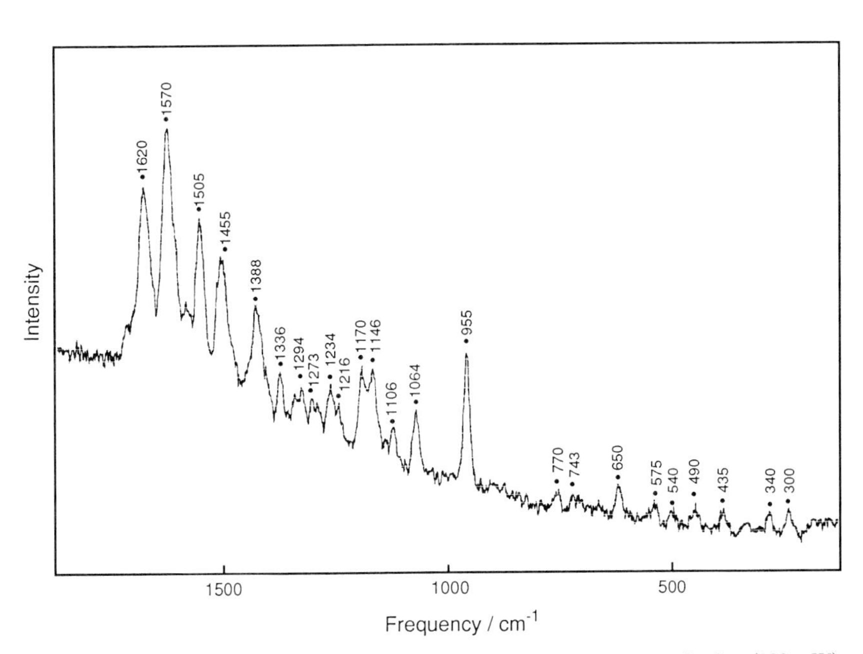

FIG. 5. Resonance Raman spectrum of **1** obtained by using 457.9 nm excitation (100 mW) (KBr disk sample).

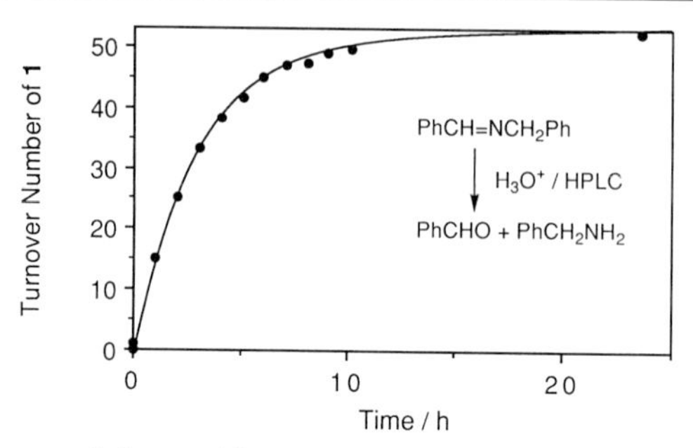

FIG. 6. Time course of the aerobic oxidation of benzylamine (100 m*M*) catalyzed by **1** (1 m*M*) in CH_3OH at room temperature under O_2 atmosphere.

Oxidation of Benzylamine by 1

The catalytic efficiency of model compound **1** was examined via the aerobic oxidation of benzylamine. The oxidation of benzylamine (100 m*M*) was carried out under O_2 atmosphere in the presence of a catalytic amount (1 m*M*) of compound **1** in CH_3OH. The ^1H NMR and IR spectra of the concentrated final reaction mixture indicated that *N*-benzylidenebenzylamine ($PhCH_2{-}N{=}CHPh$) had been formed almost quantitatively [Eq. (2)]. Figure 6 shows the time course of the oxidation reaction, which

$$\underset{(100\ \text{mM})}{PhCH_2NH_2} \xrightarrow[\text{aerobic condition}]{\mathbf{1}\ (1\ \text{mol\%}),\ \text{MeOH, rt, 30 h}} \underset{\text{quant.}}{PhCH{=}NCH_2Ph} \tag{2}$$

was obtained by monitoring the formation of the oxidation product (PhCHO) by HPLC [pump: Hitachi L-6000; UV detector: Hitachi L-3000; column: Waters Radial Compression Separation System (C_{18}); eluent: $CH_3CN/H_2O/H_3PO_4$, 100/100/1 (v/v/v), PhCHO was readily generated in the HPLC column by the acid hydrolysis of *N*-benzylidenebenzylamine]. As clearly shown, compound **1** acts as a very efficient turnover catalyst, indicating that it possesses the same chemical function of methylamine dehydrogenases. Such an efficient amine oxidation does not proceed with ordinary quinones but with another type of heterocyclic *o*-quinone cofactor, PQQ.[18,19] Detailed studies of the model systems will provide much informa-

[18] E. J. Rodriguez and T. C. Bruice, *J. Am. Chem. Soc.* **111,** 7947 (1989).

[19] S. Itoh, M. Mure, M. Ogino, and Y. Ohshiro, *J. Org. Chem.* **56,** 6857 (1991); S. Itoh, Y. Fukui, M. Ogino, S. Haranou, M. Komatsu, and Y. Ohshiro, *ibid.* **57,** 2788 (1992); S. Itoh, Y. Fukui, S. Haranou, M. Ogino, M. Komatsu, and Y. Ohshiro, *ibid.*, p. 4452.

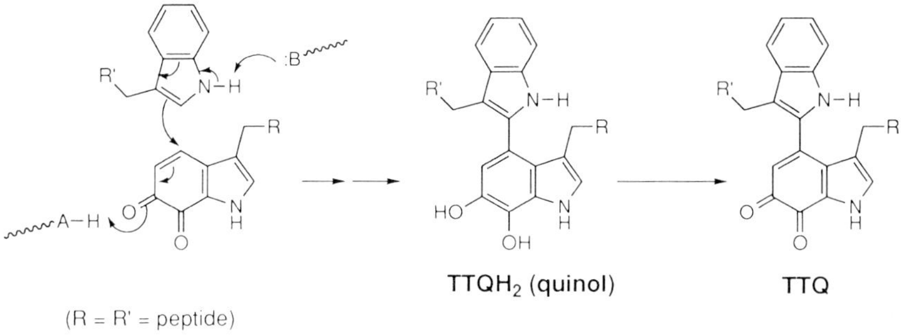

TTQH$_2$ (quinol)   TTQ

(R = R' = peptide)

SCHEME 2

tion on the amine oxidation mechanism, such as the addition position of the substrate, the structure and physicochemical properties of the intermediates, and the structure–reactivity relationships of the cofactor as has been done on PQQ and TPQ cofactors.[18–21] Preliminary results of the anaerobic reactions of model compound **1** and several amines indicated that the amine oxidation proceeds via a transamination mechanism through an iminoquinone intermediate[22] as has been proposed for the enzymatic reaction.[4,5]

Model Studies of TTQ Biosynthesis

TTQ is thought to be formed via post-translational modification of two tryptophan residues of the MADH subunit peptide chain,[7] but little is known about the mechanistic details. It has been speculated that one of the precursors is a 6,7-indolequinone derivative of the tryptophan that combines with another tryptophan residue as illustrated in Scheme 2.[1] Here we demonstrate a simple synthesis of model compound **1** by mimicking the speculated biosynthetic pathway that involves the coupling reaction between a 6,7-indolequinone derivative and an indole derivative itself.

The 6,7-indolequinone derivative of tryptophan might be produced from 6- or 7-hydroxytryptophan that could be derived via direct hydroxylation of the tryptophan residue. However, regioselective hydroxylation at the 6-

[20] S. Itoh, M. Ogino, Y. Fukui, H. Murao, M. Komatsu, Y. Ohshiro, T. Inoue, Y. Kai, and N. Kasai, *J. Am. Chem. Soc.* **115,** 9960 (1993).

[21] M. Mure and J. P. Klinman, *J. Am. Chem. Soc.* **115,** 7117 (1993).

[22] Unpublished results. Details will be reported elsewhere.

or 7-position of indoles has not been reported so far.[23] In the Fenton or Udenfriend reaction, tryptophan is unselectively converted into many products.[24,25] Direct hydroxylation of the benzene ring of indole derivatives was reported to occur in the reaction with hydrogen peroxide in superacids, but the regioselectivity was poor.[26] Enzymatic oxygenation of a 3-substituted indole such as skatole with horseradish peroxidase or chloroperoxidase was examined in 0.1 *M* phosphate buffer solution (pH 6.6) containing 20% CH_3CN at 30° according to the reported procedure,[27] and the reaction mixture was analyzed by GC-MS and ^{1}H-NMR. However, ring-opening and dimeric products were mainly obtained. These results may indicate that regioselectivity of the oxygenation is controlled to a large extent by three-dimensional construction of the MADH active site, whether this process is enzymatic or nonenzymatic.

In order to study the next step, 6- and 7-hydroxyskatoles (**8** and **9**) were employed as the model compounds of 6- and 7-hydroxytryptophans, respectively. 6-Hydroxyskatole (**8**) was obtained by deprotection (with iodotrimethylsilane) of the methoxy group of 6-methoxyskatole which was prepared by the reported procedure from *m*-anisidine.[28] 7-Hydroxyskatole (**9**) was obtained from 2-carbomethoxy-3-methyl-7-methoxyindole via ester hydrolysis, thermal decarboxylation, and deprotection of the methoxy group as in the case of compound **1** (Scheme 1). Oxidation of 6-hydroxyskatole (**8**) with Fremy's salt $[(KO_3S)_2NO\cdot]$ by the standard procedure (vide ante) gave the expected *o*-quinone derivative **10** in a 7% yield [Eq. (3)].

Me HO N H **8** —Fremy's salt→ Me O O N H **10** + Me O N H =O **11** (3)

Unexpectedly, a relatively large amount of the 2,6-dioxo derivative **11** was also produced (63%). The same treatment on **9** gave *o*-quinone **10** together with *p*-quinone **12** in 11 and 64% yields, respectively [Eq. (4)]. These prod-

[23] W. A. Remers and T. F. Spande, *in* "The Chemistry of Heterocyclic Compounds: Indoles. Part 3" (W. J. Houlihan, ed.), p. 9. Wiley, New York, 1979.

[24] K. Uchida, N. Enomoto, K. Itakura, and S. Kawakishi, *Arch. Biochem. Biophys.* **279,** 14 (1990).

[25] Z. Maskos, J. D. Rush, and W. H. Koppenol, *Arch Biochem. Biophys.* **296,** 514 (1992).

[26] C. Berrier, J.-C. Jacquesy, M.-P. Jouannetaud, and A. Renoux, *Tetrahedron Lett.* **27,** 4565 (1986).

[27] S. Colonna, N. Gaggero, G. Carrea, and P. Pasta, *J. Chem. Soc., Chem. Commun.*, p. 357 (1992).

[28] J. E. Nordlander, D. B. Catalane, K. D. Kotain, R. M. Stevens, and J. E. Haky, *J. Org. Chem.* **46,** 778 (1981).

TABLE II
ANALYTICAL DATA OF COMPOUNDS **10**, **11**, and **12**

10	mp: >300°
	MS (EI): *m/z* 161 (M^+)
	HRMS: *m/z* 161.0471 (M^+) calcd. for $C_9H_7O_2N$: 161.0477
	IR (KBr): 3180 (NH), 1632 cm^{-1} (C=O)
	^{1}H-NMR (DMSO-d_6): δ 2.07 (3 H, s, Me), 5.85 (1 H, d, *J* = 9.7 Hz, H-5), 7.09 (1 H, s, H-2), 7.45 (1 H, d, *J* = 9.7 Hz, H-4), 12.50 (1 H, br s, NH)
	^{13}C-NMR (DMSO-d_6): 9.0 (3-Me), 120.5, 122.8, 127.9, 128.7, 129.6, 137.5 (6 aromatic carbons), 166.6 and 184.2 ppm (quinone carbonyl carbons)
11	mp: >300°
	MS (EI): *m/z* 161 (M^+)
	IR (KBr): 3284 (NH), 1728, 1692, 1670, 1634 cm^{-1} (C=O and/or C=C)
	^{1}H-NMR ($CDCl_3$): δ 2.14 (3 H, s, Me), 5.80 (1 H, s, H-7), 6.20 (1 H, d, *J* = 9.9 Hz, H-5), 7.13 (1 H, br s, NH), 7.21 (1 H, d, *J* = 9.9 Hz, H-4)
12	mp: 204–205°
	MS (EI): *m/z* 161 (M^+)
	IR (KBr): 3212 (NH), 1644 cm^{-1} (C=O)
	^{1}H-NMR (DMSO-d_6): δ 2.22 (3 H, s, Me), 6.55 (1 H, d, *J* = 10.2 Hz, H-5 or H-6), 6.61 (1 H, d, *J* = 10.2 Hz, H-5 or H-6), 7.06 (1 H, s, H-2), 12.50 (1 H, br s, NH)
	^{13}C-NMR (DMSO-d_6): 10.6 (3-Me), 120.6, 122.4, 125.2, 130.4, 136.4, 137.7 (6 aromatic carbons), 176.6 and 184.2 ppm (quinone carbonyl carbons)
	Elemental Anal.: Calcd. for $C_9H_7O_2N$: C, 67.07; H, 4.38; N, 8.69.
	Found: C, 67.02; H, 4.34; N, 8.49.

9 —Fremy's salt→ **10** + **12** (4)

ucts were separated by flash column chromatography on SiO_2 and their spectral data are listed in Table II. In solution, the regioselectivities, 2,6- vs 6,7- and 4,7- vs 6,7-, may be determined by both the electron spin density of the radical intermediates that are formed during the Fremy's salt oxidation[29] and the thermal stability of the products, but in the enzymatic system the reaction course might be restricted by the steric environment of the enzyme active site to produce 6,7-indolequinone selectively.

The coupling reaction between *o*-quinone **10** (0.048 mmol) and skatole (0.048 mmol) itself was then examined in CH_3CN (5 ml) at room temperature under N_2. No reaction took place in the absence of any additive, but

[29] Details about the mechanism of Fremy's salt oxidation, see R. H. Thomson, *in* "The Chemistry of the Quinonoid Compounds, Part 1" (S. Patai, ed.), p. 111. Wiley, London, 1974.

$AlCl_3$ (5 equiv) was found to function as a good catalyst to afford the expected model compound **1** in 58% yield [^{1}H-NMR, Eq. (5)]. The primary

$$\mathbf{10} + \text{skatole} \xrightarrow[\text{2) } O_2]{\text{1) } AlCl_3 / MeCN} \mathbf{1} \quad (5)$$

product of the reaction must be reduced compound $\mathbf{1H_2}$, but it is easily oxidized by molecular oxygen to generate quinone **1** during the workup treatment. Details of the function of $AlCl_3$ are not clear now, but it could be speculated that $AlCl_3$ coordination to the quinone function activates the 4-position of **10** toward the nucleophilic addition of skatole. $AlCl_3$ might also form a complex with skatole at the N-1 position, which enhances the nucleophilicity at the 2-position of skatole.

Consequently, the model compound of novel cofactor TTQ of methylamine dehydrogenase was synthesized very simply by mimicking the postulated biosynthetic route to TTQ. The present results would suggest the probability of the hypothesis shown in Scheme 2, although the regioselective hydroxylation of 3-substituted indoles and the effective 6,7-indolequinone formation have not been achieved yet.

Further studies of the chemistry of TTQ model compounds will provide more information about the physicochemical properties, the amine oxidation mechanism, biosynthesis, structure–reactivity relationships, and so on.

[14] Detection of Intermediates in Tryptophan Tryptophylquinone Enzymes

By VICTOR L. DAVIDSON, HAROLD B. BROOKS, M. ELIZABETH GRAICHEN, LIMEI H. JONES, and YOUNG-LAN HYUN

Introduction

Properties of TTQ Enzymes

Tryptophan tryptophylquinone (TTQ) has been found thus far in two enzymes: methylamine dehydrogenase (MADH)[1] and aromatic amine de-

[1] V. L. Davidson, *in* "Principles and Applications of Quinoproteins" (V. L. Davidson, ed.), p. 73. Dekker, New York, 1993.

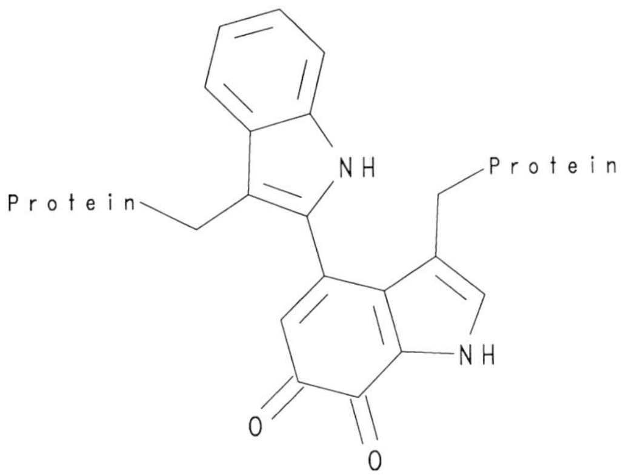

FIG. 1. Tryptophan tryptophylquinone.

hydrogenase (AADH).[2] MADHs have been isolated from several gram-negative methylotrophic and autotrophic bacteria. AADH has been isolated from *Alcaligenes faecalis*. These enzymes catalyze the oxidation of primary amines to their corresponding aldehydes plus ammonia. The two electrons which are derived from this oxidation are also transferred by the enzyme to some electron acceptor:

$$R{-}CH_2NH_2 + H_2O + 2A \rightarrow R{-}CHO + NH_3 + 2A^- + 2H^+. \quad (1)$$

When these enzymes are assayed *in vitro*, small redox-active compounds such as phenazine ethosulfate or phenazine methosulfate are routinely used as the electron acceptor.[3] The natural electron acceptor for most MADHs is a type I copper protein, amicyanin, which mediates electron transfer from MADH to *c*-type cytochromes.[4,5] The physiologic electron acceptor for AADH appears to be an azurin.[6]

TTQ is a covalently bound prosthetic group that is not acquired, but instead is formed by post-translational modification of two gene-encoded tryptophan residues of the host protein. The TTQ structure (Fig. 1) was proposed in 1991[7] and was based on mass spectroscopy and nuclear mag-

[2] S. Govindarai, E. Eisenstein, L. H. Jones, J. Sanders-Loehr, A. Y. Chistoserdov, V. L. Davidson, and S. L. Edwards, *J. Bacteriol.* **176,** 2922 (1994).

[3] V. L. Davidson, this series, Vol. 188, p. 241.

[4] M. Husain and V. L. Davidson, *J. Biol. Chem.* **260,** 14626 (1985).

[5] M. Husain, V. L. Davidson, and A. J. Smith, *Biochemistry* **25,** 2431 (1986).

[6] S. L. Edwards, V. L. Davidson, Y.-L. Hyun, and P. T. Wingfield, *J. Biol. Chem.* **270,** 4293 (1995).

[7] W. S. McIntire, D. E. Wemmer, A. Y. Christoserdov, and M. E. Lidstrom, *Science* **252,** 817 (1991).

netic resonance studies of derivatized prosthetic group-containing peptides from MADH from bacterium W3A1. This structure was verified by X-ray crystallographic studies of MADH from *Thiobacillus versutus* and *Paracoccus denitrificans*.[8] Resonance Raman spectroscopy studies[9,10] further confirmed that the MADHs from *T. versutus*, *P. denitrificans*, and bacterium W3A1 possessed identical prosthetic groups. Resonance Raman spectroscopy was also used to identify TTQ as the prosthetic group of AADH.[2,11] Both MADH[1] and AADH[2,12] possess an $\alpha_2\beta_2$ structure with TTQ present on the β subunit. Each molecule of holoenzyme possesses, therefore, two TTQs.

The most extensively studied TTQ-bearing enzyme is the MADH from *P. denitrificans*. The protocols and results described in this chapter will be primarily those obtained with this enzyme. Any similarities or differences that have been observed with other TTQ-bearing enzymes will be noted.

Summary of Reaction Mechanisms

The overall oxidation–reduction reaction of *P. denitrificans* MADH with methylamine and amicyanin is believed to proceed by the mechanism which is shown in Fig. 2. The reaction is initiated by a nucleophilic attack by the amine nitrogen on one of the quinone carbonyls. This results in formation of a carbinolamine intermediate (II) which loses water to form an imine intermediate (III). Next, an active-site nucleophile abstracts a proton from the methyl carbon, thus forming a carbanionic intermediate concomitant with the reduction of the prosthetic group. The resultant imine intermediate (IV), which now possesses a double bond between the methyl carbon and amine nitrogen, is hydrolyzed to yield the aldehyde product and the reduced aminoquinol form of TTQ (VI). Reoxidation of MADH by its physiologic electron acceptor, amicyanin, proceeds via two one-electron transfers requiring the formation of a free radical intermediate (VII). Release of ammonia occurs during or after the oxidation of this aminosemiquinone intermediate.

In order to monitor changes between intermediates in the reaction mechanism of MADH and other TTQ enzymes, it was first necessary to generate stable forms of these intermediates, or enzyme adducts which

[8] L. Chen, F. S. Mathews, V. L. Davidson, E. Huizinga, F. M. D. Vellieux, J. A. Duine, and W. G. J. Hol, *FEBS Lett.* **287,** 163 (1991).

[9] G. Backes, V. L. Davidson, F. Huitema, J. A. Duine, and J. Sanders-Loehr, *Biochemistry* **30,** 9201 (1991).

[10] W. S. McIntire, J. L. Bates, D. E. Brown, and D. M. Dooley, *Biochemistry* **30,** 125 (1991).

[11] J. Cohen, G. Backes, S. L. Edwards, S. Govindarai, V. L. Davidson, and J. Sanders-Loehr, in preparation.

[12] M. Nozaki, this series, Vol. 142, p. 650.

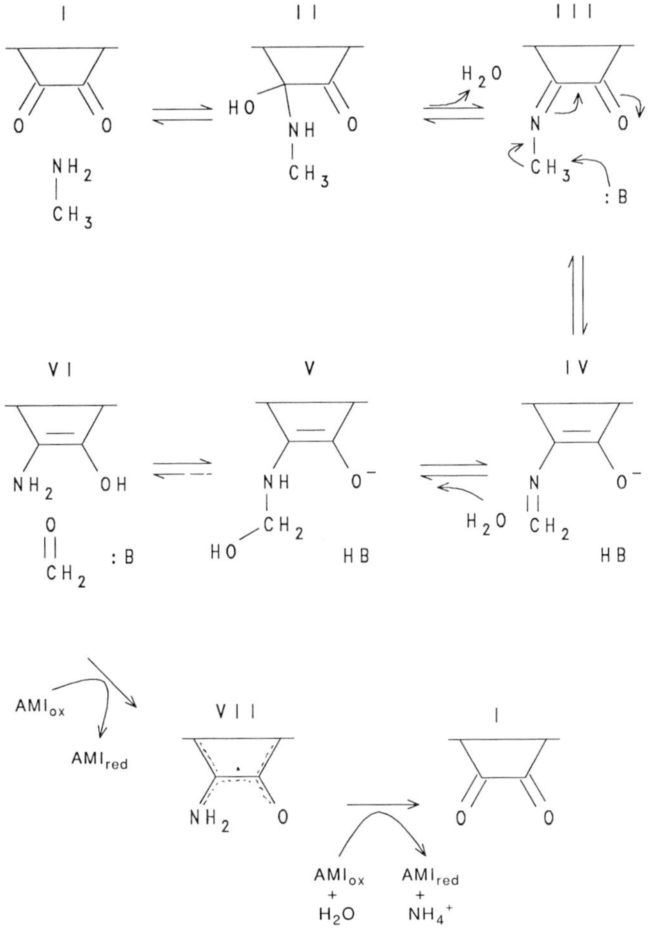

FIG. 2. Proposed reaction mechanism for MADH. This scheme describes the oxidation of methylamine by MADH and the reoxidation of MADH by two molecules of amicyanin (AMI). Only the quinone portion of the enzyme-bound TTQ is shown.

resemble these intermediates, and to characterize their spectral properties. This has been done for several forms of MADH. The different forms of MADH that have been described spectroscopically are shown in Fig. 3. The methods of preparation of these enzyme forms and their characterization are described in this chapter. Also described are rapid kinetic studies of the reduction and oxidation of MADH, which illustrate the value of this information in interpreting the transient spectral changes which accompany the reactions of this TTQ enzyme.

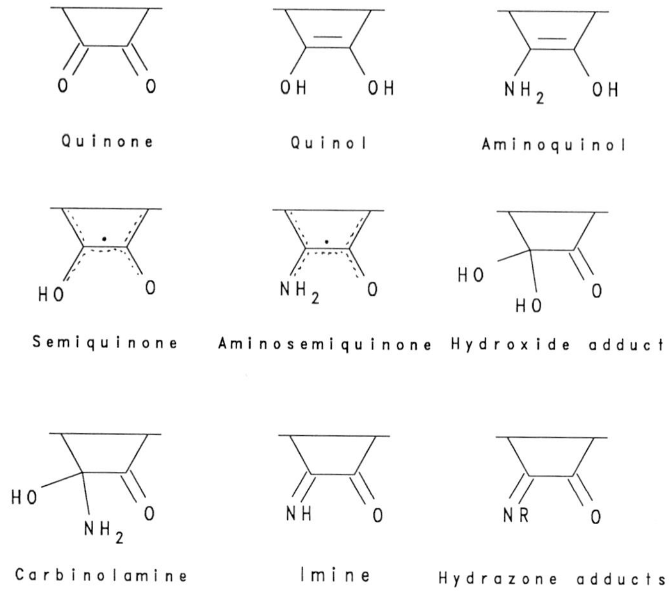

O O OH OH NH_2 OH

Quinone Quinol Aminoquinol

HO O NH_2 O HO HO O

Semiquinone Aminosemiquinone Hydroxide adduct

HO NH_2 O NH O NR O

Carbinolamine Imine Hydrazone adducts

FIG. 3. Redox forms and adducts of MADH that have been characterized. Only the quinone portion of the enzyme-bound TTQ is shown.

Formation and Characterization of Different Redox Forms of TTQ Enzymes

The study of TTQ enzymes has been facilitated by the fact that the reduced and semiquinone states of these enzymes are relatively stable to reoxidation and that the quinone, semiquinone, and quinol redox states exhibit readily distinguishable visible absorption spectra[13] (Fig. 4). These absorption spectra are also quite distinct from other quinoproteins which possess either PQQ[14] or topa[15] as a prosthetic group. The quinol and semiquinone forms of MADH may be generated by titration with sodium dithionite.[13,16] The aminoquinol form of MADH may be formed by reaction with

[13] M. Husain, V. L. Davidson, K. A. Gray, and D. B. Knaff, *Biochemistry* **26,** 4139 (1987).

[14] S. A. Salisbury, H. A. Forrest, W. B. T. Cruse, and O. Kennard, *Nature (London)* **280,** 843 (1979).

[15] S. M. James, D. Mu, D. Wemmer, A. J. Smith, S. Kaur, A. L. Burlingame, and J. P. Klinman, *Science* **248,** 981 (1990).

[16] W. C. Kenney and W. McIntire, *Biochemistry* **22,** 3858 (1983).

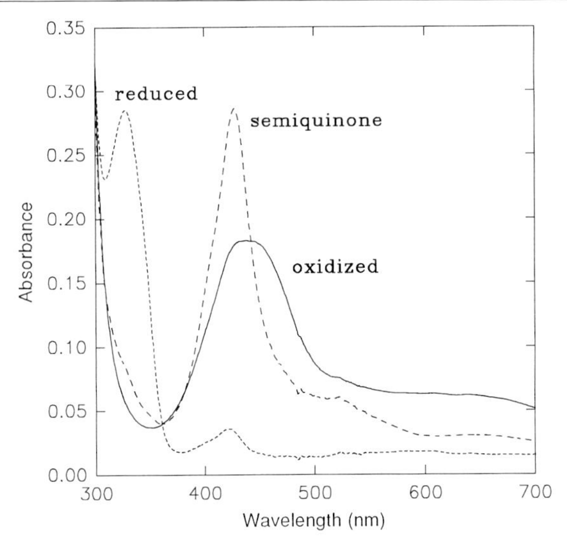

FIG. 4. Visible absorption spectra of different redox states of *P. denitrificans* MADH. The semiquinone and reduced forms were generated by the stoichiometric addition of dithionite.

a substrate.[13,16] The aminosemiquinone of MADH (see Fig. 3) may be formed via comproportionation of the aminoquinol and quinone forms.[17,18]

Formation of the MADH Quinol and Semiquinone by Reduction with Dithionite

Because dithionite is unstable in solution in the presence of oxygen, it is necessary to prepare the dithionite solution anaerobically and to add this solution to the enzyme under anaerobic conditions. Anaerobic conditions were achieved by connecting glassware and cuvettes to a Firestone rapid purge valve (Ace Glass, Vineland, NJ) connected to a vacuum pump and a source of oxygen-free argon. Using the Firestone valve it was possible to remove O_2 from the solution by alternate cycles of evacuation and flushing with argon. Such cycling was repeated 70 times to achieve anaerobiosis. Cuvettes and glassware were sealed with Suba-Seal septa and solutions were transferred with gas-tight Hamilton syringes.

The dithionite solution may be prepared by adding solid sodium dithionite to 10 m*M* potassium phosphate buffer, pH 7.5, which has been made

[17] V. L. Davidson, L. H. Jones, and M. A. Kumar, *Biochemistry* **29,** 10786 (1990).

[18] K. Warncke, H. B. Brooks, G. T. Babcock, V. L. Davidson, and J. L. McCracken, *J. Am. Chem. Soc.* **115,** 6464 (1993).

TABLE I
ABSORPTION PROPERTIES OF TTQ ENZYMES

		Absorbance maximum (nm)		
Enzyme	Source	Oxidized	Semiquinone	Reduced
MADH	*P. dentrificans*[a]	440	428	330[b]
MADH	*M. extorquens* AM1[c]	430	Nd[d]	325[e]
MADH	Bacterium W3A1[f]	429	429	331[b]
MADH	*Pseudomonas* sp. J[g]	430	Nd	330[e]
AADH	*A. faecalis*[h]	456	433	330[b]

[a] Husain and Davidson.[5]
[b] Dithionite was used as a reductant.
[c] Eady and Large.[20]
[d] Not determined.
[e] Methylamine was used as a reductant.
[f] Kenney and McIntire.[16]
[g] Matsumoto.[21]
[h] Govindarai *et al.*[2]

anaerobic by the procedure described earlier. Anaerobic reductive titration of the *P. denitrificans* MADH with dithionite proceeds through the semiquinone intermediate. The semiquinone form of MADH is formed by the addition of two molar equivalents of dithionite to oxidized MADH (one electron per TTQ). The addition of four molar equivalents of dithionite to oxidized MADH (two electrons per TTQ) will cause complete reduction to the quinol. Once formed, the semiquinone and quinol forms of MADH are relatively insensitive to reoxidation by air. No reoxidation of the quinol was observed and the half-time for reoxidation of the semiquinone on exposure to air was approximately 6 hr.[13]

Titration of AADH from *A. faecalis* with dithionite yielded essentially identical results.[2] A comparison of the absorption properties for the three redox states of AADH with MADHs is given in Table I. One difference between the MADHs and AADH was that the latter was somewhat more susceptible to reoxidation. The quinol of AADH exhibited about 10% reoxidation after 6 hr and the half-time for reoxidation of the semiquinone of AADH on exposure to air was approximately 1.5 hr.[19]

Formation of the Aminoquinol Form of MADH

The amino group which is derived from the substrate of MADH is not released from TTQ until the enzyme is reoxidized (see Fig. 2). The

[19] Y.-L. Hyun and V. L. Davidson, unpublished observations.

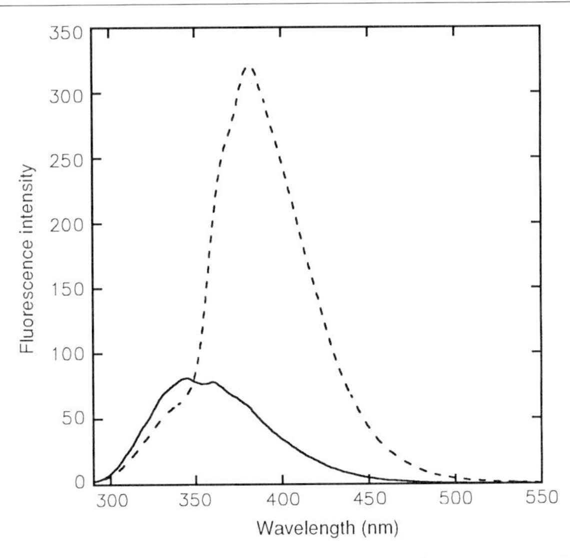

FIG. 5. Fluorescence spectra of *P. denitrificans* MADH. Fluorescence emission spectra of oxidized (solid line) and methylamine-reduced (dashed line) MADH (0.25 μM) were recorded at room temperature in 0.05 M potassium phosphate, pH 7.5, using an excitation wavelength of 280 nm.

aminoquinol form of the enzyme may, therefore, be generated by simply reducing MADH with substrate. Reductive titration of *P. denitrificans* MADH with methylamine at neutral pH proceeded directly to the fully reduced form of the enzyme without detectable formation of the semiquinone.[13] Once formed the aminoquinol is quite insensitive to reoxidation. As such, with this MADH it is not necessary to perform the reduction under anaerobic conditions. The absorption spectrum of the aminoquinol is essentially identical to that of the dithionite-generated quinol.

The addition of methylamine to oxidized *P. denitrificans* MADH also results in changes in its fluorescence spectrum (Fig. 5). Excitation of the oxidized enzyme at 280 nm yields a broad emission spectrum centered at approximately 350 nm. After the addition of methylamine, a shift in the emission spectrum to higher wavelengths and a large increase in fluorescence intensity are observed. Similar results have been reported for other MADHs.[20,21]

AADH could also be completely reduced by the amine substrate. This substrate-reduced form was much more sensitive to oxidation than the

[20] R. R. Eady and P. J. Large, *Biochem. J.* **123,** 757 (1971).
[21] T. Matsumoto, *Biochim. Biophys. Acta* **522,** 291 (1978).

aminoquinol of MADH and the dithionite-reduced forms of MADH and AADH. The half-time for reoxidation of the substrate-reduced AADH on exposure to air was approximately 3 hr.[2]

Formation of the Aminosemiquinone of MADH

The TTQ prosthetic groups of *P. denitrificans* MADH exhibit a pH-dependent redistribution of electrons from the substrate-reduced plus oxidized to semiquinone and aminosemiquinone redox forms.[17] This comproportionation reaction is only observed at pH values greater than 7.5, and this phenomenon has been exploited as a method by which to generate the aminosemiquinone of MADH. Because the comproportionation reaction occurs between quinone and aminoquinol forms of TTQ, the reaction will yield an equal mixture of aminosemiquinone and semiquinone.

The simplest method for forming this mixture is by adding one molar equivalent of methylamine to MADH (0.5 mol methylamine per TTQ) at pH 9.0. Rapid reduction of 50% of the MADH is followed by slow comproportionation. The rate of the comproportionation reaction is dependent on the MADH concentration.[17] For MADH concentrations in the micromolar range, once the comproportionation is complete at pH 9.0, no change in redox state was observed when the pH of the buffer was shifted to 7.5. However, at very high concentrations (i.e., greater than 100 μM), which are often required for biophysical studies, the aminosemiquinone was not stable at pH 7.5. If it is necessary to concentrate aminosemiquinone solutions at pH 7.5, special measures must be taken. It has been possible to minimize deleterious oxidation and disproportionation reactions during concentration by performing ultrafiltration under anaerobic conditions at $-20°$ in the presence of 50% ethylene glycol.

The absorption spectrum of the aminosemiquinone of MADH could not be recorded directly because it is present together with semiquinone, and the two species cannot be separated. Comparison of the spectrum of the product of this comproportionation reaction with that of the dithionite-generated semiquinone suggests that the spectra of the semiquinone and aminosemiquinone are very similar.

That the aminosemiquinone is truly formed as a result of the protocol described earlier has been verified by EPR analysis. The peak to trough width of the EPR spectrum of the aminosemiquinone is broadened relative to the dithionite-generated semiquinone.[22] The presence of substrate-derived nitrogen on the aminosemiquinone was verified by electron spin echo envelope modulation spectroscopy.[18]

[22] K. Warncke, H. B. Brooks, G. T. Babcock, V. L. Davidson, and J. L. McCracken, *Biophys. J.* **64,** A351 (1993).

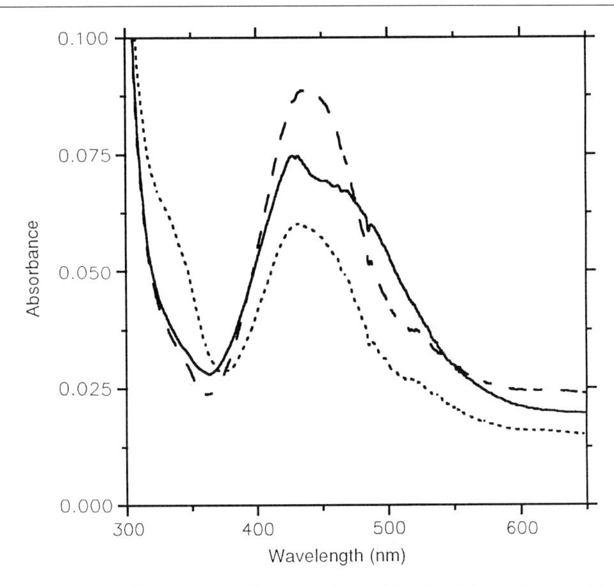

FIG. 6. Visible absorption spectra of ammonia and hydroxide adducts of *P. denitrificans* MADH. The dashed line represents the spectrum of oxidized MADH (3.3 μM) in 0.05 M potassium phosphate buffer, pH 7.5. The solid line represents the spectrum of the same MADH sample after the addition of 0.4 M NH_4Cl. The dotted line represents the spectrum of oxidized MADH (3.3 μM) in 0.05 M sodium–CHES buffer, pH 9.0.

Formation and Characterization of TTQ Adducts of MADH

Formation of Ammonia Adducts

Ammonium has been shown to be a reversible competitive inhibitor of MADHs and exhibits a K_i of approximately 20 mM.[23,24] For the MADH from bacterium W3A1, ammonium also functions as an activator at lower concentrations with a K_a of 2 mM.[23] Given the ability of ammonium to interact with MADH, it has been possible to use this compound to generate adducts which resemble reaction intermediates of MADH. The addition of millimolar concentrations of ammonium ion to MADH causes changes in the visible absorption spectrum. The addition of 0.4 M ammonium chloride to the MADHs from *P. denitrificans* and *T. versutus* caused a decrease in intensity and a shift in the absorption maximum from 440 to 425 nm, with a shoulder also forming at 470 nm (Fig. 6).[9] Resonance Raman spectroscopy of this ammonia adduct of *P. denitrificans* MADH indicated that

[23] W. S. McIntire, *J. Biol. Chem.* **262,** 11012 (1987).
[24] V. L. Davidson and L. H. Jones, *Biochim. Biophys. Acta* **1121,** 104 (1992).

these spectral perturbations reflected the formation of a carbinolamine adduct formed between ammonia and the reactive carbonyl of the TTQ prosthetic group[9] (see Fig. 3). In contrast, the addition of ammonium sulfate to MADH from bacterium W3A1 caused an increase in intensity and a shift in the absorption maximum to higher wavelengths with a peak centered at 491 nm.[16] This spectral form was attributed to the formation of an imine adduct between ammonia and TTQ.[10] It has been proposed that the differences in the reactivity of these enzymes toward ammonia are probably due to differences in the active site environment surrounding TTQ in the respective proteins.[9] No spectral changes have been observed on addition of up to 1 *M* ammonium chloride to AADH from *A. faecalis*.[19] This may be due to lack of binding because of the very poor affinity which this enzyme exhibits for small nonaromatic compounds.[12]

Formation of Hydroxide Adduct

P. denitrificans MADH may be converted to a hydroxide adduct (see Fig. 3) by incubation at alkaline pH. Although there is no evidence that this is a catalytically relevant form, it can be distinguished spectroscopically. A pH titration of oxidized *P. denitrificans* MADH revealed spectral changes which exhibited a pK_a value of 8.2.[25] As pH was increased, the absorption maximum of the enzyme decreased in intensity and shifted from a maximum of 440 nm to lower wavelengths with a shoulder forming at 460 nm (Fig. 6). The changes in the absorption spectrum that are observed at high pH vary somewhat depending on which buffer is used. Resonance Raman spectroscopy of MADH in sodium–Bicine buffer at pH 9.0 indicated that this spectral perturbation was due to the formation of a hydroxide adduct of TTQ.[9] Nearly identical spectral changes were observed with *T. versutus* MADH. In contrast, no such changes were observed with MADH from bacterium W3A1[10] or with AADH from *A. faecalis*.[19]

Formation of Inactive Adducts

All TTQ enzymes are irreversibly inactivated by a class of compounds with the general structure NH_2–NH–R. This includes hydrazine, phenylhydrazine, semicarbazide, and aminoguanidine. Hydroxylamine (NH_2–OH) also inhibits in a similar fashion. The spectral changes caused by the addition of these compounds to oxidized *P. denitrificans* MADH[24] are shown in Fig. 7. Although the spectral changes that were observed were not identical, very similar results have been obtained with other MADHs.[16,20]

It should be noted that this class of inhibitors is well known to function

[25] V. L. Davidson, *Biochem. J.* **261,** 107 (1989).

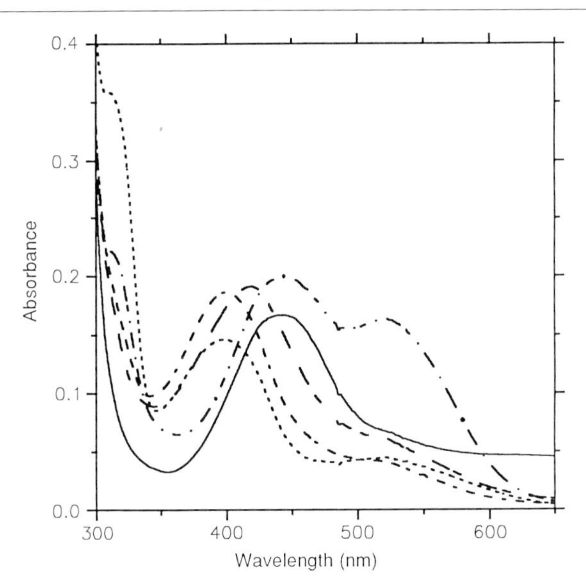

FIG. 7. Effect of inactivating carbonyl reagents on the visible absorption spectrum of *P. denitrificans* MADH. The solid line represents the spectrum of oxidized MADH (5.8 μM) in 0.05 M potassium phosphate buffer, pH 7.5. The other curves represent spectra of an equivalent amount of MADH that has been inactivated by the addition of aminoguanidine (long dashes), semicarbazide (short dashes), hydrazine (dotted), or phenylhydrazine (dash dot).

as carbonyl reagents and they are not exclusively specific for TTQ enzymes. While these TTQ adducts are not competent reaction intermediates, their structural analysis and more detailed information on the factors which influence the rate of reactivity of these agents toward specific TTQ enzymes may shed light on the precise factors which influence the initial binding of amino compounds to TTQ enzymes.

Stopped-Flow Studies of the Reduction and Oxidation of MADH

The characterization of different reaction intermediates of TTQ enzymes has allowed the study of individual steps in the overall reaction mechanism by rapid kinetic techniques. The very distinct absorption spectra of the three redox forms of MADH make it an ideal candidate for study by stopped-flow spectroscopy. Three such reactions are described below.

Methods

MADH and amicyanin from *P. denitrificans* were purified according to published procedures.[3,4] Stopped-flow experiments were performed using

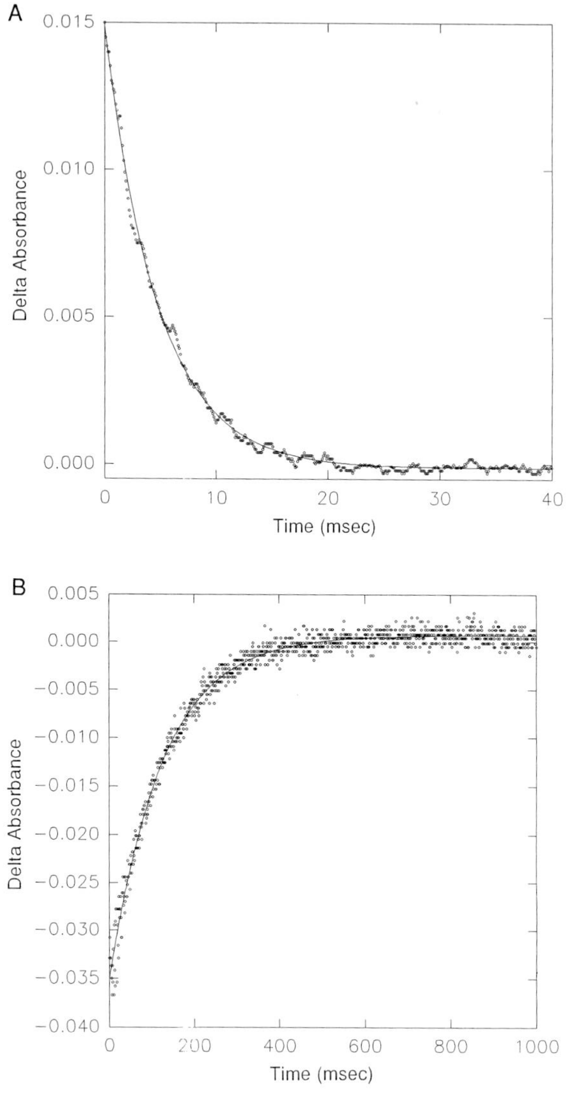

FIG. 8. Analysis of reactions of *P. denitrificans* MADH by stopped-flow spectroscopy. (A) Reduction of oxidized MADH by methylamine. Methylamine (71 μM) was mixed with MADH (1.0 μM) in 0.01 M potassium phosphate, pH 7.5, at 30°. Absorbance was monitored at 440 nm. The solid line drawn through the points represents the fit of these data to a single exponential. That fit gave a value for k_{obs} of 207 sec^{-1} for this data set. (B) Oxidation of dithionite-reduced MADH by amicyanin. Oxidized amicyanin (32 μM) was mixed with MADH (1.0 μM) in 0.01 M potassium phosphate, pH 7.5, at 25°. Absorbance was monitored at 443 nm. The solid line drawn through the points represents the fit of these data to a single exponential. That fit gave

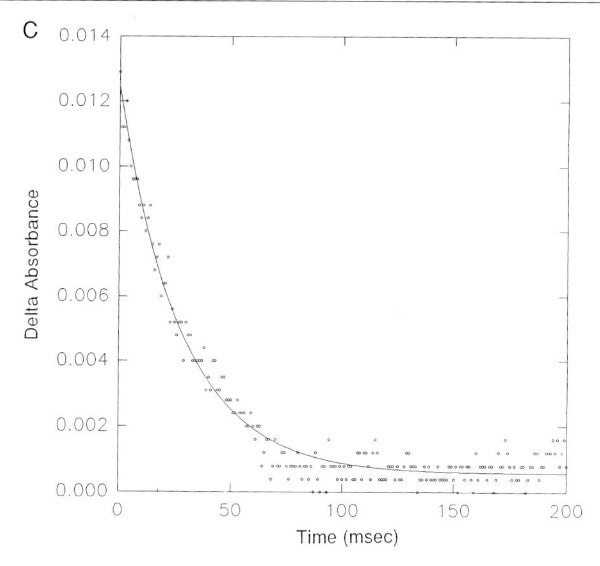

a value for k_{obs} of 8 sec^{-1} for this data set. (C) Oxidation of substrate-generated semiquinone MADH by amicyanin. Oxidized amicyanin (6 μM) was mixed with MADH (1.0 μM) in 0.01 *M* potassium phosphate, pH 7.5, at 5°. Absorbance was monitored at 428 nm. The solid line drawn through the points represents the fit of these data to a single exponential. That fit gave a value for k_{obs} of 36 sec^{-1} for this data set.

an On-Line Instrument Systems (OLIS, Bogart, GA) stopped-flow sample handling unit instrument coupled to Durrum optics. A 486 class computer controlled by OLIS software was used to collect data.

Reduction of Oxidized MADH by Methylamine

The conversion of intermediate III in Fig. 2 to intermediate IV is accompanied by a decrease in absorbance at 440 nm and an increase at 330 nm which correspond to the two electron reduction of oxidized TTQ (see Fig. 4). The observed rate constant for the reaction that is observed on addition of methylamine to oxidized *P. denitrificans* MADH (Fig. 8A) reflects the rate of that conversion and the rates of the two preceding steps. No spectral evidence was observed for intermediates II or III (see Fig. 4), suggesting that the formation of those intermediates is very rapid relative to the reductive step. Thus, in a simple kinetic model for this reaction, all steps leading to iminoquinone formation are described kinetically as a binding step. Analysis of the concentration dependence of k_{obs} for this reaction at 30° in 0.01 *M* potassium phosphate, pH 7.5, yielded values of 275 sec^{-1} for the reductive step and a kinetically determined K_d of 13 μM.[26] A similar study with MADH from bacterium W3A1 yielded a rate constant for the reductive step of 193 sec^{-1}.[27]

Oxidation of Reduced MADH by Amicyanin

The conversion of the MADH quinol to the MADH semiquinone is accompanied by distinct spectral changes (see Fig. 4). To ensure that only the one electron oxidation of the quinol is monitored, the reaction was monitored at 443 nm, which is isosbestic for the semiquinone and quinone. The physiologic reoxidant, amicyanin, does not exhibit appreciable absorbance at this wavelength. The reaction of dithionite-reduced MADH with oxidized amicyanin is shown in Fig. 8B. Analysis of this reaction at 25° in 0.01 *M* potassium phosphate, pH 7.5, yielded a value of 8 sec^{-1} for the redox step.[28] Similar studies with substrate-reduced MADH yielded a value of 24 sec^{-1} at 20°.[29]

Oxidation of Semiquinone MADH by Amicyanin

The conversion of the MADH semiquinone to the fully oxidized quinone is also accompanied by distinct spectral changes (see Fig. 4). This reaction may be monitored at 428 nm. The reaction of substrate-generated semiquinone MADH with oxidized amicyanin at 5° is shown in Fig. 8C. The reaction of dithionite-generated semiquinone is significantly faster.[30]

Acknowledgment

Work from this laboratory was supported by National Institutes of Health Grant GM-41574.

[26] H. B. Brooks, L. H. Jones, and V. L. Davidson, *Biochemistry* **32,** 2725 (1993).
[27] R. B. McWhirter and M. H. Klapper, *in* "PQQ and Quinoproteins" (J. A. Jongejan and J. A. Duine, eds.), p. 259. Kluwer Academic Publishers, Dordrecht, The Netherlands, 1989.
[28] H. B. Brooks and V. L. Davidson, *Biochemistry* **33,** 5696 (1994).
[29] H. B. Brooks and V. L. Davidson, *Biochem. J.* **294,** 211 (1993).
[30] H. B. Brooks and V. L. Davidson, *J. Am. Chem. Soc.* **116,** 11201 (1994).

[15] X-Ray Studies of Quinoproteins

By F. SCOTT MATHEWS

Introduction

This chapter focuses on the crystal structure analysis of two types of quinoproteins, methanol dehydrogenase (MEDH) and methylamine dehydrogenase (MADH), and on two crystalline complexes of MADH with its redox partners. The structure analysis of galactose oxidase (GO), also containing a redox active amino acid, is described elsewhere in this volume and will only be discussed in the context of the other two enzymes. Since this volume is principally concerned with redox-active amino acids in biology, detailed discussions of the crystallographic methodology used in the analyses will not be presented here, as these techniques have been described very well in earlier volumes of this series (see Volumes 114 and 115 of "Methods of Enzymology").

The structures of five different quinoproteins or quinoprotein complexes are presented. All are obtained from facultative or obligate methylotrophic bacteria and are located in the periplasmic space of these organisms. The structures of two forms of MADH have been determined, one from *Thiobacillus versutus* (TV-MADH)[1] and the other from *Paracoccus denitrificans* (PD-MADH).[2] A binary complex between PD-MADH and amicyanin, a cupredoxin, has been analyzed[3] as well as a ternary complex among PD-MADH, amicyanin, and a *c*-type cytochrome (cytochrome c_{551i}).[4] The fifth protein structure described is that of MEDH, from *Methylobacterium* W3A1.[5]

Methylamine dehydrogenase is a quinoprotein with MW of approximately 120,000 and containing the redox cofactor tryptophan tryptophylquinone (TTQ) (Fig. 1).[6] It is a heterotetramer, H_2L_2, consisting of two heavy

[1] F. M. D. Vellieux, F. Huitema, H. Groendijk, K. H. Kalk, J. Frank, J. A. Jongejan, J. A. Duine, K. Petratos, J. Drenth, and W. G. J. Hol, *EMBO J.* **8,** 2171 (1989).

[2] L. Chen, F. S. Mathews, V. L. Davidson, E. G. Huizinga, F. M. D. Vellieux, and W. J. G. Hol, *Proteins: Struct., Funct., Genet.* **14,** 288 (1992).

[3] L. Chen, R. Durley, B. J. Poliks, K. Hamada, Z.-W. Chen, F. S. Mathews, V. L. Davidson, Y. Satow, E. Huizinga, F. M. D. Vellieux, and W. G. J. Hol, *Biochemistry* **31,** 4959 (1992).

[4] L. Chen, R. Durley, F. S. Mathews, V. L. Davidson, M. Tegoni, C. Rivetti, and G. L. Rossi, *Protein Sci.* **2,** 147 (1993).

[5] Z.-X. Xia, W.-W. Dai, Z.-P. Hao, V. L. Davidson, S. White, and F. S. Mathews, *J. Biol. Chem.* **267,** 22289 (1992).

[6] W. S. McIntire, D. E. Wemmer, A. Chistoserdov, and M. E. Lidstrom, *Science* **252,** 817 (1991).

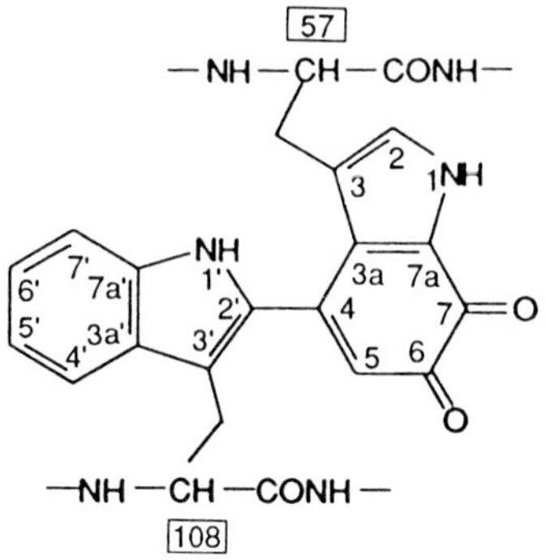

FIG. 1. Chemical structure of the tryptophan tryptophylquinone cofactor of MADH. The precursor tryptophans are located at positions 57 and 108 in the amino acid sequence of the light subunit of MADH as derived from their respective gene sequences in *T. versutus*[29] and *P. denitrificans*[11] MADH. Reproduced by permission from Huizinga *et al.*[31]

subunits (H), each of MW 45,000, and two light subunits (L), each of MW 15,000.[7] The reaction catalyzed by the enzyme is

$$CH_3NH_3^+ + H_2O \rightarrow NH_4^+ + CH_2O + 2H^+ + 2e^-.$$

Subsequently, electrons are transferred to the membrane-bound terminal oxidase, cytochrome aa_3, via a series of soluble electron carrier proteins. In several methylotrophs, the primary electron acceptor from MADH is amicyanin, a blue copper protein, of MW 12,500.[8] In the case of *P. denitrificans*, cytochrome c_{551i}, a protein of MW 17,500, can serve as the second sequential electron acceptor.[9] Both of these electron carriers, along with MADH, are induced when the bacteria are grown on methylamine as the sole carbon source.

The specificity of the interaction between amicyanin and MADH has been well documented.[10–14] The amicyanin gene is located immediately downstream of the gene for the small subunit of MADH.[10,11] As shown for *T. versutus*, disruption of the amicyanin gene results in inability of the

[7] M. Husain and V. L. Davidson, *J. Bacteriol.* **169,** 1712 (1987).

[8] M. Husain and V. L. Davidson, *J. Biol. Chem.* **260,** 14626 (1985).

[9] M. Husain and V. L. Davidson, *J. Biol. Chem.* **261,** 8577 (1987).

[10] R. M. J. van Spanning, C. W. Wansell, W. N. M. Reijnders, L. F. Itmann, and A. M. Stouthamer, *FEBS Lett.* **275,** 217 (1990).

[11] A. Y. Chistoserdov, J. Boyd, F. S. Mathews, and M. E. Lidstrom, *Biochem. Biophys. Res. Commun.* **184,** 1181 (1992).

[12] K. A. Gray, V. L. Davidson, and D. B. Knaff, *J. Biol. Chem.* **263,** 13987 (1988).

[13] V. L. Davidson and L. H. Jones, *Anal. Chim. Acta* **249,** 235 (1991).

[14] M. A. Kumar and V. L. Davidson, *Biochemistry* **29,** 5299 (1990).

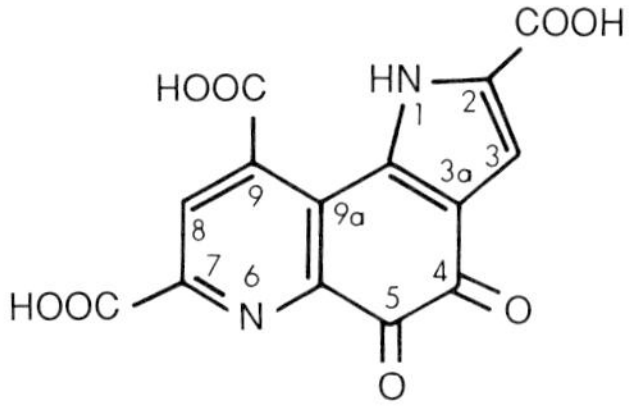

FIG. 2. Chemical structure of pyrroloquinoline quinone.

organism to grow on methylamine.[10] Formation of a complex between MADH and amicyanin isolated from *P. denitrificans* causes a −73 mV shift of the redox potential of amicyanin as well as perturbations of the visible spectrum of the TTQ cofactor of MADH.[12] Kinetic[13] and chemical cross-linking studies[14] of the proteins suggest that complex formation is stabilized in part by hydrophobic and electrostatic interactions between the light subunit of MADH and amicyanin.

Cytochrome c_{551i} is the most efficient electron acceptor of three periplasmic cytochromes induced in *P. denitrificans* during growth on methylamine.[9] It appears that a ternary complex among MADH, amicyanin, and the cytochrome must form for electron transfer to occur from MADH to the cytochrome. The cytochrome will not accept electrons from either MADH or amicyanin alone.

Methanol dehydrogenase is a periplasmic bacterial quinoprotein that catalyzes the oxidation of methanol to formaldehyde. Other primary alcohols, as well as formaldehyde, can serve as substrates. The reaction catalyzed is

$$RCH_2OH \rightarrow RCHO + 2H^+ + 2e^-.$$

The enzyme is an $\alpha_2\beta_2$ heterotetramer of approximately 140 kDa MW with subunit molecular weights of 62,000 and 8000, respectively.[15] The amino acid sequences of MEDH from several bacteria including *P. denitrificans*,[16] *Methylobacterium extorquens* AM1,[17] and *Methylophilus* W3A1[18] have been determined by DNA sequencing methods. The enzyme contains two molecules of noncovalently bound pyrroloquinoline quinone (PQQ) (Fig. 2) per tetramer as well as one or more tightly bound calcium ions.[19] The primary

[15] D. N. Nunn, D. Day, and C. Anthony, *Biochem. J.* **260,** 857 (1989).

[16] N. Harms, G. E. de Vries, K. Maurer, J. Hoogendijk, and A. H. Stouthamer, *J. Bacteriol.* **169,** 3969 (1987).

[17] D. J. Anderson, C. J. Morris, D. N. Nunn, C. Anthony, and M. E. Lidstrom, *Gene* **90,** 173 (1990).

[18] G. W. Boyd and F. S. Mathews, unpublished results (1993).

[19] I. W. Richardson and C. Anthony, *Biochem. J.* **287,** 709 (1992).

electron acceptor *in vivo* is cytochrome c_L, an acidic protein of approximately 17,000 molecular weight.[20] Electrons are then transferred to a second carrier, cytochrome c_H, a basic protein similar to soluble mitochondrial or bacterial *c*-type cytochromes, and subsequently to a membrane-bound aa_3-type cytochrome oxidase.

Methods

Crystallization

Crystals of all five proteins (or protein complexes) were grown by the vapor diffusion method. The conditions for crystallization are described in Table I. In all but one case, crystal growth began within a few days and nothing unusual occurred during the process. In the case of PD-MADH, however, crystallization did not occur unless the protein was allowed to stand for about 3 weeks at room temperature in dilute buffer solution (5 m*M* phosphate buffer, ~pH 6.0).[21] The reason for this requirement was not discovered until after the protein structure had been determined and partially refined.[2] It was then found that there was little electron density for the first 18 amino acid residues of the heavy subunit (Fig. 3). Subsequent SDS gel electrophoresis of a similarly aged protein sample indicated that it had a molecular weight about 2000 smaller than a fresh sample, showing that proteolysis had occurred during the incubation and was required for crystal growth. In crystals of the binary and ternary complexes, this segment was intact and no proteolysis was needed for crystal formation.

Structure Analyses

The structure analyses of the quinoproteins and complexes described in this chapter are summarized in Table II. The structures of TV-MADH[1] and of MEDH[5] were solved by the multiple isomorphous replacement (MIR) method. In both cases, several heavy atom derivatives were needed and anomalous scattering measurements were incorporated into the phase determination. In the case of MEDH, the structure of the enzyme isolated from the methylotrophic bacterium *Methylophilus methylotrophus* was actually solved first. Since crystals of MEDH from *Methylophilus* W3A1 are isomorphous to the *M. methylotrophus* enzyme, its structure could be obtained directly by refinement of the *M. methylotrophus* model against the observed *Methylophilus* W3A1 structure factors.

[20] C. Anthony, *Biochim. Biophys. Acta* **1099,** 1 (1991).

[21] L. Chen, L. W. Lim, F. S. Mathews, V. L. Davidson, and M. Husain, *J. Mol. Biol.* **203,** 1137 (1988).

TABLE I

CRYSTALLIZATION CONDITIONS AND UNIT CELL DATA FOR QUINOPROTEINS AND COMPLEXES

Protein	MADH[a]	MADH[b]	MADH–amicyanin[b]	MADH–amicyanin–cytochrome c_{551i}[c]	MEDH[d]
Source	*Thiobacillus versutus*	*Paracoccus denitrificans*	*P. denitrificans*	*P. denitrificans*	*Methylophilus* W3A1
Crystallization method	Hanging drop	Hanging drop	Hanging drop	Hanging drop	Sitting drop macroseeding[e]
Crystallization conditions	37–42% SAS[f] 0.1 *M* sodium acetate, pH 5.0 [MADH] = 10 mg/ml	1.8–2.0 *M* AS [2.5 m*M*, P_i[f] (Na/K) = 90:10][g] [MADH] = 15 mg/ml	[2.2–2.4 *M* Pi (Na/K) = 70:30] [MADH] = 15 mg/ml (MADH/Ami) = 1:2.5–1:3 (molar ratio)	[2.5–2.6 *M* Pi (Na/K) = 90:10] [MADH] = 15 mg/ml (MADH/Ami/Cyt) = 1:3:3 (molar ratio)	12% (w/v) PEG[f] 8000 50 m*M* Tris–HCl, pH 8.25, [MEDH] = 2.5 mg/ml
Remark	Crystallization at 4° for 3 weeks, temperature raised to 37° over 5 weeks	Crystallization at room temperature, 3-week prior incubation of protein solution at room temperature	Crystallization at room temperature	Crystallization at room temperature	Macroseeds obtained by direct mixing of protein and precipitant at 13% PEG
Unit cell data	Space group $P3_121$ a = b = 130.1 Å c = 104.0 Å 1 HL dimer/a.u.	Space group $P2_12_12_1$ a = 151.8 Å b = 135.8 Å c = 55.1 Å 1 H_2L_2 tetramer per asym. unit	Space group $P4_12_12$ a = b = 124.6 Å c = 247.3 Å 1 $H_2L_2A_2$ hexamer per asym. unit	Space group $C222_1$ a = 148.8 Å b = 68.8 Å c = 187.2 Å HLAC tetramer per asym. unit	Space group $P2_1$ a = 124.1 Å b = 62.9 Å c = 84.7 Å β = 92.9° 1 $\alpha_2\beta_2$ tetramer per asym. unit

[a] F. M. D. Vellieux, J. Frank, M. B. A. Swarte, H. Groendijk, J. A. Duine, J. Drenth, and W. J. G. Hol, *Eur. J. Biochm.* **154,** 383 (1986).

[b] L. Chen, L. W. Lim, F. S. Mathews, V. L. Davidson, and M. Husain, *J. Mol. Biol.* **203,** 1137 (1988).

[c] L. Chen, R. Durley, F. S. Mathews, V. L. Davidson, M. Tegoni, C. Rivetti, and G. L. Rossi, *Protein Sci.* **2,** 147 (1993).

[d] Z.-X. Xia, Z.-P. Hao, F. S. Mathews, and V. L. Davidson, *FEBS Lett.* **258,** 175 (1989).

[e] C. Thaler, G. Eichele, L. H. Weaver, E. Wilson, R. Karlsson, and J. N. Jansonius, this series, Vol. 114, p. 132.

[f] Abbreviations: AS, ammonium sulfate; PEG, polyethylene glycol; Pi, sodium or potassium phosphate; SAS, saturated ammonium sulfate.

[g] Phosphate buffer of composition [NaH_2PO_4:K_2HPO_4] in a given ratio.

TABLE II

STRUCTURAL ANALYSIS AND REFINEMENT OF QUINOPROTEINS AND COMPLEXES

Protein	MADH[a,b]	MADH[c,d]	MADH–amicyanin[e,f]	MADH–amicyanin–cytochrome c_{551i}[g,h]	MEDH[i,j]
Source	*Thiobacillus versutus*	*Paracoccus denitrificans*	*P. denitrificans*	*P. denitrificans*	*Methylophilus* W3A1
Method of solution	Multiple isomorphous replacement	Molecular replacement	Molecular replacement and model building	Molecular replacement and model building	Multiple isomorphous replacement
Derivatives	K_2PtI_6 Pt(ethylenediamine)Cl_2 $UO_2(C_2H_3O_2)_2$	Not applicable	Not applicable	Not applicable	$KAuCl_4$ $(NH_4)_3IrCl_6$ K_2PtCl_6
Additional data sets	None	None	MADH–apoamicyanin	MADH–apoamicyanin–cytochrome c_{551i}	MEDH from *Methylophilus methylotrophus*
Current resolution (Å)	6.0–2.25	11.0–1.75	6.0–2.6	12.0–2.4	10.0–2.4
Refinement method	TNT[k]	TNT[k]	X-PLOR[l]	X-PLOR[l]	X-PLOR[l]
Refinement results	R = 0.209 $RMSD^{m}_{bonds}$ = 0.018–0.022 Å $RMSD_{angles}$ = 3.0–3.8° No solvent molecules	R = 0.167 $RMSD_{bonds}$ = 0.007 Å $RMSD_{angles}$ = 2.3° 556 solvent molecules	R = 0.134 $RMSD_{bonds}$ = 0.016 Å $RMSD_{angles}$ = 3.3° 637 solvent molecules	R = 0.179 $RMSD_{bonds}$ = 0.017 Å $RMSD_{angles}$ = 3.6° 128 solvent molecules	R = 0.201 $RMSD_{bonds}$ = 0.009 Å $RMSD_{angles}$ = 1.1° No solvent molecules

Remarks	Model consists of X-ray sequence for H subunit and DNA sequence for L subunit. Refinement still in progress.	Refinement complete	Refinement complete	Further refinement necessary	MEDH from *M. methylotrophus* solved by MIR, MEDH from W3A1 isomorphous, refined with W3A1 sequence. Refinement not complete.

[a] F. M. D. Vellieux, F. Huitema, H. Groendijk, K. H. Kalk, J. Frank, J. A. Jongejan, J. A. Duine, K. Petratos, J. Drenth, and W. G. J. Hol, *EMBO J.* **8,** 2171 (1989).

[b] E. G. Huizinga, B. A. M. van Zanten, J. A. Duine, J. A. Jongejan, F. Huitema, K. S. Wilson, and W. G. J. Hol, *Biochemistry* **31,** 9789 (1992).

[c] L. Chen, F. S. Mathews, V. L. Davidson, E. G. Huizinga, F. M. D. Vellieux, and W. J. G. Hol, *Proteins: Struct., Funct., Genet.* **14,** 288 (1992).

[d] L. Chen, M. Doi, A. V. Chisterserdov, M. E. Lidstrom, and F. S. Mathews, unpublished results.

[e] L. Chen, R. Durley, B. J. Poliks, K. Hamada, Z.-W. Chen, F. S. Mathews, V. L. Davidson, Y. Satow, E. Huizinga, F. M. D. Vellieux, and W. G. J. Hol, *Biochemistry* **31,** 4959 (1992).

[f] R. Durley, L. Chen, and F. S. Mathews, unpublished results.

[g] L. Chen, R. Durley, F. S. Mathews, V. L. Davidson, M. Tegoni, C. Rivetti, and G. L. Rossi, *Protein Sci.* **2,** 147 (1993).

[h] L. Chen, R. Durley, F. S. Mathews, and V. L. Davidson, unpublished results (1993).

[i] Z.-X. Xia, W.-W. Dai, Z.-P. Hao, V. L. Davidson, S. White, and F. S. Mathews, *J. Biol. Chem.* **267,** 22289 (1992).

[j] S. White, G. Boyd, F. S. Mathews, Z.-X. Xia, W.-W. Dai, Y.-F. Zhang, and V. L. Davidson, *Biochemistry* **32,** 12955 (1993).

[k] D. E. Tronrud, L. F. ten Eyck, and B. W. Mathews, *Acta Crystallogr., Sect. A: Found. Crystallogr.* **A43,** 489 (1987).

[l] A. T. Brunger, J. Kuriyan, and M. Karplus, *Science* **235,** 458 (1987).

[m] RMSD is the root mean square deviation from ideal bond lengths or angles. In the case of TV-MADH, the two sets of values for $RMSD_{bonds}$ and $RMSD_{angles}$ refer to the L and H subunits, respectively.

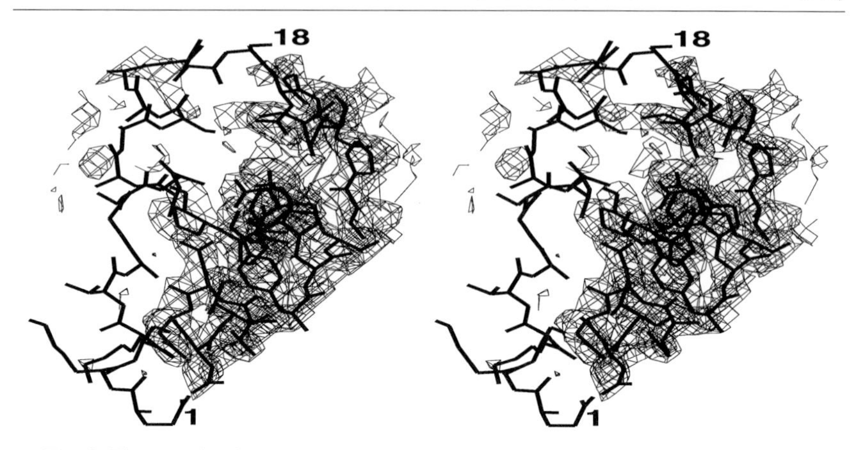

FIG. 3. Electron density map for the first 31 residues of the heavy subunit of PD-MADH after partial refinement of the model based on the TV-MADH "X-ray sequence."[2] The electron density was calculated with coefficients ($2F_o$-F_c) and phases obtained from the final cycle of least-squares refinement using the TV-MADH sequence. The electron density is very weak for residues 1–18 and strong beyond residue 18.

The structure of PD-MADH[2] was solved by the molecular replacement method using the TV-MADH structure as the search probe. The model was then refined against the observed structure factors of PD-MADH. Later, when the gene sequence of PD-MADH became available,[11] refinement was completed using the true amino acid sequence.

For both the binary[3] and ternary[4] complexes, data were recorded from crystals containing apoamicyanin as well as the copper-containing holoprotein. In each case, the MADH portion of the complex was located in the respective unit cells by molecular replacement. In the case of the binary complex, the TV-MADH model was used as a probe, after partial refinement in the PD-MADH unit cell. For the ternary complex, the correct model for PD-MADH, based on the gene sequence,[11] was used as the search model. In both cases, the partial models were further refined in the unit cells of the respective complexes. The copper of the amicyanin components was then located in isomorphous difference maps calculated with $\{[F(apo)-F(holo)]\exp[i\alpha_p]\}$ as Fourier coefficients, where F(apo) and F(holo) are the apo and holo structure factors and α_p is the phase calculated from the partial structure. For the binary complex, the amicyanin structure was traced in a difference map calculated using $[(F_o-F_c)\exp(i\alpha_p)]$ as Fourier coefficients, where F_o and F_c are the observed and calculated structure factors, respectively. In the ternary complex, the iron atom was located in an anomalous difference map using $[\Delta F^{\pm}\exp(i\alpha_p)]$ as Fourier coefficients, where $\Delta F^{\pm}$ is the anomalous difference in structure factors. The cytochrome chain was then traced in a $(F_o\text{-}F_c)$ difference map.

Amino Acid Sequences

A problem common to all the quinoprotein structures, described in this chapter, was the lack of amino acid sequence data during the initial and intermediate stages of the structure analyses. Part of the problem arose from the large size of the heavy subunits of both MADH and MEDH, making chemical sequencing difficult; furthermore, progress in cloning and DNA sequencing of the structural genes for these proteins lagged behind the crystallography. When the structure of TV-MADH was solved, the only sequence data available for MADH were for the light subunit from *M. extorquens* AM1.[22] This sequence was used as a basis for interpreting the experimental electron density. For the heavy subunit, no sequence data were available, so that an "X-ray sequence," based solely on the shape of the electron density, was derived from the map. This was sufficient to provide a chain tracing and allow identification of a few large amino acids, but the model was considerably biased. When the gene sequence became available,[23] it showed that subsequent refinement[24] had reduced bias sufficiently to improve sequence identification from 25% correct for the initial model to 50% for the refined model. The refinement process did not, however, enable a correct identification of the redox cofactor in MADH (see below).

For PD-MADH and the MADH–amicyanin complex, the X-ray sequence from TV-MADH was used for structure analysis and partial refinement. In this case the problem was more severe since the X-ray sequence corresponded to the electron density map for the *T. versutus* protein. However, use of the TV-MADH X-ray sequence in the PD-MADH structures yielded a great deal of structural information, including the structure of amicyanin in the binary complex.[3]

In the case of MEDH, the gene sequence of three other MEDHs had been published at the time the initial electron density map of the *Methylobacterium* W3A1 MEDH was calculated. One of these sequences[16] matched the density better than the others and was used for tracing the chain and for initial refinement. As in the case of MADH, a reasonably accurate chain tracing could be obtained, as well as the approximate location of the

[22] Y. Ishii, T. Hase, Y. Fukumori, H. Matsubara, and J. Tobari, *Biochemistry* **93,** 107 (1983).

[23] F. Huitema, J. Van Beeumen, G. Van Driessche, J. A. Duine, and G. W. Canters, *J. Bacteriol.* **175,** 6254 (1993).

[24] F. M. D. Vellieux, K. H. Kalk, J. Drenth, and W. G. J. Hol, *Acta Crystallogr., Sect. B: Struct. Sci.* **B46,** 806 (1990).

PQQ cofactor. However, details of the PQQ orientation and the chemical environment of the active site could not be defined accurately from these data and were only accessible after the DNA sequence of the W3A1 enzyme had been determined.[18]

Cofactor Identification

For both MADH and MEDH, difficulty was encountered in identifying the nature and/or orientation of the respective redox cofactors. This problem was especially acute in the case of MADH. At the time the crystal structure of TV-MADH was solved, the limited spectroscopic and chromatographic evidence available suggested that the cofactor of MADH was a modified form of PQQ, bound covalently to the light subunit.[25] The site of covalent attachment of the cofactor to the light subunit was known, based on its peptide sequence in *M. extorquens* AM1,[22] since two amino acids, at positions 55 and 106, behaved abnormally and could not be identified. After extensive model building and refinement of the TV-MADH structure, including assignment of the X-ray sequence, but without modeling the cofactor, an electron density map was calculated which suggested a nonplanar cofactor that could not be fitted with a PQQ model. The electron density was interpreted as a modified or precursor form of PQQ, called pro-PQQ, containing an indole ring linked to an acidic group. After the cofactor had been correctly identified as TTQ on genetic and chemical grounds and the protein-cofactor structure was refined at 2.2 Å resolution, the agreement of the TTQ model with the ($2F_o$-F_c) electron density was excellent (Fig. 4a).[26] In the case of the PD-MADH structure, the model resulting from the molecular replacement solution was refined using the TV-MADH X-ray sequence, but omitting a model for the redox cofactor. The resulting ($2F_o$-F_c) electron density differed significantly from both PQQ and the pro-PQQ model. The electron density was not interpreted further at that time since the cofactor was by then correctly identified as TTQ.[6] When TTQ was modeled to the observed electron density, the correspondence was striking (Fig. 4b).[26]

In contrast to MADH, much firmer evidence identified PQQ as the redox cofactor of MEDH.[27] At 2.6 Å resolution and using a homologous amino acid sequence for refinement, the PQQ could be identified as corresponding to a flat portion of electron density of appropriate size, but its orientation in the density was ambiguous (Fig. 5a). After the *Methylophilus*

[25] R. A. van der Meer, J. A. Jongejan, and J. A. Duine, *FEBS Lett.* **254,** 299 (1987).

[26] L. Chen, F. S. Mathews, V. L. Davidson, E. G. Huizinga, F. M. D. Vellieux, J. A. Duine, and W. G. J. Hol, *FEBS Lett.* **287,** 163 (1991).

[27] S. A. Salisbury, H. S. Forrest, W. B. T. Kruse, and O. Kennard, *Nature* (*London*) **280,** 843 (1979).

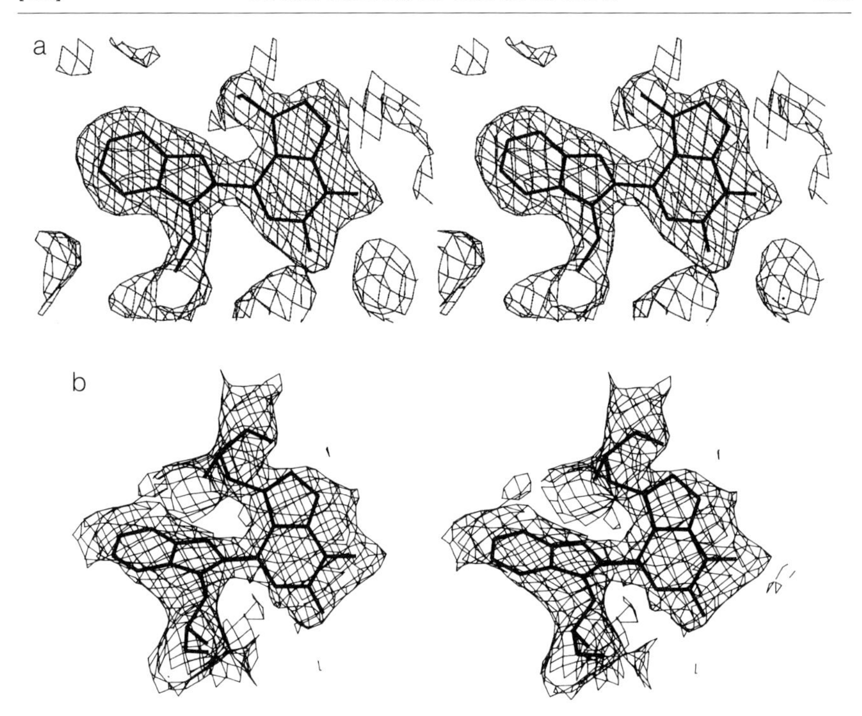

FIG. 4. (a) The refined model and electron density of the TTQ cofactor in TV-MADH. The electron density was calculated using coefficients $2F_o$-F_c and phases based on the refined "X-ray sequence" and the TTQ model. (b) PD-MADH electron density fitted with the TTQ cofactor. The electron density map was computed using coefficients $2F_o$-F_c and phases based on refinement using the TV-MADH X-ray sequence but no model for the redox cofactor.

W3A1 sequence became available[18] and refinement was extended to 2.4 Å resolution,[28] the corresponding electron density improved significantly, making it possible to define uniquely the orientation (Fig. 5b).[28a] A calcium ion, partially coordinated by PQQ, was also visible in the map.

Structures of the Quinoproteins

Methylamine Dehydrogenase

The MADH molecule is shown in Fig. 6. It consists of two heavy chains (H) and two light chains (L). The redox cofactor of MADH, TTQ (Fig.

[28] S. White, G. Boyd, F. S. Mathews, Z.-X. Xia, W.-W. Dai, Y.-F. Zhang, and V. L. Davidson, *Biochemistry* **32,** 12955 (1993).

[28a] A. Hodel, S.-H. Kim, and A. T. Brunger, *Acta Crystallogr., Sect. A: Found. Crystallogr.* **A48,** 851 (1992).

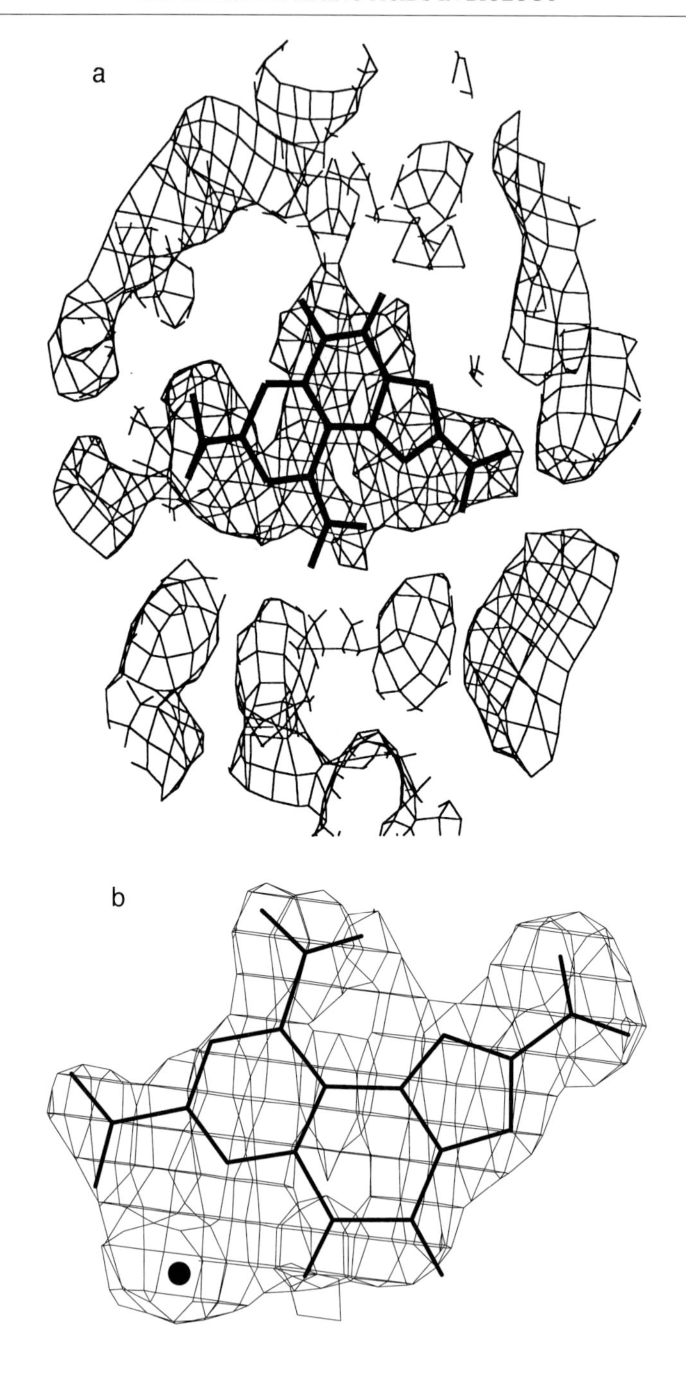
a
b

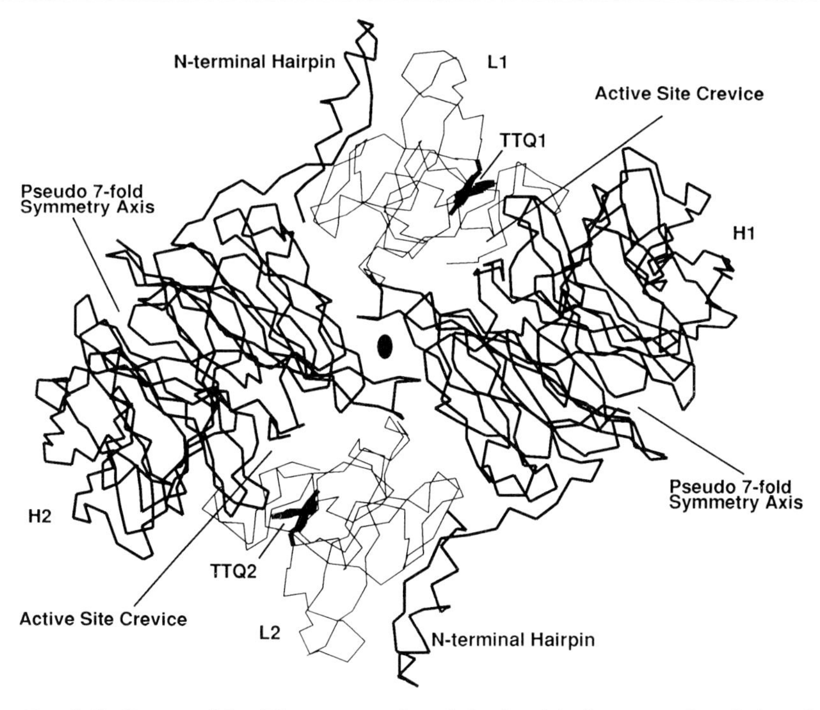

FIG. 6. C_α diagram of the H_2L_2 tetramer of methylamine dehydrogenase viewed along the molecular twofold axis.

1), is located in each of the light subunits. It is made from two tryptophan side chains, both encoded by the gene for this protein and modified by an as-yet unknown post-translational process. In both TV and PD-MADH, these tryptophans occur at positions 57 and 108 of the mature protein. The tryptophan at position 57 has been modified to tryptophylquinone (Trq), which contains an *o*-quinone moiety on atoms $C^{\eta 2}$ and $C^{\zeta 2}$ (atoms C6 and C7, respectively, Fig. 1). The two tryptophan side chains are linked by a

---

FIG. 5. (a) Electron density of PQQ in MEDH from the methylotroph W3A1 with the model superimposed. The electron density was computed with coefficients $2F_o$-F_c after refinement using the *P. denitrificans* amino acid sequence. The maps were averaged about the molecular twofold axis. PQQ was not included in the calculation. (b) Simulated annealing omit map[28a] for methanol dehydrogenase from *Methylophilus* W3A1 using the *Methylophilus* W3A1 amino acid sequence. The PQQ moiety plus all atoms within a 3-Å sphere of all PQQ atoms in both subunits were omitted from the calculation. The filled circle represents the calcium ion.

covalent bond between atom $C^{\varepsilon 3}$ of Trq-57 and $C^{\delta 2}$ of Trp-108. The two indole rings are tilted with respect to each other by a dihedral angle of 40°.

From a functional viewpoint, MADH can be considered as a pair of HL dimers, H_1L_1 and H_2L_2, related by a molecular twofold axis. The active site pocket in L1, containing the quinone oxygen atoms of TTQ, is adjacent to the H1 subunit. A channel leading into the active site is located in the H_1L_1 interface. The H_1L_1 interface is the most extensive, covering an area of about 1600 Å^2. The interactions are largely hydrophobic, involving approximately 25 residues from each subunit with an additional 20 hydrogen bonds linking the subunits. The H_1L_2 interface is slightly less extensive, covering about 1400 Å^2, but having roughly the same number of hydrophobic and hydrogen-bonding interactions. This interface involves a N-terminal hairpin of the heavy subunit (Fig. 6). The H_1H_2 interface is the smallest, at about 700 Å^2, and contains two salt bridges and 12 van der Waals interactions from each subunit. There is no contact between L_1 and L_2.

The heavy subunit of MADH contains about 370 residues and can be divided into two domains as shown in Fig. 7.[28b] The first domain is very small, consisting of a 31-residue hairpin which packs against the light subunit of the other HL dimer of the heterotetramer. The first 16 residues of this hairpin form an α-helix. As described above, 18 residues were cleaved from the PD-MADH molecule prior to crystallization.

The second domain of the H subunit consists of about 330 residues. This domain is in predominantly β structures and forms a cylinder made up of seven four-stranded antiparallel β sheets. Each sheet forms a β leaflet structure topologically in the form of the letter "W". These leaflets are arranged around a central axis with pseudo sevenfold symmetry. In the H_2L_2 tetramer, these pseudo sevenfold axes lie in a plane approximately perpendicular to the molecular twofold axis (see Fig. 6). The inside strand of each β leaflet is approximately parallel to the pseudo sevenfold axis whereas the outermost strand is inclined to it by approximately 45°. For each β leaflet, the lower sequence number starts at the innermost strand and fans outward toward the outermost strand as the sequence advances. The one exception to the simple W-like motif is sheet 7 which contains both chain termini. In this sheet, the innermost strand is formed by the C-terminal segment of the subunit whereas the three succeeding strands are formed by the N-terminal part of the chain, just after the hairpin domain. Consecutive leaflets are connected by long loops which link the outer strand of one to the inner strand of the next. The only helix in this domain is located between the fourth and fifth sheets. The heavy subunit contains one disulfide bond, within sheet 4.

[28b] P. J. Kraulis, *J. Appl. Crystallogr.* **24,** 946 (1991).

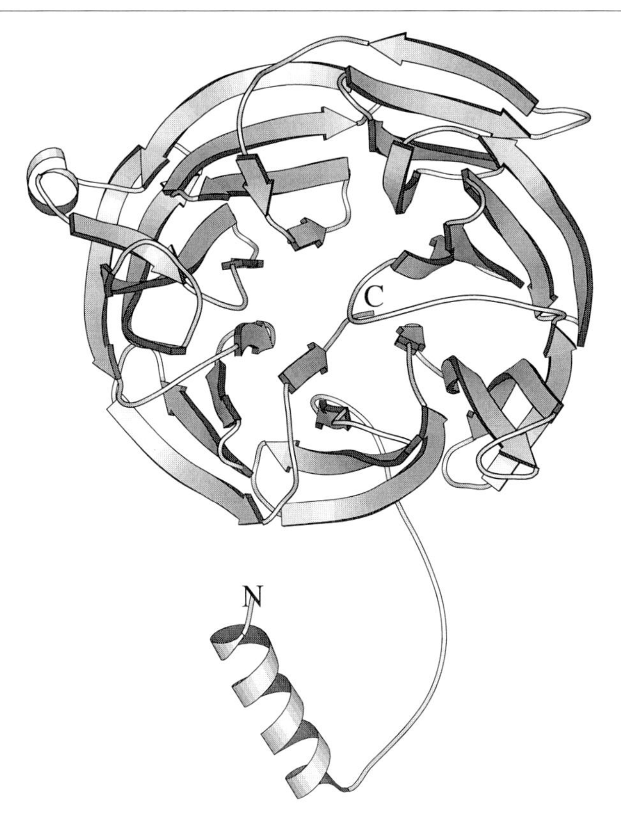

FIG. 7. Ribbon drawing of the heavy subunit of methylamine dehydrogenase. The pseudo sevenfold axis is approximately perpendicular to the diagram. The diagram was made using the program MOLSCRIPT.[28b]

The light subunits of TV-MADH[29] and PD-MADH[11] contain 130 and 131 residues, respectively, of which the first six are disordered in both crystal structures. It consists of six β strands which form two antiparallel β sheets, of two strand and three strands each. The topological connectivity of these sheets is shown in Fig. 8a. The three-stranded sheet is approximately 12 residues in length and is twisted by nearly 180° (Fig. 8b). It is irregular, with one edge of the sheet consisting of two distinct strands, 2 and 2′, separated by a three-residue excursion of the polypeptide chain. One of the attachment points of TTQ, at Trq-57, is located at the beginning of strand 2′. The TTQ cofactor is positioned between the three-residue excur-

[29] M. Ubbink, M. A. G. van Kleef, D. Kleinjan, D. Hoitink, F. Huitema, J. J. Beintema, J. A. Duine, and G. Canters, *Eur. J. Biochem.* **202,** 1003 (1991).

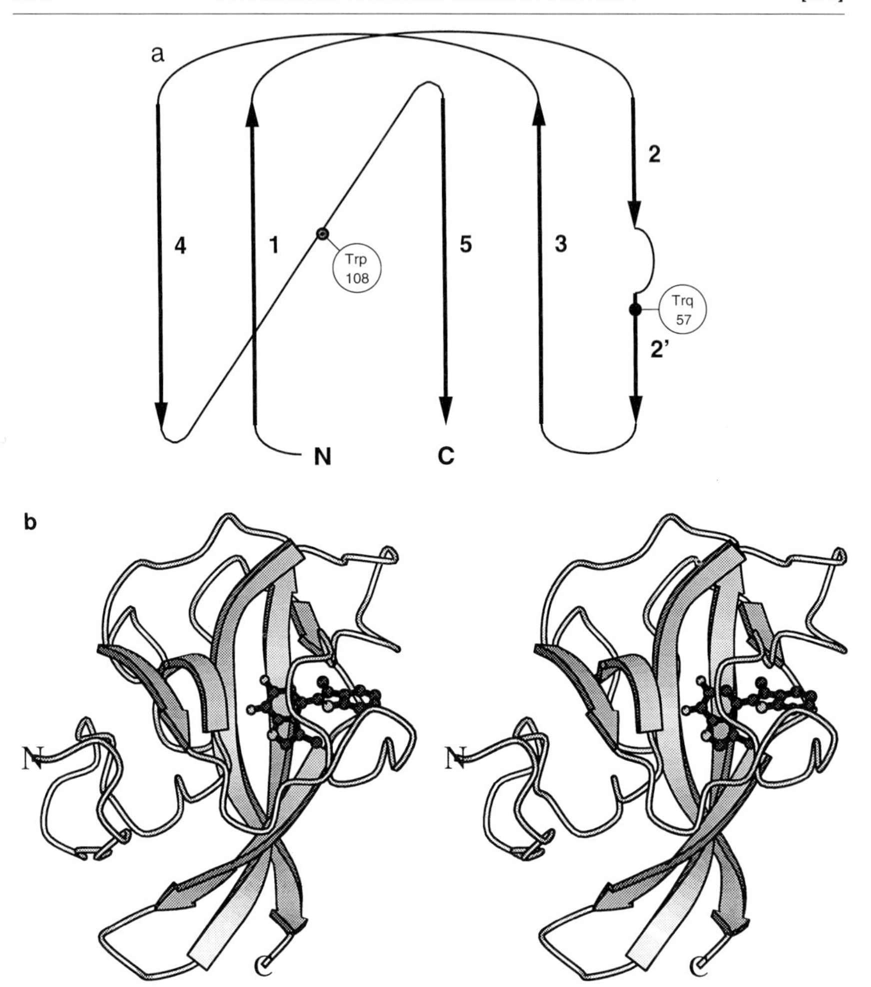

FIG. 8. (a) Topological diagram of the light subunit of methylamine dehydrogenase. There are two antiparallel β sheets, a two-stranded sheet formed by β strands 1 and 4, and a three-stranded sheet formed by β strands 2, 2′, 3, and 5. The attachment points of TTQ (Trq-57 and Trp-108) are indicated. (b) Ribbon diagram showing the light subunit of methylamine dehydrogenase. A ball and stick representation of the redox cofactor, TTQ, is also shown. The diagram was made using the program MOLSCRIPT.[28b]

sion and a portion of a long, irregular loop between strands 4 and 5, which includes position 108, the second attachment site of TTQ to the light subunit. The two-stranded sheet is located on one side of the light subunit and makes contact with the quinone-containing edge of Trq-57.

The L subunit is highly cross-linked, containing six pairs of disulfide

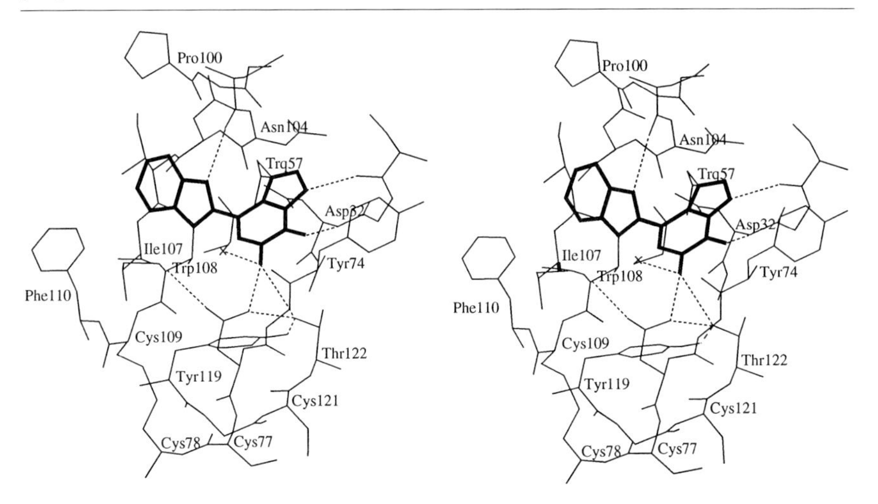

FIG. 9. Stereo diagram of the cofactor region in TV-MADH. Possible hydrogen bonds to TTQ, defined by a donor–acceptor distance shorter than 3.2 Å, are indicated by dotted lines. The cross indicates a solvent molecule or ion in the active site pocket. Reproduced by permission from Huizinga *et al.*[31]

bones. These intrasubunit cross-links are probably responsible for the thermal stability of the subunit.[30] One additional cross-link exists between the two indole rings at positions 57 and 108 which form the TTQ cofactor.

The environment of TTQ in the L subunit of MADH is shown in Fig. 9. The Trq-57 portion is part of the three-stranded β sheet located at the core of the L subunit. In the HL dimer, Trq-57 is located just above the heavy subunit, coincident with the pseudo sevenfold symmetry axis. Most of the TTQ ring system is deeply buried within the subunit, its center of gravity located about 6 Å below the molecular surface. Atoms O6 and O7 of Trq 57 (Fig. 1) are contained within a small pocket at the active site, whereas atoms C5 and C6 of the phenyl ring of Trp-108 are located on the protein surface. Three of the four heteroatoms of TTQ form hydrogen bonds to backbone atoms: Trp-108 N1 to Ala-103 O, Trq-57 N1 to Gly-31 O, and Trq-57 O7 to Asp-32 N. The fourth heteroatom, Trq-57 O6, is hydrogen bonded to Thr-122 OG1, Asp-76 OD, and a solvent molecule located in the active site pocket. The chemical nature of the solvent molecule is not known, but it could be an ammonium ion since ammonia is a competitive inhibitor of the enzyme and is present in the crystals at high concentration (Table I). The active site pocket is lined with several hydrophobic side chains together with the side chains of Asp-32 and Asp-76.

[30] S. Shirai, T. Matsumoto, and J. Tobari, *J. Biochem.* (*Tokyo*) **83,** 1599 (1978).

Atoms O6 and O7 of TTQ also protrude into the pocket. The pocket is isolated, by a phenylalanine side chain of the H subunit, from a channel between the H and L subunits leading to the protein surface. Movement of this or other side or main chain groups would be needed for the substrate to gain access to the active site.

MADH can be irreversibly inhibited by several hydrazines, compounds known to bind to carbonyl groups. The structures of two hydrazine complexes of TV-MADH have now been refined.[31] One of these complexes, with methyl hydrazine (MH), was prepared by soaking pregrown crystals of TV-MADH with the reagent. The other, with trifluoroethylhydrazine (TFEH), was prepared by cocrystallization. Both derivatives were isomorphous to the native crystals. For the TFEH derivative, the only significant features in the isomorphous difference map [using F_o(TFEH)-F_o(native) as coefficients][32] occurred at the active site. Refinement and difference Fourier analysis indicated a slight shift of the Trq-57 ring, disappearance of the solvent peak in the active site pocket, and the presence of elongated electron density attached to C6 of the Trq ring. This last feature was interpreted as the two hydrazine nitrogen atoms of TFEH. Although the first nitrogen is coplanar with the Trq ring, the second nitrogen protrudes from the plane of the ring. The isomorphous difference map for the MH derivative was considerably more noisy, but refinement and difference Fourier analysis showed changes almost identical to the TFEH derivative. In neither case were any peaks corresponding to the trifluoroethyl or methyl groups of the hydrazines found. The volume of the active site pocket was unchanged and its access was still blocked by side chains, making it difficult to model additional atoms of either the methyl or trifluoroethyl groups. A slight conformational change of the TTQ was found to occur, resulting in a 5° to 8° increase in the initial 40° dihedral angle between tryptophan rings. Two possible explanations for the absence of the methyl and trifluoroethyl groups from the refined structures are that these groups are disordered or that chemical degradation of the hydrazines had occurred during binding.

MADH–Amicyanin Binary Complex

The binary complex is a heterohexamer, containing two molecules of amicyanin per molecule of MADH.[3] Each amicyanin molecule is in contact with both the H and L subunits of a HL dimer. The surface of interaction is about twofold larger with the L subunit than with the H subunit. The

[31] E. G. Huizinga, B. A. M. van Zanten, J. A. Duine, J. A. Jongejan, F. Huitema, K. S. Wilson, and W. G. J. Hol, *Biochemistry* **31,** 9789 (1992).

[32] F_o(TFEH) and F_o(native) are the observed structure factors for the TFEH and native crystals, respectively.

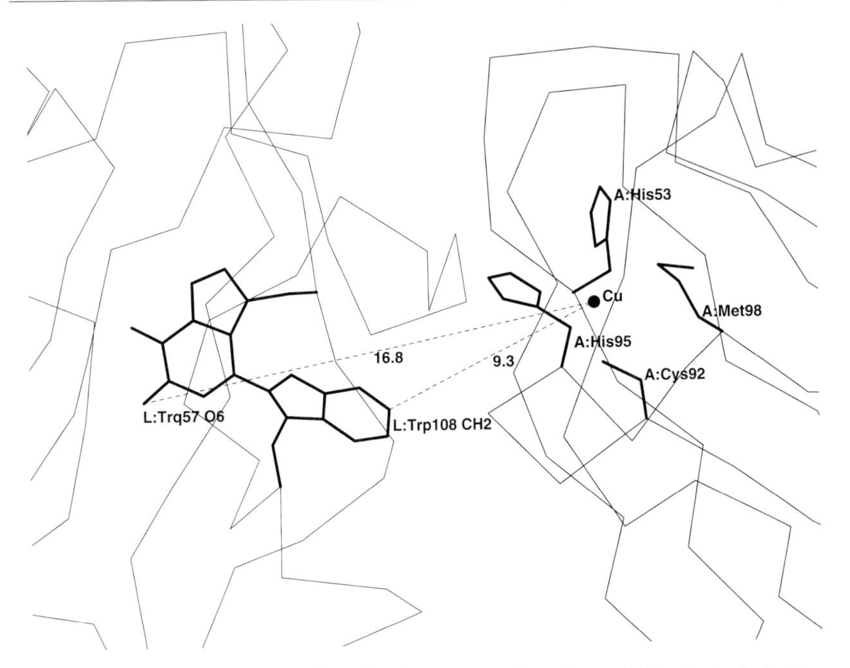

FIG. 10. Close-up view of the interface between amicyanin and MADH in their binary complex, showing the relative positions and distances between TTQ and copper. Only the side chains of the TTQ and the copper ligands are shown, with all remaining residues drawn as the C_α backbone. The H subunit of MADH is omitted for clarity. Side chains in the L subunit of MADH and amicyanin are labeled by the prefixes "L:" and "A:", respectively.

structures of both MADH and amicyanin are virtually the same in the binary complex as in the isolated molecules.[2,33]

The interface between amicyanin and the L subunit of MADH is shown in Fig. 10. The distance from the copper atom to C6 of Trq-57 (see Fig. 1), the primary site of reduction of TTQ by the substrate, is 16.8 Å whereas the closest atom of TTQ to the copper (the solvent exposed C6′ of Trp 108) is 9.3 Å. The area of the interface is approximately 700 $Å^2$ and it contains 23 residues of MADH (15 of these from the L subunit) and 19 residues of amicyanin. The interface is predominantly hydrophobic, even though portions of nine charged and nine neutral hydrophilic residues are contained within it. In addition to the hydrophobic interactions, there are two salt bridges and three water bridges connecting the two molecules.

The presence of an electrostatic component to the interaction between MADH and amicyanin is consistent with solution studies which show that

[33] R. Durley, L. Chen, L. W. Lim, F. S. Mathews, and V. L. Davidson, *Protein Sci.* **2,** 739 (1993).

the association between these proteins in low ionic strength buffer decreases on the addition of 0.2 *M* NaCl.[13] The presence of the two salt bridges connecting the H and L subunits of MADH with amicyanin is also consistent with cross-linking studies using EDC {1-ethyl-3-[3-(dimethylamino)propyl]carbodiimide}, which affects a covalent linkage between carboxylate groups and basic amino acid side chains that are in close proximity. It was shown that EDC specifically cross-links both the H and L subunits of MADH to amicyanin in solution.[14] Subsequent chemical modification studies demonstrated that the cross-linked carboxylate groups were located on MADH, implying that the cross-linked basic residues were on amicyanin.[34] Neutralization of two positive charges on amicyanin, which are in the vicinity of its redox center, may explain the 73 mV decrease in redox potential of amicyanin on complex formation with MADH.[12]

MADH–Amicyanin–Cytochrome c_{551i} Ternary Complex

The ternary complex is a heterooctomer containing two molecules of cytochrome c_{551i} per heterohexamer of the binary complex.[4] The two halves of the heterooctomer are related by an exact crystallographic twofold axis. The arrangement of the cytochrome, amicyanin, and the light subunit of MADH is shown in Fig. 11. The TTQ, copper, and heme groups are arranged in a linear fashion, placing the cytochrome and MADH on opposite sides of the amicyanin. The copper to iron distance is 24.8 Å and the distance from copper to the nearest atom of the heme is about 21 Å.

The interface between MADH and amicyanin in the ternary complex is essentially identical to that of the binary complex. The amicyanin–cytochrome interface is somewhat different from the MADH–amicyanin interface. It is smaller, covering approximately 400 $Å^2$, and considerably more polar.[35] It contains nine charged, five neutral hydrophilic, and eight hydrophobic side chains, which form one salt bridge and three hydrogen bonds; in addition, there are two solvent molecules bridging the two molecules. Thus, despite the smaller interface, the number of direct links between amicyanin and the cytochrome is greater than that between amicyanin and MADH.

Methanol Dehydrogenase

The heterotetramer of MEDH consists of a pair of heterodimers, each made up of a large subunit containing 571 amino acids and a small subunit containing about 70 amino acids. Only the first 57 residues of the small

[34] V. L. Davidson, L. H. Jones, and M. A. Kumar, *Biochemistry* **29,** 10786 (1990).
[35] L. Chen, R. Durley, F. S. Mathews, and V. L. Davidson, unpublished results (1993).

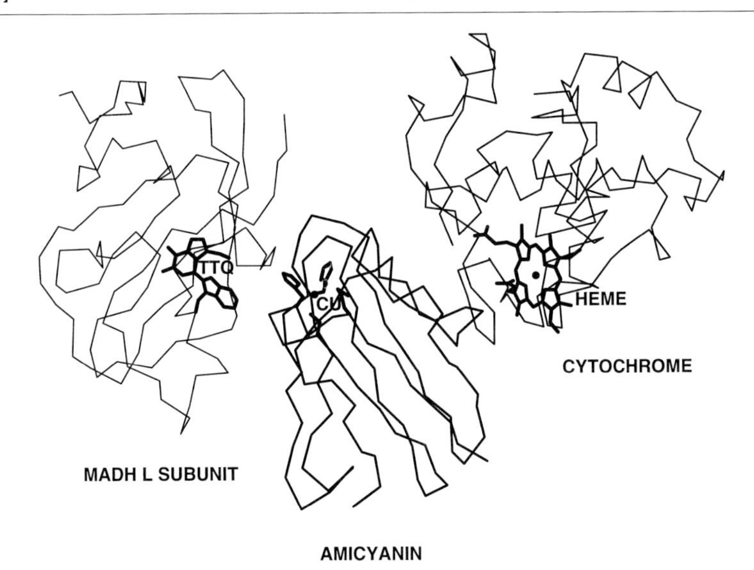

FIG. 11. The relationship of the three redox centers in the ternary complex of MADH, amicyanin, and cytochrome c_{551i}. The C_α diagrams of the L subunit of MADH, amicyanin, and cytochrome c_{551i} and the redox cofactors, TTQ, Cu^{2+}, and heme are shown.

subunit have been located in the electron density. Each large subunit consists of a single domain containing eight sets of four-stranded antiparallel β sheets, each forming a structure shaped like the letter W. This structural motif is remarkably similar to that of the H subunit of MADH and has been found in two other protein structures (see below). The eight β sheets are arranged circularly, forming an eightfold superbarrel (Fig. 12) with some additional short helices, loops, and excursions within and between some of the W's. The PQQ prosthetic group is located within the funnel-shaped central channel near one end of the superbarrel. The normal to the PQQ plane is inclined by about 20° from the pseudo eightfold axis.

The small subunit of MEDH is folded into the shape of the letter "j" and consists of an irregular N-terminal segment of about 34 residues followed by a long α helix. The conformation of the small subunit was predicted from the amino acid sequence of the small subunit of MEDH from *M. extorquens* AM1.[36] The subunit does not appear to form an independent globular domain, but extends over the outer surface of the large subunit. It has little independent tertiary structure, at least in the portion visible in the electron density map, and most of its interactions are with the large subunit.

The initial model of MEDH[5] was based on a homologous amino acid

[36] J. M. Cox, D. J. Day, and C. Anthony, *Biochim. Biophys. Acta* **1119,** 97 (1991).

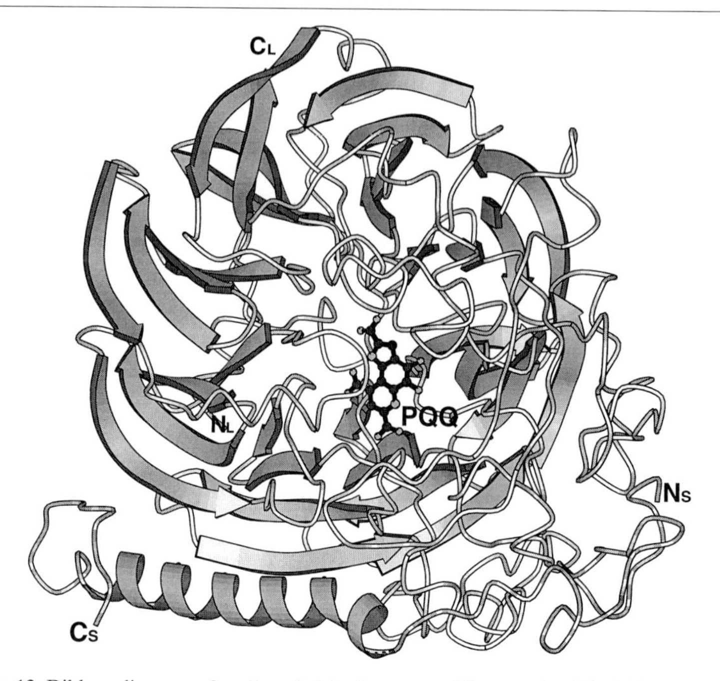

FIG. 12. Ribbon diagram of methanol dehydrogenase. The pseudo eightfold symmetry axis relating the eight β leaflets of the large subunit is approximately perpendicular to the diagram. A ball and stick representation of PQQ is shown in the center of the superbarrel. The small subunit of MEDH consisting of a helix and an extended chain is located at the bottom of the diagram. The amino and carboxyl termini of the large and small subunits are labeled N_L, C_L, Ns, and Cs, respectively. The diagram was made using the program MOLSCRIPT.[28b]

sequence and thus contained a number of inaccuracies. After the gene sequence of the large subunit of W3A1 MEDH was determined,[18] the molecular model was rebuilt and refined. This provided a more reliable model of the protein, especially at the active site; the PQQ moiety could be positioned accurately. In the process, a calcium ion, bound between the PQQ and the protein, was identified and a vicinal cystine disulfide group was located adjacent to one face of the PQQ.[28] The active site of MEDH is shown in Fig. 13a. The ring system of PQQ is in a hydrophobic environment sandwiched between Trp-237, below, and the Cys-103–Cys-104 disulfide bridge, above. The carboxyl group at position 9 forms a salt bridge with Arg-109 (Fig. 13b) and both groups are thoroughly shielded from the bulk solvent by the disulfide. This hydrophobic arrangement is in agreement with ENDOR experiments[37] which also predicted the involvement of the

[37] J. A. Duine, J. Frank, and R. de Beer, *Arch. Biochem. Biophys.* **233,** 708 (1984).

9-carboxylate group in PQQ binding. The carboxyl group at position 2 is located 2.7 Å away from a carboxyl oxygen of Glu-55. Since both acidic groups are shielded from solvent, their proximity to each other suggests that one or both of them are protonated, which would stabilize their interaction through hydrogen bond formation. The calcium ion is 6-coordinate. It is bound to atoms O5, N6, and O7A of PQQ, the two carboxyl oxygen atoms of Glu-171, and the side chain oxygen of Asn-255. Somewhat farther away are the side chains of Asp-297 and Arg-324 which are hydrogen bonded to each other. The latter residue is also hydrogen bonded to O4 and O5 of PQQ.

The functional role of Ca^{2+} in MEDH is not known at this time. Its importance is demonstrated by the fact that at least three gene products of the *mox* gene cluster are required for its incorporation into the enzyme.[19] MEDH isolated from mutants defective in any of these genes lack Ca^{2+}, but have structural properties identical to the wild-type enzyme and contain a full complement of PQQ. However, they are inactive and exhibit perturbed PQQ absorption spectra. The fully active enzyme, indistinguishable from the wild type, can be constituted from these inactive proteins by prolonged *in vitro* incubation with high concentrations of Ca^{2+}. The positive charge on the Ca^{2+} in the wild-type enzyme may have an inductive effect on the redox potential of PQQ, helping to stabilize the reduced and semiquinone forms of the enzyme. The absence of Ca^{2+} could allow alternate orientations of the PQQ ring and of some of the protein side chains in the active site, thereby destroying the catalytic ability of the enzyme.

Superbarrel Motif

The β leaflet motif found in MADH and MEDH has been observed in two other proteins: the influenza virus neuraminidase[38] and galactose oxidase (GO)[39] (see also Knowles *et al.* this volume). In each case, the four stranded W-like β sheets are arranged about a central axis of pseudosymmetry, with the strands of each sheet progressing outward from the central axis. However, the degree of pseudosymmetry differs among the four proteins, being sixfold for the neuraminidase, sevenfold for MADH and GO, and eightfold for MEDH. The topologies of these four proteins are compared schematically in Fig. 14.

The arrangement of the "last" β sheet (see Fig. 14) of each protein, which contains the N- and C-terminal strands of the domains, differ somewhat. In MEDH, the outside strand of W-8 comes from the N-terminal

[38] J. N. Varghese, W. G. Laver, and P. W. Colman, *Nature (London)* **303,** 35 (1983).

[39] N. Ito, S. E. V. Phillips, C. Stevens, Z. B. Ogel, M. J. McPherson, J. N. Keen, K. D. S. Yadov, and P. F. Knowles, *Nature (London)* **350,** 87 (1991).

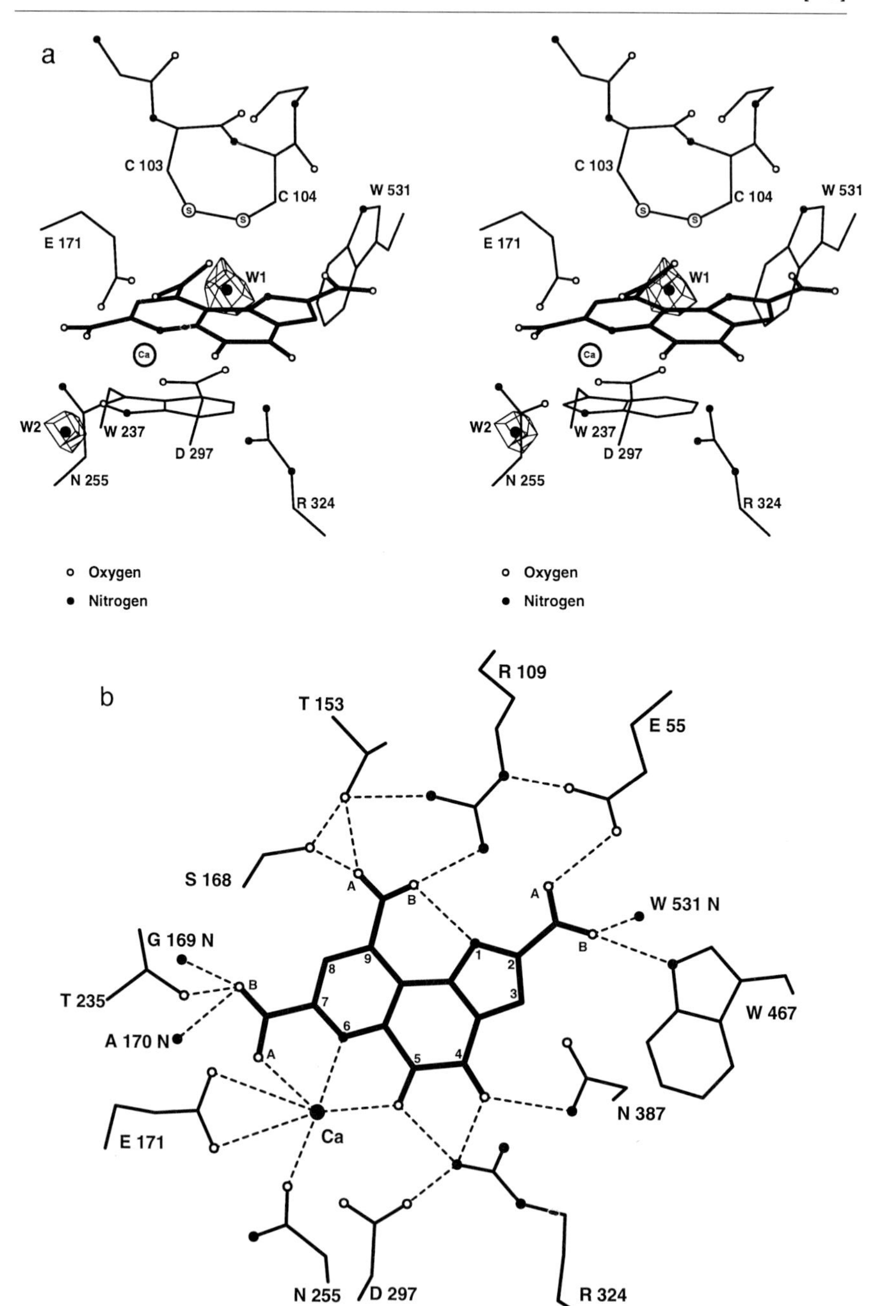
a
C 103
C 104
W 531
E 171
W1
Ca
W2
W 237
D 297
N 255
R 324
Oxygen
Nitrogen
b
R 109
T 153
E 55
S 168
A
B
W 531 N
G 169 N
T 235
A 170 N
W 467
E 171
Ca
N 387
N 255
D 297
R 324

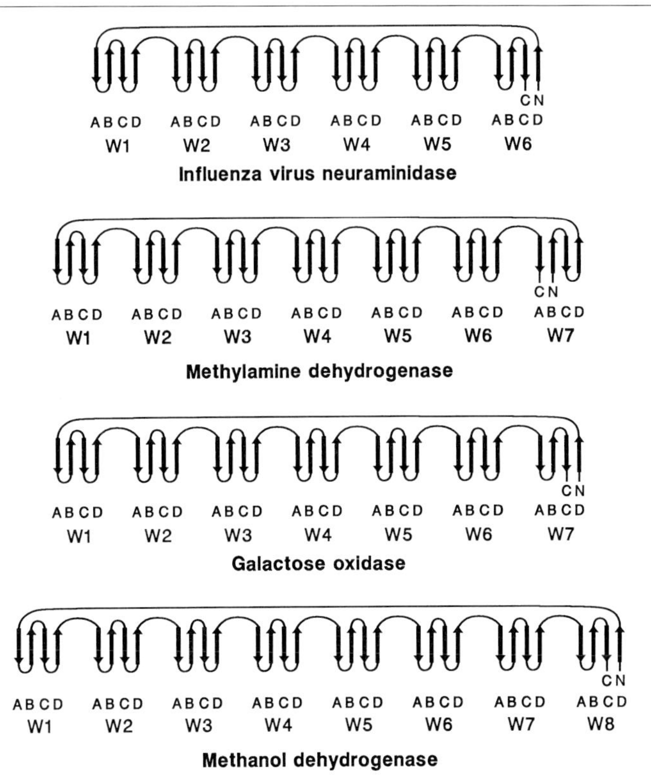

FIG. 14. The folding topology of the six-, seven-, and eightfold β leaflets of influenza neuraminidase, methylamine dehydrogenase, galactose oxidase, and methanol dehydrogenase.

segment which then goes on to form W-1. This construction is similar to that of the influenza neuraminidase and of GO but differs from that of MADH, as shown in Fig. 14. In MADH, all three outside strands of the last sheet, W-7, are formed by the N-terminal segment.

Several major structural and functional differences exist between the four protein structures. MADH is a H_2L_2 heterotetramer in which the H subunit corresponds to the superbarrel domain, except for the 30 residue

---

FIG. 13. (a) Stereoscopic view of the active site of methanol dehydrogenase from *Methylophilus* W3A1. The PQQ cofactor is seen nearly edge on. A calcium ion and two solvent molecules (W1 and W2, seen in difference electron density) are also shown. The diagram was made using the program MOLSCRIPT.[28b] (b) The active site of methanol dehydrogenase from *Methylophilus* W3A1 parallel to the plane of PQQ. All hydrogen bond interactions between PQQ and the protein, as well as the coordination bonds of the calcium ion, are shown.

N-terminal tail which protrudes from the domain. The L subunit is an independently folded domain which interacts with both H domains in the heterotetramer. MEDH is also a heterotetramer. In this case, the large subunit consists of the superbarrel domain and contains the PQQ redox cofactor. The small subunits are not folded as independent globular domains but are stretched over the large subunits and make no other intersubunit interactions in the tetramer.

The influenza virus neuraminidase is a homotetramer with circular fourfold symmetry. Each subunit of the crystalline enzyme, of approximate MW 45,000, consists of a sixfold pseudosymmetric structure, except for short appendages at the N- and C-terminal ends and a complicated excursion between two of the β sheets. The enzyme is a glycosylase, cleaving an α-ketosidic linkage between sialic acid and a sugar residue. The active site is located at the top of the superbarrel, close to the sixfold pseudosymmetric axis, in a pocket rimmed by loops from various excursions from the individual β leaflets.

Galactose oxidase is a monomeric enzyme with an approximate MW of 68,000. It consists of three domains. The central domain is a sevenfold pseudosymmetric superbarrel about 375 residues in length. The enzyme catalyzes the oxidation of a broad range of primary alcohol substrates. The active site of the enzyme contains a copper atom located at the top of the central domain on the pseudo sevenfold symmetry axis. Three of the copper ligands come from the superbarrel whereas the fourth is provided by a long loop which extends from the third domain and passes through the center of the superbarrel domain from below.

The four superbarrel structures, although quite varied with respect to each other in structure and function, share a common feature. The active sites of the enzymes are centered on the axes of pseudosymmetry. In three of the four enzymes, MEDH, GO, and neuraminidase, the active site is located in the funnel of the superbarrel, whereas in the fourth enzyme, MADH, the active site and redox cofactor are located in a separate subunit. Further investigation is needed to determine whether this similarity is a significant feature of this structural motif or merely coincidental.

Acknowledgments

This work has been supported by NSF Grant No. MCB-9119789 and by USPHS Grant No. 31611. The author thanks Drs. Rosemary Durley, Longyin Chen, and Scott White for helpful advice.

[16] Genetics of Bacterial Quinoproteins

By MARY E. LIDSTROM

Several bacterial quinoproteins have been subjected to genetic analysis, including the pyrroloquinoline quinone (PQQ)-containing enzymes methanol dehydrogenase (MEDH), alcohol dehydrogenase (ADH) and glucose dehydrogenase (GDH), and the tryptophan tryptophyl quinone (TTQ)-containing enzyme methylamine dehydrogenase (MADH) (see Table I). In all of these cases, the genetic approach has provided new insights into a variety of aspects of these enzymes, including the prosthetic groups and their synthesis, enzyme structure and function, and post-translational assembly and maturation processes. In this chapter, the methods that have been used genetically to dissect bacterial quinoproteins and their prosthetic groups will be presented with examples of how each type of approach has led to new insights into specific bacterial quinoproteins.

Gene Cloning

In most cases, major breakthroughs in genetics of bacterial quinoproteins have been made when one or more of the genes involved in the generation of active enzyme have been cloned and characterized. Two main approaches have been successful, cloning by mutant complementation or cloning by hybridization with oligonucleotide probes based on known amino acid sequence. Each of these will be considered separately.

Mutant Complementation

Genes for a specific function can be cloned using mutants deficient in that function. The most common method for doing this is to transfer a pool of plasmids comprising a genomic clone bank into each mutant and isolating those plasmids that restore the function lost in the mutant. This is termed mutant complementation. This approach requires a method for screening and identifying the mutants, an appropriate cloning vector for the strain in which the mutants are isolated, and a method for transferring the vector into the mutant strains.

This method has been used to clone genes for MEDH,[1-3] ADH,[4] and

[1] D. N. Nunn and M. E. Lidstrom, *J. Bacteriol.* **166,** 581 (1986).

[2] S. M. Machlin, P. E. Tam, C. A. Bastien, and R. S. Hanson, *J. Bacteriol.* **170,** 141 (1987).

TABLE I
QUINOPROTEINS AND BACTERIAL STRAINS DISCUSSED

Quinoprotein	Prosthetic group	Strain	Gene designations
Methanol dehydrogenase	PQQ	*Methylobacterium extorquens* AM1	*mox, pqq*
		Methylobacterium organophilum DSM 760	*mox, pqq*
		M. organophilum XX	*mox, pqq*
		Paracoccus denitrificans	*mox*
Alcohol dehydrogenase	PQQ	*Acetobacter aceti*	
Glucose dehydrogenase	PQQ	*Acetobacter calcoaceticus*	*gdh, pqq*
		Klebsiella pneumoniae	*gdh, pqq*
		Escherichia coli	*gdh, pqq*
Methylamine dehydrogenase	TTQ	*M. extorquens* AM1	*mau*
		P. denitrificans	*mau*
		Thiobacillus versutus	*mad, ami*
		Methylophilus W3A1	*mau*
		Methylobacillus flagellatum KT	*mau*

GDH[5,6] as well as genes for the synthesis of PQQ.[7–10] For the MEDH system, a positive selection method exists for *Methylobacterium* strains. Mutants deficient in MEDH activity will grow on nutrient agar plates containing 0.05% allyl alcohol in the presence of 0.5% methanol. Strains containing MEDH activity will convert the allyl alcohol to allyl aldehyde, which is toxic, and those cells will not grow. A large number of both spontaneous and mutagen-induced mutants can be readily isolated by this method.[1,3,7] However, it is only applicable to strains in which MEDH is expressed during growth in the presence of a secondary substrate. In *Paracoccus denitrificans*, MEDH is tightly regulated by repression. Therefore, both mutants and wild-types grow on these plates and this procedure does

[3] C. Bastien, S. Machlin, Y. Zhang, K. Donaldson, and R. S. Hanson, *Appl. Environ. Microbiol.* **55,** 3124 (1989).

[4] T. Inoue, M. Sunagawa, A. Mori, C. Imai, M. Fukuda, M. Takagi, and K. Yano, *J. Bacteriol.* **171,** 3115 (1989).

[5] A.-M. Cleton-Jansen, N. Goosen, K. Vink, and P. Van De Putte, *J. Bacteriol.* **170,** 2121 (1988).

[6] A.-M. Cleton-Jansen, N. Goosen, O. Fayet, and P. Van De Putte, *J. Bacteriol.* **172,** 6308 (1990).

[7] F. Biville, E. Turlin, and F. Gasser, *J. Gen. Microbiol.* **135,** 2917 (1989).

[8] C. J. Morris, F. Biville, E. Turlin, E. Lee, K. Ellermann, W.-H. Fan, R. Ramamoorthi, A. L. Springer, and M. E. Lidstrom, *J. Bacteriol.* **176,** 1746 (1994).

[9] J. J. M. Meulenberg, W. Loenen, E. Sellink, and P. Postma, *Mol. Gen. Genet.* **220,** 481 (1990).

[10] N. Goosen, D. A. M. Vermaas, and P. Van De Putte, *J. Bacteriol.* **169,** 303 (1987).

not identify mutants.[11] Mutants in ADH and GDH have been isolated by screening for lack of acid production under the appropriate growth conditions, as evidenced by lack of a halo on medium containing calcium carbonate[4] or by screening for white colonies on MacConkey plates.[10] Mutants in GDH in enteric bacteria are more difficult to obtain since these bacteria contain two systems for growing on glucose, GDH and the phosphotransferase (PTS) system. Therefore, mutants deficient in the PTS system are used as starting strains, which are then mutagenized and screened for lack of acid production on glucose.[6] However, the complemented strains do not grow well on glucose plus PQQ, and so they must be identified by selecting for the vector marker and screening individual colonies for growth and acid production. In all of these cases, if the host strain is capable of synthesizing PQQ, some of the mutants isolated will be defective in PQQ synthesis. These can be identified by testing for the appropriate phenotype in the presence and absence of PQQ, which is available from Fluka Chemical Corp. (New York, NY). A simple method for this testing is to place a sterile disk soaked in a PQQ solution onto the center of a test plate and streak sectors with test strains. Mutants can be further characterized as to the presence of apoproteins and electron acceptors, if antisera are available.

Once mutants are isolated and characterized, they can be complemented with genomic clone banks. A variety of vectors are available for use in the enteric bacteria, but the options for other strains are more limited. In the *Methylobacterium* strains, broad host range plasmids of the IncPI group are the vectors of choice[1–3,7] whereas in *P. denitrificans* and *Acinetobacter calcoaceticus*, vectors based on the IncQ systems have been used.[11] The latter vectors do not replicate in most *Methylobacterium* strains.[7] In both of these cases, the vectors are moved into the host strains by conjugation using a helper plasmid, whereas in the enteric strains both transformation and transduction have been used.

Once a clone is isolated that complements a mutant, a variety of analyses can be carried out. First, the insert in the clone is mapped with restriction enzymes. Subclones of the plasmid can then be generated and tested for complementation of the original mutant, as well as any other mutants that are available. In this way, the location of the gene defective in the original mutant can be mapped. In addition, if a number of mutants are available, complementation classes can be identified. Usually, each complementation class corresponds to at least one gene, although in some cases a single class has turned out to contain more than one gene and in other cases two classes represent different parts of a single gene.[8] Specific gene designations can

[11] N. Harms and R. J. M. Van Spanning, *J. Bioenerg. Biomembr.* **23,** 187 (1991).

only be definitively assigned after expression and/or sequencing analysis (see below). However, placing mutants in complementation classes facilitates their phenotypic analysis. Instead of carrying out extensive studies of each mutant, a few examples from each complementation class can be studied in detail.

Oligonucleotide Probing

An alternative method for gene cloning that has proven useful for isolating both GDH[12] and MADH[13–15] genes is to use hybridization to an oligonucleotide, whose sequence is based on known amino acid sequence information. In those cases in which amino acid sequence data are available for enzymes or electron acceptors, it is possible to synthesize degenerate oligonucleotides that can be labeled and used to identify specific restriction fragments in genomic DNA digests. Fragments of a similar size can be isolated, ligated into vectors and used to transform *Escherichia coli*. This partial clone bank can then be probed with the same degenerate oligonucleotide pool to identify colonies containing DNA that hybridizes to the probe. Once clones are identified, the inserts are restriction mapped and hybridization to the correct size fragment is confirmed. Although this approach is a standard gene cloning method, it is common to obtain hybridization artifacts, depending on the redundancy of the probe, and so confirmation of the correct gene by sequencing or expression (see below) is required.

Once structural genes are cloned, they in turn can be used as hybridization probes in a similar manner, to isolate homologs from other strains. This approach has been used to clone MADH genes from *P. denitrificans*[16] and *Methylophilus* W3A1.[17]

Identifying Gene Function

One of the benefits of a gene cloning approach is that it can be used to identify genes involved in an enzyme activity without prior knowledge

[12] A.-M. Cleton-Jansen, N. Goosen, K. Vink, and P. Van De Putte, *Mol. Gen. Genet.* **217,** 430 (1989).

[13] A. Y. Chistoserdov, Y. D. Tsygankov, and M. E. Lidstrom, *Biochem. Biophys. Res. Commun.* **172,** 211 (1990).

[14] R. J. M. Van Spanning, C. W. Wansell, W. N. M. Reijnders, L. F. Oltmann, and A. H. Stouthamer, *FEBS Lett.* **275,** 217 (1990).

[15] F. Huitema, J. Van Beeumen, G. Van Driessche, J. A. Duine, and G. W. Canters, *J. Bacteriol.* **175,** 6254 (1993).

[16] A. Y. Chistoserdov, J. Boyd, F. S. Mathews, and M. E. Lidstrom, *Biochem. Biophys. Res. Commun.* **184,** 1226 (1992).

[17] A. Y. Chistoserdov, W. S. McIntire, F. S. Mathews, and M. E. Lidstrom, *J. Bacteriol.* **176,** 4073 (1994).

of the function of each gene. Once genes are identified, their function can be analyzed by a variety of techniques, including expression, sequencing, and mutant characterization. Genes of unknown function are generally isolated either by mutant complementation or by their proximity to structural genes identified by hybridization approaches. In the sections below, examples will be given in which genes required for generation of active bacterial quinoproteins have been analyzed and used to obtain new information concerning those enzymes.

Expression

An isolated DNA fragment that has been identified as containing a gene of interest may contain other related genes nearby, as part of a gene cluster. One method for analyzing gene clusters is regulated expression in a heterologous system. Expression systems provide information on the direction of transcription of genes on the insert and on the number and size of encoded polypeptides. Therefore, the number of genes on a particular DNA insert can be judged much more rapidly than by DNA sequencing, especially if the insert is large. In addition, if antisera are available for known proteins, expressed polypeptides can be identified by immunoblotting techniques.[18] By using a series of subclones in the expression and immunoblot experiments, gene product relationships can be quickly established, even in the absence of known amino acid sequence information. The most common expression strain is *E. coli*, and a number of vectors are available for achieving regulated expression of insert DNA. In both the MEDH and MADH systems, the T7 expression system[19] has been used in *E. coli* to study polypeptides encoded on DNA fragments containing genes for structural polypeptides for the respective enzymes. In this vector system, the insert to be expressed is cloned immediately after a promoter from phage T7. Another compatible plasmid contains the T7 RNA polymerase, under control of an inducible promoter. Unlike bacterial RNA polymerases, the T7 RNA polymerase is insensitive to rifamycin. Therefore, expression of insert DNA can be initiated in the presence of rifamycin, and very little background from nonplasmid DNA is generated.[19] This system is especially useful for DNA of high %G+C since the T7 RNA polymerase transcribes through the hairpin structures that are common in high %G+C DNA. These structures tend to terminate transcription by *E. coli* RNA polymerase, interfering with expression.

[18] D. J. Anderson and M. E. Lidstrom, *J. Bacteriol.* **170,** 2254 (1988).

[19] S. Tabor, *in* "Current Protocols in Molecular Biology" (F. A. Ausubel, ed.), Sect. 16.2.1. Wiley (Interscience) New York, 1987.

For MEDH, expression experiments were used to show that the DNA fragment containing *mxaF*, the gene encoding the known 65-kDa subunit of MEDH, also encoded three other polypeptides. One was a 9-kDa polypeptide that routinely copurifies with MEDH.[18] This second polypeptide was subsequently shown to be a second subunit of MEDH.[20] Another encoded polypeptide was shown by immunoblots to be the cytochrome c_L, the electron acceptor of MEDH, and a fourth polypeptide of 30 kDa was encoded by a gene of unknown function. Likewise, in the MADH system, a DNA fragment known to contain the gene for the 14-kDa subunit of MADH was shown to encode a total of five polypeptides.[21] Immunoblot experiments identified one of these as the 43-kDa subunit of MADH, one was known from sequencing data to be amicyanin, the electron acceptor of MADH, and the other two were of unknown function.[21]

Sequencing

DNA sequencing provides a great deal of information concerning isolated genes. First, it provides the primary amino acid sequence for genes that have been isolated. This has been extremely useful for structural studies of the bacterial quinoproteins discussed here (see [15], this volume). Second, if related genes are clustered, sequencing on both sides of a primary isolated gene can reveal how many genes are present in the gene cluster and the deduced amino acid sequence of open reading frames can provide clues as to the function of those unknown genes. Identification of potential genes is done by searching for open reading frames, checking upstream for an appropriate ribosome binding site, and then checking each proposed open reading frame to determine whether it has the expected codon usage. If codon usage information exists for the strain from which the DNA was obtained, this can be used for direct comparison. If codon usage information is not available, general matching can be analyzed using codon usage tables for an organism with a similar %G+C content of the DNA. It is usually true that if more than one potential open reading frame is identified within a putative gene, only the correct one will show appropriate codon usage. Once potential genes are identified, deduced amino acid sequences for open reading frames can be compared to those present in current databases, and it is often possible tentatively to identify the function of the gene product based on similarity to other proteins. Genes identified solely on the basis of DNA sequencing are usually referred to as ORFs (open reading frames) and must be confirmed as functional genes by either expression or mutant studies.

[20] D. N. Nunn, D. Day, and C. Anthony, *Biochem. J.* **260,** 857 (1989).

[21] A. Y. Chistoserdov, Y. D. Tsygankov, and M. E. Lidstrom, *J. Bacteriol.* **173,** 5901 (1991).

TABLE II
GENES ASSOCIATED WITH MADH

Strain	Genes[a]
Methylobacterium extorquens AM1	*mauFBEDACJGLMN*
Paracoccus denitrificans	*mauFBEDACJ*
Thiobacillus versutus	*madB orf3 orf1 madB ami orf2*
Methylotrophus W3A1	*mauFBEDAGLMN*
Methylobaccilus flagellatum KT	*mauFBEDAGLMN*

[a] *mauB, madB* encode MADH large (β) subunit; *mauA, madA* encode MADH small (α) subunit; *mauC, ami* encode amicyanin; *orf3* is similar to *mauE, orf1* is similar to *mauD,* and *orf2* is similar to *mauJ*; *mauG* has similarity to cytochrome *c* peroxidase; and *mauM, mauN* have similarity to ferredoxins.

This approach has been used most successfully with the MADH system. Genes required for MADH activity *in vivo* have been sequenced from five different strains (see Table II). In all cases, the genes are clustered, are found in a similar order, are all transcribed in the same direction, and have high levels of similarity between homologs.[15–17,22,23] The most extensive sequencing has been carried out with *M. extorquens* AM1,[23] *Methylophilus* W3A1,[17] and *M. flagellatum* KT.[24] In these cases, 11, 9, and 9 genes have been identified, respectively. In the latter two strains the gene for amicyanin is missing from the gene cluster, consistent with the lack of amicyanin in these two strains. In addition, an adjacent gene of unknown function (*mauJ*) is missing, and this may be involved with amicyanin production.[23] An analysis of the predicted amino acid sequences has been quite revealing in these cases. First, indications of the cell compartment in which the gene product is located have been obtained. Most of the gene products are predicted to be periplasmic, except MauE and MauF, which are predicted to be membrane proteins, and MauJ, which is predicted to be cytoplasmic. Second, the sequencing has confirmed the position of the genes for the large and small subunits of MADH and for amicyanin (Table II), which had been identified for *M. extorquens* AM1 in expression studies (see above). In addition, the sequencing has shown that the gene products of all of the *mauG* genes have similarity to cytochrome *c* peroxidase and that the gene products of all of the *mauM* and *mauN* genes have similarity to ferredoxins. This

[22] M. E. Lidstrom and A. Y. Chistoserdov, *in* "Microbial Growth on C_1 Compounds" (J. C. Murrell and D. P. Kelly, eds.), p. 381. Intercept Ltd., Andover, UK, 1993.

[23] A. Y. Chistoserdov, L. V. Chistoserdova, W. S. McIntire, and M. E. Lidstrom, *J. Bacteriol.* **176,** 4052 (1994).

[24] E. Gak, A. Y. Chistoserdov, and M. E. Lidstrom, unpublished.

TABLE III
GENES INVOLVED IN PQQ SYNTHESIS

Strain	Genes[a]							
Acetobacter calcoaceticus			IV	V	I	II	III	*orfR*
Klebsiella pneumoniae		*orfX*	*pqqA*	*B*	*C*	*D*	*E*	*F*
Methylobacterium organophilum DSM 760	*pqqF*	*pqqE*	*D*	*G*	*C*	*B*	*A*	
Methylobacterium extorquens AM1	*pqqEF*		*pqqD*	*G*	*C*	*B*	*A*	

[a] All genes are similar to those directly above or below them, except that *orfR* is similar to *orfX*, and *pqqE* from *M. organophilum* and *M. extorquens* are similar.

suggests the involvement of a periplasmic peroxidase and two periplasmic ferredoxins in the MADH system.

Information has also been gained concerning PQQ synthesis using this approach. Sequence of DNA fragments isolated by complementation of PQQ mutants showed the presence of six genes in *A. calcoaceticus* (genes IV, V, I, II, and III and *orfR*) and *Klebsiella pneumoniae* (*orfX pqqABCDE*), each with considerable similarity to the homolog in the other strain,[25,26] although in slightly different order (Table III). In addition, a seventh gene (*pqqF*) was identified in *K. pneumoniae*. More recent sequencing has shown that the first three genes in the *pqq* cluster in *M. extorquens* AM1 also have high similarity to the first three *pqq* genes for *A. calcoaceticus* and *K. pneumoniae*.[8] The most surprising result of the sequencing was the identification of a peptide (gene IV of *A. calcoaceticus*, *pqqA* of *K. pneumoniae*, and *pqqD* of *M. extorquens* AM1) with conserved tyrosine and glutamate residues. It has been proposed that this peptide is the precursor of PQQ synthesis (see [17], this volume). Consistent with this hypothesis is the identification of a gene with similarity to proteases (*pqqF* of *K. pneumoniae*) and another gene with similarity to dipeptidases (*orfX* of *K. pneumoniae* and *orfR* of *A. calcoaceticus*),[26] which may be involved in processing of the peptide precursor. The sequences of the other *pqq* genes have not provided clues to their function.

Mutant Construction and Characterization

Once genes are identified by sequencing or expression, it is important to confirm their involvement in the enzyme of interest by constructing

[25] N. Goosen, H. P. A. Horsman, R. G. M. Huinen, and P. Van De Putte, *J. Bacteriol.* **171,** 447 (1989).

[26] J. J. M. Meulenberg, E. Sellink, N. H. Riegman, and P. W. Postma, *Mol. Gen. Genet.* **232,** 284 (1992).

mutants in each gene. In addition, the mutant phenotypes often reveal the function of the gene mutated. The most common method for constructing mutants in cloned genes is to generate an insertion mutation in the gene using a fragment of well-characterized DNA, often containing a selectable antibiotic resistance marker. This can be inserted into a convenient restriction site, to generate a simple insertion mutation, or it can be inserted into a site that deletes a portion of the insert, so that it generates a combination deletion/insertion mutation. If the insertion is generated in a plasmid unable to replicate in the host strain, transfer of this plasmid to the host strain and selection for the marker in the inserted fragment will identify strains in which the marker has recombined into the chromosome at the same site as in the plasmid. This is confirmed by probing chromosomal digests of the resultant mutant with vector and insert DNA. The problem with this approach is that these insertion/deletion mutations can have polar effects on downstream genes, and therefore the phenotype of insertions generated within gene clusters can be complex and difficult to interpret. In *M. extorquens* AM1, insertions of a kanamycin-resistance gene were shown to not have polar effects, if the insertion was made such that the kanamycin-resistance gene was transcribed in the same direction as the gene in which it was inserted.[23] In the case of the MEDH system in *P. denitrificans*, a novel approach was used to generate mutants, in which an insertion was generated in the normal fashion, and then a second plasmid was used to generate an unmarked deletion by recombination that was not polar on downstream genes.[27]

The phenotypic analysis of mutants generated from cloned genes has produced important information for the bacterial quinoproteins discussed here. For instance, in the case of the glucose dehydrogenase in *A. calcoaceticus*, the presence of two different forms of the enzyme and the substrate specificity of each enzyme were confirmed by studying mutants generated in each gene, after the genes were cloned.[12]

For the MEDH system, mutant analysis has provided a great deal of information. In some cases these were mutants generated from cloned genes as described earlier, and in other cases they were point mutants generated from chemical mutagenesis. Mutants in *mxaI*, the gene encoding the small subunit of MEDH, have confirmed that this polypeptide is required for MEDH activity.[3,27] Mutants in cytochrome *c* genes have demonstrated that although multiple cytochrome *c*'s can serve as electron acceptors to MEDH *in vitro*, only cytochrome c_L, encoded by *mxaG*, functions

[27] R. J. M. Van Spanning, C. W. Wansell, T. De Boer, M. J. Hazelaar, H. Anazawa, N. Harms, L. F. Oltmann, and A. H. Stouthamer, *J. Bacteriol.* **173,** 6948 (1991).

in vivo.[1,3,28] Mutants in *mxaA*, *mxaK*, and *mxaL* have been used to confirm that the Ca^{2+} in MEDH is required for activity.[29] The products of these three genes are involved in calcium insertion in MEDH.[29] Other genes (*mxaJ*, *mxdR*, and *mxdS*[3,12]) have been identified, which by mutant analysis have been shown to be required for MEDH activity but whose precise function is unknown. The phenotypes of the mutants defective in these genes and the known sequence information suggest that these genes function in regulation, stability, or assembly of MEDH,[12] pointing out the presence of still more unidentified features of the MEDH system.

Mutant analysis has also been useful in studies of PQQ biosynthesis. Mutants generated in the peptide encoded by *pqqA* of *K. pneumoniae* and gene IV of *A. calcoaceticus* have been used to demonstrate that the peptide is translated and is required for PQQ biosynthesis.[26,30] In these cases the mutants were generated in cloned genes and were used to test for PQQ production in strains lacking the ability to generate PQQ.

Mutant characterization has been instrumental in developing a model for how MADH is synthesized. As noted above, sequence information suggested that *mauG* might encode a periplasmic peroxidase. Mutants in *mauG* produce inactive MADH in which the small subunit appears to contain quinone but is apparently not cross-linked. Mutants in *mauL* have an identical phenotype. These data suggest that the products of *mauG* and *mauL* are involved in generating the cross-link in TTQ, that this step occurs in the periplasm, after transport of the small subunit polypeptide, and that it involves a peroxidase reaction.[23] Analysis of other mutants has shown that amicyanin is required for MADH activity *in vitro* in *P. denitrificans*[14] and *M. extorquens* AM1,[21] and that *mauD*, *mauE* and *mauF* are probably involved in transport of the small subunit to the periplasm and/or maturation.[21] The role of *mauM* and *mauN* are unknown, although as noted previously, their sequences suggest they are ferredoxins. Mutants in these genes grow normally on methylamine in *M. extorquens* AM1,[23] but in *M. flagellatum* KT *mauM* mutants have no detectable MADH activity.[24]

Regulatory Genes

Some of the genes identified by the approaches discussed earlier may be involved in transcriptional regulation. A common approach for assessing transcription is to generate a transcriptional fusion between a gene to be studied and a reporter gene, which encodes an enzyme for which a rapid

[28] R. J. M. Van Spanning, C. W. Wansell, W. N. M. Reijnders, N. Harms, J. Ras, L. F. Oltmann, and A. H. Stouthamer, *J. Bacteriol.* **173,** 6962 (1991).

[29] I. W. Richardson and C. Anthony, *Biochem. J.* **287,** 709 (1992).

[30] N. Goosen, R. G. M. Huinen, and P. Van De Putte, *J. Bacteriol.* **174,** 1426 (1992).

and sensitive assay exists. The transcription of this fusion is then assessed in wild-type and mutant strains by measuring the reporter gene enzyme activity. Using this approach, genes involved as positive regulators of gene expression have been identified in the MEDH system.[31–33] Gene fusions between the regions upstream of *mox* genes and either *lacZ*[31,32] or *xylE*[33] as reporter genes have been generated and the activities of the reporter genes have been assessed in various mutants. These studies have identified three regulatory genes in *P. denitrificans*,[32] *mxaX*, *mxaY*, and *mxaZ*, and six regulatory genes in *M. extorquens* AM1 and *M. organophilum* XX, *mxaB*, *mxcQ*, *mxcE*, *mxbM*, *mxbN*, and *mxbD*.[31,33] In all of these cases, mutants in these genes showed low or nondetectable levels of the reporter enzyme for at least one of the gene fusions tested. The sequence of *mxaX* and *mxaY* suggests that these are members of a two-component regulatory system,[32] but the precise role of the other genes is unknown. However, these data suggest that transcriptional regulation of the *mox* system is complex in these strains and in the *Methylobacterium* strains, and probably involves a regulatory hierarchy.[33]

[31] C. J. Morris and M. E. Lidstrom, *J. Bacteriol.* **174,** 4444 (1992).
[32] N. Harms, W. N. M. Reijnders, H. Anazawa, and C. H. N. M. Van der Palen, *Mol. Microbiol.* **8,** 457 (1993).
[33] H. H. Xu, M. Viebahn, and R. S. Hanson, *J. Gen. Microbiol.* **139,** 743 (1993).

[17] Biogenesis of Pyrroloquinoline Quinone from ^{3}C-Labeled Tyrosine

By CLIFFORD J. UNKEFER, DAVID R. HOUCK, B. MARK BRITT, TOBIN R. SOSNICK, and JOHN L. HANNERS

Introduction

Pyrroloquinoline quinone (PQQ, 2,7,9-tricarboxy-1*H*-pyrrolo[2,3f]-quinoline-4,5-dione) is one of several quinones that serve as redox-active prosthetic groups in dehydrogenases and oxidases. Described elsewhere in this volume are the covalently linked quinone cofactors topa and tryptophan tryptophylquinone which are produced by the oxidation of the side chains of tyrosyl and tryptophanyl residues encoded within the parent polypeptide

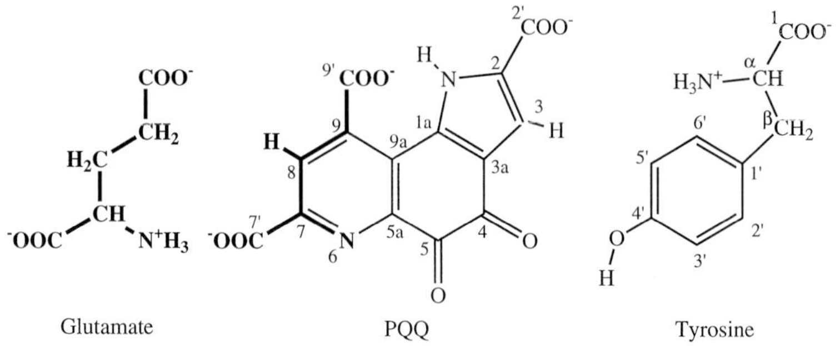

Glutamate PQQ Tyrosine

SCHEME 1

chain.[1,2] These prosthetic groups can only be removed by hydrolysis of the polypeptide backbone. PQQ, although tightly bound to the bacterial methanol and glucose dehydrogenases, can be removed under enzyme denaturation conditions.[3,4] Therefore, PQQ is distinct from the other quinone prosthetic groups in that it is biosynthesized independent of its site of action. In the presence of calcium, PQQ binds to the bacterial dehydrogenases to form an active holoenzyme.[5,6] Classical genetics studies have demonstrated that the PQQ biosynthesis pathway requires the expression of at least seven genes.[7] Using stable isotope labeling methods we demonstrated that PQQ is produced from the amino acids glutamate and tyrosine.[8,9] In the process of these studies we developed methods for specifically labeling PQQ with ^{13}C and ^{15}N. Because these specifically labeled PQQs are potentially useful to probe the chemical mechanism of PQQ-dependent dehydrogenases, the methods for their production are described here.

PQQ is biosynthesized from glutamate and tyrosine as described in Scheme 1.[8,9] Labels can be introduced into the pyrrole and quinone ring by culturing a PQQ-producing strain of bacteria in the presence of labeled

[1] S. M. Janes, D. Mu, D. E. Wemmer, A. J. Smith, S. Kaur, D. Maltby, A. L. Burlingame, and J. P. Klinman, *Science* **248,** 981 (1990).

[2] W. S. McIntee, D. E. Wemmer, A. Chistoserdov, and M. E. Lidstrom, *Science* **252,** 817 (1991).

[3] S. A. Salisbury, H. S. Forrest, W. B. T. Cruse, and O. Kennard, *Nature* (*London*) **280,** 843 (1979).

[4] J. A. Duine, J. Frank, Jzn, and P. E. J. Werwiel, *Eur. J. Biochem.* **108,** 187 (1980).

[5] I. W. Richardson and C. Anthony, *Biochem. J.* **287,** 715 (1992).

[6] S. White, G. Boyd, F. S. Mathews, Z. Xia, W. Dai, Y. Zhang, and V. L. Davidson, *Biochemistry* **32,** 12955 (1993).

[7] M. E. Lidstrom, *FEMS Microbiol. Rev.* **87,** 431 (1991).

[8] D. R. Houck, J. L. Hanners, and C. J. Unkefer, *J. Am. Chem. Soc.* **110,** 6920 (1988).

[9] D. R. Houck, J. L. Hanners, and C. J. Unkefer, *J. Am. Chem. Soc.* **113,** 3162 (1991).

tyrosine.[9] For example, we produced PQQ labeled at C-5 and C-9a by adding L-[3′,5′-$^{13}C_2$]tyrosine to the culture medium of the methylotrophic bacterium *Methylobacterium extorquens* AM1. All efforts to label PQQ directly with glutamate have failed.[9,10] We observed that glutamate added to *M. extorquens* AM1 cultures was removed rapidly from the medium. We suspect that, as is the case with succinate,[11] glutamate added to the medium induces the expression of a functional TCA cycle in *M. extorquens* AM1 and labeled glutamate is rapidly oxidized.

Because PQQ is biosynthesized by organisms which grow using methanol as their sole source of carbon, PQQ labeled uniformly with ^{13}C can be produced from [^{13}C]methanol.[8] [$^{15}N_2$]PQQ can be obtained by culturing a methylotrophic strain using [^{15}N]ammonium salts as the sole nitrogen source. Finally, growth of *M. extorquens* AM1 using [1-^{13}C]ethanol as the sole source of carbon yields PQQ that is enriched in the three carboxyl groups (C-2′, C-7′, and C-9′) and at C-5, C-5a, and C-9a.[9]

Synthesis of Labeled Tyrosine

We prepared specifically labeled L-tyrosines from L-serine and phenol using a procedure based on that developed by Walker and co-workers.[12] This synthesis involves the tyrosine phenol-lyase catalyze condensation of phenol and serine and allows for introduction of specific ^{13}C and ^{15}N labels into tyrosine. For example, L-[3′,5′-$^{13}C_2$]tyrosine is prepared from the condensation of L-serine and [2,6-$^{13}C_2$]phenol. Similarly, L-[β-^{13}C]tyrosine is prepared from L-[3-^{13}C]serine and phenol, and L-[α-^{15}N,2′,6′-$^{13}C_2$]tyrosine is prepared from L-[α-^{15}N]serine and [3,5-$^{13}C_2$]phenol. To ensure good isotopic yields, unlabeled substrates are added in a twofold molar excess. Detailed protocols for the preparation of ^{13}C-labeled phenols from *p*-nitrophenol via the condensation of [1,3-$^{13}C_2$]acetone with nitromalonaldehyde have been reported.[12,13] L-[3-^{13}C]Serine was produced by microbial transformation of [^{13}C]CH_3OH and glycine using *M. extorquens* AM1 using a procedure developed in our laboratory.[14]

Tyrosine phenol-lyase is expressed in high concentrations in the bacterium *Erwinia herbicola* (ATCC 21434) when cultured as follows. The culture medium is prepared by adding per 1.5 liters of deionized water: L-tyrosine, 3.0 g; KH_2PO_4, 3.0 g; $MgSO_4$, 1.5 g; ferric citrate, 1.8 mg;

[10] M. A. G. vanKleef and J. A. Duine, *FEBS Lett.* **237,** 91 (1988).

[11] I. J. Taylor and C. Anthony, *J. Gen. Microbiol.* **93,** 99 (1976).

[12] T. E. Walker, C. Matheny, C. B. Storm, and H. Hayden, *J. Org. Chem.* **51,** 1175 (1986).

[13] V. Viswanatha and V. J. Hruby, *J. Org. Chem.* **44,** 2892 (1979).

[14] J. Hanners, R. Gibson, K. Velarde, J. Hammer, M. Alvarez, J. Griego, and C. J. Unkefer, *J. Labelled Compd. Radiopharm.* **16,** 781 (1991).

pyridoxine · HCl, 150.0 mg; glycerol, 7.1 ml; succinic acid, 7.5 g; D,L-methionine, 1.5 g; D,L-alanine, 3.0 g; glycine, 0.75 g; L-phenylalanine, 1.5 g; yeast extract (Difco), 15.0 g; meat extract (Difco), 7.5 g. The ingredients are mixed and the pH of the resulting suspension is adjusted to pH 7.5 with the addition of ammonium hydroxide. The culture medium is sterilized in an autoclave. Then 200 ml of sterile culture medium is transferred to each of six 1-liter fluted culture flasks. Using a sterile loop, *E. herbicola* cells are transferred from a nutrient agar plate (1.5% w/v) to one of the culture flasks. The organism is cultured at 30° on a rotary shaker (200 rpm) for 24 hr. Five milliliters of this culture is used to inoculate each of the five remaining flasks which are then incubated at 30° on a rotary shaker at 200 rpm. After 24 hr the cells are harvested from the culture broth by centrifugation (10,000*g* for 10 min). For the production of labeled tyrosine the cells are resuspended in labeling medium that is prepared by adding [2,6-$^{13}C_2$]phenol (5 g, 52 mmol), L-serine (10 g, 95 mmol), and ammonium acetate (5 g) to 500 ml of water and adjusting the pH to 7.5 with the addition of ammonium hydroxide. The suspension of cells in the labeling medium is then placed in a 500-ml jacketed spinner flask (Bellco) and stirred using a magnetic stirrer. The temperature in the spinner flask is maintained by circulating water from a 37° constant temperature bath through the jacket. After 24 hr, a copious amount of precipitated tyrosine forms. The precipitate and the cells are pelleted by centrifugation (10,000*g* for 10 min). The precipitate is resuspended in 6 *N* HCl and the cells are removed by centrifugation (10,000*g* for 10 min). The supernatant is concentrated to a white solid *in vacuo* using a rotary evaporator. The solid L-[3′,5′-$^{13}C_2$]tyrosine · HCl is dissolved in a small amount of water, and the solution is neutralized with the addition of ammonium hydroxide. L-[3′,5′-$^{13}C_2$]Tyrosine crystallizes from this solution as its zwitter ion. Using this procedure, 8.5 to 9.0 g of L-[3′,5′-$^{13}C_2$]tyrosine is recovered (90–96% yield).

Production of Labeled PQQ

Like many methylotrophic bacteria, *M. extorquens* AM1 (*Pseudomonas* AM1, ATCC 14718) excretes PQQ into the growth medium when cultured using methanol or ethanol as its sole source of carbon.[15] PQQ is produced by culturing the organism in a 5-liter New Brunswick BioFlow III fermentor at 30°. The organism is cultured in mineral medium[14] that contains the following per liter of deionized water: $(NH_4)_2SO_4$, 0.2 g; NH_4Cl, 1.6 g; KH_2PO_4, 2.72 g; $Na_2HPO_4 \cdot 7H_2O$, 5.6 g; $MgSO_4 \cdot 7H_2O$, 100 mg; $CuSO_4$, 70 μg; $MnSO_4 \cdot H_2O$, 35 μg; $ZnCl_2$, 23.7 μg; $CaCl_2$, 1 mg; $CoCl_2$, 18 μg;

[15] M. A. G. vanKleef and J. A. Duine, *Appl. Environ. Microbiol.* **55,** 1209 (1989).

H_3BO_4, 7 μg; $(NH_4)_6Mo_7O_{24} \cdot 4H_2O$, 60 μg; $FeSO_4 \cdot 7H_2O$, 550 μg; and citric acid, 600 μg. Methanol (4 g/liter) is added to the sterile culture medium as a carbon source. Labeled tyrosine (90 mg/liter) is added to the culture medium as a solid immediately prior to inoculation. The culture is initiated with a 10% inoculum. Oxygen is maintained in the culture by bubbling air (2 liters/min) and stirring (150 rpm). The pH of the culture is maintained by the automatic addition of a 1 *M* NaOH solution using the fermentor's pH controller. Methanol in the culture medium is monitored by gas chromatography as follows. A sample (1 ml) is taken from the fermentor and centrifuged to remove the bacteria. The supernatant (20 μl) is diluted with water (180 μl) that contains ethanol (56 μl ethanol/100 ml H_2O) as an internal standard. The sample (1 μl) is injected onto a Porapac QS column (Alltech Associates, 80–100 mesh, 2 mm ID × 4 ft, glass) at 128° and monitored using a flame ionization detector. When the methanol is exhausted from the medium (24–48 hr, ~4.0 OD at 560 nm), the culture is harvested by centrifugation (10,000*g* for 10 min); PQQ is isolated from clearified culture broth.

Using this same protocol [U-^{13}C]PQQ can be produced by substituting [^{13}C]methanol for unlabeled methanol and omitting the tyrosine. Similarly, substitution of [^{15}N]ammonium salts in the medium yields [$^{15}N_2$]PQQ. Finally, substitution of [1-^{13}C]ethanol (4 g/liter) for methanol yields PQQ that is selectively enriched at the three carboxyl groups (C-2′, C-7′, and C-9′) and at C-5, C-5a, and C-9a.

Isolation of PQQ

The isolation of PQQ from the clarified culture broth is based on published procedures[14] and is carried out in our laboratory as follows. The clarified culture broth is passed over a DEAE Sephadex-G25 column (10 ml resin per 5 liters of culture broth) and discarded. A step gradient of NaCl in 5 m*M* potassium phosphate buffer (pH 7) is used to elute PQQ (75 ml each of 0.1, 0.2, 0.4, 0.6, 0.8, and 1.0 *M* NaCl). Fractions containing PQQ (HPLC assay below) are pooled and passed over a reversed-phase column (Baker Bakerbond spe C_{18}, sorbent wt. 1.0 g); under these conditions (pH 7), PQQ does not bind to the resin and is in the column eluent. The eluent solution is acidified to pH 2 with HCl and is loaded on a second reversed-phase column (Baker Bakerbond spe C_{18}, sorbent wt. 1.0 g); under acidic conditions, PQQ binds to the resin. The column is then washed with 10 ml of 5 m*M* HCl. PQQ is eluted with 4 ml 50% aqueous methanol and dried *in vacuo*. The yield of PQQ is typically 1 mg/liter of culture broth. Using this procedure, ^{13}C-labeled tyrosine was found to be incorporated into PQQ with minimal dilution (Table I).

TABLE I
BIOSYNTHESIS OF STABLE ISOTOPE-LABELED PQQ

	Precursor		PQQ	
Biosynthetic precursor	Labeled positions	Enrichment	Labeled positions	Enrichment
Tyrosine	C-3′, C-5′,	99.2%	C-5, C-9a	88%[a]
	C-β	49.6%	C-3	42%[a]
	C-2′, C-6′	99.2%	C-4, C-1a	88%[a]
	α-N	98%	N-1	15.7%
Ethanol	C-1	92%	C-2′	82%
			C-7′	59%
			C-9′	82%
			C-5	27%
			C-5a	54%
			C-9a	35%

[a] The label was expected to be diluted 10% by the inoculum.

HPLC Quantitation of Labeled PQQ

Solution A: add tetrabutylammonium phosphate (1.188 g, 3.5 mmol) and K_2HPO_4 (0.348 g, 2.0 mmol) to liter of HPLC grade water and filter through a 0.45-μm filter.

Solution B: add 100 ml of tetrahydrofuran (THF) to 100 ml of filtered (0.45 μm) HPLC grade water. The THF is freshly distilled under argon from a sodium–potassium alloy using the blue ketyl of benzophenone as an indicator.[16]

Samples (10 μl) are injected onto a C_{18} reversed-phase column (5 μm C_{18} 4.6 mm $\times$ 25 cm; Ranin C18 80-225-C5) equilibrated with 95.5% solution A:4.5% solution B and flowing at 0.8 ml/min. A linear gradient is used to increase the fraction of solution B to 20% in 6 min. The eluant is then maintained at 20% solvent B for 5 min. PQQ is monitored by its absorbance at 248 nm and by its fluoresence. The fluorescence is monitored with a filter fluorimiter (Spectrovision FD-100) using an excitation filter with maximum transmission at 365 nm and an emission filter that transmits light above 400 nm. PQQ elutes 9 min after injection.

Biogenesis of Radiolabeled PQQ

Because PQQ has only two nonexchangeable hydrogens, biosynthesis of high specific activity tritiated PQQ would be difficult. We prepare ^{14}C-

[16] D. D. Perrin and W. L. F. Armarego, "Purification of Laboratory Chemicals," 3rd ed. Pergamon, Oxford, 1988.

TABLE II
CHEMICAL SHIFT ASSIGNMENTS OF PQQ AND ADDUCTS OF PQQ[a]

	δ (ppm)				
Carbon	PQQ	**I**	**II**	**III**	**IV**
1a	136.7	135.7	135.3	133.6	
2	127.6	138.6	135.8		
2′	161.3	162.4	167.6		
3	113.8	108.2	110.6		
3a	123.4	121.2	120.8		
4	173.4	173.8	189.5	186.5	
5	179.2	183.2	94.6	94.1	74.8
5a	148.1	144.8	157.3		
7	146.5	156.7	151.8		
7′	165.4	167.0	171.6		
8	130.3	128.8	126.5		
9	142.2	145.4	145.0		
9′	167.2	168.5	173.8		
9a	126.1	123.4	119.3	123.5	119.6
CN					117.9

[a] Fourier transform ^{13}C-NMR spectra were acquired at 25° using Bruker AF-250 and AMX-500 NMR spectrometers. Proton-decoupled ^{13}C-NMR spectra were obtained using a 45° pulse and a total recycle time of 5 sec. After the transform, the digital resolution in the spectra was 0.30 Hz/point (AF-250) or 0.47 Hz/point (AMX-500). Phase-sensitive COSY spectra were obtained using Bruker's implementation of the time proportional phase incrementation method.

labeled PQQ by adding commercial ^{14}C-labeled tyrosine to the *M. extorquens* AM1 cultures.[17] At the time of inoculation, 250 μCi of L-[U-^{14}C]tyrosine (500 μCi/μmol) is added to a 1-liter fermentor containing standard culture broth and methanol as a carbon source. Using the standard isolation protocol, 616 μg of PQQ is recovered. The biosynthesis of tyrosine was found to dilute the specific activity of the radiolabel significantly to 0.43 μCi/μmol.

^{13}C-NMR of Labeled PQQ and PQQ Adducts

Table II contains the ^{13}C chemical shift assignments of PQQ, its tri(tetrabutylammonium) salt (**I**) and of adducts formed when PQQ is treated

[17] C. R. Smidt, C. J. Unkefer, D. R. Houck, and R. B. Rucker, *Proc. Soc. Exp. Biol. Med.* **197,** 27 (1991).

with methanol (**II**), benzyl alcohol (**III**), or cyanide (**IV**) (Scheme 2). The assignments for the 14 ^{13}C-NMR signals of PQQ were determined from the ^{1}H–^{13}C coupling patterns ($^{1}J_{CH}$ and $^{3}J_{CH}$) and carbon–carbon spin–spin correlations. Carbon–carbon couplings were observed using samples of [U-^{13}C]PQQ (99+ %^{13}C) isolated from cultures grown on [^{13}C]methanol (99.7%). Unambiguous assignments for PQQ were reported previously and were achieved by selecting for one-bond ^{13}C coupling interactions ($^{1}J_{CC}$=55 Hz) in ^{13}C COSY experiments.[9] This approach has been used to make the complete carbon chemical shift assignments for the tri(tetrabutylammonium) salt of PQQ (**I**) and the C-5 methyl hemiketal of the tri(tetrabutylammonium) salt of PQQ (**II**). The partial chemical shift assignments reported for the cyannohydrin (**IV**) and benzyl ketal (**III**) adducts

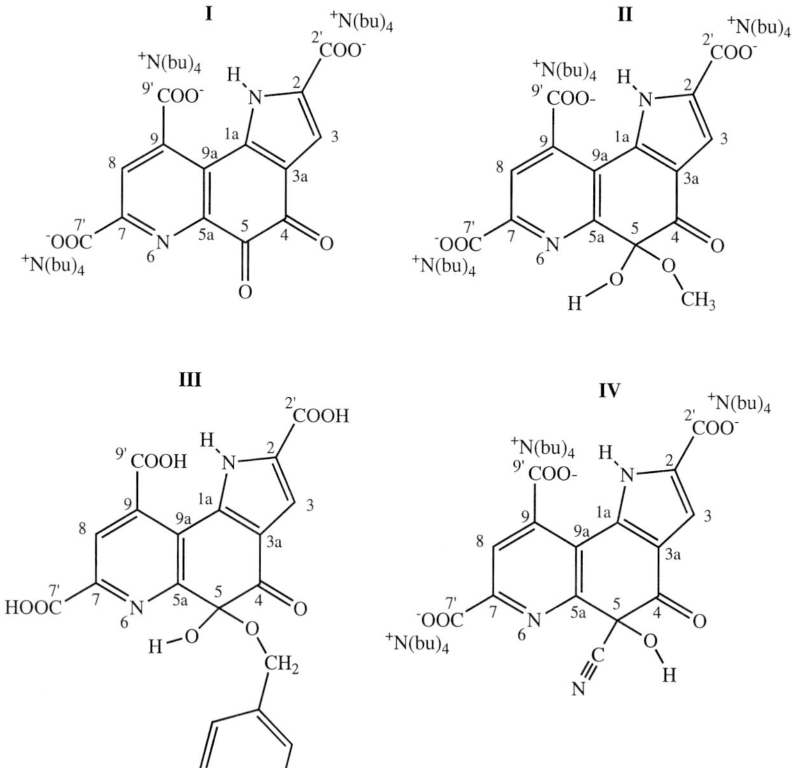

Scheme 2

of PQQ are based on observation of signals from specifically labeled PQQ and are also unambiguous.

Sample Preparation

PQQ samples are prepared for NMR spectroscopy in several ways. ^{13}C-NMR spectrum of the acid form of PQQ is obtained by dissolving it directly in DMSO-d_6 (2–5 mg/0.4 ml). The benzyl C-5 hemiketal of PQQ (**III**) is formed by adding benzyl alcohol (250 μl) to a solution of the acid form of PQQ in DMSO-d_6 (2–5 mg/0.25 ml). The tri(tetrabutylammonium) salt of PQQ (**I**) is prepared by neutralizing an aqueous solution of the acid form of PQQ with tetrabutylammonium hydroxide. The solution is then lyophilized to remove water. For NMR analysis the tri(tetrabutylammonium) salt of PQQ (**I**) is dissolved in DMSO-d_6. The methyl C-5 hemiketal of PQQ (**II**) was formed by dissolving the tri(tetrabutyl)ammonium salt of PQQ in methanol -*methyl*-d_3. The cyanide adduct of PQQ (**IV**) is formed by adding an equimolar ratio of tetrabutylammonium [^{13}C]cyanide to a DMSO-d_6 solution of the tri(tetrabutyl)ammonium salt of PQQ.

[18] X-Ray Crystallographic Studies of Cofactors in Galactose Oxidase

By NOBUTOSHI ITO, PETER F. KNOWLES, and SIMON E. V. PHILLIPS

Introduction

Galactose oxidase (GOase; EC 1.1.3.9) is one of a group of copper-containing oxidative enzymes which includes monoamine oxidase, laccase, ascorbate oxidase, tyrosinase, and dopamine monooxygenase. Within this group, galactose oxidase has the simplest structure which consists of a single polypeptide chain of 639 amino acid residues[1] (MW = 68,000) and contains a single copper with no other dissociable prosthetic group.[2]

GOase catalyzes the oxidation of primary alcohols (e.g., the hydroxyl group at the C6 position in D-galactose) to aldehydes, accompanied by the reduction of molecular oxygen to hydrogen peroxide:

$$R{-}CH_2OH + O_2 \rightarrow R{-}CHO + H_2O_2.$$

[1] M. J. McPherson, Z. B. Ogel, C. Stevens, K. D. S. Yadav, J. Keen, and P. F. Knowles, *J. Biol. Chem.* **267,** 8146 (1992).

[2] D. J. Kosman, M. J. Ettinger, R. E. Weiner, and E. J. Massaro, *Arch. Biochem. Biophys.* **165,** 456 (1974).

Substrate specificity is very broad, ranging from small alcohols (e.g., propanediol) to polysaccharides with D-galactose at the nonreducing terminus.[3,4] The best substrate reported so far is dihydroxyacetone, which is more than three times better than D-galactose. Despite this, GOase is strictly stereospecific at Carbon 6, and does not oxidize either D-glucose or L-galactose.

A major issue with respect to the catalytic mechanism of GOase is how, with only one Cu(II) atom and no other dissociable prosthetic groups, it can catalyze a two-electron redox reaction. Is there a secondary cofactor? This question remained unanswered for nearly 30 years and has been a major obstacle in understanding the catalytic mechanism.

It has been known for some time that enzymic activity is significantly increased in the presence of horseradish peroxidase or oxidants such as $K_3Fe(CN)_6$ or H_2IrCl_6. Hamilton[5] found that native GOase, as isolated, is a mixture of an active, EPR-silent form of the enzyme with an inactive, EPR-detectable one that is the result of a one-electron reduction of the active form. From these observations, the presence of Cu(III) in the active enzyme was suggested, which would be subject to a two-electron reduction to Cu(I) during the turnover.

Whittaker and Whittaker[6] reported conditions for preparation of the active and inactive forms and concluded, from a variety of spectroscopic measurements, that the copper was divalent in both. They proposed that a protein free-radical in the GOase active site is antiferromagnetically coupled with the copper. This coupling could explain the absence of the EPR signal from both the copper ion and free radical in the active form of the enzyme. Further resonance Raman scattering experiments led them to propose that a copper-coordinated tyrosine is complexed with an unidentified but probably aromatic free-radical, X, in the redox-activated enzyme.[7]

The "EPR invisible" free-radical of GOase, which became "visible" when the apoenzyme was treated with $K_3Fe(CN)_6$,[8] showed an EPR spectrum characteristic of an aromatic-free radical. Identification of the aromatic species was established by incorporating [β,β-D_2]tyrosine into the enzyme. In the modified enzyme, the EPR signal corresponding to the free radical

[3] D. J. Kosman, *in* "Copper Proteins and Copper Enzymes" (R. Lontie, ed.), Vol. 1, pp. 1–26. CRC Press, Boca Raton, FL, 1984.

[4] P. F. Knowles and N. Ito, *Perspect. Bioinorg. Chem.* **2,** 208 (1993).

[5] G. A. Hamilton, P. K. Adolf, J. de Jersey, G. C. DuBois, G. R. Dyrkacz, and R. D. Libby, *J. Am. Chem. Soc.* **100,** 1899 (1978).

[6] M. M. Whittaker and J. W. Whittaker, *J. Biol. Chem.* **263,** 6074 (1988).

[7] M. M. Whittaker, V. L. DeVito, S. A. Asher, and J. W. Whittaker, *J. Biol. Chem.* **264,** 7104 (1989).

[8] M. M. Whittaker and J. W. Whittaker, *J. Biol. Chem.* **265,** 9610 (1990).

is grossly perturbed, which identifies tyrosine as a component of the radical site.

The crystallographic studies reported in this chapter were undertaken in order to obtain the complete structure of the enzyme, including characterization of any cofactors. The high resolution structure is consistent with the presence of a tyrosine free-radical in the active site, provides further structural information on the active site, and also suggests a possible model for substrate binding. These and other properties of galactose oxidase have been reviewed.[4] A more complete account of the crystallography has been published.[9]

Purification and Crystallization of the Enzyme

The fungus *Dactylium dendroides*[10] was grown using the method reported by Tressel and Kosman.[11] Temperature stability (~25°) is crucial to the fungal growth and, therefore, to the production of GOase in the medium. The addition of a concentrated solution of trace metals (Mg^{2+}, Mn^{2+}, Cu^{2+}, Fe^{2+}, Zn^{2+}) after 3 days of growth stimulates secretion of the enzyme and stabilizes it.

The purification procedure also followed that reported by Tressel and Kosman[11] except that ultrafiltration was used to concentrate the initial culture medium. Following anion-exchange chromatography on DEAE-cellulose, the solution was subjected to affinity chromatography on Sepharose 6B and was then concentrated, again by ultrafiltration.

The crystals were grown by the hanging drop method, where the well solution consisted of 300–400 μl of a saturated aqueous solution of ammonium sulfate diluted to 1.0 ml with 800 m*M* acetate buffer (pH 4.3–4.7). The drops contained 15 μl of protein solution (3 mg/ml in 50 m*M* ammonium acetate buffer at pH 7.0) together with 10 μl of the well solution. Crystals were grown for 2 to 3 weeks at 18 ± 2°. A typical crystal of GOase is a very thin plate with dimensions of ~1.0 × 0.4 × 0.1 mm^3 and diffracts X-rays beyond 2 Å resolution.

The space group is *C*2, with unit cell parameters a = 98.0 Å, b = 89.4 Å, c = 86.7 Å, and β = 117.8°. The b axis is parallel to the longest axis of the crystal whereas a lies in the flat plane. The asymmetric unit contains one GOase molecule. V_M (volume per unit molecular weight) is 2.47

[9] N. Ito, S. E. V. Phillips, K. D. S. Yadav, and P. F. Knowles, *J. Mol. Biol.* **238,** 794 (1994).

[10] *D. dendroides* has recently been reclassified to be *Fusarium*. Z. B. Ogel, D. Brayford, and M. J. McPherson, *Mycol. Res.* **98,** 474 (1994).

[11] P. Tressel and D. J. Kosman, *Anal. Biochem.* **105,** 150 (1980).

$Å^3$/D, indicating that crystals contain 50% solvent if a value of 0.74 cm^3/g is assumed for the partial specific volume of the protein.[12]

Crystals were transferred to a stabilizing buffer (25% PEG 6000) in 100 m*M* acetate buffer (pH 4.5), which was also used for the preparation of heavy-atom derivatives. The pH 7 form was prepared by soaking a crystal in 25% PEG 6000, 100 m*M* PIPES buffer (pH 7.0). Another crystal was soaked in 25% PEG 6000, 20 m*M* sodium diethyldithiocarbamate, 100 m*M* PIPES buffer (pH 7.0) to give a crystal of apoenzyme (copper-depleted enzyme). The azide derivative was prepared by the addition of 100 m*M* sodium azide to the crystal growth medium; the crystals were transferred for crystallographic study to 25% PEG 6000, 100 m*M* PIPES buffer (pH 7.0) containing 100 m*M* sodium azide.

Active site mutants of GOase (W290H and C228G) have been constructed and expressed in *Aspergillus nidulans.*[13] The secreted mutant enzymes were purified and crystallized by procedures similar to those described earlier for the wild-type enzyme.

Data Collection and Phase Determination

All the diffraction data sets used in this study were collected at room temperature using a Xentronics/Siemens multiwire area detector X-100A with a three-axis goniometer mounted on a Rigaku RU200 rotating anode X-ray generator. CuK_α radiation monochromatized by a graphite monochromator was used. The crystals are stable to the X-ray radiation, allowing a full data set to be collected from a single crystal. The oscillation angle per detector data frame was 0.2°.

The raw frame data were integrated by the program XDS[14] and unmerged data were then converted to the LCF format for the subsequent calculations with the CCP4 program suite.[15] The diffraction intensities from the data frames were divided into batches, each corresponding to 5° of rotation, and then scaled to each other. For heavy-atom derivatives, where anomalous scattering data were required, Bijvoet pairs in the same or adjacent batches (i.e., reflections whose ω angles lie within 5–10°) were used to calculate the anomalous differences.

Data collection statistics are shown in Table I.

[12] B. W. Matthews, *J. Mol. Biol.* **33,** 491 (1968).

[13] M. J. McPherson, C. Stevens, A. J. Baron, Z. B. Ogel, K. Seneviratne, C. M. Wilmot, N. Ito, I. Brocklebank, S. E. V. Phillips, and P. F. Knowles, *Biochem. Soc. Trans.* **21,** 752 (1993).

[14] W. Kabsch, *J. Appl. Crystallogr.* **21,** 916 (1988).

[15] CCP4, "Collaborative Computing Project No. 4. A Suite Program for Protein Crystallography." Distributed from Daresbury Laboratory, Warrington, UK, 1979.

TABLE I
STATISTICS OF DATA COLLECTION

Crystal	Resolution (Å)	N_{obs}[a]	N_{unique}[a] (% of total possible)	R_{sym}[b] (%)
Native A (pH 4.5)	20–2.2	70,069	29,711 (87.8)	5.6
Native B (pH 4.5)	10–1.7	117,980	57,092 (79.1)	3.9
	[1.8–1.7	11,365	5,908 (52.6)][c]	
Native C (pH 7.0)	10–1.9	83,387	46,302 (89.5)	5.0
	[2.0–1.9	7,378	5,589 (75.8)][c]	
Apoenzyme	20–2.2	49,151	27,690 (82.4)	4.2
	[2.3–2.2	2,843	1,764 (42.0)][c]	
K_2PtCl_4	20–2.3	44,990	22,292 (75.8)	5.4
H_2IrCl_6	20–2.2	54,428	27,817 (82.4)	4.4
$Pb(NO_3)_2$	20–2.2	54,932	27,703 (81.6)	4.9

[a] N_{obs} and N_{unique} are the total number of reflections observed and the number of unique, nondegenerate reflections, respectively.
[b] $R_{sym} = \Sigma_h \Sigma_i |I(h)_i - \langle I(h) \rangle| / \Sigma_h \Sigma_i |I(h)_i|$, where h are unique reflection indices.
[c] These numbers are for the highest resolution shell.

Heavy-Atom Derivatives and Phase Refinement

An extensive search was made for heavy-atom derivatives suitable for phasing by the isomorphous replacement method. Derivatives were prepared by soaking native GOase crystals in solutions of various heavy-atom compounds. The soaking conditions and phase statistics for the derivatives used in structure determination are shown in Table II. The first useful derivative was with K_2PtCl_4 (10 mM). This solution was freshly made for the soak because the compound is unstable in the soaking buffer. The

TABLE II
SOAKING CONDITIONS AND PHASING STATISTICS FOR HEAVY-ATOM DERIVATIVES

Compound	K_2PtCl_4	H_2IrCl_6	$Pb(NO_3)_2$
Concentration (mM)	10	2	200
Soaking time (days)	1	1	13
Completeness (20–2.5 Å)			
Isomorphous part (%)	86.0	93.2	92.2
Anomalous part (%)	65.4	65.6	69.4
No. of binding sites	1	3	10
Isomorphous difference (20–2.5 Å) (%)	16.1	14.9	13.9
FHLE refinement (20–3.0 Å)			
$R_{centric}$ (%)	54	45	56

[a] $\Sigma_h |F(h)_{DER} - F(h)_{NAT}| / |F(h)_{NAT}|$.

3-Å isomorphous difference Patterson map shows a clear single peak on the Harker section ($v = 0$) with height about 25% of that of the origin peak. The 3-Å anomalous difference Patterson map also shows a peak of lower height (about 12% of the origin peak) at the same position, demonstrating the good quality of the anomalous difference data. Since the unit cell origin is arbitrary in the y direction for space group $C2$, the y coordinate of the platinum site was set to zero. Since the arrangement of the platinum atoms in the crystal is centrosymmetric, the correct enantiomorph for subsequent derivatives could be determined directly from difference Fourier maps phased from isomorphous and anomalous data for the platinum derivative.

H_2IrCl_6 (2 mM) also gave a useful derivative. A cluster of peaks can be clearly seen in the 3-Å isomorphous difference Patterson map, with similar but weaker features in the anomalous difference Patterson map. A cluster of two major and one minor sites was deduced from these peaks, with the help of the difference Fourier map calculated with phases from the platinum derivative (including the anomalous scattering contribution). The 6-Å MIRAS (multiple isomorphous replacement with anomalous scattering) electron density map calculated with α_P(Pt,Ir) (phases from the two derivatives) clearly shows less dense regions, corresponding to the solvent channels in the crystal lattice. The molecular envelope is recognizable in the 3-Å map, although the solvent regions are less flat.

A serious problem with the first two derivatives, however, is that the three iridium atoms have almost the same y coordinate as the platinum atom (i.e., zero). The phase ambiguity inherent in isomorphous replacement, therefore, is resolved almost solely by the anomalous scattering. If the anomalous scattering was too weak, the MIRAS map would be a mixture of the true electron density with its fake mirror image across the $y = 0$ plane.

The presence of fake mirror symmetry was confirmed when data from the apoenzyme (copper-depleted enzyme) were processed. The difference Fourier map between the holo- (native) and apoenzyme with α_P(Pt,Ir) shows a very strong peak with fractional coordinates (0.40, 0.09, 0.39) corresponding to the copper atom, but also a significant mirror-related peak (0.40, −0.09, 0.39). Another derivative containing a heavy-atom not lying on $y = 0$ was necessary to resolve this phase ambiguity. Although the apoprotein itself could be seen as a derivative with the copper of negative occupancy, the copper proved too light to contribute significantly to the phasing.

The problem was finally solved when a third derivative, $Pb(NO_3)_2$ (200 mM), was found. The difference Fourier map for this derivative, phased with α_P(Pt,Ir), indicated the presence of 10 Pb sites, many of which are not close to the $y = 0$ plane. The final MIRAS phases [α_P(Pt,Ir,Pb)] calculated with all three derivatives have an average figure of merit of 0.58 for the resolution range of 20–2.5 Å. Although the improvement in the figure of merit was small, the ratio between heights of the true copper peak and its fake mirror image in the difference Fourier map for apoenzyme increased by about 70% with α_P(Pt,Ir,Pb). The mirror image peak for the copper was almost at noise level, suggesting that the addition of the third derivative had solved the phase ambiguity.

Ten cycles of solvent flattening[16] were applied to the MIRAS map in the resolution range of 20–2.5 Å. The molecular envelope determined from the MIRAS map at cycle 1 was used throughout the cycles. At the end of this phase refinement process, the average figure of merit of the combined phases was 0.80.

Model Building and Refinement

Primary Sequence

N-terminal amino acid sequence data from galactose oxidase and various protease (V-8, endoproteinase Lys-c) and cyanogen bromide-derived peptides were determined and used to design redundant primers for the polymerase chain reaction.[1] A unique DNA fragment of 1.4 kb was amplified from *Dactylium dendroides* DNA and used as a homologous probe to isolate a galactose oxidase clone from the genomic library. The DNA sequence of the galactose oxidase gene is consistent with all available peptide sequence data (approximately 42% of the whole protein) with the exception of peptides covering the region 251–276 where Tyr-272 could not be detected in the protein sequence; the significance of this result will become clear later. Translation of the DNA sequence produces a mature protein of 639 amino acid residues (Fig. 1).

Initial Model Building

The polypeptide chain was fitted to the 2.5-Å MIRAS solvent-flattened map (referred to as the "SF map") using the program FRODO[17] on an Evans and Sutherland PS330 computer graphics system.

The SF map was of such excellent quality that the main-chain carbonyl oxygens in β strands were readily discerned and the polarity of the chain could be determined solely from the electron density. Many of the side chains could be recognized from the size and shape of the density.

The polypeptide chain was traced from the N terminus to the C terminus, with the exception of a few surface loops. The connectivity in these loops,

[16] B. C. Wang, this series, Vol. 115, p. 90.
[17] T. A. Jones, *J. Appl. Crystallogr.* **11,** 268 (1978).

```
          MKHLLTLALCFSSINAVAVTVPHKAVGTGIPEGSLQFLSLR
          -41                                     -1
ASAPIGSAISRNNWAVTCDSAQSGNECNKAIDGNKDTFWHTFYGANGDPK
1                                               50
PPHTYTIDMKTTQNVNGLSMLPRQDGNQNGWIGRHEVYLSSDGTNWGSPV
51                                             100
ASGSWFADSTTKYSNFETRPARYVRLVAITEANGQPWTSIAEINVFQASS
101                                            150
YTAPQPGLGRWGPTIDLPIVPAAAAIEPTSGRVLMWSSYRNDAFGGSPGG
151                                            200
ITLTSSWDPSTGIVSDRTVTVTKHDMFCPGISMDGNGQIVVTGGNDAKKT
201                                            250
SLYDSSSDSWIPGPDMQVARGYQSSATMSDGRVFTIGGSWSGGVFEKNGE
251                                            300
VYSPSSKTWTSLPNAKVNPMLTADKQGLYRSDNHAWLFGWKKGSVFQAGP
301                                            350
STAMNWYYTSGSGDVKSAGKRQSNRGVAPDAMCGNAVMYDAVKGKILTFG
351                                            400
GSPDYQDSDATTNAHIITLGEPGTSPNTVFASNGLYFARTFHTSVVLPDG
401                                            450
STFITGGQRRGIPFEDSTPVFTPEIYVPEQDTFYKQNPNSIVRVYHSISL
451                                            500
LLPDGRVFNGGGGLCGDCTTNHFDAQIFTPNYLYNSNGNLATRPKITRTS
501                                            550
TQSVKVGGRITISTDSSISKASLIRYGTATHTVNTDQRRIPLTLTNNGGN
551                                            600
SYSFQVPSDSGVALPGYWMLFVMNSAGVPSVASTIRVTQ
601                                   639
```

FIG. 1. The amino acid sequence of GOase and its probable leader peptide obtained by sequencing the *gaoA* gene. The N-terminal amino acid of the mature protein (639 amino acids) is indicated by the number 1, with the region of the protein that has been confirmed by peptide sequencing shown by shading. Adapted from McPherson *et al.*[1]

however, was easily established from the partial primary sequence information then available and subsequently published.[1]

As the sequencing of the cloned GOase gene was progressing in parallel with the model building, alanine residues were initially used to fit the density for the residues without sequence information. Residues whose side chains were not well-defined in the map were also substituted with alanine. The SF map implied that one of the copper ligands, Tyr-272, formed a covalent bond with the sulfur of the adjacent Cys-228. To confirm the existence of this unexpected and unusual bond, i.e., to avoid model bias in the map, Cys-228 was replaced with an alanine residue, even though very strong electron density was clearly present at the position corresponding to S_γ.

The initial model subjected to refinement consisted of 638 residues, of which 538 had complete side chains, and one cupric ion. The crystallographic

R factor for the 20- to 3-Å resolution range was 32% before refinement, suggesting a very high quality model.

Refinement

One hundred seventy-three cycles of Hendrickson–Konnert restrained least-squares refinement[18] were performed on this initial model. The refinement was started at relatively low resolution (20.0–3.0 Å, cycles 1–15), but the resulting R factor was so low (22%) that the resolution range was immediately expanded to 2.5 Å (cycles 16–21) then to 2.2 Å (cycles 22 and thereafter) without manual rebuilding of the model. During these cycles, the B factors of individual atoms were kept constant (20.0 $Å^2$), and only the overall B factor was refined. After cycle 27, the model was manually rebuilt for all 639 residues, using the primary sequence data that had become available.[1] Those with ambiguous side chains, and Cys-228, however, were still treated as alanine residues. Further cycles of refinement reduced the R factor to 28%.

Restrained atomic isotropic B-factor refinement was carried out from cycle 36. At this stage, S_γ of Cys-228 was included in the model since the strong density of the sulfur without its corresponding atomic model had caused a shift of the phenol ring of Tyr-272 toward it. The van der Waals' interactions between Cys-228 and Tyr-272 were omitted from the model restraints, but no other special restraints (e.g., bond length) were applied to the S_δ–C_ε bond. This allowed the geometrical relation between the two residues to be determined by the diffraction data. From cycle 44, very low resolution reflections (below 10 Å) were omitted from the refinement as they include a strong component from the bulk solvent in the crystal lattice. Although these constitute a very small portion (~0.1%) of the data, the improvement in the R factor was significant. The model at this stage (cycle 50) consisted of 639 residues (622 with the correct side chains) with a R factor of 22.8% for all data between 10 and 2.2 Å.

Solvent atoms were then added to the model. Most of the extra density peaks not corresponding to protein atoms were of a globular shape and lay near polar groups, so they were assigned as water molecules. However, there were two cases where the extra density could not be accounted for by water, and they were interpreted as an acetate ion and a sodium ion, which were added to the model.

During the refinement, a new data set from a larger native crystal to higher resolution (native B) became available, which made more precise structure determination possible. The resolution of the refinement was

[18] W. A. Hendrickson, this series, Vol. 115, 252.

immediately extended to 1.7 Å (cycle 109), the limit of the new data set. Inclusion of higher resolution data improved the map in several ways, including (i) the conformation of the four residues at the N terminus, (ii) the assignment of the external ligand to the copper as an acetate ion, and (iii) the conformation of the side chains of some residues, especially aliphatic amino acids.

The assignment of the acetate ion deserves more description. The presence of an exogenous ligand to the copper was obvious even in the 2.5-Å SF map. Initially the ligand was refined as a water molecule at 2.2 Å resolution, but the density in the $2F_O - F_C$ map became elongated in one direction and appeared to suggest a binuclear species. When 1.7 Å data became available, the ligand was refined as molecular oxygen, but the subsequent $2F_O - F_C$ map showed the density to be triangular in shape, and it was replaced by acetate. This assignment is consistent with the report that the copper site of GOase is easily (~5 m*M*) saturated by acetate,[19] which was present at high concentrations in the crystallization medium.

The final model includes 4830 protein atoms, one copper ion, one sodium ion, two acetate ions, and 316 water molecules. The R factor is 17.7% for all observed data between 10 and 1.7 Å. Refinement statistics are shown in Table III.

Five other isomorphous structures (i.e., native at pH 7.0, apoenzyme, azide derivative, and mutant enzymes) were determined by the difference Fourier method. The copper ion and water molecules in its vicinity were omitted in the initial model of the apoenzyme. The refinements for these structures were carried out in a similar way to that described earlier, with some statistics shown in Table III.

Accuracy and Reliability of the Model

The model of GOase has been refined according to fairly strict stereochemical restraints, and the small RMS deviations from ideal values indicate that the bond length, bond angle, and planarity restraints are well observed in the structure. Extra care was also taken to ensure that all dihedral angles are reasonable.

The Ramachandran plot for the 1.7-Å refined structure shows nearly 90% of the nonglycine residues in the "core" region and all but two residues within the "allowed" regions.[20] These exceptions are Lys-60, which lies at

[19] D. J. Kosman, M. J. Ettinger, R. D. Bereman, and R. S. Giordano, *Biochemistry* **16,** 1597 (1977).

[20] A. L. Morris, M. W. MacArthur, E. G. Hutchinson, and J. M. Thornton, *Proteins: Struct., Funct., Genet.* **12,** 345 (1992).

TABLE III
STATISTICS FOR MODEL REFINEMENT

	Native (pH 4.5)	Native (pH 7.0)	Apoprotein
Resolution (Å)	10–1.7	10–1.9	10–2.2
No. of reflections			
All observed	57,092	46,302	27,365
$\|F\| > 1\sigma$	52,404	43,159	25,713
Crystallographic R factor (%)			
All observed	17.7	17.0	15.6
$\|F\| > 1\sigma$	15.8	15.3	14.2
RMS deviation from stereochemical ideality (Å)			
Bond	0.018	0.019	0.016
Angle 1–3	0.048	0.051	0.038
Planar 1–4	0.054	0.060	0.050
Average B factor (Å^2)			
Protein atoms	23.9	24.2	27.3
No. of nonhydrogen atoms in the model			
Total	5,156	5,142	5,114
Protein	4,830	4,830	4,830
Copper ion	1	1	0
Sodium ion	1	1	1
Acetate ion	8	0	0
Water	316	310	238

a turn in the polypeptide chain, and Ser-432, which forms an intermolecular contact with an adjacent molecule in the crystal. Profiles for side chain conformation are in good agreement with empirically defined distributions.[21]

The Luzzati plot,[22] which is a plot of R factor against resolution, gives the upper limit of the mean positional error for the atoms as 0.16 Å, assuming that the differences between F_O and F_C are solely due to positional errors in the model.

Galactose Oxidase Structure

General Description of the Structure

Galactose oxidase consists of three domains predominantly formed from the β structure (Fig. 2)[22a] with only a short single α-helix. The preponder-

[21] J. W. Ponder and F. M. Richards, *J. Mol. Biol.* **193,** 775 (1987).
[22] P. V. Luzzati, *Acta Crystallogr.* **5,** 802 (1952).
[22a] P. Kraulis, *J. Appl. Crystallogr.* **24,** 946 (1991).

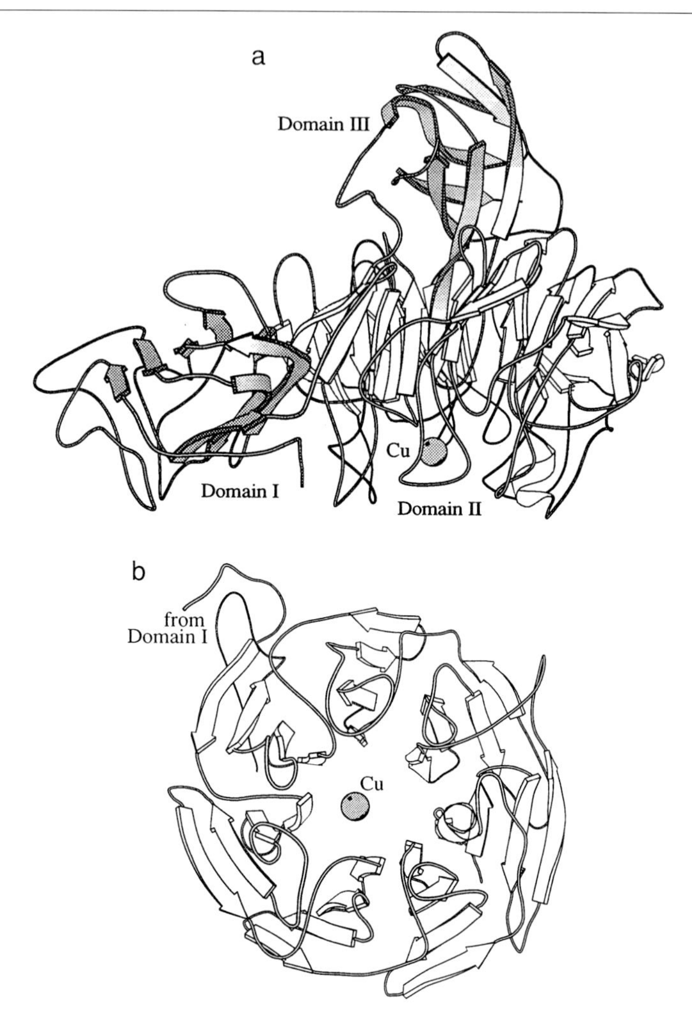

FIG. 2. Overall three-dimensional structure of GOase as ribbon diagrams drawn using the program MOLSCRIPT.[22a] (a) Side view of the molecule with domains I and III shaded. (b) View of domain II approximately along the pseudo sevenfold axis. The Cu is shown as a shaded sphere in both cases.

ance of the β structure probably contributes to the remarkable stability of the enzyme, which is stable in 6 *M* urea.[2] The first domain (residues 1–155) has a β sandwich structure. It is linked to the second domain by a well-ordered stretch of polypeptide chain, which is resistant to proteolytic attack (P. F. Knowles, unpublished observation). The second and largest domain (residues 156–532) has pseudo sevenfold symmetry with an overall appearance of a seven-petalled flower, each petal consisting of a four-stranded antiparallel β sheet. Similar arrangements have been reported for methylamine dehydrogenase[23] and neuraminidase,[24] although the latter contains only six sheets instead of seven. The copper lies on the solvent-accessible surface of this domain close to the pseudo sevenfold axis. The third domain (residues 533–639) lies on the opposite side of the second domain from the copper, and two of its seven β strands pierce the middle of the second domain along the pseudo sevenfold axis to provide a ligand (His-581) to the copper. Free space between this "finger" and the second domain is filled with approximately 40 well-ordered water molecules.

All electron density observed can be accounted for by the known primary structure of the protein and by the solvent and ions present in the solution. No electron density is present for either pyrroloquinoline quinone (PQQ) or carbohydrate. Although carbohydrate was reported to be present at about 1% in the enzyme (i.e., about four monosaccharides per protein molecule),[2] it may be disordered or simply absent. PQQ, which has been suggested to be bound to the enzyme covalently as a cofactor,[25] should have appeared in the density map if present. As the copper site is fairly close to an intermolecular contact in the crystal, there is no physical space at the active site to accommodate PQQ in the model. The crystallographic model specifically excludes the possibility of a PQQ cofactor. Topa (6-hydroxydopa), which has been reported to be present in bovine serum amine oxidase,[26] is also inconsistent with the electron density map.

Copper Coordination and Site Geometry

The copper site on the second domain lies in a region extremely rich in aromatic residues. Side chains of three tyrosines, three phenylalanines, three histidines, and one tryptophan all lie within 7 Å of the copper. Some seem to be involved in the formation and stabilization of the free radical whereas others form a hydrophobic wall for the putative substrate binding pocket (see below).

In the structure at pH 4.5 (acetate buffer), the copper site has square pyramidal coordination (Fig. 3). O_η of Tyr-272, $N_{\varepsilon 2}$ of His-496 and His-581, and an acetate ion form an almost perfect square, with coordination distances of 1.94, 2.11, 2.15, and 2.27 Å, respectively. O_η of Tyr-495 provides

[23] F. M. D. Vellieux, F. Huitema, H. Groendijk, K. H. Kalk, J. Jzn. Frank, J. A. Jongejan, J. A. Duine, K. Petratos, J. Drenth, and W. G. J. Hol, *EMBO J.* **8,** 2171 (1989).

[24] J. N. Varghese, W. G. Laver, and P. M. Colman, *Nature (London)* **303,** 35 (1983).

[25] R. A. van der Meer, J. A. Jongejan, and J. A. Duine, *J. Biol. Chem.* **264,** 7792 (1989).

[26] S. M. Janes, D. Mu, D. Wemmer, A. J. Smith, S. Kaur, D. Maltby, A. L. Burlingame, and J. P. Klinman, *Science* **248,** 981 (1990).

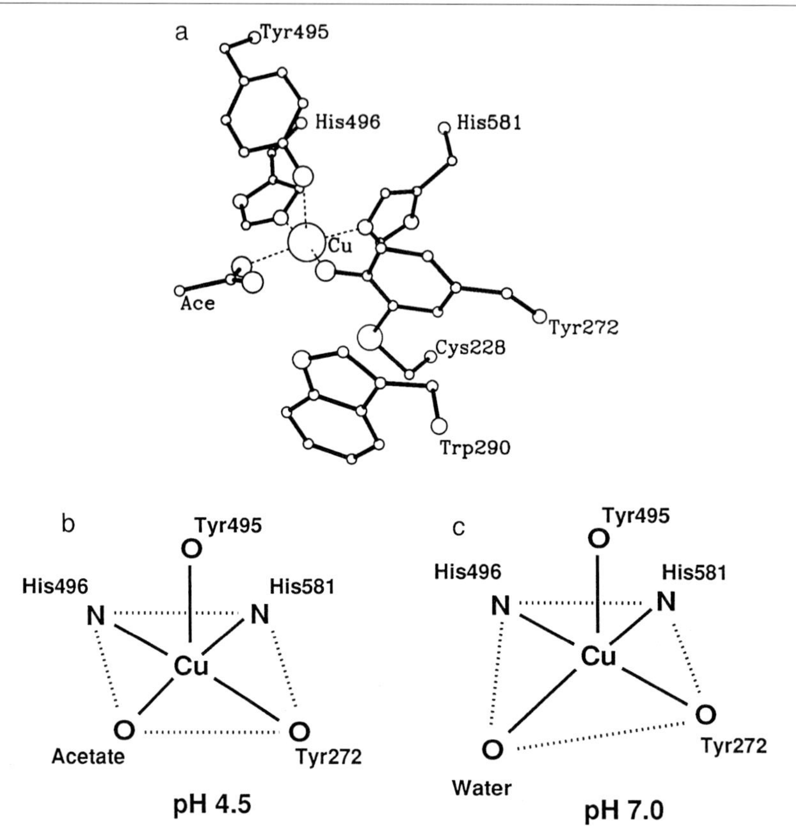

FIG. 3. (a) Crystallographic model for the copper site in GOase at pH 4.5 (b) Schematic diagram of copper coordination at pH 4.5. (c) Schematic diagram of copper coordination at pH 7.0.

a fifth, axial ligand at 2.69 Å from the copper, suggesting relatively weak coordination of this ligand. Some of the geometrical parameters describing copper coordination are listed in Table IV.

The relative orientations of the equatorial plane of the copper coordination and the planes of each ligand (i.e., the phenol ring of Tyr, the imidazole ring of His, and the carboxylate of acetate) are interesting (Fig. 3). Tyr-272 and His-581 are almost perpendicular to the equatorial plane (73 and 71°, respectively) whereas His-496 and the acetate ion are almost parallel to it (21 and 17°). The relation with Tyr-272 is particularly important since this residue is most likely to be the free radical site and, therefore, is probably involved in electron transfer with the copper center.[7,27]

[27] M. M. Whittaker and J. W. Whittaker, *Biophys. J.* **64,** 762 (1993).

TABLE IV

SOME GEOMETRICAL PARAMETERS OF THE COPPER SITE IN THE pH 4.5 STRUCTURE[a]

Bond length (Å)		
Cu–O_η(Y272)	1.94	(1.92)[b]
Cu–$N_{\varepsilon 2}$(H496)	2.11	(2.22)
Cu–$N_{\varepsilon 2}$(H581)	2.15	(2.20)
Cu–O_η(Y495)	2.69	(2.61)
Cu–O(Ace)	2.27	(—)
Cu–O(Wat)	—	(2.81)
Bond angle (degrees)		
O_η(Y272)–Cu–$N_{\varepsilon 2}$(H581)	99.1	(100.0)
$N_{\varepsilon 2}$(H496)–Cu–$N_{\varepsilon 2}$(H581)	88.6	(90.7)
$N_{\varepsilon 2}$(H496)–Cu–O(Ace)	89.5	(—)
$N_{\varepsilon 2}$(H496)–Cu–O(Wat)	—	(92.8)
O_η(Y495)–Cu–$N_{\varepsilon 2}$(H581)	97.2	(104.2)
O_η(Y495)–Cu–$N_{\varepsilon 2}$(H496)	105.7	(102.5)
O_η(Y495)–Cu–$N_{\varepsilon 2}$(Y272)	75.2	(79.0)
O_η(Y272)–Cu–O(Ace)	82.8	(—)
O_η(Y272)–Cu–O(Wat)	—	(75.8)
O_η(Y495)–Cu–O(Ace)	90.6	(—)
O_η(Y495)–Cu–O(Wat)	—	(88.8)
Cu–O_η(Y272)–C_ζ(Y272)	130.7	(127.3)
Cu–O_η(Y495)–C_ζ(Y495)	104.7	(108.2)
Cu–$N_{\varepsilon 2}$(H496)–$C_{\varepsilon 1}$(H496)	126.1	(124.5)
Cu–$N_{\varepsilon 2}$(H581)–$C_{\varepsilon 1}$(H581)	128.2	(127.1)
Cu–O(Ace)–C(Ace)	122.6	(—)

[a] Parameter errors are difficult to estimate in protein crystallography. It would be reasonable to expect average errors of the order of 0.13 Å in the bond length in this table.

[b] Values are for the pH 7.0 structure.

The copper site of GOase is very specific, and there are no reports of any metal ion other than Cu(II) able to bind to it. The high specificity is probably partly achieved by the rigidity of the site. The orientation of the imidazole ring of the two ligand histidines is fixed by hydrogen bonding through $N_{\delta 1}$ (His-496 to the main chain oxygen of Gly-513 and His-581 to O_γ of Thr-580). The rigidity of this and the other copper ligands is shown by the observation that these residues rarely change position when the copper is removed.

In the absence of acetate (i.e., the structure in PIPES buffer at pH 7) the acetate ion at the copper site is replaced by a smaller molecule, most probably water. This ligand lies 2.8 Å from the copper, which is very long for direct coordination. The removal of the copper in the crystal leaves the position of this ligand unchanged. It is present even in the structure of the apoenzyme, where it forms a hydrogen bond with O_η of Tyr-272. This suggests that the water is coordinated to the copper only very weakly, if at all, and anionic inhibitors such as F^-, CN^-, and N_3^- would easily replace it, inhibiting the binding of substrate, i.e., alcohol and/or molecular oxygen. The crystal structure of the azide derivative[28] reveals that indeed azide binds at the equatorial site occupied by acetate in the native structure (Cu–N = 2.45 Å); the only other significant structural change in this derivative is that the axial ligand, Tyr-495, is displaced to a slightly longer coordination distance (2.95 Å).

Since this water in the pH 7.0 structure is too distant from the copper for coordination, the copper site geometry can be effectively described as distorted tetrahedral. It has been argued[29] that this would facilitate the redox change from Cu(II) to Cu(I) which follows the addition of an alcohol substrate. However, none of the spectroscopic studies previously carried out on galactose oxidase in solution at pH 7.0 seem consistent with this geometry.[3,5,6,30–33] EXAFS studies[34] on polycrystalline samples of galactose oxidase at 10K in 2 *M* urea indicate that the pH 4.5 and 7.0 structures are closely similar, with the oxygen from an equatorial ligand 1.95 Å from the copper. Further work will be necessary to reconcile the differences between the crystal structures and the data from spectroscopic methods.

The EXAFS studies[34] further suggest that the structure of the "inactive" and "redox-activated" forms of galactose oxidase are closely similar, again with an equatorial oxygen ligand at 1.95 Å. Contrary to this view, it has been concluded from optical spectral studies that a structural change occurs at the copper site when the tyrosine-free radical is created by the oxidation of the enzyme; the active form is proposed to have more tetragonal structure than the inactive form.[6] There has been some concern, therefore, whether the crystallographic structure at pH 4.5 represents the active or inactive form or a mixture of the two. It is not possible to visualize a free radical directly by protein crystallography, but a crystal of GOase grown in the

[28] I. Brocklebank, N. Ito, P. F. Knowles, S. E. V. Phillips, and C. Wilmot, unpublished results (1994).

[29] N. Ito, S. E. V. Phillips, C. Stevens, Z. B. Ogel, M. J. McPherson, J. N. Keen, K. D. S. Yadav, and P. F. Knowles, *Nature (London)* **350,** 87 (1991).

[30] M. J. Ettinger and D. J. Kosman, *in* "Copper Proteins" (T. G. Spiro, ed.), p. 220. Wiley, New York, 1981.

[31] B. J. Marwedel, D. J. Kosman, R. D. Bereman, and R. J. Kurland, *J. Am. Chem. Soc.* **103,** 2842 (1981).

[32] D. J. Kosman, J. Peisach, and W. B. Mims, *Biochemistry* **19,** 1304 (1980).

[33] J. W. Whittaker, *in* "Bioinorganic Chemistry of Copper" (K. D. Karlin and Z. Tyeklar, eds.), p. 447. Chapman & Hall, London, 1993.

[34] P. F. Knowles, R. D. Brown, S. F. Koenig, S. Wang, R. A. Scott, M. A. McGuirl, D. E. Brown, and D. M. Dooley, *Inorganic Chem.*, (in press) (1995).

presence of 100 mM $K_3Fe(CN)_6$ (which would convert the enzyme to the active form[6]) shows no significant change in the copper site. This would indicate that the reported structural change may be very small and below the accuracy of protein crystallography (~0.2 Å) or that all the crystal forms are already fully oxidized.

Based on spectroscopic data, the copper site has been predicted to be in roughly square planar coordination with an extra axial ligand.[3,6] The crystallographic studies confirm this geometry, as well as identifying the ligands. The main difference between the studies is that spectroscopic data[3] suggest an external water molecule at the axial site whereas an endogenous protein ligand (the phenol oxygen of Tyr-495) is found in the crystal structure. The assignment of the external ligand at the axial site was mainly based on $^{19}F^-$ NMR studies.[31] This could be reconciled with the crystal structure by assuming two fluoride sites, one equatorial (K_d ~0.1 M) and the other axial (K_d ~50 M). It is likely that the fluoride ion at high concentration would replace the oxygen from Tyr-495 under the experimental conditions of the NMR studies.

Novel Thioether Bond between Tyr-272 and Cys-228: "Tyr–Cys Bridge"

The most surprising finding in the structure of GOase is that one of the copper ligands, Tyr-272, is covalently bonded at $C_{\varepsilon 1}$ to the sulfur atom of Cys-228 (Fig. 4).[34a] There is strong evidence to indicate that Tyr-272 is the site of the radical species.[6–8,27]

As mentioned earlier, the model refinement was carried out without any special restraint between $C_{\varepsilon 1}$ of Tyr-272 and S_γ of Cys-228 except that the van der Waals' interactions between the two side chains were omitted. Both the S_γ–$C_{\varepsilon 1}$ bond length (1.84 Å) and bond angles at $C_{\varepsilon 1}$ (119.4° for $C_{\delta 1}$–$C_{\varepsilon 1}$–S_γ and 120.6° for C_ξ–$C_{\varepsilon 1}$–S_γ) are almost ideal for the *o*-substitution on the phenol ring. The bond angle at S_γ [C_β(Cys228)–S_γ–$C_{\varepsilon 1}$] is 105.1°. The *cis* conformation of the bond (the torsion angle $C_{\delta 1}$–$C_{\varepsilon 1}$–S_γ–C_β is 7°) results in the formation of a large planar group and suggests that the S_γ–$C_{\varepsilon 1}$ bond may have partial double-bond character (Fig. 5).

Confirmatory Evidence for the Thioether Bond

The identification of the thioether bond described in this chapter is based solely on the crystallographic data and the primary sequence of the protein. It may be stated unequivocally that the crystallographic evidence

[34a] N. Ito, unpublished results.

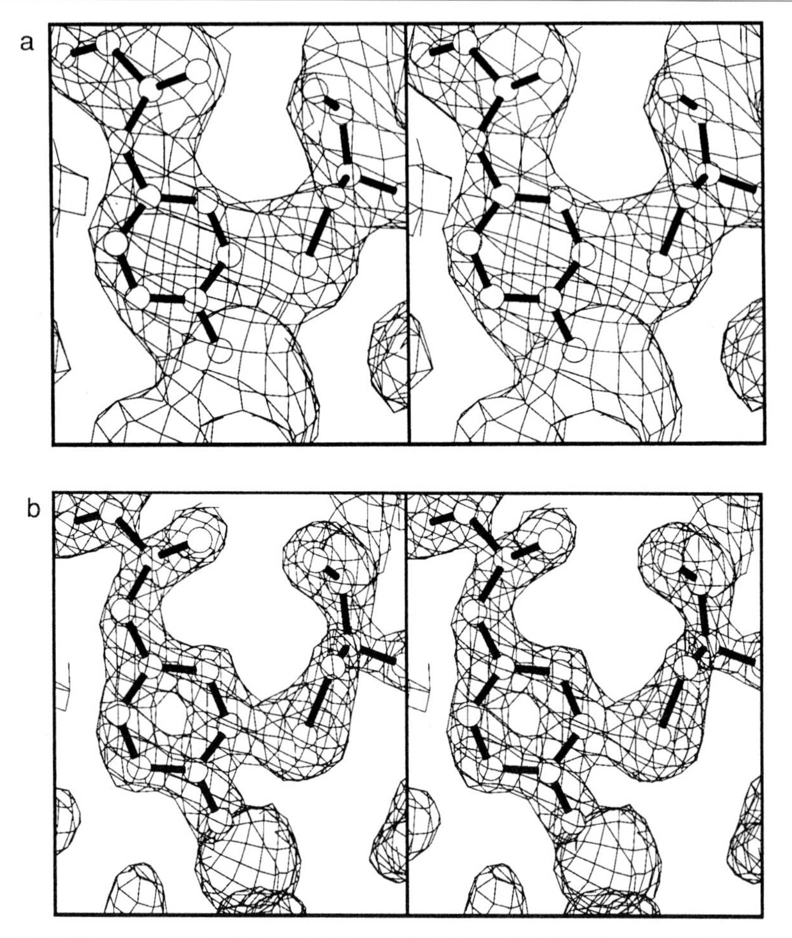

FIG. 4. The unusual thioether bond between Tyr-272 and Cys-228, drawn using the program SKULD.[34a] (a) Refined model with the 2.5-Å experimental MIR electron density map after solvent flattening, which is not biased by the molecular model. (b) The same view of the bond but with the refined 1.7-Å $2F_O - F_C$ map.

is direct and unambiguous, as shown by the rigorous analysis of the data described earlier. There is, however, much independent evidence that supports the existence of such a bond.

1. Titration of unfolded and fully reduced GOase shows that there are five free cysteine residues in the protein[2] whereas the peptide sequence

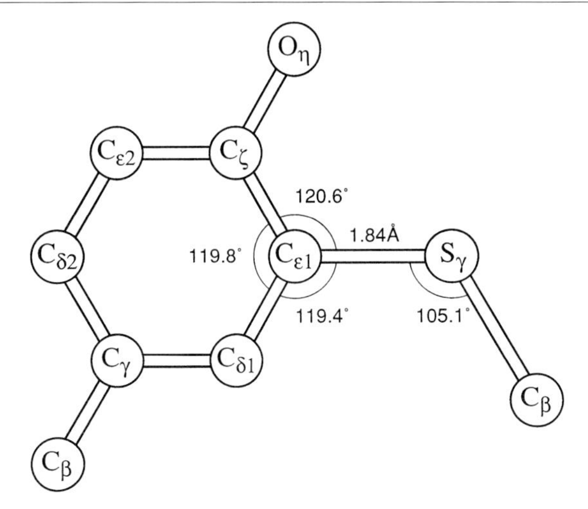

FIG. 5. Refined geometrical parameters for the thioether bond between Tyr-272 and Cys-228. Only the side chains of the two residues are shown. The bond lengths and angles shown here were refined from the diffraction data and were not artificially restrained to their ideal values. The torsion angles are: C_δ–C_ε–S_γ–C_β = 7°, χ_1 (Cys-228) = −83°, χ_2 (Cys-228, using C_ε of Tyr-272 as the fourth atom) = −99°, χ_1 (Tyr-272) = −68°, and χ_2 (Tyr-272) = −166°.

predicted from the DNA sequence indicates six.[1] One cysteine must, therefore, be modified such that it is not detected by conventional cysteine titration.

2. Amino acid sequencing[1] has repeatedly failed to identify Tyr-272 from the peptide which should contain it. This could be explained if Tyr-272 is modified in some way. One would expect by similar reasoning that Cys-228 would also be undetectable, although no peptides covering the region 200–249 have been isolated and sequenced (see Fig. 1).

3. Amino acid analysis on acid-hydrolyzed GOase reveals an unusual peak in the aromatic region[35] which was attributed to *o,o*-dityrosine. This peak might have originated instead from the *S*-linked dipeptide.

4. Resonance Raman data for GOase show systematically and significantly lower frequencies for all the ring modes of a copper-bound tyrosine than found previously in other copper-containing enzymes.[7] This can be explained by the substitution of one of the ring hydrogens with the heavier sulfur atom.

5. Simulation of the EPR spectrum for the free radical in the apoenzyme

[35] P. Tressel and D. J. Kosman, *Biochem. Biophys. Res. Commun.* **92,** 781 (1980).

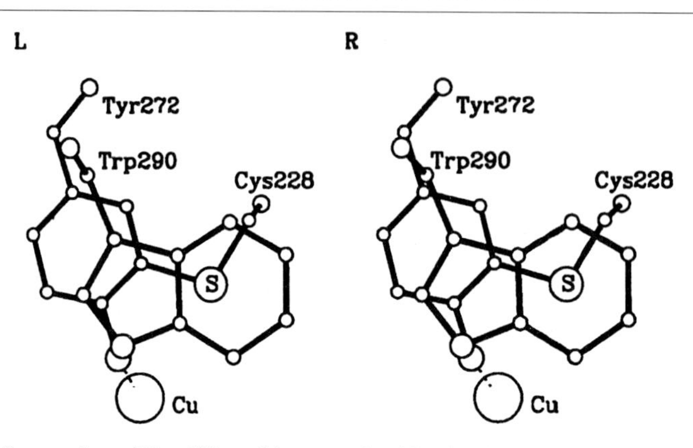

FIG. 6. Stereo view of Trp-290 stacking over the thioether bond between Tyr-272 and Cys-228, in which the six-membered ring of the indole lies exactly over the sulfur of Cys-228.

has been reported to indicate that one ring hydrogen was missing.[36] This is consistent with the loss of one ring hydrogen replaced by the thioether bond.

The role of the thioether bond and its biosynthesis are discussed later.

Stacking Interaction of Trp-290

Another striking feature of the GOase copper site is that the indole ring of Trp-290 is stacked on the "pseudo side chain" formed by the Tyr–Cys bridge (Fig. 6). The other side of the indole ring is exposed on the protein surface, protecting the thioether bond from the solvent.

The interactions involved did not seem to be simple π–π stacking of the two aromatic side chains (Tyr-272 and Trp-290). Viewed from the top, it is evident that the indole ring of Trp-290 has more interaction with Cys-228 than with Tyr-272, with its six-membered ring located exactly above the sulfur atom (the distances between the sulfur and these six carbons of Trp-290 are all 3.84 $\pm$ 0.04 Å). The "best" plane defined by all the side chain atoms (including C_β) of Tyr-272 and Cys-228 makes an angle of 1.8°, nearly perfectly parallel, with the plane of the side chain atoms (the ring atoms and C_β) of Trp-290. The two groups can be fitted to two parallel planes 3.44 Å apart.

The interactions between the thioether bond and the indole ring suggest

[36] G. T. Babcock, M. K. El-Deeb, P. O. Sandusky, M. M. Whittaker, and J. W. Whittaker, *J. Am. Chem. Soc.* **114,** 3727 (1992).

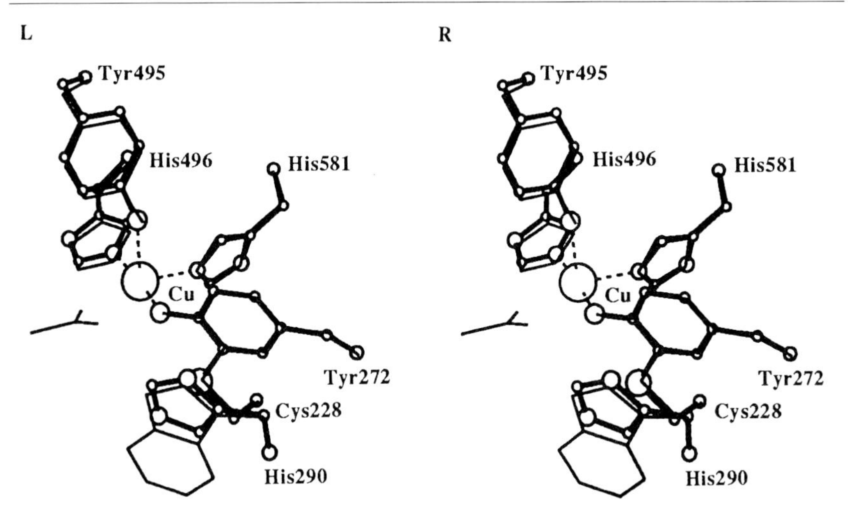

FIG. 7. Stereo view of the active site region of the W290H mutant enzyme shown as a ball and stick representation. The native structure, shown with thin lines, is superimposed.

involvement of Trp-290 in the free radical stabilization and/or other aspects of the catalytic mechanism. Although there is no evidence for the wild-type enzyme to substantiate this suggestion, it is supported by studies of a mutant form of galactose oxidase in which Trp-290 has been replaced by His (see next section).

X-Ray Crystallographic Studies on the Mutant Forms of Galactose Oxidase[13,37,38]

$W_{290}H$ Mutant (Substitution of the Stacking Tryptophan)

The crystal structure of the active site of the W290H mutant enzyme has been solved at 2.1 Å and refined to $R = 15.7\%$ (Fig. 7). It is essentially identical to that of the wild-type enzyme. The histidine ring is positioned exactly where the five-membered ring of the tryptophan indole had been; the only significant structural differences in the mutant are that O_γ of the hydroxyl of Ser-291 has rotated to point toward the missing electron density and that an additional acetate ion is present. The specific activity of W290H is 2000-fold lower than the wild type, indicating that Trp-290 plays a role, at present not completely clear, in the catalytic mechanism.

[37] A. J. Baron, C. Stevens, C. M. Wilmot, P. F. Knowles, S. E. V. Phillips, and M. J. McPherson, *Biochem. Soc. Trans.* **21,** 319S (1993).

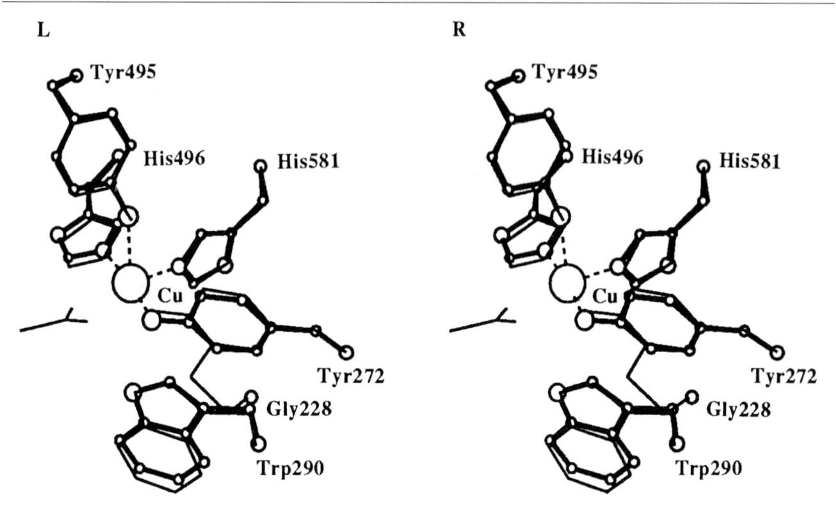

FIG. 8. Stereo view of the active site region of the C228G mutant enzyme shown as a ball and stick representation. The native structure, shown with thin lines, is superimposed.

$C_{228}G$ Mutant (Disruption of the Thioether Bond)

The crystal of C228G diffracted less well than the wild type or the W290H mutant, and good data were only obtained to 2.6 Å. The crystal structure was refined to $R = 14.9\%$. Inspection of the active site region (Fig. 8) revealed no change in main chain positions or the position of the copper and its ligands, including Tyr-272. However, the six-membered ring of Trp-290 and the ring of Phe-227 have moved into the space created by the loss of the cysteine side chain, and the side chain of Phe-194 has moved to occupy the space created by the movement of Phe-227. The specific activity of the C228 mutant is 10,000 times less than the wild type.

It has further been observed[38] that both the W290H and C228G mutant proteins in their redox-activated forms lack the 810-nm absorbance peak present in the native enzyme. This peak has been interpreted[7,27] as resulting from a mixture of Tyr $\rightarrow$ Cu charge transfer and charge resonance excitation between the aromatic π systems involved in the charge transfer complex; its absence in the two mutants provides the first evidence that the stacking tryptophan and the thioether bond are involved in generating and stabilizing the free radical.

[38] A. J. Baron, C. Stevens, C. M. Wilmot, K. D. Seneviratne, V. Blakeley, D. M. Dooley, S. E. V. Phillips, P. F. Knowles and M. J. McPherson, *J. Biol. Chem.* **269,** 25095 (1994).

Current Issues

Biosynthesis of the Thioether Linkage

The mechanism by which this unusual thioether is formed is not yet clear. One possibility is that it is created in the first turnover of the enzyme. It is evident in the case of ribonucleotide reductase[39] that, with the help of a metal center, a free radical can be formed on normal (i.e., unmodified) tyrosine under physiological conditions. If a tyrosine free-radical is created in the vicinity of a free cysteine, it might well be reactive enough to form a covalent thioether bond. In such a reaction, *o*-substitution, as observed here, would be preferable to *m*-substitution because $C_{\varepsilon 1}$ is one of the carbon atoms where larger spin density of the tyrosine radical in ribonucleotide reductase is reported to reside.[40] Another possibility is that the formation of the thioether is catalyzed by another enzyme as a post-translational modification. This seems unlikely for the fully folded protein because the thioether bond is not accessible to the solvent, but could occur with the partially folded protein.

Relationship between the Tyr–Cys–Trp Complex and the Free Radical

The unusual amino acid complex, Tyr-272–Cys-228–Trp-290, found in this study must be intimately involved in the creation and stabilization of the free radical. An extended aromatic system could aid the delocalization of the spin density, thereby stabilizing the free radical.

One of the characteristic features of GOase, among other free radical proteins, is the unusually low redox potential of the radical. The redox potential of the free radical in GOase determined by redox titration is 0.41 V (at pH 7.5),[5] compared to 0.94 V reported for a free tyrosine.[41] The thioether linkage probably contributes to lowering the redox potential of the radical.

The requirement of the cupric ion for radical generation is not yet clear. The apoenzyme treated with $K_3Fe(CN)_6$ has been reported to show an EPR signal from a free radical,[8] in contrast to ribonucleotide reductase, where the binuclear iron center is essential for radical formation. The GOase used in that study was, however, prepared as the holoenzyme, then converted to the apoenzyme by removing the copper. It would, therefore, probably have the Tyr–Cys bridge already formed. It may be that the

[39] P. Nordlund, B.-M. Sjöberg, and H. Eklund, *Nature* (*London*) **345,** 593 (1990).

[40] C. J. Bender, M. Sahlin, G. T. Babcock, B. A. Barry, T. K. Chandrashekar, S. P. Salowe, J. A. Stubbe, B. Lindström, L. Petersson, A. Eherenberg, and B.-M. Sjöberg, *J. Am. Chem. Soc.* **111,** 8076 (1989).

[41] M. R. DeFelippis, C. P. Murthy, M. Faraggi, and M. H. Klapper, *Biochemistry* **29,** 4847 (1989).

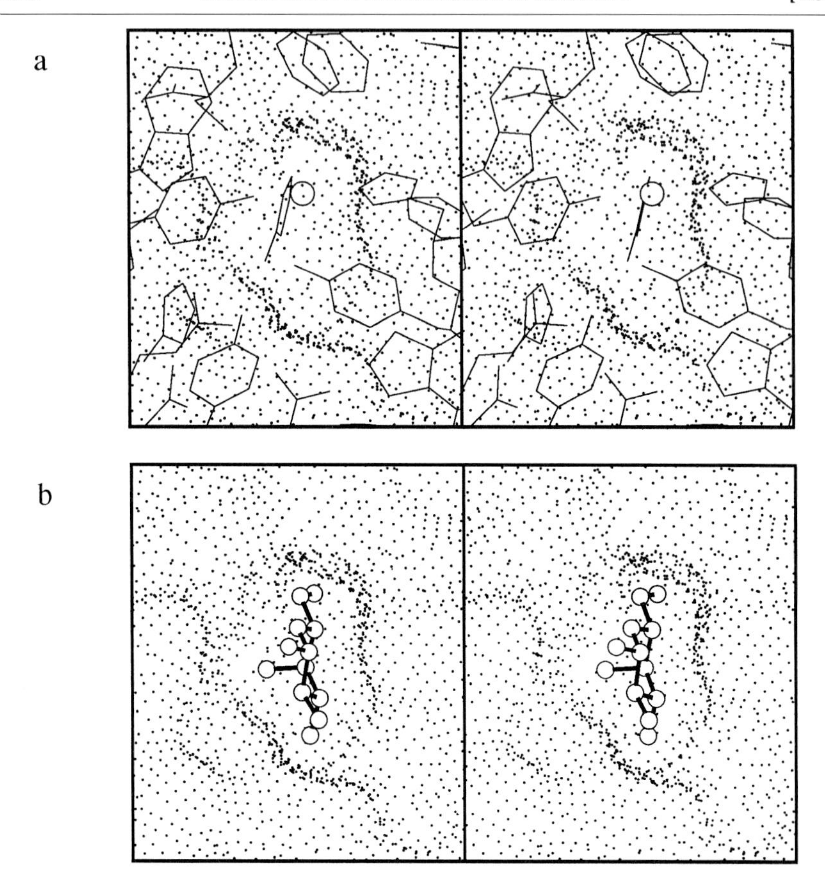

FIG. 9. Proposed substrate-binding site. (a) View looking down into the copper site with the Connolly surface[45] shown dotted. The sphere is the copper ion. (b) The same view as in (a) but with D-galactose in the pocket. (c) The modeled D-galactose molecule from a view orthogonal to (b) and looking along the molecular surface. (d) Same as (c) but with D-glucose in the pocket. In (c) and (d), the protein lies to the right hand and the solvent region is to the left.

enzyme synthesized in the fungal cell requires the copper ion for the generation of the free radical during the first turnover. This radical would generate the Tyr–Cys bridge, which would reduce the redox potential and make it possible to create the free radical without the copper center thereafter. The copper is still essential for catalytic activity, however, including its possible role in binding the alcohol substrate (see below).

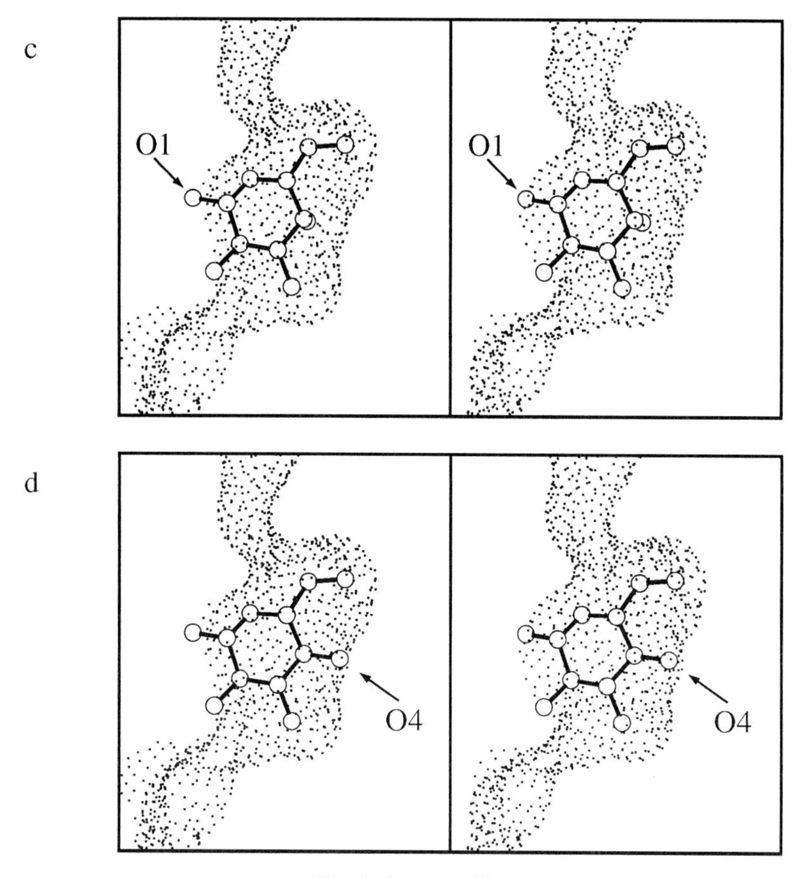

FIG. 9. (*continued*)

"Built-in" Cofactors

The ability of certain enzymes to utilize their own amino acid residues as a "built-in" cofactor does not seem limited to GOase. Bovine serum amine oxidase, another cooper-containing oxidase, has been reported to contain 6-hydroxydopa (or 3,4,6-trihydroxyphenylalanine), a hydroxylated derivative of tyrosine.[26] NADH peroxidase from *Streptococcus faecalis* is another example where an amino acid residue is modified to act as a redox cofactor. In this FAD enzyme, Cys-42 has been modified to form a stabilized cysteine–sulfenic acid.[42] Tyrosinase from *Neurospora crassa* has been re-

[42] L. B. Poole and A. Claiborne, *J. Biol. Chem.* **264,** 12330 (1989).

ported, from biochemical studies, to contain a thioether between Cys-94 and His-96,[43] although neither the involvement of the His–Cys bridge in the catalytic activity nor its relation to the binuclear copper center is known. A combination of side chain modification and covalent cross-linking of two side chains to generate a novel cofactor, 2,4′-bitryptophan-6′,7′-dione (or tryptophan tryptophylquinone), has been reported in methylamine dehydrogenase.[44] In this cofactor, an oxidized tryptophan makes a cross-link with a second unmodified tryptophan.

These reports suggest that such "built-in" cofactors occur more widely in enzymes. It is tempting to speculate that certain metalloenzymes have evolved a novel mechanism for circumventing the requirement for exogenous secondary cofactors by modifying their own amino acid residues.

Substrate Binding

The structure of GOase suggests a model for binding alcohol substrates. Calculations of the water-accessible surface[45] show a pocket at the copper site, which is filled by the ligand acetate ion and one water molecule in the native structure (Fig. 9a). A simple manual docking experiment using computer graphics (without either energy minimization or adjustment of protein side chains) shows that this pocket is structurally complementary to D-galactose in its "chair" conformation (Figs. 9b and 9c).

In addition to shape complementary, the model suggests favorable interactions between the enzyme and substrate (Fig. 10). O6 of the galactose is directly coordinated to the copper at the equatorial site occupied by the acetate ion in the native structure. The torsion angle O5–C5–C6–O6 is *trans*, which is reported to be the most favorable conformation.[46] O4 and O3 of the substrate form hydrogen bonds with Arg-330, and perhaps O2 with Gln-406. In addition to these polar interactions, part of the backbone (C6, C5, and C4) makes hydrophobic interactions with the side chains of two phenylalanine residues which form one wall of the pocket.

The model can also explain the substrate specificity of the enzyme. For instance, D-glucose, which is not a substrate, differs from D-galactose only in the configuration at C4, that is, the direction of O4. The movement of O4 would not only break the hydrogen bond with Arg-330 but also cause serious steric hindrance with Tyr-495 (Fig. 9d). Since Tyr-495 is one of the copper ligands, it would seem difficult to accommodate D-glucose in the pocket. Similar, or worse, steric hindrance would occur with other hexose

[43] K. Lerch, *J. Biol. Chem.* **257,** 6414 (1982).
[44] W. S. McIntire, D. E. Wemmer, A. Chistoserdov, and M. E. Lidstrom, *Science* **252,** 817 (1991).
[45] M. J. Connolly, *J. Appl. Crystallogr.* **16,** 23 (1983).
[46] A. Maradufu, G. M. Cree, and A. S. Perlin, *Can. J. Chem.* **49,** 3429 (1971).

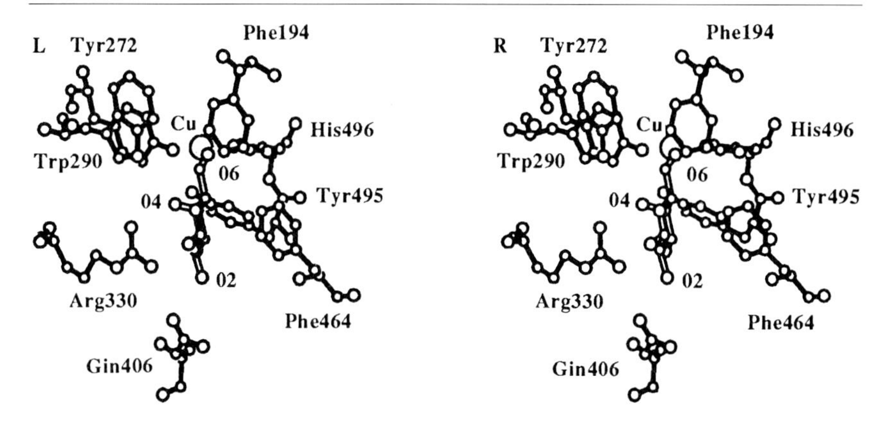

FIG. 10. Stereo view of the model for bound substrate (D-galactose). In this model, O6 of D-galactose is directly coordinated to the copper whereas O4, O3, and possibly O2 are involved in hydrogen bonding to the enzyme. C6, C5, and C4 make hydrophobic interactions with two phenylalanine residues.

derivatives such as L-galactose. On the other hand, the observation that polysaccharides with D-galactose at the nonreducing terminus could be substrates is consistent with the model since O1, where the next saccharide unit would be attached in such polysaccharides, is totally exposed to the solvent (Fig. 9c).

This model, where D-galactose binds at an equatorial site on the copper, contradicts one proposed on the basis of other experimental methods, where the alcohol substrate is bound to the copper at an axial site, and one equatorial site is occupied by molecular oxygen.[3] This proposal is based on spectroscopic experiments,[31] which suggested the presence of two external ligand binding sites, and a steady-state kinetics study,[47] which suggests an ordered mechanism for the GOase-catalyzed reaction.

Unfortunately, the active site of GOase is very close to a symmetrically related molecule in the crystal lattice and there is no physical space for substrates as large as D-galactose to bind in the crystal. Smaller primary alcohols such as 1,3-propanediol are reported to be poor substrates. Probably because of this, attempts to obtain substrate (or inhibitor) complexes of GOase by soaking crystals in solutions of such molecules have been unsuccessful. Cocrystallization also does not seem hopeful because acetate ion has a much higher affinity for the copper site than any substrate reported and the crystallization medium requires high concentrations (at least 0.6 *M*) of acetate. A new crystal form in which the copper site faces a large

[47] L. D. Kwiatkowski, M. Adelman, R. Pennelly, and D. J. Kosman, *J. Inorg. Biochem.* **14,** 209 (1981).

solvent region would probably be needed for crystallographic studies of substrate binding. There is, however, experimental evidence to support our proposed model for substrate binding; nuclear magnetic relaxation dispersion studies carried out with the reductively inactivated [Cu(II)–Tyr] form of the enzyme have shown that the substrate dihydroxyacetone displaces the equatorial water coordinated to the copper.[34]

The results reported in this chapter demonstrate the value and future potential of X-ray crystallographic studies of novel "built-in" cofactors. The availability of crystallographic structures offers a sound basis for spectroscopic and other studies of the properties of such cofactors.

Acknowledgments

The authors thank Dr. C. M. Wilmot for providing the data for Figs. 7 and 8. We thank the UK Science and Engineering Research Council (SERC) for financial support. S. E. V. P. is an International Research Scholar of the Howard Hughes Medical Institute and SERC Senior Fellow.

[19] Spectroscopic Studies of Galactose Oxidase

By James W. Whittaker

The structural probes that are useful in bioinorganic studies cover an enormous energy spectrum spanning more than 12 orders of magnitude (Fig. 1). These methods include probes of the electronic ground state [paramagnetic nuclear magnetic resonance (NMR) and electron paramagnetic resonance (EPR) and magnetic susceptibility (χ_{sus})], as well as techniques that explore electronic excited states [optical absorption, circular dichroism (CD), magnetic circular dichroism (MCD), and X-ray absorption (XAS) spectroscopies] and even nuclear excited states (Mössbauer spectroscopy). All of these, including the methods most useful in studies of free radical-containing enzyme (EPR and optical absorption spectroscopies), directly probe the *electronic structure* of an enzyme, extending beyond the atomic level of resolution available from a crystal structure and forming a bridge to *chemical reactivity*. Since each approach yields information on a specific aspect of structure, no single approach is capable of a giving a complete description of the active site, and a combination of these methods generally gives the most insight.

In free radical metalloenzymes the radical site represents a special feature in the electronic structure of the protein that results in unique chemistry

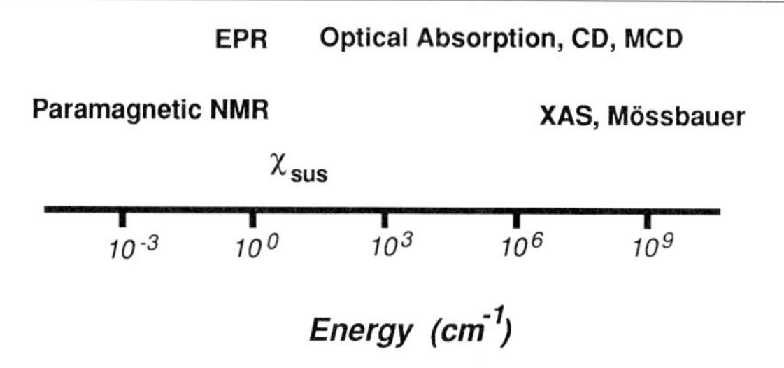

FIG. 1. The energy spectrum of bioinorganic structure probes.

and spectra. These spectra open a window on the active site for studies of ligand interactions and catalysis. Free radical metalloenzymes are often intensely colored as a result of low energy absorption bands associated with localized electronic excitations (e.g., $\pi \rightarrow \pi^*$ in a phenoxyl radical or $d \rightarrow d$ for a metal center). These and other spectra contain information that can be used at several levels of sophistication. At the most elementary level, absorption bands provide a signal that reflects quantitatively the amount of a species present. This quantitative aspect makes spectroscopy important as an analytical tool—detecting, resolving, and measuring components in a complex sample. The sensitivity of these active site spectra to substrate and other ligand interactions means that they can be used to monitor the progress of a titration, providing thermodynamic information (ligand affinities) independent of a complete knowledge of the origin of the spectrum. At a deeper level, assignment of the spectra gives information on the electronic structure of the active site. This ultimate analysis requires correlation among protein studies, modeling experiments, and theory. This overview of spectroscopic studies on galactose oxidase attempts to illustrate how these three levels of analysis can be developed in studies of a free radical metalloenzyme.

Active Site Structure and Chemistry

Galactose oxidase is a fungal copper metalloenzyme that catalyzes redox electron transfer from hydroxylic substrates (including galactose) to dioxygen, with formation of hydrogen peroxide and an aldehyde[1]:

$$RCH_2OH + O_2 \rightarrow RCHO + H_2O_2$$

Both half reactions involve two electron changes, yet the enzyme contains a single metal ion, a mononuclear copper center for which one-electron reactivity would be expected. The apparent mismatch in redox properties

[1] D. Amaral, F. Kelly-Falcoz, and B. L. Horecker, this series, Vol. 9, p. 87.

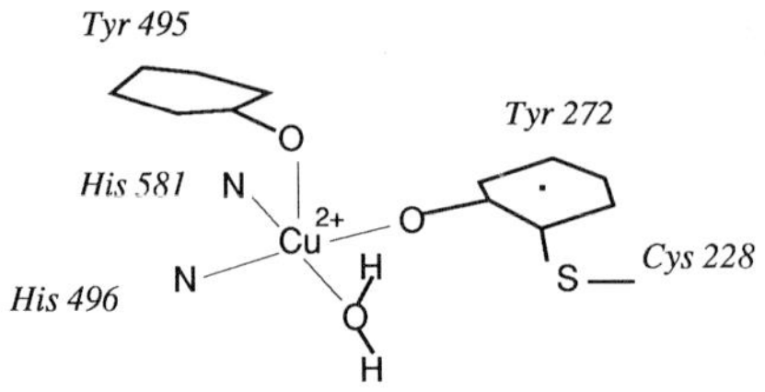

FIG. 2. Active site structure for galactose oxidase (based on Ito *et al.*[3]; see [18] by Ito and Knowles).

has been resolved by the discovery that galactose oxidase contains a protein-derived free radical localized on an amino acid side chain in the catalytically active form, and that this radical directly participates in catalysis. The actual catalytic complex is of an entirely new type in which the free radical interacts with a cupric metal center to form a free radical-coupled copper active site.[2] The most likely suspect for the radical-forming site in the enzyme, based on spectroscopic studies, is a sulfur-linked tyrosine–cysteine dimer in the protein that has been revealed by crystallographic studies[3] on galactose oxidase. This novel post-translational covalent modification of active site residues involves a cross-link between a tyrosine residue (Tyr-272) and a cysteinyl residue from a remote part of the polypeptide chain (Cys-228) as shown in the active site (Fig. 2). A second, unmodified tyrosine (Tyr-495) is also coordinated to the metal ion, as are two histidine imidazoles (His-496 and -591). Water coordinated to the copper indicates the position of the exogenous ligand binding site. While this atomic resolution protein crystal structure answers the questions: Where are the atoms? What is the geometric structure?, spectroscopy brings a focus to new questions: Where are the electrons? How does it work?

EPR Probes the Electronic Ground State

EPR spectroscopy is the electronic analog of NMR spectroscopy and is specifically sensitive to the presence of unpaired electrons in a sample.[4] This special sensitivity makes EPR extremely valuable as a specific probe of the metal and radical sites in enzymes since the bulk of the sample is transparent in this technique. As few as 10^{12} unpaired electrons (spins $S = 1/2$) may be detected in a sample with commercial instrumentation.

[2] M. M. Whittaker and J. W. Whittaker, *J. Biol. Chem.* **263,** 6074 (1988).

[3] N. Ito, S. E. V. Phillips, C. Stevens, Z. B. Ogel, M. J. McPherson, J. N. Keen, K. D. S. Yadev and P. F. Knowles, *Nature* (*London*) **350,** 87 (1991).

[4] H. Beinert, W. H. Orme-Johnson and G. Palmer, this series, Vol. 54, p. 111.

The number of unpaired electrons in a molecule can determine whether or not EPR spectroscopy will successfully detect the spins. An odd number of unpaired electrons in a molecule (half-integer spin S = 1/2, 3/2, ...) leads to ground state degeneracies (Kramers degeneracies) that split in an applied magnetic field ($\Delta E = g\beta H$, where g is the experimental spectroscopic splitting factor, β is the Bohr magneton, and H is the applied magnetic field) resulting in resonance and therefore EPR detectability. For an even number of electrons (integer spin S = 1, 2, 3 ...) a non-Kramers ground state occurs that in general lacks all degeneracies. EPR spectra are often difficult to observe in this case, even for a paramagnetic species ($S > 0$). Weak interactions between unpaired electrons can complicate this picture further through dipolar or exchange splittings in the ground state. In these cases, other methods such as magnetic susceptibility need to be applied since conventional EPR spectroscopy is not well suited to measuring the magnitude of these interactions.

In biological samples, unpaired electrons are generally associated with transition metal ions having a partly filled d shell and with organic free radicals. Signals from these two types of sites can be easily distinguished by the magnitude of the g shifts, the deviation of the g value at resonance from the free electron g value, $g_e = 2.0023$. The orbital contribution to electron paramagnetism in transition ion complexes results in relatively large g shifts (typically $g - 2 > 0.1$) whereas significantly smaller g shifts ($g - 2 \approx 0.002$) are characteristic of EPR spectra for organic free radicals.

These points are well illustrated by studies on galactose oxidase. The EPR spectrum of the reductively inactivated form of galactose oxidase,[2] recorded at cryogenic temperatures to improve sensitivity, is shown at the top of Fig. 3. The observed spectrum has a large g shift ($g - 2 = 0.2$) and is recognizable as the spectrum of a cupric (Cu^{2+}) ion. This simple analysis confirms the presence of copper in the enzyme sample, but contains much more additional information. For example, EPR spectroscopy can be used quantitatively to determine the amount of paramagnetic species contributing to a spectrum. In the spectrum shown in Fig. 3, the amount of EPR-detectable copper is evaluated by double integration of the EPR signal. Double integration is required since the normal EPR spectrum is actually the first derivative of the absorption spectrum. The first integral computes the absorption spectrum (Fig. 3, middle) which for an isolated EPR transition will be positively signed and returns to baseline above the transition. The second integral evaluates the area under the absorption band, which is proportional to the number of spins undergoing resonance. Comparison of the value of the integrated intensity (I) between an unknown sample and a suitable reference (a spin standard) yields a quantitative estimate of the amount of material contributing to the observed EPR signal, usually

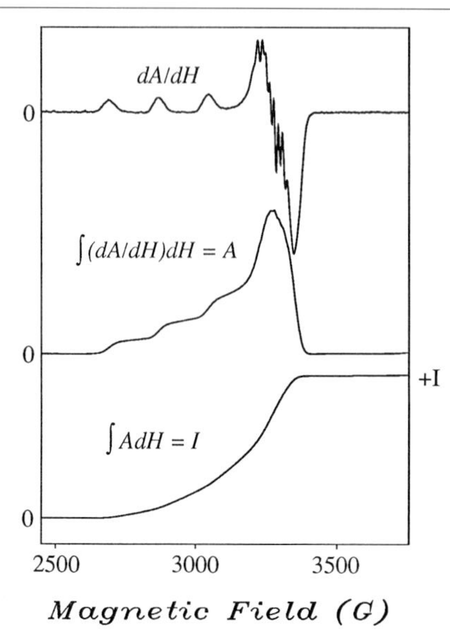

FIG. 3. Quantitative EPR measurements. (Top) First derivative EPR spectrum of reductively inactivated galactose oxidase. (Middle) First step integration over the full range of the resonance spectrum computes the absorption envelope. (Bottom) Second step integration evaluates the area under the absorption curve, yielding the integrated intensity, I. Spectra recorded on a Varian E109 X-band EPR spectrometer equipped with an Oxford Instruments E910 pumped helium cryostat for cryogenic temperature control. Instrumental parameters: microwave frequency, 9.223 GHz; microwave power, 10 μW; modulation amplitude, 10 G; temperature, 30 K. Spectrum shown is the result of coaddition of five scans. Protein concentration is 16 mg/ml in 50 m*M* $NaHPO_4$ buffer, pH 7.0.

accurate to $\pm 10\%$. A variety of spin standards have been reported for quantitative EPR studies, the general rule being that an appropriate standard should exhibit an EPR spectrum resembling that of the unknown in terms of resonance magnetic field and line width to avoid integration errors. The standard must also conform to analytical specifications of purity and homogeneity. To obtain reliable measurements, Aasa–Vänngård correction factors[5] must be applied in comparing the integrated intensities of spectra that differ significantly in *g* value. In work on galactose oxidase, copper perchlorate ($[Cu(OH_2)_6](ClO_4)_2$), prepared by digestion of electrolytic grade copper metal in concentrated HCl followed by dilution with sodium perchlorate, has proven valuable as a spin standard.[2] [*CAUTION!* Perchlorates may detonate without warning and should be handled with extreme care.] The most common errors in quantitation arise from analysis of inade-

[5] R. Aasa and T. Vänngård, *J. Magn. Reson.* **19,** 308 (1976).

quate data, where the signal is comparable in magnitude with instrument noise or baseline artifacts. These errors can be recognized in displays of the first and second integrals as nonclosure problems, for example, as an absorption spectrum that does not return to or overshoots the baseline, resulting in a second integral that does not converge.

The value of these quantitative applications of EPR spectroscopy becomes particularly clear when they are related to independent analytical methods, such as atomic absorption spectroscopy,[6] that provide an estimate of the total metal content. Differences between values obtained by the two approaches reveals that for galactose oxidase, a fraction of the copper is EPR nondetectable in the native form of the enzyme isolated from culture filtrates.[2] Careful correlation between the amount of EPR nondetectable Cu and the catalytic activity of the enzyme leads to the conclusion that in the most active enzyme the Cu is in an EPR silent, oxidized complex. This result, by itself, is consistent with the presence of a trivalent copper (Cu^{3+}) site in the enzyme, in the singlet ($S = 0$) ground state expected for a d^8 metal ion in low symmetry. However, when the results of other experiments are also considered (see below), it becomes clear that the metal retains the cupric oxidation state in the active enzyme complex, leading to the indirect identification of a free radical site in the protein.

On the basis of this indirect evidence for a free radical, a $S = 1/2$ free radical EPR signal should be detectable near the free electron g value ($g_e = 2.0023$). In fact, in oxidant-treated enzyme, a sharp free radical EPR signal *is* observed in this region.[2] In these spectra the free radical EPR signal appears disproportionately large compared to other features as a result of a relatively narrow line width and small g shifts. Quantitation of this signal indicates that it is actually a minority species present at less than 1% of the protein concentration and so is not even clearly assignable as arising from the enzyme. (However, see below.) EPR signals for both copper and free radical are therefore missing from the spectrum of the active enzyme. Where are the signals? The presence of both cupric and free radical centers in the active site suggests that interactions between the two paramagnetic centers might be responsible for the EPR silent character of the active enzyme complex. Magnetic susceptibility,[7] which is better suited to address this problem, confirms this picture, revealing a 200-cm^{-1} antiferromagnetic exchange splitting stabilizing a diamagnetic ($S_T = 0$) EPR silent sublevel lowest in the ground state.

It is possible to unmask the free radical EPR signal in galactose oxidase by removing the copper, forming a metal-free apoenzyme which can be

[6] C. Veillon and B. Vallee, this series, Vol. 54, p. 446.

[7] T. H. Moss, this series, Vol. 54, p. 379.

oxidized under mild conditions to generate a new EPR signal near the free electron g value ($g - 2 \approx 0.003$) associated with a stable organic free radical.[8] Experimental conditions for acquiring EPR spectra for free radicals are generally different from those used for recording metal ion spectra. The relatively long spin-lattice relaxation times for free radicals demand the use of relatively high temperatures and low microwave powers to avoid saturation of the EPR signal. Power saturation artifacts complicate quantitative measurements by reducing the observed signal intensity in the steady state, leading to correspondingly low estimates of spin concentration for a partly saturated sample. Saturation may be detected as a deviation from the square law for signal intensity ($I \propto \sqrt{P}$) when the microwave power (P) is varied systematically. More extensive characterization of the free radical EPR signal requires accurate measurement of the g values from experimental data. In general, factory instrument calibration is insufficient for measuring free radical spectra, and additional calibration is required. Magnetic field calibration may be accomplished by either gaussmeter measurements or by measurement of a resonance field for a sample having a known g value [for example, 2,2′-diphenyl-1-picrylhydrazyl (DPPH) free radical, $g = 2.0043$] when the microwave frequency is accurately known. Frequency calibration is most readily accomplished using a frequency counter. A complete description of the spectrum requires accurate simulation to obtain estimates of the components of the g tensor and the magnitudes and multiplicities of hyperfine coupling terms.[9]

The spectrum of the free radical formed in apogalactose oxidase is qualitatively the same as the free radical signal detectable in the oxidant-treated holoenzyme but quantitation shows that the signal in the oxidized apoenzyme is present at much higher levels (up to 0.4 spins/protein, representing approximately 40% of the active sites). This information is useful and provides a plausible interpretation of the free radical EPR signal in the oxidant-treated enzyme as arising from the small fraction of active sites lacking a bound metal ion. Quantitation of this signal can also be used to identify the optical spectrum of the radical site in oxidized apogalactose oxidase: reduction of the radical by a titrant like ascorbate may be monitored by EPR spectroscopy in parallel with measurements of the optical absorption spectrum. If the amount of reductant that is required to eliminate the free radical EPR signal matches the spin quantitation for the signal, it is possible to identify the component of the optical spectrum that has changed as arising from the free radical as well.[8] This type of quantitative

[8] M. M. Whittaker and J. W. Whittaker, *J. Biol. Chem.* **265,** 9610 (1990).

[9] G. T. Babcock, M. K. El-Deeb, P. O. Sandusky, M. M. Whittaker and J. W. Whittaker, *J. Am. Chem. Soc.* **114,** 3727 (1992).

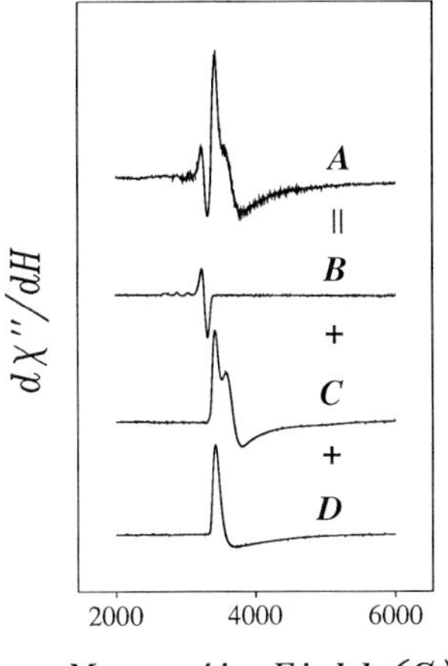

FIG. 4. Resolution of a complex EPR spectrum. (A) Spectrum of reductively inactivated galactose oxidase exposed to nitric oxide. (B) Spectrum of galactose oxidase before exposure to NO. (C) EPR spectrum of fast-relaxing nitric oxide species isolated by power saturation techniques. (D) Residual spectrum, identical to NO frozen in ice. Spectra recorded on a Bruker ER300 X-band EPR spectrometer equipped with an Oxford Instruments E900 helium flow cryostat for cryogenic temperature control. Instrumental parameters: microwave frequency, 9.460 GHz; microwave power, 1 mW; modulation amplitude, 10 G; temperature, 10 K. Spectrum shown is the result of coaddition of 10 scans. Protein concentration is 43 mg/ml in 50 m*M* $NaHPO_4$ buffer, pH 7.0.

correlation between techniques is extremely valuable in sorting out the complex behavior of the galactose oxidase active site.

A more complicated application of EPR approaches is illustrated in a tutorial example of an analysis of the spectrum of the nitric oxide complex of galactose oxidase (Fig. 4A). This spectrum is clearly complex, with overlapping signals from several distinct paramagnetic species present in the sample. Resolution of the component spectra is possible since the superposed spectrum is simply the sum of signals from each site. Standard spectra for the cupric site in the enzyme (B) and a pure component isolated by power saturation at low temperature (C) can be sequentially stripped off, leaving a residual signal (D) that can be recognized as the spectrum of NO in ice, an impurity signal in this sample. Each of the resolved signals may be quantitated, providing the basis for analysis of NO interactions with the enzyme.[10]

A variety of related spectroscopic approaches including electron nuclear double resonance[9] and electron spin echo envelope modulation spectro-

[10] M. M. Whittaker and J. W. Whittaker, *Biophys. J.* **64,** (1993).

scopies are valuable in higher resolution studies of the ground state spectra, specifically in evaluating nuclear hyperfine components that are often not clearly resolved in conventional EPR.

Optical Absorption Probes Excited States

The green color of redox-activated galactose oxidase, one of its most striking properties, results from low energy electronic excitations in the protein associated with the free radical site and the metal center. In the visible-near infrared spectral region, approximately 300–1200 nm, the bulk of the protein does not contribute to the absorption, so spectra directly relate to the special components of the active site (the free radical and the metal ion). Characterization of the enzyme in each of its accessible redox forms by optical spectroscopies makes use of this special active site selectivity, using absorption features arising from groups in the active site as reporters of ligand interactions and redox changes (Figs. 5 and 6). As for EPR spectroscopy, this information may be used at several different levels of sophistication.

Fundamental information contained in a simple absorption measurement are the energy and intensity of an optical absorption band. The *energy* of the transition is related to λ_{max}, the wavelength of the absorption maximum, by the relation $E(cm^{-1}) = 10^7/\lambda_{max}(nm)$. The *intensity* of the

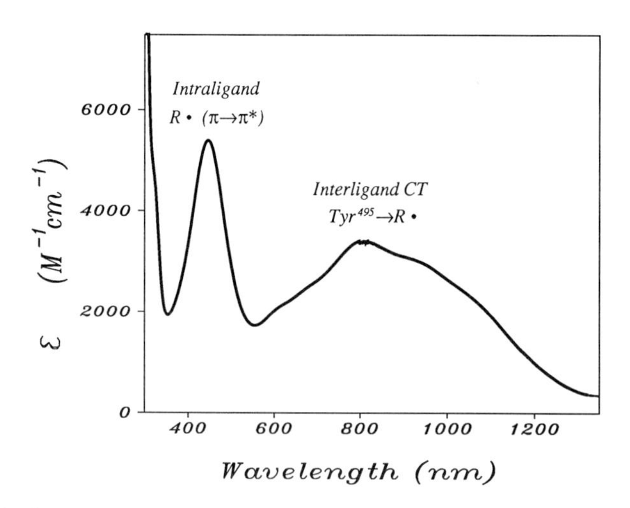

FIG. 5. Optical absorption spectrum for Redox-activated galactose oxidase. Assignments for major components of the spectrum are indicated. Spectrum recorded on a Cary 5 UV/VIS absorption spectrometer. Protein concentration is 11 mg/ml in 50 m*M* $NaHPO_4$ buffer in D_2O, pD = 7.0; 1-cm light path.

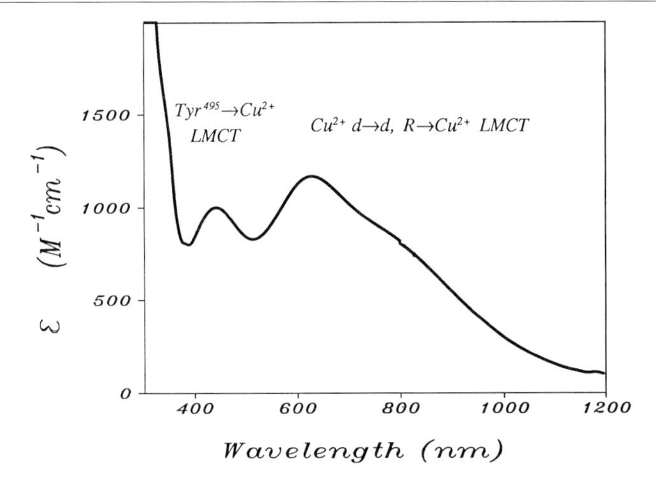

FIG. 6. Optical absorption spectrum for reductively inactivated galactose oxidase. Assignments for major components of the spectrum are indicated. Spectrum recorded on a Cary 5 UV/VIS NIR absorption spectrometer. Protein concentration is 27 mg/ml in 50 mM $NaHPO_4$ buffer in D_2O, pD = 7.0; 1-cm light path. Sample prepared by ferrocyanide reduction as described in Whittaker and Whittaker.[2]

absorption, measured as the molar extinction ε of an absorption band, is computed from the peak optical absorption A_{max} and the molar concentration c ($\varepsilon = A_{max}/c$). The intensity of the optical transition derives mainly from the magnitude of the charge displacement in the electronic excitation, and therefore varies characteristically over different types of excitation. Charge transfer (CT) transitions appear relatively strong in absorption as a result of the large electric dipole moment associated with the charge displacement. For example, the ligand-to-metal charge transfer (LMCT) excitations in metal complexes generally exhibit molar absorptivities in the range of 10^3–10^4 $M^{-1}cm^{-1}$. The energy of a LMCT transition relates to the type of ligand involved and to the metal oxidation state, and relatively easily oxidized ligands like tyrosine and cysteine give rise to the lowest energy LMCT spectra. Metal-centered d $\rightarrow$ d transitions are relatively weak in absorption ($\varepsilon \approx 10^2$–10^3 $M^{-1}cm^{-1}$) and the energies and splittings in the d $\rightarrow$ d spectra reflect the geometry of the metal complex and ligand interactions. d $\rightarrow$ d spectra will only occur for metal ions having a partly filled valence shell, so for copper only Cu^{2+} absorption is typically observed. Clearly, for copper, d $\rightarrow$ d spectra provide an important probe of the metal ion oxidation state and ligand environment.

At the simplest level, optical spectroscopy can be used to characterize the green-to-blue color changes associated with redox transformations of galactose oxidase, providing a basic signal for monitoring the progress of the redox reaction. Limiting spectra for reduced and oxidized samples can be used as quantitative factors for analysis of spectra for mixed, heterogeneous samples, permitting estimates of the fraction of each species involved. Using this approach, it is possible to quantitate the oxidized and reduced components in a heterogeneous mixture such as the native enzyme.[2] Alternatively, resolution of components is possible by making use of the distinct reactivities of these forms. The oxidized, catalytically active enzyme is reduced by substrates in the absence of O_2, generating a cuprous (Cu^+) form lacking absorption bands in the visible region of the spectrum. Any residual absorption in the optical spectrum of this substrate-reduced galactose oxidase arises from inactive enzyme, providing a straightforward method for determination of this species independent of EPR spin quantitation.[2]

Ligand interactions in the active site perturb its structure, resulting in changes in the optical spectra. The interpretation of these changes generally requires a detailed understanding of the origins of the spectral features, but in some cases may yield direct insights. For example, on binding azide to galactose oxidase, the absorption spectra shift and a new feature appears near 390 nm in the near UV region for both active and inactive enzyme forms as illustrated in the spectra on the left in Fig. 7. On the basis of comparison with known cupric azide model complexes, it is possible to assign the 390-nm absorption to the low energy π component of the azide-to-Cu^{2+} LMCT transition for an equatorially coordinated azide ligand. The coincidence of the energies observed for this LMCT band in complexes with both active and inactive forms of galactose oxidase is strong evidence for a cupric (Cu^{2+}) center in the redox-activated enzyme, based on the known sensitivity of the energies of LMCT spectra to metal oxidation state.[2] (Cu X-ray absorption edge spectroscopy is a sensitive probe of metal ion oxidation state[11] that confirms the Cu^{2+} assignment for the metal in the unligated enzyme.[12])

As a direct probe of the active site, optical spectroscopy is a valuable tool for monitoring the progress of ligand binding reactions. Changes in the absorption spectra reflect conversion to the ligand–bound complex, with intermediate spectra being a simple superposition of the spectra of the components. For single step (two-state) binding reactions, isosbestic points will occur between successive spectra recorded during the titration. The most appropriate wavelength for monitoring exogenous ligand interac-

[11] S. I. Chan and R. C. Gamble, this series, Vol. 54, p. 323.

[12] K. Clark, J. E. Penner-Hahn, M. M. Whittaker and J. W. Whittaker, *J. Am. Chem. Soc.* **112,** 6433 (1990).

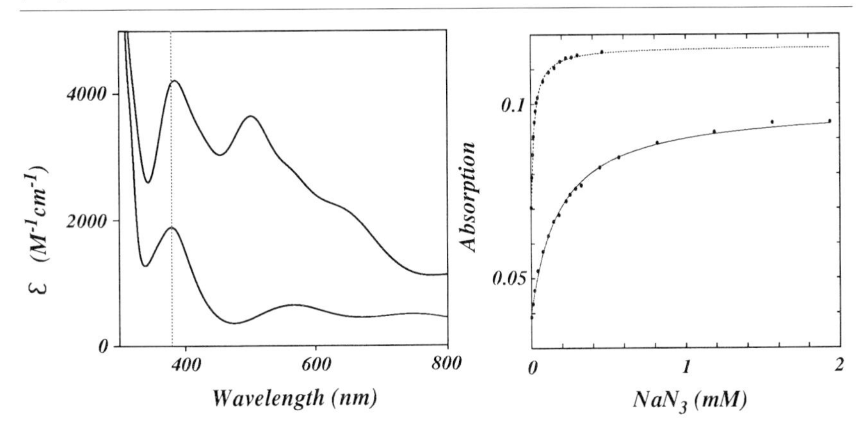

FIG. 7. Ligand interactions with galactose oxidase. (Left) Absorption spectra for azide complexes of active (upper) and inactive (lower) galactose oxidase. The broken line identifies the 390-nm bands in both spectra. (Right) Titration curves obtained by monitoring 390 nm absorption during serial addition of sodium azide to active (upper) and inactive (lower) enzyme samples. K_d's for azide binding are evaluated by fitting the experimental data with theoretical binding isotherms (smooth curves). For the active enzyme, $K_d(N_3^-) = 20\ \mu M$; for the inactive enzyme, $K_d(N_3^-) = 190\ \mu M$. Spectra recorded on a Cary 5 UV/VIS NIR absorption spectrometer. Protein concentration is 5.3 mg/ml (active) or 10.6 mg/ml (inactive) in 50 mM $NaHPO_4$ buffer in D_2O, pD = 7.0; 1-cm light path. NaN_3 added to 1.1 mM (active) or 10 mM (inactive) to quantitatively convert enzyme complexes.

tions may be determined by comparing initial spectrum of unligated enzyme with the spectrum of the complex formed by saturating levels of exogenous ligand and identifying the point at which the maximum ΔA displacement occurs. Monitoring absorption changes at this wavelength (not necessarily a maximum or minimum of either spectrum) will provide the greatest sensitivity for acquiring titration data. Thermodynamic data (ligand dissociation constants, K_D) are extracted by fitting the titration data with a theoretical binding isotherm (Fig. 7, right).

Although absorption spectra may be roughly assigned by analogy with known models, detection and identification of free radicals in proteins by optical spectroscopy are problematical. In certain cases, for example, for the tyrosine phenoxyl free radical in ribonucleotide reductase, the optical spectrum is sufficiently characteristic to provide convincing evidence for the identity of the radical.[13] However, in general this is not the case. In galactose oxidase, the intense absorption features of the redox-activated enzyme may be qualitatively identified as being associated with the presence of a free radical in the active site, but there is no obvious analogy with

[13] B.-M. Sjöberg, P. Reichard, A. Gräslund and A. Ehrenberg, *J. Biol. Chem.* **253,** 6863 (1978).

other radical spectra aside from similarity to the benzene dimer radical.[14] Even for the oxidized apoenzyme, where the optical spectrum is expected to more directly relate to the isolated free radical site, modeling studies[15] and calculations based on thc crystallographic tyrosine–cysteine dimer structure have been required to resolve the spectroscopic assignments. The detailed assignments indicated in Figs. 5 and 6 are the result of a combination of spectroscopic, chemical, and biochemical studies which together form the basis for a characterization of the complex spectra.[2,8,10,14,15]

Polarization Methods Resolve Spectra

Polarization spectroscopies (CD and MCD) contribute to spectroscopic characterization of galactose oxidase by providing methods for detection, resolution, and quantitation of spectral components. The basis for CD spectroscopy is the differential absorption of left- and right-handed light ($\Delta A_{\text{L-R}}$) in an absorption band. The requirements for observing CD spectra are therefore optical absorption (obvious!) and chirality (essentially guaranteed for biological complexes). In contrast to optical absorption, CD is a signed quantity which may be either positive or negative, depending on whether left or right hand absorption occurs preferentially. The signed character of CD intensity makes this approach useful in separating broad and overlapping absorption bands which may not be resolved in absorption.

The net circular polarization in optical spectra of biomolecules is generally small, on the order of 1% or less of the total absorption. CD spectroscopy therefore requires special instrumentation and is subject to special polarization artifacts. Sample concentrations for CD tend to be relatively high compared to absorption experiments, on the order of 0.1–1 m*M* for measurements on galactose oxidase, a basic criterion being that the maximum sample absorption should be approximately 1 OD. Simply detecting a CD signal may be useful as a method of monitoring a titration or redox reaction, particularly if the modifying reagent has absorption bands in the visible region of the spectrum that interfere with the analysis of optical titration data. This type of application for CD spectroscopy is illustrated in the redox interconversion of active and inactive forms of galactose oxidase in the presence of ferri-/ferrocyanide redox buffers.[2] The redox agent ferricyanide absorbs strongly in the visible spectrum, with absorption bands that overlap the spectra of the enzyme, making the analysis of the enzyme

[14] M. M. Whittaker, V. L. DeVito, S. A. Asher and J. W. Whittaker *J. Biol. Chem.* **264,** 7104 (1989).

[15] M. M. Whittaker, Y.-Y. Chuang, and J. W. Whittaker, *J. Am. Chem. Soc.* **115,** 10029 (1993).

spectra difficult. The achiral redox agents make no contribution in CD, however, permitting the conversion to be followed quantitatively. As an analytical tool, CD has certain advantages over absorption spectroscopy and is a valuable complementary technique for characterizing mixtures. As a specific example, inactive, native, and active galactose oxidase all have distinct absorption spectra lacking isosbestic points that might be used to define the number of distinct species, contributing to the difficulty of characterizing these enzyme forms. In CD, however, *isodichroic* points occur between the spectra of pure active and inactive enzyme (Fig. 8) and the observation that the CD spectrum of native galactose oxidase intersects the spectra of the limiting reduced and oxidized forms at precisely the same points conclusively demonstrates that native enzyme is not a distinct form but simply a mixture of partly inactivated enzyme.[2]

As for absorption and EPR measurements, quantitative CD studies demand proper instrument calibration and exclusion of artifacts. CD calibration is typically accomplished by recording the spectrum of a reference sample of known molar ellipticity (e.g., (1S)-(+)-10-camphorsulfonic acid) following standard procedures. The most common artifact in CD spectroscopy is due to depolarization of the circularly polarized light in birefringent media. Spectroscopic glassware used in CD spectroscopy should be examined for depolarization artifacts which may result from stresses within the optical windows by measuring the spectrum of a CD-active sample placed before and after the cell. Coincidence of the two spectra indicates lack of

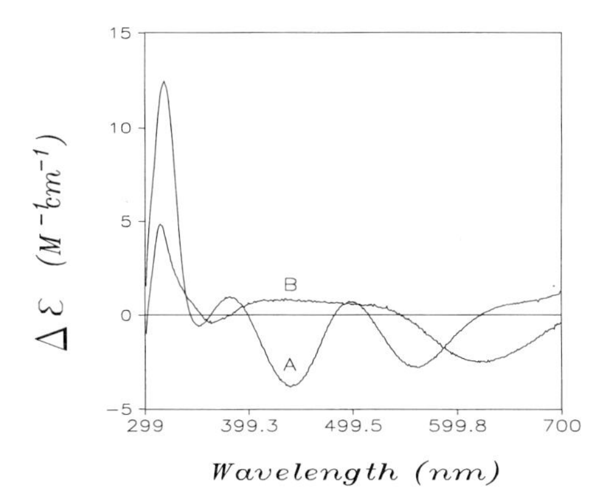

FIG. 8. Circular dichroism spectra for active (A) and inactive (B) galactose oxidase. Spectra recorded on an AVIV Model 41DS UV/VIS NIR spectropolarimeter. Instrumental parameters: time constant, 2.5 sec; band width, 2.5 nm; sampling interval, 1 nm. Protein concentration is 11 mg/ml (A) or 5.4 mg/ml (B) in 50 m*M* $NaHPO_4$ buffer in D_2O, pD = 7.0; 1-cm pathlength.

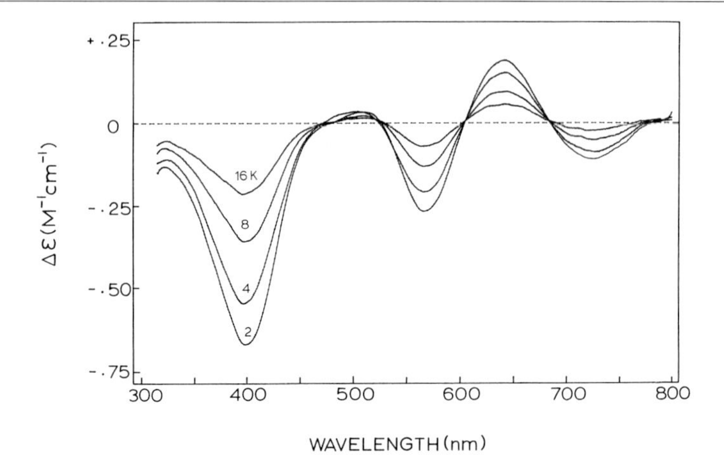

FIG. 9. Magnetic circular dichroism spectrum for azide complex of inactive galactose oxidase. Spectra recorded on an AVIV Model 41DS UV/VIS NIR spectropolarimeter equipped with an Oxford Instruments SM4-6T magnetocryostat. Instrumental parameters: magnetic field strength, 5.5 T; time constant, 0.5 sec; band width, 2 nm; sampling interval, 1 nm. Sample temperature for each scan is recorded on the spectrum. Protein concentration is 88 mg/ml in 50 mM $NaHPO_4$ buffer in D_2O, pD = 7.0; 50% d_3-glycerol; 0.1-cm pathlength. NaN_3 (5 mM) was added to form the anion complex.

significant depolarization.[16] The standard units of intensity for CD spectroscopy vary with application. Ellipticity [θ (m°)] is commonly the default unit of measurement on commercial CD instrumentation and may be converted to units more closely related to absorption measurements by the relation $\Delta A = \theta(\text{m}°)/32{,}900$ and to molar units ($\Delta\varepsilon$) by further dividing by the molar concentration.

Intensities in CD spectra may also be useful in making spectroscopic assignments. While intraligand and charge transfer transitions dominate the absorption spectra, d $\rightarrow$ d spectra tend to be relatively strong in CD. By comparing the anisotropy ($\Delta\varepsilon/\varepsilon$) over the CD spectrum it is sometimes possible to identify features arising from a metal center. For example, in the active form of galactose oxidase, intraradical and interligand charge transfer absorption bands dominate the visible spectrum, whereas in CD a relatively strong band associated with the cupric center may be detected.[2]

Magnetic circular dichroism (MCD) spectroscopy is superficially related to CD in terms of instrumentation requirements, but represents a distinct approach, not limited as CD is to chiral chromophores.[17] MCD results from

[16] W. R. Browett, A. F. Fucaloro, T. V. Morgan, and P. J. Stephens, *J. Am. Chem. Soc.* **105,** 1868 (1983).

[17] B. Holmquist, this series, Vol. 130, p. 270.

circular polarization induced in optical absorption bands in the presence of a magnetic field. This magnetic field-induced circular polarization arises from interaction of electronic spin and orbital magnetic moments in ground and excited electronic states with the applied magnetic field. For paramagnetic samples, the dominant intensity mechanism leads to temperature-dependent Faraday C terms which increase in intensity at low temperature. As a result, for greatest sensitivity, spectra are typically recorded on frozen samples at cryogenic temperatures (2–100 K) and strong magnetic fields (ca. 6 T). MCD experiments require optical transmission at low temperature which for biological samples can generally be achieved by using aqueous buffers containing 50% glycerol. High quality glasses are produced from this solvent by slow cooling of 1-mm-thick samples to cryogenic temperatures. In terms of quantity of protein, the sample requirements for MCD experiments are similar to those for CD spectroscopy. The sample volume is reduced, but the shorter path length for the MCD sample requires an order of magnitude increase in concentration, to the 1–5 m*M* range. Samples of this type may be readily prepared for the paramagnetic, redox-inactivated form of galactose oxidase for characterization of the isolated cupric active site metal center (Fig. 9). The results of the MCD experiment can be interpreted in terms of theory and by correlation with results from synthetic models as a powerful basis for spectroscopic assignments.

Summary

A combination of powerful spectroscopic approaches is now available for detailed characterization of active sites in free radical metalloenzymes. As illustrated for galactose oxidase, a single approach such as EPR or absorption spectroscopy is not sufficient to characterize a sample completely, and only by combining results from complementary techniques can the information in each approach be used most effectively.

Acknowledgments

The spectroscopic studies on the free radical coupled copper active site in galactose oxidase described here have been supported by the National Institutes of Health (GM-46749).

[20] Use of Rapid Kinetics Methods to Study the Assembly of the Diferric–Tyrosyl Radical Cofactor of *E. coli* Ribonucleotide Reductase

By J. MARTIN BOLLINGER, JR., WING HANG TONG, NATARAJAN RAVI, BOI HANH HUYNH, DALE E. EDMONDSON, and JOANNE STUBBE

Introduction

Since the early 1990s, studies made possible by advances in molecular biological and physical biochemical methods have demonstrated that the essential prosthetic groups of a number of enzymes are generated via metal- and O_2-mediated post-translational modifications of common amino acids.[1–3] Our laboratories have focused on understanding one such case, the mechanism of assembly of the tyrosyl radical–μ-oxodiiron(III) cofactor essential for nucleotide reduction in *Escherichia coli,* mammalian, and herpes viral ribonucleotide reductases (RNRs). Specifically, we have been investigating the mechanism by which the R2 protein of *E. coli* RNR self-assembles its cofactor (Fig. 1)[3a] when metal-free (apo) R2 is treated with Fe^{2+} and O_2.[4–8]

Our studies[7] and those of Elgren *et al.*[9] and Ochiai *et al.*[10] suggest that an "extra" reducing equivalent, in addition to the three obtained by

[1] S. M. Janes, D. Mu, D. Wemmer, A. Smith, S. Kaur, D. Maltby, A. L. Burlingame, and J. P. Klinman, *Science* **248,** 981 (1990).

[2] A. F. V. Wagner, M. Frey, F. A. Neugehauer, W. Shater, and J. Knappe, *Proc. Natl. Acad. Sci. U.S.A.* **89,** 996 (1992).

[3] J. Stubbe, *Curr. Opin. Struct. Biol.* **1,** 788 (1991).

[3a] P. Nordlund and H. Eklund, *J. Mol. Biol.* **232,** 123 (1993).

[4] J. M. Bollinger, Jr., D. E. Edmondson, B. H. Huynh, J. Filley, J. R. Norton, and J. Stubbe, *Science* **253,** 292 (1991).

[5] J. M. Bollinger, Jr., J. Stubbe, B. H. Huynh, and D. E. Edmondson, *J. Am. Chem. Soc.* **113,** 6289 (1991).

[6] N. Ravi, J. M. Bollinger, Jr., B. H. Huynh, D. E. Edmondson, and J. Stubbe, *J. Am. Chem. Soc.* **116,** 8007 (1994).

[7] J. M. Bollinger, Jr., N. Ravi, Tong, W. H. Tong, B. H. Huynh, D. E. Edmondson, and J. Stubbe, *J. Am. Chem. Soc.* **116,** 8015 (1994).

[8] J. M. Bollinger, Jr., N. Ravi, Tong, W. H. Tong, B. H. Huynh, D. E. Edmondson, and J. Stubbe, *J. Am. Chem. Soc.* **116,** 8024 (1994).

[9] T. E. Elgren, J. B. Lynch, C. Juarez-Garcia, E. Münck, B.-M. Sjöberg, and L. Que, Jr., *J. Biol. Chem.* **266,** 19265 (1991).

[10] E.-I. Ochiai, G. J. Mann, A. Gräslund, and L. Thelander, *J. Biol. Chem.* **265,** 15758 (1990).

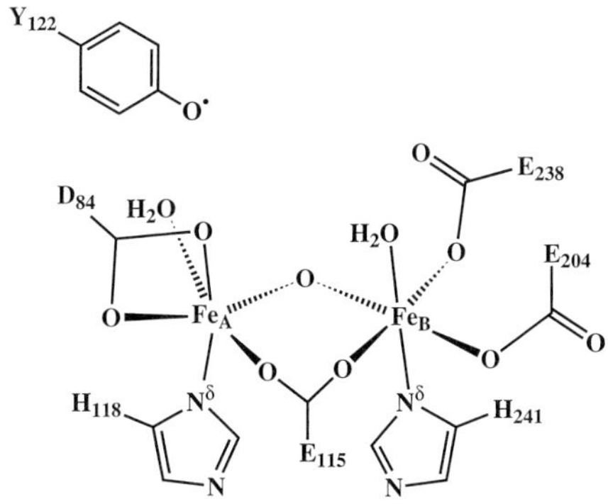

FIG. 1. Structure of the diferric–tyrosyl radical cofactor of RNR (adapted from Nordlund and Eklund[3a]).

oxidation of tyrosine 122 (Y122) and 2 Fe^{2+}, is required in the assembly reaction to balance the four electron reduction of O_2:

$$2Fe^{2+} + Y122 + O_2 + H^+ + e^- \rightarrow Fe^{3+}\text{-}O^{2-}\text{-}Fe^{3+} + Y122\bullet + H_2O. \quad (1)$$

This reducing equivalent can be supplied by a third Fe^{2+}. We have presented evidence that the reaction partitions between two pathways and that the partition ratio depends on the availability of the extra reducing equivalent, which is determined by the Fe^{2+}/apo R2 ratio employed in the reaction. This chapter focuses on experiments in which the Fe^{2+}/apo R2 ratio is ~5, under which conditions the extra reducing equivalent is readily available from Fe^{2+}. As will be seen, useful properties of the reaction under these conditions are its first-order kinetic behavior and the independence of its rate constant on Fe^{2+} and O_2 concentrations. The postulated mechanism for the reaction under these conditions is shown in Scheme I.

Three methods were employed to examine the kinetics of the assembly process: stopped-flow UV/VIS absorption spectroscopy (SF-Abs), rapid freeze-quench EPR spectroscopy (RFQ-EPR), and rapid freeze-quench Mössbauer spectroscopy (RFQ-Möss). Evidence will be presented that these three methods, despite their inherent differences, give rate constants that are in remarkably good agreement. Furthermore, the EPR and Möss methods have provided insight into the structure of a novel diferric-radical

$$\text{apo R2} + 3Fe^{2+} + O_2 \xrightarrow{k_1} (Fe^{3+})_2\,L\bullet + \text{“}Fe^{3+}\text{”} \xrightarrow{k_2} Fe^{3+}\text{-}O^{2-}\text{-}Fe^{3+}\ (Y122\bullet)$$

$$(Fe^{3+})_2\,L\bullet = \text{intermediate X}$$

SCHEME I

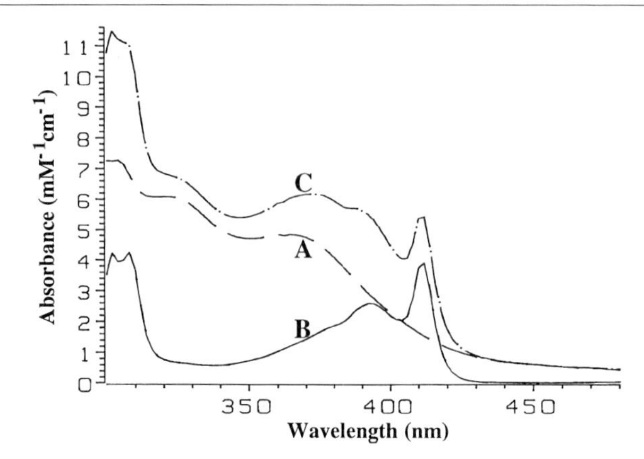

FIG. 2. UV/visible light absorption spectra of the diferric cluster (A) and the tyrosyl radical (B). Spectra were obtained by subtraction as described in Methods. Spectrum C is the sum of A and B.

intermediate that is kinetically competent to oxidize Y122 to the stable tyrosyl radical (Y122•).[5–7]

Reaction of Apo R2 with Excess Fe^{2+} and O_2 Probed by SF-Abs Spectroscopy

Three criteria are essential for use of SF-Abs to monitor a reaction. First, at least one of the components must have a UV/VIS light absorption spectrum. The R2 cofactor meets this criterion, as both the diferric cluster (Fig. 2A, with absorption features at 325 and 360 nm) and the tyrosyl radical (Fig. 2B, with a sharp absorption feature at 410 nm) have distinctive spectra. Second, the rate constant for the reaction must be sufficiently low to be measurable ($<1000\ sec^{-1}$). The assembly reaction carried out by mixing apo R2 with saturating Fe^{2+} and O_2 has an overall rate constant of $\sim 1\ sec^{-1}$ at 5°, and, hence, this criterion is met as well. Third, sufficient quantities of the reactants must be available. Application of molecular biological methods[11] has allowed preparation of R2 in gram quantities, and the pioneering work of Atkin *et al.*[12] made preparation of the apo protein straightforward. The ability to obtain grams of apo R2 was particularly important in this study, since it was considered important to carry out the SF-Abs and RFQ-Möss experiments (described subsequently) under identical reaction conditions, and Mössbauer spectroscopy is extremely

[11] S. P. Salowe and J. Stubbe, *J. Bacterol.* **165,** 363 (1986).
[12] C. L. Atkin, L. Thelander, P. Reichard, and G. Lang, *J. Biol. Chem.* **248,** 7464 (1973).

insensitive (requiring Fe concentrations in excess of 500 μM). It is clear that the cofactor assembly reaction is an excellent system to monitor by SF-Abs. The caveat is that, when this work was initiated, very little was known about the types of intermediates likely to be involved in the assembly process and the physical properties (such as UV/VIS spectra) that they might exhibit.

Apparatus

SF-Abs experiments were carried out on two different apparatus. Our dirth of knowledge concerning the absorption spectra of likely intermediates made multi-wavelength monitoring desirable. This was accomplished by using a Hewlett Packard 8452A diode array spectrometer in conjunction with an Applied Photophysics RX.1000 rapid kinetics spectrometer accessory. With this apparatus, both the rapid mixing accessory and the spectrometer were actuated manually, resulting in a long and somewhat variable dead time (estimated to be 100–250 msec). Experiments that required increased time resolution or dead time reproducibility were carried out on an Applied Photophysics DX.17MV sequential stopped-flow spectrofluorimeter.

Protocol

All three types of kinetics experiments in this study (SF-Abs, RFQ-EPR, and RFQ-Möss) were carried out according to the same general protocol: apo R2 (550–640 μM) in 100 mM HEPES, pH 7.7, was mixed with an equal volume of an Fe^{2+} stock solution containing 4.9–5.1 molar equivalents Fe^{2+}. [Additional SF-Abs experiments were carried out with lower concentrations of apo R2 (40–120 μM) and Fe^{2+}/R2 ratios of 5–10].[13] The Fe^{2+} stocks were prepared as follows. A 1.5- to 14-mg piece of Fe metal was dissolved anaerobically in 2 N H_2SO_4. The volume of H_2SO_4 added was such that 4 molar equivalents of H^+ relative to Fe^0 was initially present. The Fe^0 and H_2SO_4 were heated at 60° to speed dissolution, which took 12–24 hr. This stock was diluted for storage with H_2O to 5 mM Fe^{2+}. An aliquot of the 5 mM Fe^{2+} solution was diluted either with H_2O or with dilute H_2SO_4 in order to give the desired concentration of Fe^{2+} in 2–3 mM H_2SO_4. (It was assumed that 2 mol H^+ was converted to H_2 per mol Fe^0 oxidized to Fe^{2+}.) In order to ensure that O_2 was in excess, both the protein solution and the Fe^{2+} solution were saturated with 1 atm of O_2 prior to their being loaded into the stopped-flow apparatus. Aliquots of the stock

[13] The rate constants obtained for the assembly process were independent of the apo R2 concentrations from 20 to 640 μM and of the Fe^{2+}/R2 ratios from 5 to 40.

solutions were placed in separate 100-ml tonometers. The tonometers were connected to a house vacuum and to a source of humidified O_2. The tonometers were gently evacuated by opening the house vacuum. After several seconds, the vessels were refilled with H_2O-saturated O_2. This vacuum/O_2 cycle was repeated six to eight times, and the solutions were incubated on ice for 20–30 min. This routine of six to eight cycles of vacuum/O_2 followed by incubation on ice for 20–30 min was repeated three to four times. The solutions were then loaded into the stopped-flow or rapid freeze-quench syringes.

Results

When apo R2-wt is mixed at 5° with excess Fe^{2+} (5 molar equivalents),[13] development of the absorption spectrum characteristic of the tyrosyl radical–μ-oxodiiron(III) cofactor is complete within 30 sec. The time-dependent spectra of the reaction (Fig. 3) indicate that an intermediate rapidly accumulates prior to formation of the product cofactor. Since nothing was known about the UV/VIS spectra of potential intermediates, the HP8452A/RX1000 apparatus was initially used to monitor the assembly process. In the dead time + 0.2-sec spectrum of this representative experiment, 47% of the final absorbance at 360 nm was developed, whereas only 10% of the

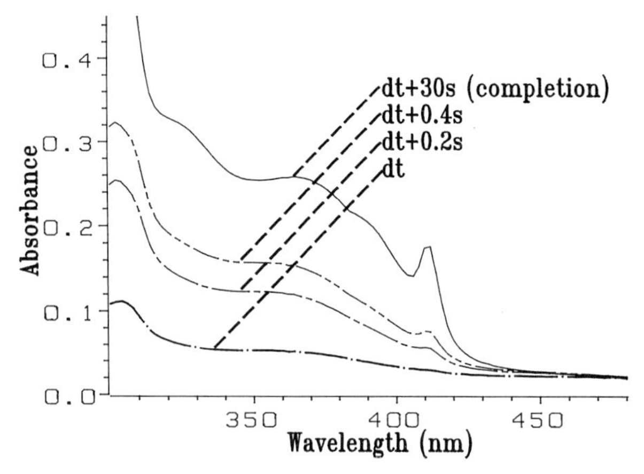

FIG. 3. Development of the absorption spectrum of the tyrosyl radical–diferric cluster cofactor on mixing of apo R2-wt with excess Fe^{2+} in the presence of O_2. The reaction conditions (after mixing) were 27 μM apo R2-wt, 250 μM Fe^{2+}, 50 mM HEPES (air-saturated), pH 7.6, 5°. Spectra were acquired on the HP8452A diode array apparatus. The time after mixing at which each spectrum was acquired is indicated, with the dead time of the apparatus denoted dt. The contribution to each spectrum from the protein has been subtracted away for clarity. Similar results were obtained with 300 μM apo R2-wt and Fe^{2+}/R2 ratio of 5.

final Y122• was formed. Although this early spectrum seems to have the ~360-nm broad band associated with the μ-oxodiiron(III) (diferric) cluster, the 325-nm shoulder of the diferric cluster is absent. Therefore, this spectrum must arise from some species other than the diferric cluster. As the spectrum at completion is that characteristic of the product cofactor (with the appearance of the more intense 365- and 325-nm bands of the diferric cluster), the species that exhibits the ~360-nm band must decay during the reaction, which suggests that it may be a kinetically competent intermediate. This deduction is substantiated by the RFQ-EPR and RFQ-Möss data as described below.

To examine the assembly reaction with better time resolution, the Applied Photophysics DX.17MV SF spectrofluorimeter was used. Spectra were generated by acquiring absorbance vs time traces at different wavelengths. Kinetic traces were acquired every 5 nm between 320–400 nm and 420–460 nm, and every 3 nm between 404 and 416 nm. This information was converted to time-dependent spectra, which showed the same general features shown in Fig. 3.

The feature of Y122• at 410 nm, which is much sharper than any other spectral feature, provides a unique opportunity for quantitation of this species. When the absorbance at 410 nm of a sample of R2 is corrected to the line defined by the absorbance at 404 nm and the absorbance at 416 nm, only the sharp feature of the tyrosyl radical contributes since the broader features of the cofactor spectrum are well approximated as straight lines over this 12-nm wavelength range. Thus, this "dropline-corrected" absorbance at 410 nm ($A_{410,\mathrm{dropline}}$) is proportional to the Y122• concentration, providing a simple method for quantitation of Y122• either at completion or as a function of reaction time.

The $A_{410,\mathrm{dropline}}$ versus time trace of the reaction illustrates that, after a significant lag phase, formation of Y122• is approximately first order. Nonlinear least-squares fitting of the equation for a first-order growth to the region of this curve between 0.5 and 5 sec (Fig. 4) gives an observed first-order rate constant (k_{obs}) of 0.75 $\pm$ 0.04 sec^{-1} (Table I). (This value represents the average of results from six separate experiments, with the results of three to six individual trials averaged in each experiment. The quoted uncertainty is the difference between the mean and extreme values.) As illustrated in Table I, in the range of apo R2 concentrations ([apo R2]) that was investigated (22–290 μM), the k_{obs} is independent of [apo R2]. The Y122•/R2 ratio at completion (calculated from the magnitude of $A_{410,\mathrm{dropline}}$ in each experiment) is also independent of [apo R2] in the reaction. The mean value is 1.2 Y122•/R2.

The lag phase in the $A_{410,\mathrm{dropline}}$ versus time curve is consistent with the accumulation of an intermediate prior to formation of Y122•. In order to

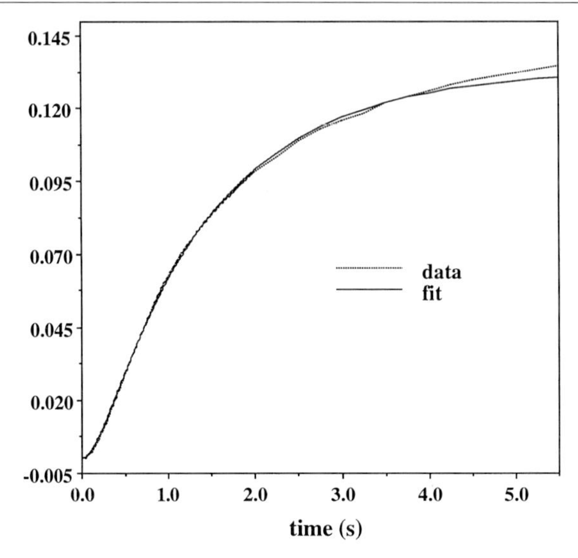

FIG. 4. The $A_{410,\mathrm{dropline}}$ used to assess the time course of tyrosyl radical production in the reaction of apo R2-wt with excess Fe^{2+}. The reaction conditions (after mixing) were 0.29 m*M* apo R2-wt, 1.4 m*M* Fe^{2+}, 50 m*M* HEPES (O_2-saturated), pH 7.6, 5°. Data were acquired on the Applied Photophysics DX.17MV apparatus. The experimental trace was constructed as A_{410}-$(A_{404} + A_{416})/2$ by averaging three trials for each wavelength. The theoretical curve generated by fitting the equation for a first-order growth to the 0.5- and 5.0-sec region of the trace is also shown.

TABLE I
SUMMARY OF RATE CONSTANTS FOR Y122• FORMATION AND OF Y122•/R2 RATIOS AT COMPLETION FOR REACTION OF Apo R2-wt WITH EXCESS Fe^{2+}

Experiment	[Apo R2] (μM)	Fe^{2+}/R2	k_{obs}[a] (sec^{-1})	$(Y122•/R2)_\infty$
1	22	4.9	0.78	1.20
2	27	9.5	0.76	1.12
3	54	10.	0.74	1.23
4	290	5.0	0.78	1.27
5	295	5.0	0.71	—
6	295	4.9	0.72	1.15
Average	—	—	0.75 ± 0.04	1.19 ± 0.08

[a] Fits were made to 0.5–5 sec of the experimental traces.

calculate a rate constant for formation of this intermediate, the 0- to 5-sec regions of traces from three of the experiments of Table I were analyzed according to Eq. (2). This equation gives $A_{410,\text{dropline}}$ as a function of time $[A_{410,\text{dropline}}(t)]$ for the hypothetical reaction sequence of Scheme I, in which an intermediate species intervenes between the reactants (Fe^{2+}, apo R2, and O_2) and the product (native R2, containing the diferric cluster and Y122•). $A_{410,\text{dropline}}(t)$ is related to the concentration of Y122• at completion ($[\text{Y122•}]_\infty$), the dropline-corrected molar absorptivity at 410 nm of Y122• ($\varepsilon_{410,\text{dropline}}$), and the rate constants, k_1 and k_2. The assumption is made in Eq. (2) that both reactants and intermediate have $\varepsilon_{410,\text{dropline}} = 0$. A representative fit from this analysis is shown in Fig. 4 and the results are summarized in Table II. The data are consistent with a rate constant (k_1) of 4.9–7.5 sec^{-1} for formation of the inferred intermediate, and a rate constant (k_2) of 0.72–0.77 sec^{-1} for formation of Y122•:

$$A_{410,\text{dropline}}(t) = [\text{Y122•}]_\infty \varepsilon_{410,\text{dropline}} \left(1 + \frac{k_1 \exp(-k_2 t) - k_2 \exp(-k_1 t)}{k_2 - k_1}\right). \tag{2}$$

In a further attempt to assess the validity of Scheme I and to determine values for k_1 and k_2, multicomponent, multivariate analysis (Applied Photophysics Global Analysis) was applied to the time-dependent absorption spectra of the reaction. When k_2 was held fixed at 0.75 sec^{-1} (the mean value calculated from the experiments of Tables I and II), a value of 5.4 sec^{-1} was calculated for k_1, and the spectra calculated for the three components were well resolved. (In other words, the spectrum of the reactant was featureless and that of the intermediate resembled the early spectra of Fig. 3). Thus, assuming a rate constant of 0.75 sec^{-1} for conversion of the intermediate into the product cofactor, the time-dependent spectra are consistent with a rate constant (k_1) of 5.4 sec^{-1} for formation of the

TABLE II
SUMMARY OF RATE CONSTANTS CALCULATED FROM NONLINEAR LEAST-SQUARES FITTING OF Eq. (2) TO $A_{410,\text{dropline}}$ VERSUS TIME TRACES FROM THE REACTION OF Apo R2-wt WITH EXCESS Fe^{2+}

Experiment[a]	[Apo R2] (μM)	Fe^{2+}/R2	k_1 (sec^{-1})	k_2 (sec^{-1})
3	54	10.	4.9	0.77
5	295	5.0	7.5	0.72
6	295	4.9	7.1, 5.9	0.73, 0.75

[a] Experiments are the same as those of Table I.

intermediate. The value of k_1 agrees well with that obtained by analysis of the $A_{410,\mathrm{dropline}}$ versus time curves (Table II).

Reaction of Apo R2 with Excess Fe^{2+} and O_2 Monitored by RFQ-EPR Spectroscopy

The SF-Abs data suggest the accumulation of an intermediate that is kinetically competent to be on the reaction pathway. The multicomponent analysis suggests that decay of the intermediate leads to the simultaneous production of Y122• and the diferric cluster. Thus, the intermediate should be oxidized by one electron relative to the diferric cluster, which would imply that it must be paramagnetic and might be detectable by EPR spectroscopy. Furthermore, the kinetic constants derived from the SF data suggest that ~0.9 equiv of this odd electron species should accumulate, making it readily detectable. EPR spectroscopy was therefore used to obtain additional evidence for and to probe the nature of the intermediate postulated in Scheme I. The relatively short lifetime of this intermediate made it necessary to use the rapid freeze-quench method to trap it.[14,15]

Protocol

The RFQ method involves mixing several reactants rapidly and efficiently, allowing the reaction to proceed for a desired length of time (the time required for the mixture to pass through an aging hose of appropriate volume), freeze-quenching the reaction mixture by squirting it into isopentane cooled to $-150°$, and packing the resulting ice crystals into a receptacle appropriate for spectroscopic analysis.[14] The apparatus used in preparation of RFQ samples consists of an Update Instruments (Madison, WI) Ram unit and Model 705A computer controller, and a home-built quenching bath. A schematic diagram of the quenching bath is shown in Figs. 5A and 5B. The bath holds ~8 liters isopentane which is cooled with liquid N_2 and maintained at $-150°$ with a Bayley Instruments Company (Danville, CA) precision temperature controller. The temperature of the reaction was maintained at $5 \pm 1°$ by filling the Ram unit with enough cold water to immerse the drive syringes and by periodically adding ice. The aging hose was submerged in the water bath for several minutes before each sample was prepared in order to allow the hose to reach $5 \pm 1°$. Immediately

[14] R. C. Bray, *in* "Rapid Mixing and Sampling Techniques in Biochemistry" (B. Chance, R. H. Eisenhart, Q. H. Gibson, and K. K. Lonberg-Holm, eds.), p. 195. Academic Press, New York, 1964.

[15] D. P. Ballou and G. Palmer, *Anal. Chem.* **46,** 1248 (1974).

before the Ram unit was actuated, the aging hose was removed from the cold water and held over the quenching bath.

Calibration of RFQ Sampling Apparatus. In the rapid freeze-quench method, the reaction time at which a given sample is quenched is the sum of two quantities: the time elapsed after the reactants are mixed until the reaction mixture traverses the aging hose, and the time elapsed after the reaction mixture contacts the cold isopentane until it is cooled to the temperature at which no further reaction occurs.[14] The latter quantity (quenching time) can vary from one reaction system to another[15] and is difficult to determine. The quenching times of many reactions are short (5–10 msec)[15] compared to the time scale of the R2 reconstitution reaction ($t_{1/2}$ for Y122• formation ~1 sec), and quenching time was therefore neglected in this study. The former quantity depends on the volume of the aging hose, the volume expelled by the drive syringes for a given displacement of the Ram drive, and the velocity of the Ram drive. In these experiments, the velocity of the Ram drive was assumed to be that quoted by the manufacturer. The volume expelled by the drive syringes for a given displacement and the volumes of the aging hoses were measured experimentally. One drive syringe was filled with distilled H_2O, and the Ram unit was actuated for a programmed displacement. The water expelled from the drive syringe was collected in a previously weighed microcentrifuge tube. The weight of the water collected was determined, and this was converted to volume. The volume expelled by the second drive syringe was determined in identical fashion. The volume of each aging hose was determined by connecting the empty hose to one or both of the H_2O-filled drive syringes. The Ram drive was actuated for a given displacement, and the water expelled from the aging hose was collected in a previously weighed microcentrifuge tube. The volume of the aging hose was calculated as the difference between the volume expelled by the drive syringe(s) for that displacement and the volume collected in the tube. In some experiments it was necessary to couple two hoses together to obtain a desired volume. In these cases, the volume of the coupled system was determined as above.

EPR Spectroscopy and Analysis. EPR spectra of the time course samples were recorded on a Brüker ER 200D-SRC spectrometer equipped with an Oxford Instruments ESR 910 continuous flow cryostat. Double integration of EPR spectra, subtraction of EPR spectra, and simulation of EPR spectra were all carried out with programs written in the laboratory of B. H. Huynh. Simulation of EPR spectra was also carried out with the ESR[a] program of Calleo Scientific Software Publishers (Fort Collins, CO). Analysis of all the kinetic data was carried out with the Git and Gear nonlinear regression programs of Drs. R. J. McKinney and F. J. Wiegert, Central Research and Development Department, E. I. du Pont de Nemours and Co.

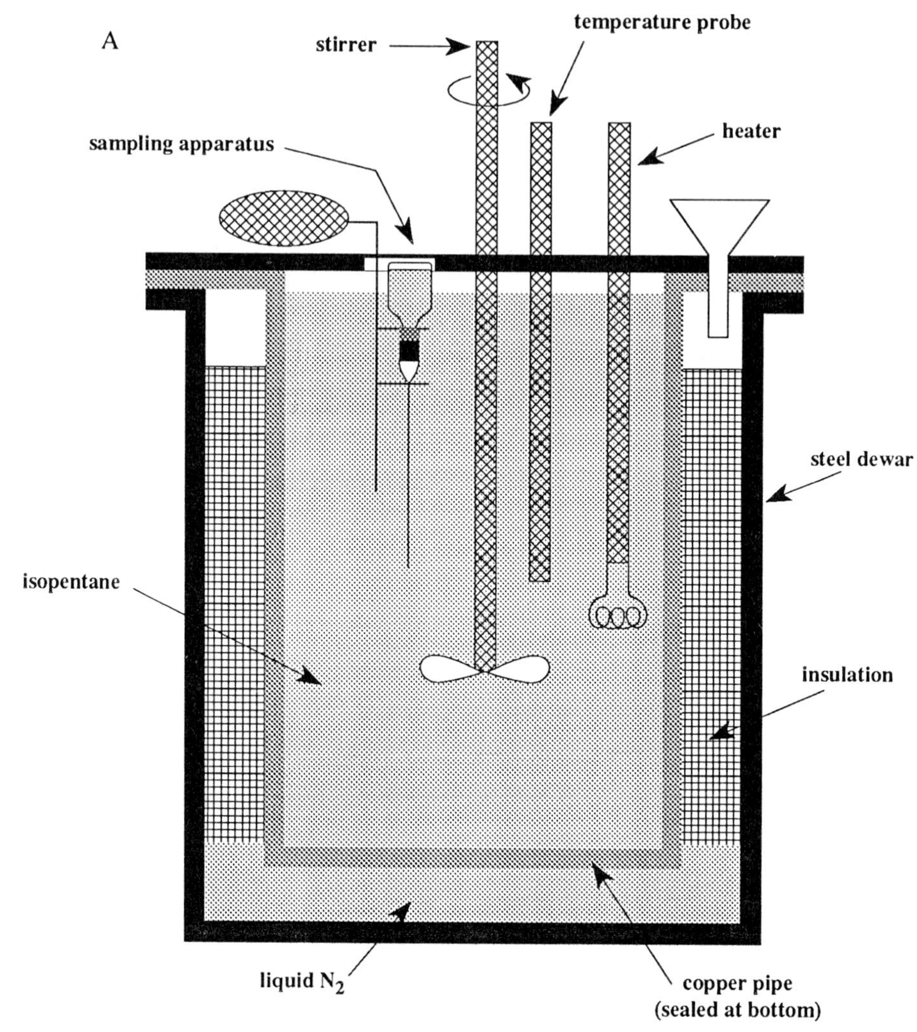

FIG. 5. Rapid freeze-quench apparatus. (A) A schematic diagram of the isopentane freeze-quenching bath with the EPR sampling apparatus inside. (B) A more detailed schematic diagram of the sampling apparatus. In preparation of the samples, immediately before the Ram drive was actuated, the nozzle through which the reaction mixture was sprayed was positioned less than 1 cm above the level of the isopentane inside the funnel.

Control for Rapid Freeze-Quenching of R2

In order to ascertain whether R2 is adversely affected by the freeze-quenching procedure, a control was carried out in which native R2 was

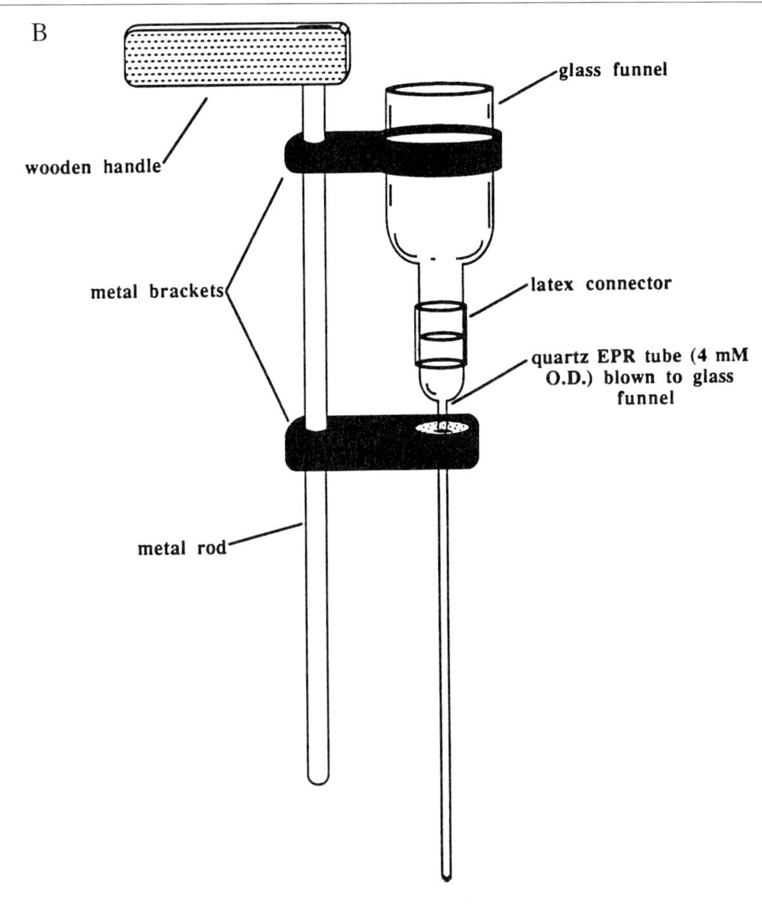

FIG. 5. (*continued*)

squirted into liquid isopentane in contact with the frozen solid. The absorption spectrum of a 25 μM solution of native R2 [in 100 mM HEPES (pH 7.7)] was acquired. A 500-μl aliquot of this R2 solution was expelled forcefully from a 1-ml gas-tight syringe into 2 ml of partially frozen isopentane in a 5-ml test tube. The resulting ice crystals were allowed to melt, the protein solution was recovered, and the absorption spectrum was recorded. The concentration of Y122• before and after the mock freeze-quenching procedure was determined by the magnitude of $A_{410,\mathrm{dropline}}$ and was found to be the same. Thus, the procedure is not deleterious to the cofactor of R2.

Results

When apo R2-wt is mixed at 5° in the presence of O_2 with excess Fe^{2+}, an EPR-active intermediate rapidly accumulates. The intermediate exhibits a sharp, isotropic singlet with a g value of 2.00 (Fig. 6A).[4] (When the field width of the spectrum was increased, no additional features were observed with the exception of a weak signal at $g = 4.3$, which is generally associated with rhombic, high-spin ferric species.) With increasing time, the singlet of the intermediate decays (hereafter designated **X**), while the doublet characteristic of Y122• develops (Figs. 6B–6D and Table III).

Quantitative Analysis of Time-Dependent EPR Spectra. An EPR time course was also carried out with a site-directed mutant R2 subunit, R2-

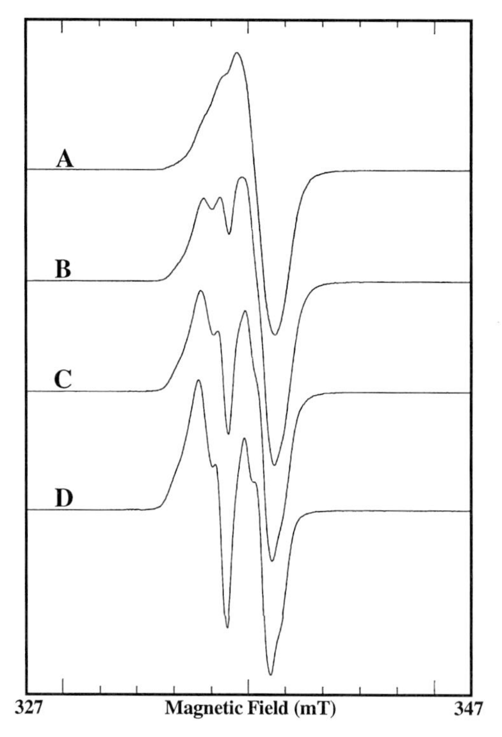

FIG. 6. Time course of the reaction of apo R2-wt with excess Fe^{2+} as monitored by EPR spectroscopy. Apo R2-wt (0.59 m*M*) in O_2-saturated 100 m*M* HEPES buffered at pH 7.7 was mixed at 5° with an equal volume of 2.94 m*M* Fe^{2+} in 2.5 m*M* H_2SO_4. The reaction was freeze-quenched at 0.15 sec (A), 0.63 sec (B), 1.5 sec (C), or 60 sec (D). The spectra were recorded at 20 K with a microwave power of 2 μW, a frequency of 9.43 GHz, a modulation frequency of 100 kHz, a modulation amplitude of 4 G, a time constant of 200 msec, a scan time of 200 sec, and a receiver gain of 4×10^4.

TABLE III

SUMMARY OF RELATIVE AND ABSOLUTE QUANTITIES OF **X** AND Y122• AS DETERMINED BY EPR IN THE TIME COURSE SAMPLES FOR THE REACTION OF Apo R2-wt WITH EXCESS Fe^{2+}

Time (sec)	Integral	% **X**	% Y122•	X/R2	Y122•/R2
0.061	7.0	85	15	0.54	0.095
0.16	11.6	82	18	0.87	0.19
0.22	12.9	80	20	0.95	0.23
0.31	12.4	77	23	0.87	0.26
0.44	16.4	67	33	1.01	0.50
0.63	14.7	61	39	0.82	0.53
1.02	11.9	45	55	0.49	0.60
1.52	13.7	28	72	0.35	0.90
2.22	8.8	19	81	0.15	0.65
3.02	13.3	10	90	0.12	1.10
4.02	13.6	5	95	0.06	1.18
5.02	13.1	2	98	0.02	1.18
60	13.1	—	100	—	(1.20)

Y122F, in which Y122• cannot form. As in the wild-type protein (R2-wt), the isotropic singlet of **X** rapidly develops, but the line shape of the spectrum does not change with time, suggesting that the intermediate **X** is the only EPR active species which forms. Therefore, the R2-Y122F reaction provides a reference EPR spectrum for **X** that can be used to estimate the relative amounts of **X** and Y122• present in the time course samples from the R2-wt reaction. The relative quantities of the two species were estimated for each sample from Table III by iterative addition of the spectra of **X** and Y122• in varying ratios, until the experimental spectrum was satisfactorily reproduced. At all times during the reaction the experimental spectrum can be accounted for as the sum of the spectra of **X** and Y122•. Figure 7 shows a representative example of this analysis. The relative quantities of **X** and Y122• in each sample are listed in Table III, along with the double integral for each. By using the values from Table III and the 60-sec sample as a radical concentration standard (Y122•/R2 at completion = 1.2), the absolute amounts of **X** and of Y122• in each sample were estimated according to Eqs. (3) and (4):

$$\frac{\mathbf{X}}{\text{R2}} = 1.2\,\frac{I(t)}{I(60)}\,F_{\mathbf{X}}(t) \tag{3}$$

$$\frac{\text{Y122}\bullet}{\text{R2}} = 1.2\,\frac{I(t)}{I(60)}\,F_{\text{Y122}\bullet}(t) \tag{4}$$

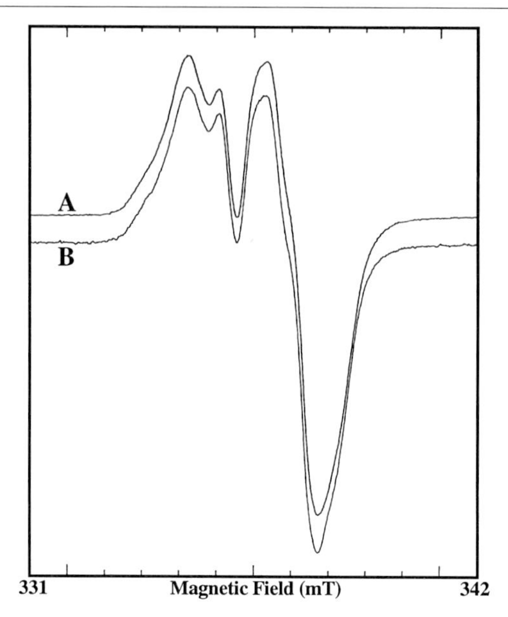

FIG. 7. Example of deconvolution of an EPR spectrum from the reaction of apo R2-wt with excess Fe^{2+} to determine the relative quantities of **X** and Y122• present. Spectrum A was acquired on the 1.0-sec sample from the experiment depicted in Fig. 6. Spectrum B was constructed by summation of the spectra of **X** and Y122• in a ratio of 0.45 : 0.55 (**X** : Y122•). The reference spectrum for Y122• was acquired on the 60-sec sample from the same experiment. The spectrometer settings for the experimental and reference spectra were as in Fig. 6.

where $I(t)$ is the double integral of the $g = 2$ region for a given sample, $F_{\mathbf{X}}(t)$ and $F_{Y122\cdot}(t)$ are the fractions of **X** and Y122• present in that sample, and I(60) is the double integral for the 60-sec sample. These data were analyzed according to the kinetic model of Scheme I. They are consistent with a rate constant (k_1) of 10 sec^{-1} for formation of **X** and a rate constant (k_2) of 0.95 sec^{-1} for its decay and the concomitant formation of Y122• (Fig. 8). The concentrations of **X** and Y122• as functions of time suggest that decay of **X** and formation of Y122• are concomitant processes. This result is consistent with a mechanism in which **X** generates Y122•.

Summary

The SF-Abs and RFQ-EPR data for the reaction of apo R2 with excess Fe^{2+} and O_2 can readily be interpreted within the mechanism of Scheme I. The rapid accumulation of **X**, characterized by the $g = 2.00$ isotropic singlet, correlates in time with the development of the 360-nm broad absorption band. The rate constant for formation of **X** as determined by RFQ-

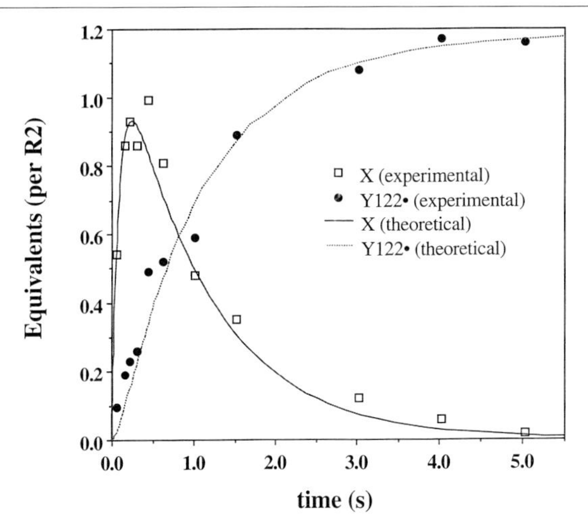

FIG. 8. Nonlinear least-squares analysis of quantities of **X** and Y122• as functions of time in the reaction of apo R2-wt with excess Fe^{2+}. The equations appropriate for the kinetic model of Scheme I were fit to the data from Table III. The curves generated in the analysis are shown, corresponding to $k_1 = 10\ sec^{-1}$, $k_2 = 0.95\ sec^{-1}$, and $(Y122\bullet)_{final} = 1.2$.

EPR ($k_{obs} = 10\ sec^{-1}$) is in fair agreement with the rate constant determined for formation of the intermediate by SF-Abs ($k_{obs} = 5\ sec^{-1}$). The decay of the EPR singlet due to **X** and the development of the doublet characteristic of Y122• correspond temporally with the development of the light absorption spectrum characteristic of the product cofactor. The rate constant for decay of **X** and for formation of Y122• as determined by RFQ-EPR ($k_{obs} = 0.95\ sec^{-1}$) is in fair agreement with the rate constant for formation of Y122• ($k_{obs} = 0.72$–$0.77\ sec^{-1}$) determined by SF-Abs. Thus, the SF-Abs and rapid freeze-quench EPR data together provide strong evidence for Scheme I.

Reaction of Apo R2 with Excess Fe^{2+} and O_2 Monitored by RFQ-Möss: Characterization of a Novel Diferric Radical Species which Generates Y122•

The rapid freeze-quenching apparatus described above was adapted for preparation of Mössbauer samples. It is evident that the time-resolved Mössbauer spectra of the cofactor assembly reaction might provide a wealth of mechanistic information, as the concentration at any time of each iron-containing reactant, intermediate, and product might be determined, provided that the spectra could be deconvoluted as superpositions of appropriate reference spectra.

Apparatus and Protocol

The apparatus used to prepare the Mössbauer samples (Fig. 9) is a derivative of the EPR sampling apparatus (see Fig. 5B). It was designed and fabricated by Mr. Bud Puckett in the Department of Physics at Emory University. Sample preparation was carried out as follows. The apparatus was assembled, placed in its holder, and filled with isopentane. It was inserted into the cold quenching bath (containing isopentane at $\sim -150°$) and allowed to reach thermal equilibrium. The spray nozzle at the end of

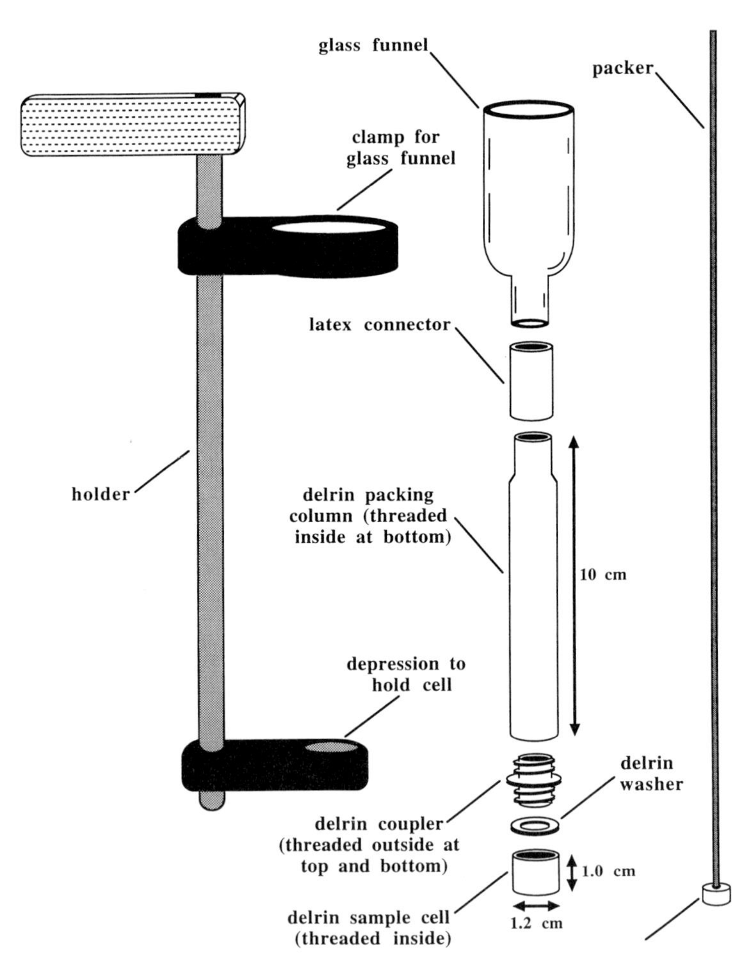

FIG. 9. Apparatus used to prepare rapid freeze-quenched samples for Mössbauer spectroscopy. It was designed and fabricated by Mr. Bud Puckett in the Department of Physics at Emory University.

the aging hose was held less than 1 cm above the level of the isopentane in the glass funnel, and the reactants were mixed and expelled into the cold isopentane by actuation of the Ram drive. Approximately 500 μl of reaction mixture was required to generate a satisfactory sample. In preparation of all the freeze-quenched Mössbauer samples the velocity of the Ram drive was 1 cm/sec. This relatively low velocity seemed to be required to obtain ice crystals that were coarse enough to settle properly. The ice crystals were allowed to settle for approximately 1 min, and then were packed into the Mössbauer cell. The sampling apparatus was removed from the quenching bath and from its holder, and the Mössbauer cell was submerged in a slush of partially frozen isopentane contained in a wide-mouth stainless-steel Dewar. With the packer still in the packing column, the latex connector and glass funnel were removed. The sample was lifted from the isopentane slush just long enough for the threaded Mössbauer cell to be manually loosened from the threaded coupler which connects it to the packing column. The sample cell was quickly reinserted into the isopentane slush bath, and the packing column and threaded coupler were unscrewed and removed from the sample cell. A hex-nut style, delrin sample cap (not depicted in Fig. 9) was inserted into the isopentane slush bath on the end of a hex-key wrench. The cooling of the cap on the wrench had the effect of temporarily fixing the two together. The sample cell in the slush bath was grasped with precooled tweezers, and the cap was threaded into the sample cell. By repeated, brief manual tightening, the sample was compressed to the minimum thickness and highest density achievable. When the cap was tight, the sample was removed from the end of the hex-key wrench, very briefly wiped free of isopentane, and then stored in liquid N_2. Acquisition of replicate spectra over a period of months has shown that the samples are stable in liquid N_2.

Preparation of Diferric Cluster Reference Sample. For analysis of the time course spectra, reference Mössbauer spectra were required. Although parameters have been reported for the diferric cluster (in native and met R2)[12,16] and for the ferrous form of R2,[16] it was deemed desirable to have experimental reference spectra for these species. In preparation of the reference sample for the diferric cluster (native R2), 140 μl of 4.38 m*M* $^{57}Fe^{2+}$ in 4.5 m*M* H_2SO_4 containing 14 m*M* ascorbic acid was added at 2 ± 2° to 250 μl of 0.98 m*M* apo R2-wt in air-saturated 100 m*M* HEPES, pH 7.7. The resulting solution, which contained 0.64 m*M* apo R2-wt, 1.6 m*M* ^{57}Fe (Fe/R2 = 2.5), and 4.9 m*M* ascorbate, was incubated (open to air) on ice for 45 min. It was then transferred to a Mössbauer cell and frozen in liquid N_2.

[16] J. B. Lynch, C. Juarez-García, E. Münck, and L. Que, Jr., *J. Biol. Chem.* **264,** 8091 (1989).

Preparation of Ferrous-R2 Reference Sample. A 1-ml aliquot of 0.98 m*M* apo R2-wt in 100 m*M* HEPES (pH 7.7) and a 1-ml aliquot of 5.1 m*M* $^{57}Fe^{2+}$ in 5 m*M* H_2SO_4 were made O_2-free by several cycles of gentle evacuation–refilling with argon-incubation on ice. A 207-μl aliquot of the apo R2 was mixed in an anaerobic chamber with 200 μl of the 5.1 m*M* $^{57}Fe^{2+}$ solution. The resulting Fe^{2+}–R2 solution was transferred to a Mössbauer cell, which was placed in a septum-stopped flask. The flask was removed from the anaerobic chamber and submerged in liquid N_2 until the sample in the Mössbauer cells was frozen.

Preparation of "Ferrous Ion in HEPES" Reference Sample. A reference sample for $^{57}Fe^{2+}$ in HEPES buffer was prepared by mixing (at 5°) equal volumes of O_2-saturated 100 m*M* HEPES (pH 7.7) with O_2-saturated 1.5 m*M* $^{57}Fe^{2+}$ stock and freeze-quenching the mixture at 0.44 sec.

Preparation of Diferric Radical Species ***(X)*** *Reference Sample.* The diferric radical species can be trapped in high yields in a site-directed mutant of R2 in which the oxidizable Y122 is replaced with F (R2-Y122F). A reference sample for **X** was prepared by mixing 1.5 m*M* apo R2-Y122F in O_2-saturated 100 m*M* HEPES buffered at pH 7.7 (at 5°) with an equal volume of O_2-saturated 4.5 m*M* $^{57}Fe^{2+}$ stock in 14 m*M* H_2SO_4 containing 5 m*M* ascorbic acid and freeze-quenching the mixture at 0.31 sec.[6]

Acquisition of Spectra. Mössbauer spectra of time course and reference samples were acquired at 4.2 K with a magnetic field of 50 mT applied parallel to the γ beam.

Quantitative Analysis of the Time-Dependent Spectra. Analysis of the Mössbauer spectra of the freeze-quenched samples to determine the quantities **X** and diferric cluster present in each was carried out using a subtraction program written by R. Zimmermann. The program computes the total area under the absorption spectrum (which is proportional to the quantity of ^{57}Fe present in the sample) and allows a theoretical or experimental reference spectrum to be superimposed on and then subtracted from the original spectrum. The superimposed spectrum can be scaled to represent any percentage of the integrated intensity of the original spectrum. By iteratively superimposing varying percentages of the reference spectrum for a given component of the reaction (ferrous R2, **X**, or the diferric cluster) on the spectrum of a time course sample, it was possible to determine the fraction of the ^{57}Fe in the sample that was present as that component and to assign error limits to this value. The known, total quantity of ^{57}Fe present in the sample served as an internal standard, which allowed the absolute quantity of each component to be calculated from the fraction of ^{57}Fe in that component.

For ferrous R2, **X**, and the diferric cluster, both experimental and theoretical spectra were used as references for analysis of the time course samples. Experimental spectra for ferrous R2, **X**, and the diferric cluster were acquired on the samples prepared as described earlier. Theoretical spectra for ferrous R2 and for the diferric cluster were generated either by nonlinear least-squares fitting of two quadrupole doublets to the data or by simulation (using parameters from the least-squares fit or from Lynch *et al.*[16]). For **X**, the theoretical spectrum was generated by analysis of the paramagnetic component of the experimental spectrum. The paramagnetic component was shown to be a spin-coupled system involving two high-spin ferric ions (with parameters shown in Table IV) and a free radical.[5,6] Because this spectrum of **X** shows imperfect agreement with the experimental spectrum in the central region, when an objective of the analysis was to subtract away the spectrum of **X** in order to identify other components of the reaction, an experimental spectrum was used as the reference so as not to distort the central region of the subtraction spectrum.

Results

Reference Spectra of Ferrous R2, X, and the Diferric Cluster. All Mössbauer spectra recorded at 4.2 K of the ferrous R2 samples containing 2.5 (A) or 5.0 (C) Fe^{2+}/R2 are shown in Fig. 10 along with the results of least-squares fits of two quadrupole doublets to the data (solid line). The parameters from the fit (Table IV) are in good agreement with those

TABLE IV
PARAMETERS FOR REFERENCE SPECTRA OF FERROUS R2, FERROUS ION IN HEPES BUFFER, THE DIFERRIC CLUSTER, AND **X**

Species	Site	ΔE_Q (mm/sec)	δ (mm/sec)	η	$A/g_n\beta_n$ (T)
Ferrous R2	1	3.24 ± 0.06^a	1.31 ± 0.03^a	—	—
	2	2.92 ± 0.06^a	1.20 ± 0.03^a	—	—
		$(3.13)^b$	$(1.26)^b$		
Ferrous ion in HEPES buffer	1	3.39 ± 0.06^a	1.40 ± 0.03^a	—	—
	2	3.03 ± 0.06^a	1.36 ± 0.03^a	—	—
Diferric cluster	1	1.64 ± 0.06^a	0.54 ± 0.03^a	—	—
		$(1.62 \pm 0.02)^c$	$(0.55 \pm 0.01)^c$		
	2	2.41 ± 0.06^a	0.45 ± 0.03^a	—	—
		$(2.44 \pm 0.03)^c$	$(0.45 \pm 0.01)^c$		
Diferric radical	1	−1.0	0.55	0.5	−52.5
(**X**)[d]	2	−1.0	0.36	1.0	+24.0

[a] Obtained by a least-squares fit of two quadrupole doublets to the data.
[b] Average parameters for the two sites quoted by Lynch *et al.*[16]
[c] Parameters from Lynch *et al.*[16]
[d] Simulation from Ravi *et al.*[6]

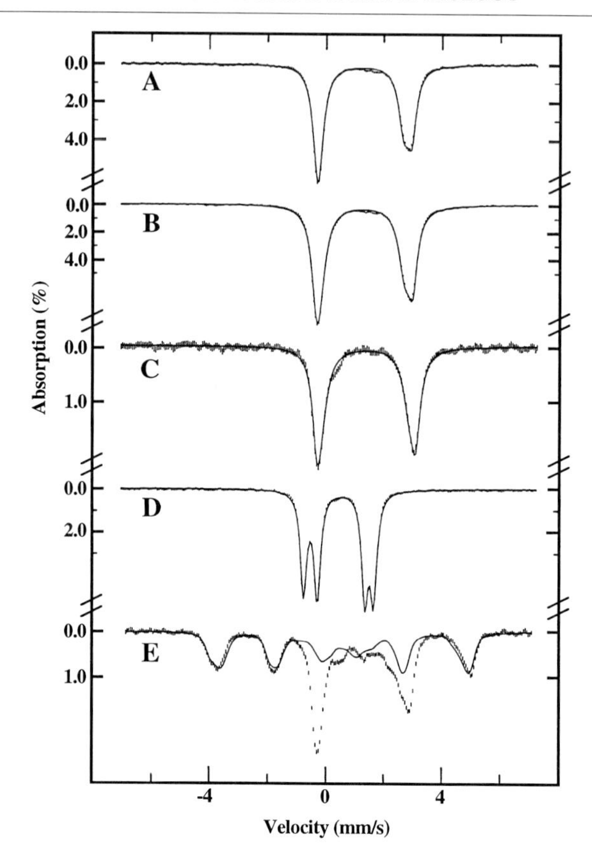

FIG. 10. Reference spectra for ferrous-R2 (A, B at Fe^{2+}/R2 of 2.5 and 5.0, respectively), ferrous ion in HEPES buffer (C), the diferric cluster (D), and the diferric-radical intermediate (E). The solid lines in A–C were obtained by least-squares fitting of two quadrupole doublets to the data. The parameters obtained in these fits are listed in Table IV. The solid line in D is the simulation described in Bollinger *et al.*[5,6] The parameters for this spectrum are also shown in Table IV. Details of sample preparation are given in Protocols.

reported by Lynch *et al.*[16] The spectrum of the reference sample containing Fe^{2+} in HEPES buffer is shown in Fig. 10C, along with the result of a least-squares fit of two quadrupole doublets to the data (solid line). The spectrum of the sample of reconstituted R2 is shown in Fig. 10D. Plotted over the experimental data is the result of a least-squares fit of two quadrupole doublets to the data (solid line). The parameters from this fit (Table IV) also agree with those reported by Lynch *et al.*[16] Figure 10E shows the spectrum of a freeze-quenched sample from the reaction of R2-Y122F, which contains 70% **X**. Plotted over the experimental data is the theoretical spectrum obtained by using the parameters reported in Table IV.[5,6] As mentioned earlier, the inner region of the theoretical spectrum shows imperfect agreement with the experimental data. The theoretical spectrum is adequate to quantify **X**, but the experimental spectrum is preferable for subtraction. The discrepancy between the theoretical and the experimental spectra of **X** is not understood, but the differences do not affect the kinetic analysis.

Time Course of the Reaction by RFQ-Möss. The time-dependent Mössbauer spectra of the reaction of apo R2-wt with excess Fe^{2+} and O_2 (Fig. 11) clearly reflect the progress of the reaction. In the first time point that was taken (0.061 sec, Fig. 11A), the spectrum is dominated by unreacted ferrous ion. Although Fe^{2+} is the predominant species at this first time point, a significant quantity of **X** has accumulated, whereas the features of the diferric cluster are not yet detectable. At somewhat longer reaction times (0.31 sec, Fig. 11B), the relative quantity of Fe^{2+} present has decreased significantly, whereas the contribution due to **X** has increased. In addition, the spectrum of the diferric cluster is now detectable. With an increasing reaction time (1.0 sec, Fig. 11C), the relative contributions from Fe^{2+} and **X** decrease, whereas that from the diferric cluster increases. Finally, at completion of the reaction (60 sec, Fig. 11C), the spectrum is dominated by the features of the diferric cluster. Figure 11 clearly demonstrates the utility of the rapid freeze-quench Mössbauer method for monitoring the reconstitution reaction, as the progression from reactant (Fe^{2+}), to intermediate **(X),** to product can be seen on inspection.

Quantitative Analysis of the RFQ-Möss Spectra. In addition to illustrating the progress of the reaction, Fig. 11 also shows that both **X** and the diferric cluster can be quantified as functions of time from the Mössbauer spectra. For **X**, the highest and lowest energy lines of site 1 (see Table IV) and the lowest energy line of site 2 are well resolved from other spectral features (see Fig. 10). Likewise, for the diferric cluster, both lines of the outer quadrupole doublet (site 2 of Table IV) are sufficiently resolved from other spectral features to allow for reliable quantitation (see Figs. 11B–11D). To estimate the quantity of **X** or the diferric cluster present in a given time course sample, the reference spectrum of the species was superimposed on the experimental spectrum of that sample. The percentage of the reference spectrum (relative to the integrated intensity of the experimental spectrum) that was plotted was varied until agreement was achieved. The proper percentage of this species was subtracted away, and the correct quantity of the next species was determined by analysis of the subtraction spectrum. In many cases, the analysis was repeated in opposite order to provide a check. Limits of error for the measured quantities of the species were also estimated. The quantitation and error estimates are summarized

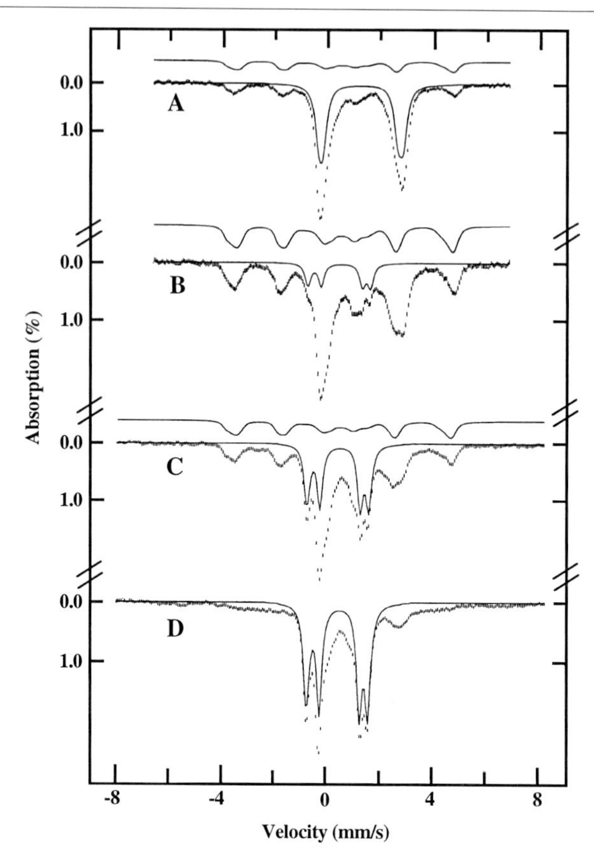

FIG. 11. Time course of the reaction of apo R2-wt (590 μM) with excess Fe^{2+} (2.97 mM) and O_2 as monitored by Mössbauer spectroscopy. The reaction was quenched at 0.061 sec (A), 0.31 sec (B), 1.0 sec (C), or 60 sec (D). The solid line plotted over the data in A is a simulation of the ferrous R2 spectrum that was obtained by using the parameters reported by Lynch *et al.*[16] It is scaled to 45% of the integrated intensity of the experimental spectrum. The solid line plotted just above the data in A is the theoretical spectrum of **X** scaled to 20% of the intensity of the experimental spectrum. In B and C, the solid line above the data is the theoretical spectrum of **X** (36% in B and 23% in C), and the line plotted over the data is the theoretical spectrum of the diferric cluster (12% in B and 32% in C). The solid line in D is the theoretical spectrum of the diferric cluster (55%). The diferric cluster reference spectrum is a simulation based on the values of ΔE_Q and δ from the least-squares fit of C and on the values of η reported by Lynch *et al.*[16] (The signs of the values of ΔE_Q (−) were taken from Lynch *et al.*[16]

TABLE V
SUMMARY OF QUANTITATION OF **X** AND THE DIFERRIC CLUSTER AS FUNCTIONS OF REACTION TIME

Expt.	Reaction time (sec)	% **X**	Equiv **X**	% Diferric cluster	Equiv diferric cluster
1	0.061	21 ± 4	0.53 ± 0.11	0	0
	0.16	33.5 ± 3.5	0.84 ± 0.10	3 ± 1	0.08 ± 0.03
	0.22	35 ± 2	0.88 ± 0.07	6 ± 1	0.15 ± 0.03
	0.31	39 ± 3	0.98 ± 0.09	9.5 ± 1.5	0.24 ± 0.04
	0.38	30 ± 3	0.75 ± 0.09	17 ± 1	0.43 ± 0.05
	0.44	33 ± 2	0.83 ± 0.07	14 ± 1	0.35 ± 0.04
	0.63	27 ± 3	0.68 ± 0.09	20 ± 2	0.50 ± 0.06
	1.0	22 ± 3	0.55 ± 0.09	32 ± 1	0.80 ± 0.05
	1.5	15 ± 2	0.38 ± 0.06	38 ± 3	0.95 ± 0.10
	2.2	6.5 ± 1.5	0.16 ± 0.04	50 ± 2	1.25 ± 0.08
	3.0	3 ± 1	0.08 ± 0.03	49 ± 2	1.23 ± 0.07
	5.0	—	—	53 ± 2	1.33 ± 0.08
	60	—	—	55 ± 2	1.38 ± 0.08
2	0.061	18 ± 4	0.46 ± 0.11	1 ± 1	0.03 ± 0.03
	0.16	32 ± 3	0.82 ± 0.09	5 ± 3	0.13 ± 0.08
	0.22	36 ± 3	0.89 ± 0.09	8 ± 3	0.21 ± 0.09
	0.31	40 ± 3	1.02 ± 0.12	10 ± 3	0.26 ± 0.08
	0.44	38 ± 3	0.97 ± 0.12	15 ± 3	0.39 ± 0.08

in Table V. The quantity of **X** as a function of time exhibits the rise–fall behavior expected of an intermediate, whereas the quantity of the diferric cluster rises smoothly. The ratio of diferric cluster/R2 at completion (1.38 ± 0.08) is in reasonable agreement with the Fe^{3+} content of native R2 as determined in our laboratory (2.8–3.2 Fe^{3+}/R2 assuming $\varepsilon_{280} = 131$ mM^{-1} cm^{-1}). The ratio also is only slightly different from the Y122•/R2 ratio (1.2 ± 0.1) observed on completion of the reaction.

The general equations for two consecutive, first-order reactions[17] (Scheme I) were fit to the measured quantities of **X** and Y122• as functions of time. The data are consistent with this simple kinetic model, and the analysis gives a rate constant (k_1) of 7.3 ± 1 sec^{-1} for formation of **X** and a rate constant (k_2) of 1.03 ± 0.1 sec^{-1} for decay of **X** and concomitant formation of the diferric cluster (Fig. 12). As would be predicted from Scheme I, there is a noticeable lag phase in formation of the diferric cluster.

[17] P. W. Atkins, "Physical Chemistry." Freeman, New York, 1986.

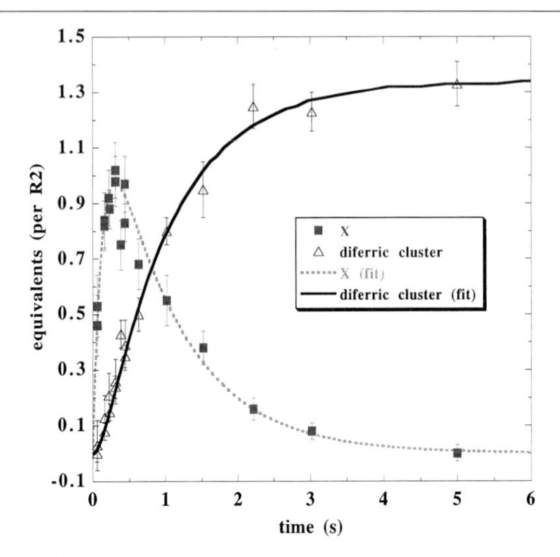

FIG. 12. Nonlinear least-squares fitting of the kinetic model of Scheme I to the measured quantities of **X** and diferric cluster as functions of time. The fit lines correspond to $k_1 = 7.3$ sec^{-1}, $k_2 = 1.03$ sec^{-1}, and a final diferric cluster/R2 ratio of 1.34.

Summary

The SF-Abs, RFQ-EPR, and RFQ-Möss data on the R2 reconstitution reaction are all consistent with the mechanism of Scheme I, in which the intermediate **X** is the immediate precursor to the product cofactor, and illustrate how the continuous SF approach and the discontinuous RFQ methods can be complementary. Given the inherent differences in the methods, it should not be taken for granted that data from the two will be consistent. A number of problems can be associated with the RFQ approach. For example, isopentane could conceivably interfere with or alter the chemistry to be studied. A second potential problem involves temperature-dependent equilibria among different intermediate species. This problem has been encountered by Dooley *et al.*[18] with the 6-hydroxydopa-requiring protein, plasma amine oxidase and was previously observed with the adenosylcobalamin-dependent ribonucleotide reductase by Blakley and co-workers.[19] This potential complication should be considered when discrepancies arise between SF and RFQ data and in low temperature structural studies of reactive intermediates in general.

[18] D. M. Dooley, M. A. McGuirl, D. E. Brown, P. N. Turowski, W. S. McIntire, and P. F. Knowles, *Nature* (*London*) **349,** 262 (1991).

[19] W. H. Orme-Johnson, H. Beinert, and R. L. Blakley, *J. Biol. Chem.* **249,** 2338 (1974).

Each of the three methods employed can yield time-resolved quantitation of reaction components. In this regard, SF-Abs has the disadvantage of poor resolution, such that quantitation of individual components most often requires sophisticated mathematical analysis. Obvious advantages to the RFQ-Möss method are the presence of an internal standard (the known amount of ^{57}Fe being proportional to the total absorption area) and the spectroscopic activity of all reaction components which contain iron. In our hands, quantitation by RFQ-EPR was most problematic and least reproducible. This irreproducibility most likely relates to heterogeneity among samples in terms of volume and density. As discussed in detail by Ballou and Palmer,[15] the packing factor, which relates to the fraction of a sample made up by the reaction solution (the remainder being frozen isopentane), is dependent on the investigator. Given this caveat, it is not surprising that the RFQ-EPR data had the greatest uncertainty in our hands. Placing a chemically unreactive, EPR active standard in each reaction mixture could help alleviate this problem.

Time-resolved Möss methods can be extremely powerful if excellent, nonoverlapping reference spectra of starting materials, products, and intermediates are available. All of the iron centers can be examined simultaneously. The problems associated with Möss arise from its extreme insensitivity. It takes millimolar solutions of proteins and several days for data collection of each time point. The millimolar concentrations of protein required could result in altered kinetics, in comparison with more dilute protein solutions used in SF and RFQ-EPR methods, and side reactions in metal-based reactions due to disproportionation. Fortunately, in our case this was not a problem.

Although there are problems associated with any methods used, the power of comparing three independent and complementary kinetic methods is readily apparent from the results reported herein. These data, in total, provide convincing evidence that **X** is a kinetically competent intermediate in the oxidation of Y122 to Y122•.

[21] Tyrosyl Radicals in Photosystem II

By BRIDGETTE A. BARRY

Introduction

Photosynthesis in plants, green algae, and procaryotic cyanobacteria involves the concerted action of two chlorophyll (chl)-containing membrane proteins; these reaction centers cooperate to transfer electrons from water to $NADP^+$. The energy to drive this process comes from light absorption. Photosystem II (PSII) is the reaction center that carries out the oxidation of water and reduction of bound plastoquinone. The path of electron transfer in photosystem II involves multiple prosthetic groups. P_{680} is the primary chlorophyll donor. After light absorption and excitation, P_{680}^* transfers an electron to a bound pheophytin molecule, which in turn reduces a bound plastoquinone, Q_A. Q_A reduces Q_B, which is a two electron acceptor and can act as an electron carrier in the membrane. On the "donor" side of photosystem II, the chlorophyll cation radical, P_{680}^+, oxidizes a cofactor, Z, which in turn oxidizes the manganese cluster. This metal cluster is the catalytic site of water oxidation. The oxidation of water requires four sequential charge separations and subsequent oxidations of the metal cluster. These sequentially oxidized forms of the manganese site are called S_n states, where n refers to the number of oxidizing equivalents that have been stored. S_1 is the dark stable state. When S_4 is produced, the system resets to S_0 with the concomittant release of O_2 (reviewed in Debus[1]).

Photosystem II is a complex, multisubunit enzyme that is composed of both hydrophobic and extrinsic subunits. The D1 and D2 subunits are believed to form the heterodimer core of the reaction center and to bind most of the prosthetic groups of photosystem II.[2] D1 and D2 show sequence homology to the L and M subunits of the reaction center from purple, nonsulfur bacteria.[3] The structure of the bacterial reaction center is known at the atomic level.[3] Although there is no crystal structure of photosystem II, two structural models for this enzyme have been proposed. These models are based on existing homologies with the bacterial reaction center.[4,5] However, although there are functional similarities between "acceptor" side electron transfer in these two proteins,[3] the bacterial reaction center does not oxidize water, so the accuracy of the "donor side" structural models is difficult to evaluate at the present time.

The donor side of photosystem II contains several redox-active amino acids (reviewed in Barry[6]). The Z cofactor has been shown to be a redox-active tyrosine residue through electron paramagnetic resonance (EPR) spectroscopy and isotopic labeling.[7] Photosystem II also contains another

[1] R. J. Debus, *Biochim. Biophys. Acta* **1102,** 269 (1992).

[2] O. Nanba and K. Satoh, *Proc. Natl. Acad. Sci. U.S.A.* **84,** 109 (1987).

[3] J. Deisenhofer and H. Michel, *Science* **245,** 1463 (1989).

[4] B. Svensson, I. Vass, E. Cedergren, and S. Styring, *EMBO J.* **7,** 2051 (1990).

[5] S. V. Ruffle, D. Donnelly, T. L. Blundell, and J. H. A. Nugent, *Photosynth. Res.* **34,** 287 (1992).

[6] B. A. Barry, *Photochem. Photobiol.* **57,** 179 (1993).

[7] R. J. Boerner and B. A. Barry, *J. Biol. Chem.* **268,** 17151 (1993).

redox active tyrosine, D, that forms a stable neutral radical, D•.[8] The function of D is still unknown, although mutagenesis experiments suggest an interaction between D• and Z.[9] Site-directed mutagenesis experiments also suggest that D is tyrosine 160 of the D2 polypeptide and Z is tyrosine 161 of the D1 polypeptide.[10–14] In site-directed mutants where nonredox-active substitutions are made at the putative Z and D sites, another radical, M^+, can be observed in some photosystem II preparations.[9,14] Isotopic labeling experiments have shown that M^+ is also a tyrosine radical with unusual spectroscopic properties.[15]

The oxidation and reduction kinetics of D and Z are different, allowing them to be easily distinguished from each other. In oxygen-evolving preparations, tyrosine Z is oxidized by $P680^+$ in the nanosecond time scale[16] and is reduced in the 10- to 1000-msec time scale with a rate that depends on the oxidation state of the metal cluster.[17–19] When the manganese cluster is removed, both the oxidation[20] and reduction[21–23] kinetics of Z are slowed by three orders of magnitude. In manganese-depleted preparations, the EPR signal of the neutral tyrosine radical, Z•, can be observed under steady-state illumination.[7,21,22] The radical is reduced rapidly in the dark and, therefore, is not observed under these conditions. Spin quantitation shows that there is up to one Z• spin per reaction center under illumination.

On the other hand, tyrosine D forms a dark-stable tyrosine radical; there is up to one D• spin per reaction center. The reduction kinetics of

[8] B. A. Barry and G. T. Babcock, *Proc. Natl. Acad. Sci. U.S.A.* **84,** 7099 (1987).

[9] R. J. Borner, K. A. Bixby, A. P. Nguyen, G. H. Noren, R. J. Debus, and B. A. Barry, *J. Biol. Chem.* **268,** 1817 (1993).

[10] R. J. Debus, B. A. Barry, G. T. Babcock, and L. McIntosh, *Proc. Natl. Acad. Sci. U.S.A.* **85,** 427 (1988).

[11] R. J. Debus, B. A. Barry, I. Sithole, G. T. Babcock, and L. McIntosh, *Biochemistry* **27,** 9071 (1988).

[12] W. F. J. Vermaas, A. W. Rutherford, and O. Hansson, *Proc. Natl. Acad. Sci. U.S.A.* **85,** 8477 (1988).

[13] J. G. Metz, P. J. Nixon, M. Rogner, G. W. Brudvig, and B. A. Diner, *Biochemistry* **28,** 6960 (1989).

[14] G. H. Noren and B. A. Barry, *Biochemistry* **31,** 3335 (1992).

[15] R. J. Boerner and B. A. Barry, *J. Biol. Chem.* **269,** 134 (1994).

[16] S. Gerken, K. Brettel, E. Schlodder, and H. T. Witt, *FEBS Lett.* **237,** 69 (1988).

[17] R. E. Blankenship, G. T. Babcock, J. T. Warden, and K. Sauer, *FEBS Lett.* **51,** 287 (1975).

[18] G. T. Babcock, R. E. Blankenship, and K. Sauer, *FEBS Lett.* **61,** 286 (1976).

[19] J. P. Dekker, J. J. Plijter, L. Ouwehand, and H. J. van Gorkom, *Biochim. Biophys. Acta* **767,** 176 (1984).

[20] M. Boska, K. Sauer, W. Buttner, and G. T. Babcock, *Biochim. Biophys. Acta* **722,** 327 (1983).

[21] G. T. Babcock and K. Sauer, *Biochim. Biophys. Acta* **376,** 315 (1975).

[22] G. T. Babcock and K. Sauer, *Biochim. Biophys. Acta* **376,** 329 (1975).

[23] J. P. Dekker, H. J. van Gorkom, M. Brok, and L. Ouwehand, *Biochim. Biophys. Acta* **764,** 301 (1984).

D• are on the order of hours.[24] D• is reduced by the manganese cluster during the S_0 to S_1 transition.[25] Similarly, D can be oxidized either by the manganese cluster (for example, by reducing S_2 to S_1) or by P_{680}^{+}.[14,26,27] In the presence of the manganese cluster, the rise time for the D• signal has been measured to be 1 sec.[26] Whatever the role of D• in the reaction center, site-directed mutagenesis experiments have shown that it is not required for oxygen evolution.[10,12] Although the EPR line shapes of D• and Z• are similar in control preparations, the spectra are not identical when these tyrosine radicals are labeled at the 3,5 positions with deuterium[7] (see below).

M^+ is observed under steady-state illumination in up to one spin per reaction center.[9,14,15] The EPR line shape of M^+ can be clearly distinguished from that of D• and Z•. Indeed, the EPR line shape of M^+ is different from all previously characterized tyrosine radicals.[6] The EPR spectra of tyrosine radicals are known to be sensitive to the interaction of the radical with its protein environment (reviewed in Barry[6]). There are three types of structural factors that can affect the EPR line shape.[15] First, a change in the geometry at the methylene position would affect the spectrum, since the coupling to these β-protons depends on the dihedral angle between the proton and the p_z orbital on the adjacent ring carbon.[8,28] Second, a change in the strength of a hydrogen bond to the phenol oxygen would change the EPR spectrum, since both the spin density distribution and the g_x component of the g tensor would be altered.[29] Third, a covalent modification of the side chain would also dramatically change the EPR line shape.[30,31] These possibilities are not mutually exclusive, as described previously.[15]

The oxidation and reduction kinetics of the M species have not yet been measured. Since M^+ is observed when nonredox-active amino acids are substituted at either the putative Z• or D• sites,[9,14,15] M is likely to be yet a third redox-active tyrosine residue in photosystem II. It has been proposed that M can be oxidized either by Z• or by $P680^+$.[9] Since this radical is not observed in nonoxygen-evolving preparations that have been stripped of peripheral photosystem II polypeptides,[13] the M^+ residue may not be located on the core polypeptides of photosystem II. Alternatively,

[24] G. T. Babcock and K. Sauer, *Biochim. Biophys. Acta* **325,** 504 (1973).

[25] S. Styring and A. W. Rutherford, *Biochemistry* **26,** 2401 (1987).

[26] G. T. Babcock and K. Sauer, *Biochim. Biophys. Acta* **325,** 483 (1973).

[27] C. A. Buser, L. K. Thompson, B. A. Diner, and G. W. Brudvig, *Biochemistry* **29,** 8977 (1990).

[28] B. A. Barry and G. T. Babcock, *Chem. Scri.* **28A,** 117 (1988).

[29] G. J. Gerfen, B. F. Bellew, S. Un, J. M. Bollinger, J. Stubbe, R. G. Griffin, and D. J. Singel, *J. Am. Chem. Soc.* **115,** 6420 (1993).

[30] N. Ito, S. E. V. Phillips, C. Stevens, Z. B. Ogel, M. J. McPherson, J. N. Keen, K. D. S. Yadav, and P. F. Knowles, *Nature (London)* **350,** 87 (1991).

[31] G. T. Babcock, M. K. El-Deeb, P. O. Sandusky, M. M. Whittaker, and J. W. Whittaker, *J. Am. Chem. Soc.* **114,** 3727 (1992).

the conformation of the reaction center may be an essential factor in promoting the oxidation of the M residue. This is not unprecedented, since D and Z are no longer both oxidized in nonoxygen-evolving D1–D2 reaction center preparations.[2]

This chapter describes the use of EPR and infrared spectroscopy in the study of redox-active tyrosines in photosystem II. Isotopic labeling of tyrosine is an important component of this work and can be performed through the use of the cyanobacterium *Synechocystis* sp. PCC 6803. Spectroscopic observation of Z• and M^+ requires the use of a purified photosystem II preparation. Accordingly, a method to purify photosystem II from *Synechocystis* is described.

Isotopic Labeling of Tyrosine in the Cyanobacterium *Synechocystis* sp. PCC 6803

Synechocystis sp. PCC 6803 is grown in liquid BG-11 or on solid BG-11 media[32,33] supplemented with 5 m*M* *N*-tris(hydroxymethyl)methyl-2-aminoethanesulfonic acid (TES)–NaOH, pH 8.0.[33] The strain is glucose tolerant, as described.[33] Cells are grown in constant light (25–50 $\mu E\ m^{-2}\ sec^{-1}$ for liquid cultures; 5–15 $\mu E\ m^{-2}\ sec^{-1}$ for plates) at 30° and 60–70% humidity. Light intensity is measured with a LI-189 light meter equipped with a LI-190SA quantum detector (Li-Cor, NE). Liquid cultures are bubbled with sterile-filtered air. Where appropriate, cultures are supplemented with glucose or aromatic amino acids, which are added by sterile filtration through 0.22-μm filters.

Isotopic labeling of tyrosine residues in high yield can be accomplished by feedback inhibition of the biosynthetic pathway for the three aromatic amino acids. Such inhibition can be accomplished by the addition of 0.5 m*M* L-phenylalanine to liquid cultures. In the absence of the other two aromatic amino acid products, cells cannot grow under these conditions; inhibition occurs at an early point in the biosynthetic pathway. Cultures that also receive 0.25 m*M* L-tyrosine and 0.25 m*M* L-tryptophan grow in the presence of 0.5 m*M* phenylalanine, but they are dependent on import of these three amino acids for growth.[8] Under these conditions, all tyrosine residues in every cellular protein can be labeled in high yield by supplying 0.25 m*M* of the appropriate tyrosine isotopomer to the culture. A number of ^{13}C- and ^{2}H-labeled tyrosines (i.e., L-4-hydroxyphenyl-d_4-alanine-2,3,3-d_3 99.6%) are available from commercial sources (i.e., Cambridge Isotopes,

[32] R. Rippka, J. Derulles, J. B. Waterbury, M. Herdman, and R. Stanier, *J. Gen. Microbiol.* **111,** 1 (1979).

[33] J. G. K. Williams, this series, Vol. 167, p. 766.

MA). Control cultures are grown as described earlier, i.e., 0.5 mM phenylalanine, 0.25 mM tryptophan, and 0.25 mM tyrosine. Mass spectral analysis of plastoquinone, naphthoquinone, and chlorophyll shows little isotopic incorporation into any of these prosthetic groups from labeled tyrosine.[8,34]

To reduce the likelihood of contamination with other microorganisms, which has the effect of decreasing the amount of isotopic incorporation, an antibiotic-resistant strain of *Synechocystis* can be employed.[7,15] Kanomycin-resistant control and mutant cells have been grown on 5 μg/ml kanomycin and labeled successfully.[7,15]

Photosystem II Preparation from Control and Isotopically Labeled Cyanobacteria

Observation of the light-induced radicals, M^+ and Z•, requires purification of photosystem II away from other membrane components, particularly photosystem I. A method of purification and an amendment for use with labeled cells are described below.[7,35] All manipulations should be performed in the range from 0 to 4°.

Thylakoid Membrane Preparation

Photoautotrophically or photoheterotrophically grown cells from 15-liter glass carboys ($OD_{730\ nm}$ approximately 1.0) are resuspended in 1.5 liter of "break" buffer (0.8 M sucrose, 50 mM 2-(N-morpholino)ethanesulfonic acid (MES)–NaOH, pH 6.0) and spun for 10 min at 5000 rpm in a Sorvall GS-3 rotor. The pellet is resuspended in approximately 300 ml of break buffer. The cells are allowed to incubate approximately 60 min on ice in the dark before being pelleted by a 10-min spin at 5000 rpm in a Sorvall SS-34 rotor. The pellet is resuspended in a small amount (<50 ml) of break buffer. To these cells, 500 μl from a 10-mg/ml phenylmethylsulfonyl fluoride (PMSF) stock solution in ethanol (Sigma, St. Louis, MO), 250 μl from a 6-mg/ml N^α-tosyl-L-phenylalanine chloromethyl ketone (TPCK) stock (Calbiochem, San Diego, CA), and 250 μl from a 1-mg/ml pepstatin A stock (Calbiochem) are added. The stocks of protease inhibitors are stored at −20°. The cells are then added to a prechilled Beadbeater (Biospec Products, OK) chamber that is two-thirds full with glass beads (0.1-mm diameter, Biospec Products) and also contains 1.25 g bovine serum albumin (Fraction V, protease free, Sigma) and 0.5 mg DNase I (bovine pancreatic type IV, Sigma). The chamber is then completely filled with break buffer,

[34] G. M. MacDonald, K. A. Bixby, and B. A. Barry, *Proc. Natl. Acad. Sci. U.S.A.* **90,** 11024 (1993).

[35] G. H. Noren, R. J. Boerner, and B. A. Barry, *Biochemistry* **30,** 3943 (1991).

so as to exclude air. The jacket of the bead beater is filled with an ice slurry. The cells are then broken in seven cycles in the dark. A cycle consists of 30 sec on and 15 min off; an intervalometer (Dimco-Gray, OH) is used to time the on and off steps. This treatment leads to greater than 90% cell breakage. After the last cycle, the cells and beads are decanted. The beads are allowed to settle, and the broken cells are separated from the beads, by four successive washes (with the break buffer described above) of approximately 200 ml each. The broken cells are then spun for 30 min at 7000 rpm in a Sorvall GSA rotor. The pellet, which is not green, is discarded. To the supernatant, 1 *M* $CaCl_2$ to is added to bring the final concentration to 40 m*M* $CaCl_2$. The supernatant is then centrifuged at 12,500 rpm in the GSA rotor for 30 min to pellet the aggregated membranes.

The pellet from this high calcium spin is made up of "thylakoid membranes" and is resuspended in approximately 30 ml of "freeze" buffer [25% glycerol (Ultrapure, BRL, MD), 20 m*M* MES–NaOH, pH 6.0, 20 m*M* $CaCl_2$, 20 m*M* $MgCl_2$]. The membranes are then pelleted again with a 19,000 rpm spin for 30 min in the SS34 rotor, in order to remove contaminating phycobilisomes. The pelleted membranes are frozen in liquid nitogen and stored in a $-80°$ freezer. For labeled cultures, this thylakoid preparation can be scaled down to account for the decreased volume of cells.[7]

To prepare the 0.1-mm glass beads, the beads are soaked overnight in 1 *M* HCl, then washed five or six times with water to a pH > 5. The beads are washed once with 1 m*M* EDTA and then five to six times with water until the pH ≥ 7. The beads are dried in a Buchner funnel.

Anion-Exchange Column Chromatography at pH 6.0

Thylakoid membranes, up to 20 mg chl, are thawed on ice. A Fast Flow Q (Pharmacia, Uppsala, Sweden) column (30 × 1.5 cm) is equilibrated with 300 ml of high salt buffer [25% (v/v) glycerol, 20 m*M* MES–NaOH, pH 6.0, 0.05% (w/v) lauryl maltoside, 20 m*M* $CaCl_2$, 120 m*M* $MgCl_2$, 400 m*M* NaCl] and 400 ml of low salt buffer [25% (v/v) glycerol, 20 m*M* MES–NaOH, pH 6.0, 0.05% (w/v) lauryl maltoside, 20 m*M* $CaCl_2$, 15 m*M* $MgCl_2$]. This procedure is used to equilibrate a new column, as well as to reequilibrate the Fast Flow Q column after use. Each packed column can be used four times and then must be discarded. The detergent is purchased from Anatrace (OH).

PMSF is added to the membranes to give a final concentration of 0.01 mg/ml from a 10-mg/ml PMSF stock solution in ethanol. A 5% (w/v) lauryl maltoside stock is made up in freeze buffer (see above) and is then added to the membrane suspension, dropwise, to a final detergent concentration of 1% (w/v). The final chlorophyll concentration is 0.6 mg chl/ml. The membrane suspension is swirled vigorously on a rotary platform during addition of the detergent. The membranes are solubilized for 10 min in the dark at 0°, at which time the sample is transferred to prechilled ultracentrifuge tubes and spun at 140,000*g* for 30 min. These steps must be accurately timed. The resultant supernatant is then loaded immediately onto the preequilibrated Fast Flow Q column.

A 400-ml gradient is applied to the column consisting of 200 ml of low salt buffer (see above) and 200 ml of elution buffer [25% (v/v) glycerol, 20 m*M* MES–NaOH, pH 6.0, 0.05% (w/v) lauryl maltoside, 20 m*M* $CaCl_2$, 120 m*M* $MgCl_2$]. The column is run at 1 ml/min, and 5-ml fractions are collected. The eluant is monitored at 280 nm using a Pharmacia UV-M monitor.

An absorption spectrum of each fraction is recorded from 720 to 600 nm through the use of a Cary 210 UV-VIS spectrophotometer, and the fractions corresponding to the most blue shifted chlorophyll Q_y bands are pooled (usually four to six tubes are pooled). These fractions are located in the second half of the first chlorophyll-containing peak and eluate at $MgCl_2$ concentrations of approximately 50 m*M*. The partially purified photosystem II (FFQ particles) in these pooled fractions are precipitated through the addition of an equal volume of 30% polyethylene glycol (PEG) buffer (30% PEG-8000, 20 m*M* MES, pH 6.0, 20 m*M* $CaCl_2$) and collected by a 30-min spin at 19,000 rpm in a SS-34 rotor. The pellet is frozen at −80°.

Anion-Exchange Column Chromatography at pH 6.5

The FFQ particles (see above) are resuspended in Mono Q buffer [25% (v/v) glycerol, 50 m*M* MES–NaOH, pH 6.5, 0.05% lauryl maltoside, 20 m*M* $CaCl_2$, 15 m*M* NaCl]. The solution is filtered through a Gelman 0.45-μm Acrodisk filter and then 1 mg is loaded onto a Pharmacia 5/5 Mono Q column using a Pharmacia FPLC system.

The column is washed with 5 ml of Mono Q buffer, and then a 20-ml gradient of NaCl in the same buffer is applied. The final NaCl concentration is 250 m*M*. The flow rate is 0.4 ml/min, and 0.5-ml fractions are collected. The eluant is monitored at 280 nm through the use of a Pharmacia UV-M monitor. The absorption spectrum of each fraction is then recorded as described earlier. Fractions with Q_y chlorophyll bands that have absorption maxima less than 675 nm are pooled and precipitated with an equal volume of 30% PEG buffer (see above). Typically, five fractions are pooled; these fractions elute at approximately 150 m*M* NaCl. The column is then washed with 5 ml of Mono Q buffer that contains 500 m*M* NaCl. This treatment elutes any remaining material on the column, which can be reequilibrated with the original Mono Q buffer. Because of the reproducible elution

TABLE I
PURIFICATION OF OXYGEN-EVOLVING PHOTOSYSTEM II PARTICLES FROM *Synechocystis* sp. PCC 6803[a]

	O_2 evolution[b]	Total chlorophyll[c]	Yield[d]
Thylakoids	470[e]	20.0	100
Fast Flow Q	1200[f]	2.5	12
Mono Q	2400[f]	0.4	2

[a] Adapted with permission from Noren *et al.*[35]
[b] μmol O_2/mg chl-hr.
[c] mg chl.
[d] Percent chl.
[e] 1 m*M* 2,6-DCBQ and 1 m*M* potassium ferricyanide.
[f] 400–600 μM 2,6-DCBQ and 1 m*M* potassium ferricyanide.

patterns using the FPLC system, multiple columns can be run, and the photosystem II-containing fractions can be pooled.

The precipitated PSII particles are resuspended to a concentration of approximately 0.5 mg chl/ml in Mono Q buffer and are stored at −80°. This method works for labeled cells grown photoheterotrophically and for unlabeled cells.

Modification for Labeled Cells Grown Photoautotrophically

Because of their higher photosystem II content, labeled cells grown photoautotrophically can be purified by a procedure that involves only one column chromatography step on a Pharmacia Mono Q column.[7] The column buffers contain 25% glycerol, 20 m*M* MES–NaOH, pH 6.0, 0.05% lauryl maltoside, 20 m*M* $CaCl_2$, and 15 mM $MgCl_2$. A 25-ml gradient of $MgCl_2$ is used to elute the column. The final $MgCl_2$ concentration is 125 m*M*. All other manipulations are as described previously.

Oxygen Evolution Measurements

Oxygen evolution measurements are used to monitor the effectiveness of each purification (Table I). Oxygen release is monitored directly through the use of a Clark electrode, a Gilson water-jacketed sample holder, and a Yellow Springs Instrument (YSI Inc., OH) oxygen meter. Temperature is controlled at 25° with a Lauda RM6 water bath. Air-saturated water is used as an oxygen standard. The electrode is also periodically calibrated through the use of a catalase assay.[36] The chlorophyll concentration is 10–

[36] W. J. Wingo and G. M. Emerson, *Anal. Chem.* **47,** 351 (1975).

20 μg chlorophyll in 1.5 ml. In order to measure the maximum rate, 1 m*M* potassium ferricyanide and 400–1000 μM 2,6-dichlorobenzoquinone (DCBQ) (Kodak) are added. The DCBQ must be recrystallized from methanol before use. Similar oxygen evolution rates are observed with recrystallized 2,5-DCBQ. Saturating light is provided by a fiber-optic light source (Dolan-Jenner, Model 180). The sample compartment is equipped with a red filter and a first surface aluminum mirror that increases the amount of light incident on the sample. The oxygen assay buffer for thylakoid membranes, FFQ particles, and PSII particles contains 1 *M* sucrose, 50 m*M* MES–NaOH, pH 6.5, 25 m*M* $CaCl_2$, and 10 m*M* NaCl. Chlorophyll *a* is quantitated by extraction into methanol and use of the extinction coefficient at 665.2 nm.[37] Typical oxygen evolution rates are in the range from 2400 to 2600 μmol O_2/mg chl/hr (Table I).

EPR Analysis of Tyrosine Radicals in Photosystem II

EPR spectra at room temperature are recorded at the X-band through the use of a Varian E-4 spectrometer equipped with a Varian TE cavity. The spectrometer is interfaced to a Macintosh IIcx computer through a digital voltmeter (Keithley Model 195A, Cleveland, OH) and a Mac488A bus controller (Iotech, Cleveland, OH). The data acquisition program is a gift of Professor John Golbeck (University of Nebraska). Data analysis is performed on the Macintosh computer using the graphics software program, IGOR (Wavemetrics, Lake Oswego, OR). Illumination of the sample in the EPR cavity is performed through the use of a fiber-optic light source (Dolan-Jenner, Model 180). Fremy's salt, $K_2(SO)_3NO_3$, is used as a standard for spin quantitation under nonsaturating conditions.[38,39] Typical chlorophyll concentrations are in the range from 0.4 to 0.6 mg chl/ml. Manganese can be removed by alkaline Tris or by hydroxylamine treatment.[35,40]

Representative EPR spectra of D• and Z• in *Synechocystis* photosystem II particles are shown in Fig. 1. The dashed lines were obtained in the light, and the solid lines were obtained after illumination in the dark. The samples contain 1 m*M* potassium ferricyanide, which acts as an electron acceptor. In Figs. 1A and 1B, the solid line corresponds to the EPR signal of D•. There is one spin of D• per reaction center or one spin per approximately

[37] H. K. Lichtenthaler, this series, Vol. 148, p. 350.

[38] G. T. Babcock, D. F. Ghanotakis, B. Ke, and B. A. Diner, *Biochim. Biophys. Acta* **723,** 276 (1983).

[39] J. E. Wertz and J. R. Bolton, "Electron Spin Resonance." Chapman & Hall, New York, 1986.

[40] G. M. MacDonald and B. A. Barry, *Biochemistry* **31,** 9848 (1992).

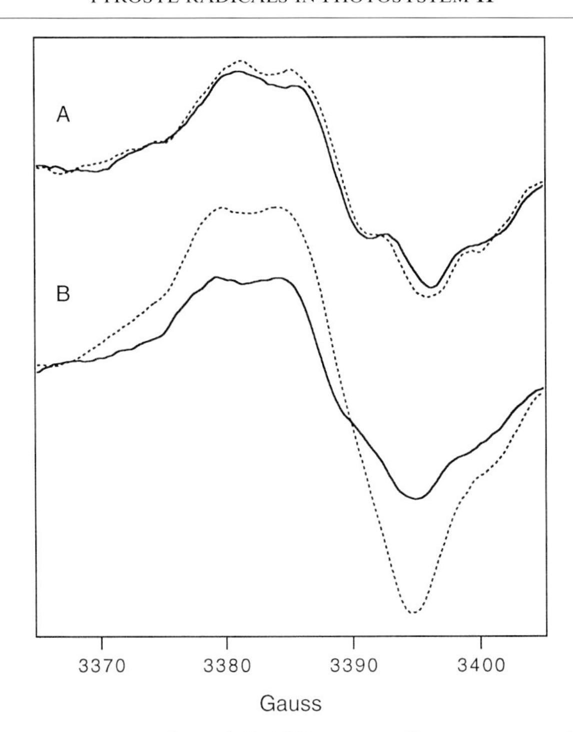

FIG. 1. Room temperature D• and Z• EPR spectra. Data were recorded on purified photosystem II particles from wild-type cultures of the cyanobacterium *Synechocystis* 6803. (A) Spectra were recorded on oxygen-evolving, manganese-containing samples either under illumination (dashed line) or in the dark following illumination (solid line). (B) Spectra were recorded on manganese-depleted samples either under illumination (dashed line) or in the dark following illumination (solid line). The samples contained 1 m*M* potassium ferricyanide. Spectral conditions: frequency, 9.5 GHz; power, 3.5 mW; scan time, 4 min; time constant, 2 sec; modulation amplitude, 3.2 G. The isotropic *g* value of D• is 2.0046.[41] Adapted with permission from Noren and Barry.[14]

50–60 chlorophylls.[7,9,14,15,42,43] When the majority of centers contain an active manganese cluster, there is very little increase in the amplitude of this signal in the light (Fig. 1A, dashed line). However, when the manganese cluster is purposely removed from these particles, a light-induced increase in the signal is observed, which is caused by the accumulation of Z• (Fig. 1B, dashed line). Up to one spin per reaction center of Z• can be generated in the light under these conditions.[14,35] This light spectrum is the composite

[41] A.-F. Miller and G. W. Brudvig, *Biochim. Biophys. Acta* **1056,** 1 (1991).

[42] R. J. Boerner, A. P. Nguyen, B. A. Barry, and R. J. Debus, *Biochemistry* **31,** 6660 (1992).

[43] G. M. MacDonald, R. J. Boerner, R. M. Everly, W. A. Cramer, R. J. Debus, and B. A. Barry, *Biochemistry* **33,** 4393 (1994).

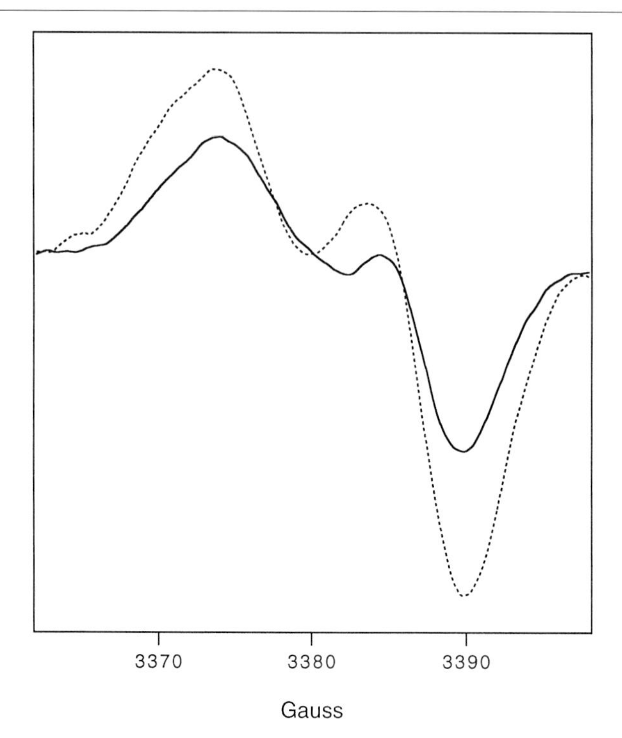

Fig. 2. Room temperature EPR spectra of 3,5-deuterated D• and Z•. Data were recorded on purified photosystem II particles from wild-type cultures of the cyanobacterium *Synechocystis* 6803, which were grown in the presence of 3,5-deuterated tyrosine (L-4-hydroxyphenyl-3,5-d_2-alanine 98% labeled). The preparations were then manganese depleted. Spectra were recorded under illumination (dashed line) or in the dark following illumination (solid line). The samples contained 0.5 m*M* potassium ferricyanide and 0.5 m*M* potassium ferrocyanide. Spectra conditions are described in Fig. 1. Adapted with permission from Boerner and Barry.[7]

of the D• and Z• EPR line shapes. The light (dashed line) and dark (solid line) spectra shown in Fig. 1B are similar.

However, a difference between the EPR line shapes of D• and Z• can be observed when the tyrosine radicals are 3,5 deuterated (Fig. 2). These doublet spectra are dominated by a large coupling to one methylene proton.[44] Since the difference that is observed between the light (Fig. 2, dashed line) and dark (Fig. 2, solid line) spectra is small and could be caused by several structural factors,[7] difference infrared spectroscopy was employed to obtain more information about the structural differences between D• and Z• (see below).

[44] B. A. Barry, M. K. El-Deeb, P. O. Sandusky, and G. T. Babcock, *J. Biol. Chem.* **265,** 20139 (1990).

The EPR signal of M^+ is shown in Fig. 3A. M^+ can be observed when a phenylalanine is substituted for tyrosine 161 (putative Z) in the D1 polypeptide or when either a phenylalanine or a tryptophan is substituted for tyrosine 160 (putative D) in the D2 polypeptide.[9,14] M^+ is a less stable radical than Z•. In wild-type preparations, spin quantitation shows that an hour of repeated illumination at room temperature leads to loss of 40% of the original Z• signal. In the YF161D1 mutant preparations, the amplitude

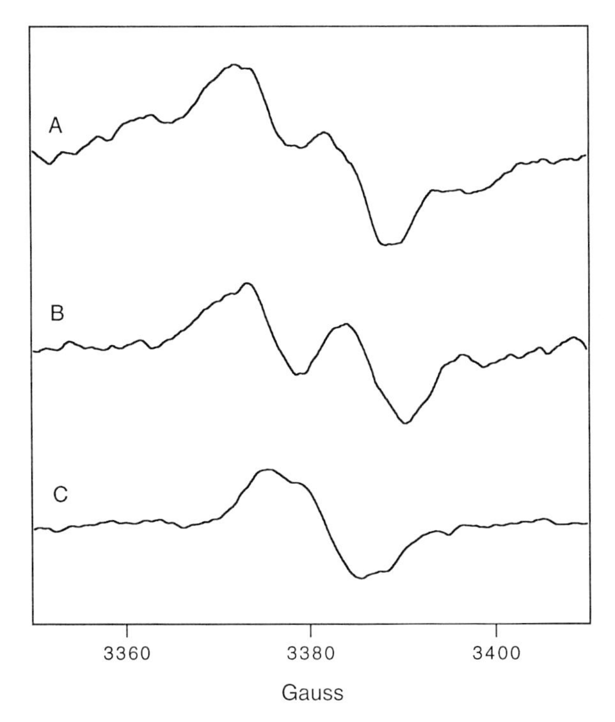

FIG. 3. Room temperature M^+ EPR spectra recorded under illumination. These difference spectra were recorded on purified photosystem II particles from the YF160D2 mutant of *Synechocystis* 6803. (A) Photosystem II was isolated from cultures grown in the presence of control, unlabeled tyrosine. (B) Photosystem II was isolated from cultures grown in the presence of 3,5-deuterated tyrosine (L-4-hydroxyphenyl-3,5-d_2-alanine 98.0% labeled). (C) Photosystem II was isolated from cultures grown in the presence of perdeuterated tyrosine (L-4-hydroxyphenyl-d_4-alanine-2,3,3-d_3 99.6% labeled). The samples contained 1 m*M* potassium ferricyanide. Since the M^+ radical is unstable and decays irreversibly in 30 min in this mutant, its line shape was obtained by subtracting the chlorophyll radical signal remaining after M^+ decay from the first spectrum recorded under illumination. The remaining chlorophyll radical signal arises from a small amount of contaminating photosystem I.[9] Spectral conditions are as in Fig. 1, except that the modulation amplitude in B and C was 2.0 G. Adapted with permission from Boerner and Barry.[15]

of the M^+ signal is decreased by 60% by the same treatment.[14] In the YF160D2 and YW160D2 mutants, M^+ decays irreversibly in 30 min.[9]

The EPR line shape of M^+ can be obtained by subtraction (Fig. 3A).[9,14] Similar subtraction procedures with control preparations do not yield the characteristic four-line signal of M^+, which is clearly distinguishable from that of Z• and D• in control preparations (Fig. 1). Upon incorporation of perdeuterated tyrosine into photosystem II, the M^+ signal narrows (Fig. 3C). However, this signal is broader than the 7-G singlet that is observed on perdeuteration of either Z• or D•.[7,8,15] Also, the M^+ doublet signal, which is observed on 3,5 deuteration, is isotropic and exhibits an 11-G splitting to one methylene proton (Fig. 3B), compared to an 8-G coupling to a methylene proton in D•.[45] This last result suggests that the geometry of the M^+ radical at the methylene position is different from the geometry of the D• radical.[15] Other structural differences may also exist and are under investigation.[15]

Difference Infrared Studies of Tyrosine Radicals in Photosystem II

Fourier transform infrared (FT-IR) spectra are obtained through the use of a Nicolet 60SXR spectrometer equipped with a Harrick sample holder and temperature controller. The temperature is $-9°$, and dehydrated samples are employed. The spectral resolution is 4 cm^{-1}, a Happs-Ganzel apodization function is used, and the mirror velocity is 1.57 cm/sec. Typically, 1000 scans are coadded for each interferogram. Samples contain 25–45 μg chl. The sample is sandwiched between a CaF_2 window and a germanium window, which is used to block He–Ne laser irradiation of the sample from the optical bench. The absorbance in the amide I band of the sample is always below 0.7 absorbance units. Samples also contain 1.5 m*M* potassium ferricyanide and 1.5 m*M* potassium ferrocyanide. The sample is illuminated with a Dolan Jenner Model 180 illuminator equipped with an annular illuminator, a red filter, and a heat filter. An Alpha Products (Darien, CT) RS-232 to A-BUS adapter and a ST-143 digital output driver are used to interface the spectrometer to the illumination system.

Difference or "light minus dark" infrared spectroscopy can be used to obtain the vibrational difference spectrum associated with the oxidation of Z and D.[34,40] This can be accomplished by removing manganese from the reaction center (see above). Under these conditions, illumination produces the state Z•D•Q_A^- and an infrared spectrum is recorded. A spectrum is also recorded in the dark after charge recombination between Z• and Q_A^- has occurred. The difference spectrum is constructed by subtraction of a dark spectrum from a spectrum recorded under illumination. If the length

[45] C. W. Hoganson and G. T. Babcock, *Biochemistry* **31,** 11874 (1992).

of the dark adaptation after illumination is short (8 min), then D• does not have time to decay. Under these conditions, the light minus dark difference spectrum reflects the oxidation of Z and reduction of Q_A. On the other hand, if there is a long (90 min) dark adaptation between illumination and the dark scans, then the light minus dark spectrum reflects the oxidation of Z *and* D. EPR spectroscopy is performed under identical conditions on dehydrated samples to assess the number of spins per reaction center that are generated in each case.[34,40]

^{2}H (L-4-hydroxyphenyl-d_4-alanine-2,3,3-d_3 99.6%, Cambridge Isotopes) and ^{13}C (L-4-hydroxyphenyl-$^{13}C_6$-alanine 99%, Cambridge Isotopes) labeling of tyrosine has been used to assign vibrational modes to the oxidized and reduced forms of D and Z. This allows us to clearly distinguish the contributions of quinone and tyrosine since plastoquinone is not labeled by this treatment. The EPR line shapes of Z• and D• are also recorded on these labeled samples in order to verify that labeling has occurred and is complete.[34,40]

Figure 4 shows the vibrational difference spectra obtained with short dark adaptations. Z• makes a positive contribution to this spectrum whereas Z makes a negative contribution. Q_A^- and Q_A also contribute to this spec-

FIG. 4. Difference (light minus dark) FT-IR spectrum of photosystem II particles isolated from wild-type cultures of the cyanobacterium *Synechocystis* 6803. The tick marks on the *y* axis correspond to 2×10^{-4} absorbance units. Spectral conditions are given in the text.

trum; vibrational modes of the quinone have not yet been identified by isotopic labeling. However, isotopic labeling *has* shown that Z tyrosine residue modes are located at 1657, 1522, and in the region between 1264 and 1231 cm^{-1}.[34] The Z• radical contributes to the spectral feature at 1477 cm^{-1}, which has been tentatively assigned to the C–O vibration of the radical. Significantly, the vibrational spectrum associated with the oxidation of D is different from the spectrum associated with the oxidation of Z.[34] In particular, the putative C–O vibration of the D• radical is downshifted with respect to the putative C–O of Z•. This difference is consistent with a difference in hydrogen bonding to each phenol oxygen.[34] This hypothesis has recently been confirmed by deuterium exchange.[47] A change in hydrogen bonding to the radicals could help to account for the difference in the midpoint potentials of the two redox-active tyrosines.[13,46]

The putative C–O vibration of the stable tyrosine radical in the enzyme, ribonucleotide reductase (RNR), has a frequency of 1498 cm^{-1}.[48] This radical is not hydrogen bonded.[48,49] Interestingly, a high frequency EPR study[29] has shown that the g tensors of D• and the RNR tyrosine radical are not identical. This EPR study implies that the RNR radical has approximately 30% more spin density at the phenol oxygen, when compared to D•.[29] Our vibrational study[34] is in qualitative agreement with this prediction since the lower frequency of the D• C–O vibration is consistent with a decrease in spin density at the phenol oxygen.

Conclusions

The factors that control the function of redox-active tyrosines need to be elucidated. Photosystem II is an excellent system in which to investigate this question since it contains three redox-active amino acids. Two of these tyrosines, D and Z, are well characterized. Our vibrational studies have shown that there are small structural differences between these two radicals. This difference may be due to a change in the strength of a hydrogen bond to the phenol oxygen of each redox-active residue. This structural difference could be important in controlling the function of the redox-active amino acid, i.e., through adjustment of midpoint potential. Interestingly, structural

[46] A. Boussac and A. L. Etienne, *Biochim. Biophys. Acta* **766,** 576 (1984).

[47] M. T. Bernard, G. M. MacDonald, A. P. Nguyen, R. J. Debus, and B. A. Barry, *J. Biol. Chem.* **270,** 1589 (1995).

[48] G. Backes, M. Sahlin, B.-M. Sjöberg, T. M. Loehr, and J. Sanders-Loehr, *Biochemistry* **28,** 1923 (1989).

[49] C. J. Bender, M. Sahlin, G. T. Babcock, B. A. Barry, T. K. Chandrashekar, S. P. Salowe, J. A. Stubbe, B. Lindström, L. Petersson, A. Ehrenberg, and B.-M. Sjöberg, *J. Am. Chem. Soc.* **111,** 8076 (1989).

models of the donor side of photosystem II predict that the environments of the D and Z tyrosines are different in terms of their hydrophobicities and in terms of the strength of a hydrogen bond to each tyrosine.[4,5] Photosystem II also contains a third redox-active tyrosine, M, with unique spectroscopic characteristics. This radical is observed when substitutions are made either at the D• or at the Z• sites. The location of this tyrosine is under investigation, as is the role of M in electron transfer in photosystem II.

Acknowledgments

Work in my laboratory is supported by NIH GM 43273. I thank all of my students and co-workers: Mary Bernard, Kathryn Bixby, Renee Boerner, Gina MacDonald, George Noren, David DeSantis, Ned Van Eps, Jacqueline Steenhuis, and Linda Eisenschenk.

[22] Role of Tryptophans in Substrate Binding and Catalysis by DNA Photolyase

By SANG-TAE KIM, PAUL F. HEELIS, and AZIZ SANCAR

Introduction

DNA photolyase catalyzes the monomerization of cyclobutane pyrimidine dimer (Pyr<>Pyr) to normal mononucleotides by utilizing the light energy of near-ultraviolet (UV) and visible wavelengths. Photolyases utilize two noncovalently bound cofactors as photosensitizers.[1] One is flavin adenine dinucleotide ($FADH^-$),[2,3] which is the essential catalytic cofactor. The other is either methenyltetrahydrofolate (MTHF)[4] or 8-hydroxy-5-deazaflavin (8-HDF),[5,6] which functions as the major light-energy harvesting cofactor. The physiologically relevant form of the enzyme contains the flavin in a two-electron reduced state, most likely as the anionic form ($FADH^-$) of dihydroflavin. However, during purification under aerobic conditions, this cofactor is oxidized to the flavin semiquinone form ($FADH^\circ$) which is not active in catalysis unless reduced chemically or

[1] A. Sancar, *Biochemistry* **33,** 2 (1994).

[2] A. Sancar and G. B. Sancar, *J. Mol. Biol.* **172,** 223 (1984).

[3] A. M. P. Eker, P. Kooiman, J. K. C. Hessels, and A. Yasui, *J. Biol. Chem.* **265,** 8009 (1990).

[4] J. L. Johnson, S. Hamm-Alvarez, G. Payne, G. B. Sancar, K. V. Rajagoplan, and A. Sancar, *Proc. Natl. Acad. Sci. U.S.A.* **85,** 2046 (1988).

[5] A. P. M. Eker, J. K. C. Hessels, and J. V. Velde, *Biochemistry* **27**(1), 758 (1988).

[6] K. Malhotra, S. T. Kim, C. T. Walsh, and A. Sancar, *J. Biol. Chem.* **267,** 15406 (1992).

photochemically.[1] Steady-state and time-resolved spectroscopic studies have revealed that absorption of light by either MTHF or 8-HDF results in the generation of a singlet excited state of $FADH^-$.[7,8] The photoexcited $FADH^-$ initiates the splitting of Pyr<>Pyr by an electron transfer process.[9,10]

In addition to the two cofactors, photolyases are rich in tryptophans, which can function as catalytic chromophores in the far UV. The two main primary photoprocesses of tryptophan in both aqueous and nonaqueous solvents are electron ejection and H-atom release. In the presence of an adequate electron acceptor such as Pyr<>Pyr, the electron transfer from photoexcited tryptophan to the Pyr<>Pyr is the main decay pathway of the photoexcited tryptophan, which results in the regeneration of the original two pyrimidines.[11,12] This suggests that the tryptophan(s) in photolyase can also function as a catalytic chromophore. Furthermore, aromatic amino acids are known to be involved in binding flavins in flavoproteins. Thus, the tryptophans, together with other aromatic amino acids in photolyase, could be involved in providing an aromatic shell around the chromophore to facilitate electron transfer such as seen in photosynthetic reaction centers[13] and in bacteriorhodopsin.[14] In fact, it has been demonstrated that the efficiency of pyrimidine dimer splitting in covalently linked synthetic dimer sensitizers is highly dependent on the solvent and that a nonpolar environment is required for high efficiency dimer repair in these model systems.[15,16] The dependence of splitting efficiency on solvent polarity was rationalized on the basis of retardation of back electron transfer due to Marcus-inverted behavior of the charge-separated species. Photolyases might achieve their high efficiency of dimer splitting in part by employing a hydrophobic environment provided by an "aromatic shell" at the active site.

In order to evaluate the functional and structural roles of tryptophans in photolyase, we have investigated how these residues in *Escherichia coli* photolyase contribute to binding and catalysis. Using site-directed mutagen-

[7] S. T. Kim, P. F. Heelis, T. Okamura, Y. Hirata, N. Mataga, and A. Sancar, *Biochemistry* **30,** 11262 (1991).

[8] R. S. A. Lipman and M. S. Jorns, *Biochemistry* **31,** 786 (1992).

[9] T. Okamura, A. Sancar, P. F. Heelis, N. Mataga, and A. Sancar, *J. Am. Chem. Soc.* **113,** 3143 (1991).

[10] S. T. Kim, A. Sancar, C. Essenmacher, and G. T. Babcock, *J. Am. Chem. Soc.* **114,** 4442 (1992).

[11] J. R. Van Camp, T. Young, R. F. Hartman, and S. D. Rose, *Photochem. Photobiol.* **45,** 365 (1987).

[12] S. T. Kim and S. D. Rose, *Photochem. Photobiol.* **47,** 725 (1988).

[13] J. Deisenhofer and H. Michel, *EMBO J.* **8,** 2149 (1989).

[14] T. Mogi, T. Marti, and H. G. Khorana, *J. Biol. Chem.* **164,** 14197 (1989).

[15] S. T. Kim, R. F. Hartman, and S. D. Rose, *Photochem. Photobiol.* **52,** 789 (1990).

[16] S. T. Kim and S. D. Rose, *J. Photochem. Photobiol. B: Biol.* **12,** 179 (1992).

esis, we replaced each of the 15 tryptophan residues in this enzyme by phenylalanine and by other natural amino acid residues. Characterization of the mutant proteins by time-resolved optical and resonance spectroscopic methods revealed that (1) Trp-277 is involved in DNA binding by intercalation and thus replacement of this residue with nonaromatic amino acid residues drastically alters the substrate affinity of the enzyme. (2) Trp-277 repairs Pyr<>Pyr by a photoinduced electron transfer process, independent of the other two cofactors. (3) Trp-306 donates an electron to the excited-state $FADH^{\circ}$ to generate $FADH^{-}$, the catalytically active form of the cofactor. These two tryptophan residues (Trp-277 and Trp-306 in *E. coli*) are exclusively conserved in all microbial DNA photolyases and, in all likelihood, perform similar functions.

General Methods

Purification of DNA Photolyase

Photolyases are not abundant proteins and thus purification to homogeneity requires overexpression of the cloned gene. From such genetically engineered strains, several milligrams of pure enzyme can be obtained from a liter of bacterial culture. We have typically used two different purification methods for *E. coli* photolyase which yield 95–99% pure enzyme in two or three chromatographic steps. These two methods are the modifications of the originally reported seven-step purification procedure.[17] The analysis of the various purification fractions of *E. coli* (Folate class) and *Anacystis nidulans* (Deazaflavin class) photolyases by SDS–polyacrylamide gel electrophoresis is shown in Fig. 1.

Growing and Harvesting the Cells. A seed culture is prepared by growing an overnight culture from a single colony picked from a plate of freshly transformed cells. The overnight culture is then diluted 100-fold into Luria broth containing proper antibiotics. When the culture reaches A_{600} = 0.6–0.8 (1.6-cm path length glass tube), isopropyl β-D-thiogalactoside (IPTG) is added to a final concentration of 1 m*M* and the culture is incubated for an additional 10–12 hr. The cells are collected by centrifugation at 10,000*g* for 20 min at 4° and are carefully resuspended in lysis buffer containing 50 m*M* Tris–HCl, 100 m*M* NaCl, 1 m*M* EDTA, 10 m*M* mercaptoethanol, and 10% (v/v) sucrose. The cell suspension is frozen in a dry ice/ethanol bath and can be stored at −80° indefinitely. The following is the purification procedure for cells from 5 to 10 liters of culture.

[17] A. Sancar, F. W. Smith, and G. B. Sancar, *J. Biol. Chem.* **259,** 6028 (1984).

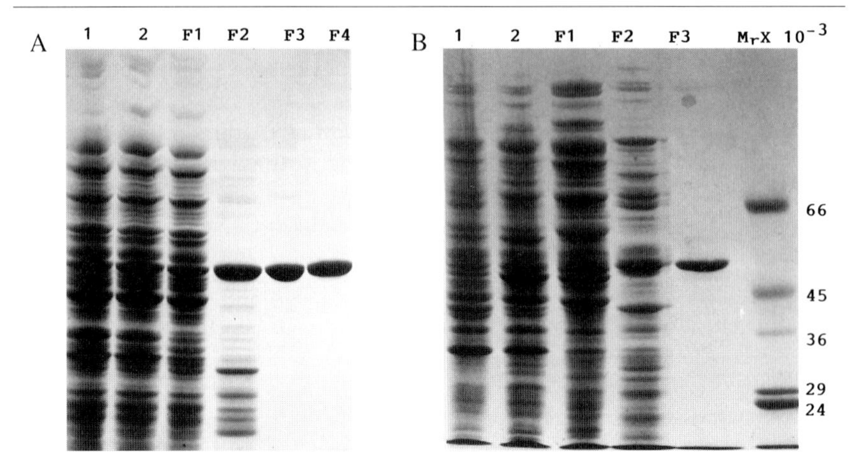

FIG. 1. Purification of (A) *E. coli* and (B) *A. nidulans* photolyases. Samples from each step of purification were analyzed on 10% SDS–polyacrylamide gels that were stained with Coomassie blue. (A) Lanes 1 and 2 contain cells from 200 μl of overnight cultures of CSR603K/pMS969 that were grown without and with 1 m*M* IPTG induction, respectively; lane F1, crude extract, 50 μg of protein; lane F2, peak fractions after Blue-Sepharose column, 10 μg of protein; lane F3, peak fractions after size exclusion column, 13 μg; lane F4, peak fractions after hydroxylapatite column, 11 μg of protein. (B) Lanes 1 and 2 contain cells from 200 μl of overnight cultures of CSR603K/pUNC1993 that were grown without and with 1 m*M* IPTG induction, respectively; lane F1, crude extract, 27 μg of protein; lane F2, peak fractions after Blue-Sepharose column, 50 μg of protein; lane F3, peak fractions after DNA cellulose column, 5 μg of protein. From Sancar *et al.*[9] and Malhotra *et al.*[6]

Cell Lysis. The frozen cell suspension is thawed on ice. Once the cells are completely thawed, lysozyme (100 μg/ml) is added to the suspension and is incubated at 4° for 30 min. Cell lysis is completed by sonication with a Branson sonicator. The sonicate is centrifuged at 100,000*g* for 1 hr to remove cell debris.

Ammonium Sulfate Precipitation (Fraction 1). Ammonium sulfate is added to the supernate with stirring over a 30-min period (65% saturation). The stirring is continued for an additional 30 min. The precipitate is collected by centrifugation at 10,000*g* for 20 min. The precipitate is redissolved in 10–20 ml of buffer A containing 50 m*M* Tris–HCl, pH 7.4, 1 m*M* EDTA, 10 m*M* mercaptoethanol, and 20% (v/v) glycerol and is dialyzed overnight against 2 liters of buffer A.

Blue-Sepharose Chromatography (Fraction 2). Fraction 1 is loaded onto a 50-ml Blue-Sepharose (Reactive Blue 2, Sepharose CL-6B, Sigma) column prequilibrated with buffer A. The column is washed with 150 ml of 10 m*M* ATP and 100 m*M* KCl in buffer A. Photolyase is eluted with 2 *M* KCl in buffer A. Photolyase is located by SDS–PAGE. Fractions containing the enzyme are pooled, concentrated, and dialyzed against buffer A.

Size Exclusion Chromatography (Fraction 3a). Fraction 2 is loaded onto a 200-ml size exclusion column (Bio-Gel P-100 Gel, medium, Bio-Rad) preequilibrated with buffer A at a flow rate of 0.3 ml/min. Fractions containing photolyase are pooled, concentrated, and dialyzed against buffer A.

Hydroxylapatite Chromatography (Fraction 4a). Fraction 3 is loaded onto a 15-ml hydroxylapatite (Bio-Gel HT Hydroxylapatite, Bio-Rad) column equilibrated with buffer containing 67 m*M* potassium phosphate, pH 6.8, 1 m*M* EDTA, 10 m*M* mercaptoethanol, and 20% (v/v) glycerol. The column is washed with 45 ml of the same buffer and then a 120-ml linear gradient of 67–330 m*M* phosphate is applied. Photolyase elutes at about 200 m*M* K-phosphate. Fractions containing photolyase are pooled, concentrated, and dialyzed against storage buffer containing 50 m*M* Tris–HCl, pH 7.4, 100 m*M* NaCl, 1 m*M* EDTA, 5 m*M* dithiothreitol, and 50% (v/v) glycerol.

DNA-Cellulose Chromatography (Fraction 3b). Alternatively, fraction 2 can be applied to a 20-ml DNA-cellulose column preequilibrated with 50 m*M* Tris–HCl pH 7.4, 100 m*M* potassium chloride, 1 m*M* EDTA, 10 m*M* mercaptoethanol, and 20% glycerol. The column is washed with 60 ml of the same buffer and developed with a 140-ml gradient of 0.1–1.0 *M* KCl. Photolyase elutes at about 250 m*M* KCl. Fractions containing photolyase are pooled, concentrated, and dialyzed against storage buffer.

Preparation of Apoenzyme

In both folate and deazaflavin class photolyases, the two chromophores are tightly bound to the apoenzymes. However, the cofactors can be dissociated from the apoenzymes without entirely denaturing the protein.[18,19] In the case of *E. coli* photolyase, the apoenzyme can be prepared by dialysis of photolyase against 2 *M* KBr at pH 4–5 for 10–20 days. This procedure usually yields apoenzyme free of MTHF and with a FAD content of 1–5%. The residual flavin-containing enzyme can be removed by subsequent hydroxylapatite column purification. Under these conditions, the enzyme containing flavin binds to the resin in 67 m*M* potassium phosphate, pH 6.8, but the apoenzyme does not.

Reconstitution

Reconstitution experiments were performed by incubating the apoenzyme with a two- to threefold excess of MTHF and/or FAD on ice for 1–

[18] M. Husain and V. Massey, this series, Vol. 53, p. 429.

[19] G. Payne, M. Wills, C. Walsh, and A. Sancar, *Biochemistry* **29,** 5706 (1990).

6 hr.[19,20] Binding can be monitored by the shift in the position of the absorption peak (MTHF) or a change in the shape of the absorption band (FAD). Unbound chromophores are usually removed by passage through a Sephadex G-50 column (Pharmacia LKB Biotechnology, Inc.) followed by dialysis.

Substrates

Two types of substrates were used: (i) A thymine dimer in a synthetic duplex. This 49-bp duplex was prepared by J.-S. Taylor and colleagues by using a thymidine dimer building block in a DNA synthesizer.[21,22] The 49-mer was 5′-end labeled with γ-^{32}P-ATP and T4 polynucleotide kinase and annealed to the complementary strand to obtain a duplex with a centrally located T<>T at a *Mse*I site (TTAA). This substrate was used either in gel retardation experiments or in a coupled enzyme assay where the susceptibility of the enzyme to *Mse*I is tested following photoreactivation. (ii) Thymine dimers in oligo$(dT)_{15}$. This substrate was prepared by acetone photosensitized irradiation.[23] The oligomer was dissolved in 10% aqueous acetone at a concentration of 1–2 mg/ml. Portions (3 ml) were placed in a quartz cuvette held at 5°, flushed with argon, and irradiated with a 313-nm light from a 75-W xenon lamp equipped with a f/4 grating monochrometer (1200 L/mm; 300-nm blaze). The fluence rate at the inner surface of the incident beam side of the cuvette was 23 Jm^{-2} sec^{-1} as monitored by an AMKO-Quantacount (Quantacount-Photon Technology International, Inc.) calibrated by ferrioxalate actinometry.[24] The sample was irradiated with 5×10^5 Jm^{-2} and the solvent was removed by flash evaporation. The residue was resuspended in diethyl ether and centrifuged to collect the product. About 75% of the thymine was converted to T<>T as determined by the decrease in thymine absorption at 260 nm.

Enzyme Assays

Three types of assays were used. A gel retardation assay was employed to measure enzyme-substrate binding and either a coupled enzyme assay or a spectrophotometric assay was used to quantify repair.

[20] M. S. Jorns, B. Wang, S. P. Jordan, and L. P. Chanderkar, *Biochemistry* **29,** 552 (1990).
[21] J.-S. Taylor, I. R. Brockie, and C. L. O'Day, *J. Am. Chem. Soc.* **109,** 6735 (1987).
[22] J.-S. Taylor and I. R. Brockie, *Nucleic Acids Res.* **16,** 5123 (1988).
[23] S. T. Kim and A. Sancar, *Biochemistry* **30,** 8623 (1991).
[24] C. G. Hatchard and C. A. Parker, *Proc. R. Soc. London, Ser A* **235,** 518 (1956).

1. The gel retardation assay is conducted as follows.[25] A 50-μl reaction mixture contained 50 m*M* Tris–HCl, pH 7.4, 100 m*M* NaCl, 10 m*M* dithiothreitol, 1 m*M* EDTA, 4% glycerol, 0.5–5 n*M* substrate, and the indicated amounts of enzyme. The mixture was incubated in the dark for 30 min and then loaded on a 5% polyacrylamide gel and subjected to electrophoresis at 75 V for 90 min. Following electrophoresis, the free and enzyme-bound DNA are located by autoradiography, excised from the gel, and quantified by Cerenkov counting.

2. In the coupled enzyme assay for repair, the 49-mer duplex is mixed with photolyase in the reaction buffer and is exposed to photoreactivating light from an appropriate source.[26,27] Following irradiation, the DNA is extracted with phenol/chloroform and precipitated with ethanol. The DNA is then dissolved in the appropriate buffer, digested with *Mse*I, and separated on a 12% polyacrylamide sequencing gel. The 49-mer band corresponding to unrepaired DNA and the 21-mer band resulting from *Mse*I incision of repaired DNA were excised and quantified by Cerenkov counting.

3. The spectrophotometric assay measures the absorbance change at 265 nm as a result of splitting of the cyclobutane ring and restoration of the 5–6 double bond of thymines.[23] The reaction mixture in 350 μl reaction buffer containing 0.02 m*M* photolyase and 5 m*M* substrate was placed in an anaerobic cuvette, and the enzyme was photoreduced by irradiation with filtered camera flashes ($\lambda > 500$ nm) and was then exposed to photoreactivating light of 366 nm. Repair was calculated for the absorbance increase at 265 nm taking $\Delta\varepsilon[\varepsilon(T - T) - \varepsilon(T<>T)] = 19{,}000\ M^{-1}\ cm^{-1}$.

Preparation of Deuterated Tryptophans

The ring-deuterated tryptophans (indole-d_5; indole-2,5-d_2) were prepared from tryptophan by refluxing in 0.2 *M* DCl/D_2O for 3 to 20 days under an argon-saturated atmosphere.[16] The DCl/D_2O solution was changed several times over the course of the experiment. The progress of the deuterium incorporation was monitored by 400-MHz ^1H-NMR. For tryptophan-d_2, deuterium was incorporated up to 98 and 96% into C-2 and C-5 positions, respectively. For tryptophan-d_5, deuterium incorporation was up to 99% in each position of the indole ring. Assignment of the deuterated positions was made on the basis of ^1H- and ^{13}C-NMR, and the deuterium contents were estimated by the integration of the signals in the aromatic region in the 400-MHz ^1H-NMR.

[25] I. Husain and A. Sancar, *Nucleic Acids Res.* **15,** 1109 (1987).

[26] S. T. Kim, K. Malhotra, C. A. Smith, J.-S. Taylor, and A. Sancar, *Biochemistry* **32,** 7065 (1993).

[27] G. Payne and A. Sancar, *Biochemistry* **29,** 7715 (1990).

Instrumentation

DNA Repair Quantum Yield Measurements

Quantitative photoreactivation was conducted on an integrated monochromator-actinometer (Quantacount-Photon Technology International, Inc., Princeton, NJ).[27] The instrument houses a 75-W xenon lamp focused onto a f/4 grating monochrometer (1200 L/mm; 300-nm blaze) equipped with an electronically operated shutter. Monochromatic (4-nm band pass/mm slit width) light exits the monochromator and enters the sample compartment where it is split into the sample beam (90%) and the reference beam (10%) by the beam splitter. Behind the reference and sample compartment are detectors, each of which consists of a solution of totally absorbing rhodamine B and a large-area silicon photocell for monitoring the rhodamine B fluorescence. An electrical signal from each silicon cell is delivered to the Quantacount electronics where it is amplified and converted to a voltage signal. The electronic component continuously substracts the sample detector signal from the reference detector signal and integrates over time. The instrument is set so that the reference signal minus the sample is equal to zero before sample irradiation. A completely absorbing screen is placed between the sample cuvette and the sample detector so that light is prevented from reaching the sample detector. As a result, the quanta counter accumulates counts proportional to the intensity of the sample beam, integrated over the time of irradiation. The counter was calibrated with ferrioxalate actinometry so that the counts can be converted to the total fluence to which the sample is exposed.

Flash Photolysis

The nanosecond laser flash photolysis was based on a JK Lasers System 2000 Nd^{3+}YAG laser emitting pulses at 532 nm with energies in the 10–200 mJ and a pulse duration of 20 nsec.[28] For measurements on a millisecond time scale, a DC-powered quartz halogen lamp of constant intensity was used as the analyzing source. Picosecond transient absorption spectra were measured with a microcomputer-controlled mode-locked Na^{3+}YAG laser emitting pulses at 355 nm of 1-mJ energy and 22-psec duration.[9] A pulsed Xe lamp, which produced light of constant intensity for 400 μsec after the laser pulse, was used as the analyzing light source for nanosecond laser photolysis. Measurements of transient changes in the absorbance at wavelengths in the range of 300–700 nm were recorded on a Phillips PM3311 digital oscilloscope and then transferred to an L51/2 computer and stored

[28] P. F. Heelis and A. Sancar, *Biochemistry* **25,** 8163 (1986).

on disc for further analysis. Pulse-to-pulse variations in the laser intensity were corrected for by monitoring the integrated laser intensity on each pulse. Samples were contained in quartz flow cells of 1-cm optical path length. Appropriate filters were placed in the analyzing light path to reduce photolysis. Transient quantum yields were determined by the comparative technique. The ruthenium tris(bipyridyl) triplet-state absorption was employed as a standard with $\varepsilon_{360} = 10^4\ M^{-1}\ cm^{-1}$ and $\phi = 1.0$.[28a] Conventional flash photolysis experiments were carried out with a Starblitz 2400 TS photographic flash unit on manual mode and are termed camera flash experiments in order to distinguish them from the laser flash photolysis experiments.

EPR Measurements

EPR spectra were taken at 298 K in a quartz flat cell with a Bruker ER-200D spectrometer operating at a microwave power of 6.3 mW, a microwave frequency of 9.2226 GHz, and a modulation amplitude of 2.8G (IG = 10^4 T). Magnetic field strength and microwave frequency were monitored continuously using a Bruker ER-035M gaussmeter and a Hewlett-Packard 5255A frequency, respectively. Transient EPR spectra were obtained by using gated integration techniques with a delay of 4 μsec.[29] The ac coupling (v > 10 Hz) was used to discriminate against any stable FADH° signals present. The time-resolved spectra were accumulated and averaged to improve the signal/noise ratio.

Results and Discussion

Role of Tryptophan in Specific Binding of Photolyase to T<>T

Intercalation of aromatic residues into DNA is an important driving force for many DNA-binding proteins such as M13 gene 5, *E. coli* SSB, and T4 gene 32 proteins.[30] Hélène and co-workers have shown that the Lys–Trp–Lys tripeptide binds to DNA by ionic interaction of Lys residues with the backbone followed by tighter binding which involves intercalation of the tryptophan.[31] Site-specific mutagenesis studies on T4 endonuclease V (an enzyme which like photolyase binds to Pyr<>Pyr) suggest that the Trp–Tyr–Lys–Tyr–Lys pentapeptide is important for substrate recognition

[28a] M. Rougee, T. Ebbesen, F. Ghetti, and R. V. Bensasson, *J. Phys. Chem.* **86,** 4404 (1982).
[29] C. W. Hoganson and G. T. Babcock, *Biochemistry* **27,** 5848 (1988).
[30] J. W. Chase and K. R. Williams, *Annu. Rev. Biochem.* **55,** 103 (1986).
[31] T. Behmoaras, J.-J. Toulme, and C. Hélène, *Proc. Natl. Acad. Sci. U.S.A.* **78,** 926 (1981).

TABLE I
BINDING, SPECTROSCOPIC, AND CATALYTIC PROPERTIES OF MUTANT PHOTOLYASES

			Relative fluorescence at[b]		Relative repair ϕ at[c]	
Photolyase	$K_A(M^{-1})$	$K_N(M^{-1})$[a]	480 nm	505 nm	366 nm	384 nm
WT	2.2×10^9	2.8×10^5	1.0	0.78–1.0	1.0	1.0
W277R	8.2×10^6	1.5×10^6	1.03	0.97	0.9	0.9
W277E	1.3×10^6	2.4×10^5	0.98	1.24	0.9	1.0
W277F	1.7×10^9	ND[d]	1.04	ND	0.9	ND
W277H	1.6×10^9	ND	0.76	ND	0.8	ND
W277Q	2.8×10^8	ND	1.01	0.72	ND	ND

[a] Expressed per base pair. K_A is the specific binding constant for DNA with a single cyclobutane thymine dimer. K_N is the nonspecific binding constant to DNA without the photodimer. In the case of nonspecific binding, every base pair is considered to constitute a binding site.

[b] The relative fluorescence is expressed relative to wild type. Fluorescence measurements were conducted twice with the wild-type enzyme on two different preparations. The two values are given. The fluorescence of the mutants is expressed relative to the higher of the two values obtained for the wild-type enzyme.

[c] The quantum yields are expressed relative to wild type.

[d] Not determined.

by this enzyme.[32] Similarly, site-specific mutagenesis studies with the *E. coli* photolyase revealed that Trp-277[33] (and its homolog in yeast photolyase, Trp-387[34]) was an essential residue for specific binding.

The *E. coli* W277R mutant has about 300-fold lower affinity to T<>T compared to the wild-type enzyme and a 5-fold higher affinity for undamaged DNA, presumably due to replacement of a residue which confers specificity with a residue that can interact with the backbone by making salt bridges.[33] The mutant enzyme retains native conformation as evidenced by spectroscopic (absorption, fluorescence, CD) properties nearly identical to those of the wild-type enzyme.

To understand the mode of interaction between W277 and DNA (H-bonding, van der Waals interactions, intercalation), the mutant enzymes W277H, W277Q, W277E, and W277F were generated by site-specific mutagenesis and their binding, catalysis, and photochemical properties were studied (Table I).[33] The photochemical and photophysical properties of the mutants are identical to the wild type. However, the substrate-binding

[32] M. L. Dodson and R. S. Lloyd, *Mutat. Res.* **218,** 49 (1989).

[33] Y. F. Yi and A. Sancar, *Biochemistry* **29,** 5698 (1990).

[34] M. E. Baer and G. B. Sancar, *J. Biol. Chem.* **268,** 16717 (1993).

properties change in unique ways. Introduction of a positive change (W277R) increased the nonspecific binding (binding to DNA without a T<>T) while reducing the specific affinity (binding to DNA with a T<>T) of the enzyme. In contrast, a negative charge at this position (W277E) reduced both specific and nonspecific affinity of photolyase to DNA. The W277Q mutation decreased the specific affinity without affecting the nonspecific binding. Most significantly, the binding constants of W277H and W277F mutants were identical with that of the wild-type enzyme, within experimental error of these measurements. As phenylalanine does not have the H-bonding properties of tryptophan, these observations are most consistent with the proposal that W277 contributes to substrate binding by stacking (intercalation) or van der Waals interactions.

Photosensitized Repair of Pyrimidine Dimers by a Tryptophan in DNA Photolyase

It has been shown that the tripeptide Lys–Trp–Lys catalyzes photodimer reversal[31] and that antibodies raised against pyrimidine dimers and which effectively reverse pyrimidine dimers appear to contain a photosensitizing tryptophan at the binding site.[35] It is believed that these tryptophans achieve dimer cycloreversion by photoinitiated electron transfer from the photoexcited tryptophan to the dimer to produce the dimer radical anion, the intermediate that actually splits into monomers. Photolyases, as a class, are rich in aromatic amino acids, especially tryptophan. Therefore, the identification of a Trp residue at the substrate-binding site of photolyase raised the possibility that this residue may directly split Pyr<>Pyr upon excitation with the appropriate wavelength as has been shown in model systems.

Near and Far-UV Photoreactivation. Photoreactivation was originally defined as the reversal of the effect of UV by exposure to longer wavelength radiation. However, the absorption spectrum of the cofactors extends into the short wavelength region of the spectrum. In fact, it has been shown that the two cofactors in *E. coli* photolyase function as photosensitizers in the far-UV as well as in the near-UV-visible region.[27] However, the contribution to photorepair in the far UV (200–300 nm) by the aromatic residues of the apoenzyme was not clear due to uncertainty introduced by the high extinction coefficients of the noncovalent chromophores. Since the cofactors interfere with the quantitative analysis of photoreactivation by the apoenzyme in the far-UV region, we have used the W306F mutant enzyme (which cannot be photoreduced; see below) to evaluate the catalytic

[35] A. G. Cochran, R. Sugasawara, and P. G. Schultz, *J. Am. Chem. Soc.* **110,** 7888 (1988).

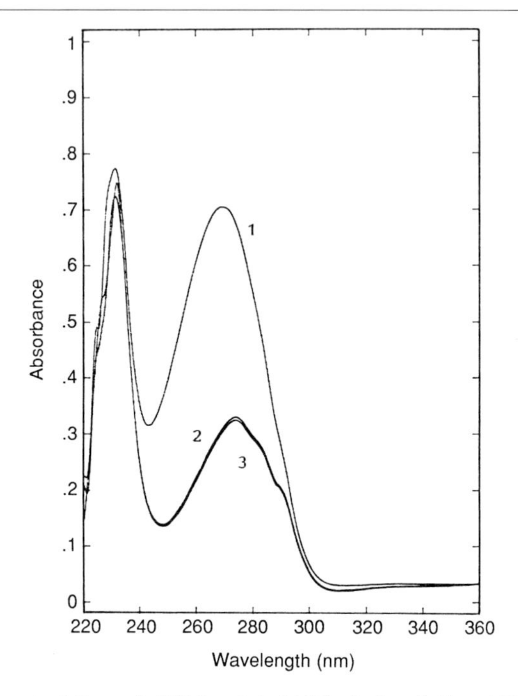

FIG. 2. Photorepair of dimers in UV-irradiated $(dT)_{15}$ by irradiation at 280 nm with enzyme containing nonphotoreducible FADH°. Curve 2: absorption spectrum of 3 μM photolyase (W306F, 3 μM Enz-FADH°) plus UV-irradiated (dT)15 (100 μM dimer) phosphate buffer (50 mM NaCl, 10 mM potassium phosphate, pH 7.0) before irradiation. Curve 3: after a 20-min irradiation at 366 nm (absorption spectrum was superimposed onto curve 1). Curve 1: after an additional 20-min irradiation at 280 nm. From Kim *et al.*[36]

role of tryptophan(s) in photolyase. Although apoenzyme devoid of flavin does not bind to Pyr<>Pyr, W306F binds to Pyr<>Pyr but has catalytically inactive FADH° (see below). The only catalytically active form of the flavin in photolyase is the fully reduced form (Enz-$FADH^-$). Therefore, W306F photolyase is suitable for investigating repair by the photolyase apoenzyme without the interference of the very efficient repair carried out by the two cofactors.

Photolyase with the W306F mutation failed to repair T<>T at 366 nm which is the absorption maximum of the enzyme-bound flavin in the near UV; however, irradiation at 280 nm resulted in complete repair of T<>T in oligo$(dT)_{15}$ substrate as evidenced by the absorbance increase at 265 nm due to regeneration of the 5,6-double bonds of the pyrimidines (Fig. 2).[36]

[36] S. T. Kim, Y. F. Li, and A. Sancar, *Proc. Natl. Acad. Sci. U.S.A.* **89,** 900 (1992).

The repair observed at 280 nm was not due to wavelength dependence of photoreduction of FADH° to FADH since preillumination at 280 nm to affect partial photorepair followed by extensive irradiation with 366 nm did not lead to any further increase at the 265-nm absorbance. Therefore, these results are consistent with direct repair of T<>T by a photolyase amino acid(s).

Identification of the Photosensitizing Tryptophan. To identify the amino acid(s) responsible for direct photolysis, we determined the action spectrum of W306F photolyase (Enz-FADH°) in the 250- to 550-nm range using apoenzyme as a control (Fig. 3).[36] The action spectrum of photolyase W306F was superimposable on the absorption spectrum of tryptophan with a photolytic cross section $\varepsilon\phi = 3250\ M^{-1}\ \text{cm}^{-1}$. This photolytic cross section is consistent with either a single Trp residue splitting T<>T with a quantum yield of 0.56 or splitting by several tryptophans with a lower and perhaps variable quantum yield. Fluorescence quenching by a substrate provided the answer to this question. Photolyase was mixed with a substrate in the absence of a reducing agent to prevent photoreduction of and repair by flavin and excited with a 295-nm light; emission was monitored at 334 nm. At saturating concentrations of substrate, 7% of total fluorescence was quenched in W306F and W306Y mutants which are incapable of flavin-mediated repair under these conditions (see below). Since these mutants contain 14 Trp residues, data were consistent with total quenching of fluorescence of a single Trp residue (1/14) by the substrate. When the fluores-

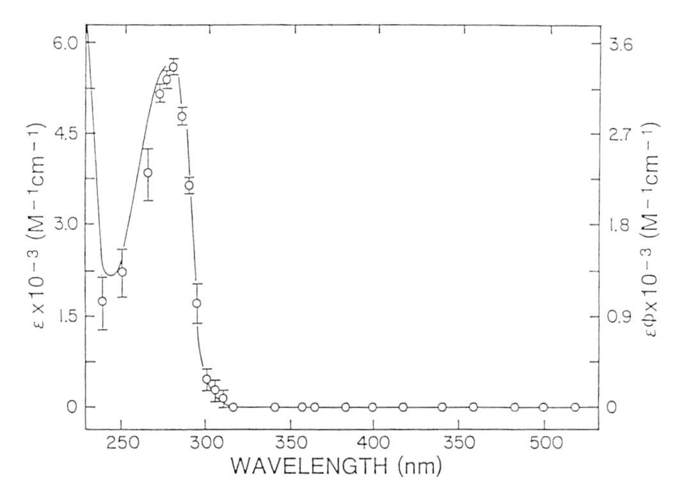

FIG. 3. Absolute action spectrum (data points) of photolyase in the form of Enz-FADH° (W306F) superimposed on the absorption spectrum of Trp (absolute scale). The purified mutant photolyase (W306F) contains nonphotoreducible FADH°. From Kim *et al.*[36]

cence quenching was conducted with photolyases mutated at each of the 14 Trp residues, the substrate quenched fluorescence by the same magnitude in all but the W277F mutant. This mutant binds the substrate with the same affinity as the wild-type enzyme and therefore a lack of fluorescence quenching with this mutant indicates that W277 is the solo Trp residue in photolyase which repairs T<>T by electron transfer.[36]

Flavin Radical Photoreduction by a Tryptophanyl Residue in DNA Photolyase

Photolyase contains flavin in the two-electron reduced form *in vivo*.[37–39] However, during purification the flavin is oxidized to the semiquinone form which is catalytically inert.[37] Exposure of this form (Enz-FADH°-MTHF) to light results in activation of the enzyme by photoreduction of the flavin even in the absence of external electron donors.[28,37] Flash photolysis and EPR spectroscopy were employed to identify the electron donor as a specific Trp residue in the apoenzyme.

Camera Flash Photolysis. The absorption spectrum of the purified enzyme (Enz-FADH°-MTHF) from *E. coli* is shown in Fig. 4.[38] The absorption at longer wavelengths ($\lambda > 400$ nm) is due to the neutral flavosemiquinone, while the absorption peak at 380 nm represents the combined absorption of the radical and the MTHF. Camera flash experiments under aerobic conditions produced a progressive loss of the long-wavelength absorption due to the radical with increasing numbers of flashes. However, after a 30-min storage in the dark at 4°, a near complete return to the original spectrum is observed (not shown). These observations are consistent with a one-electron photoreduction of the Enz-FADH° to the Enz-$FADH^-$. The nature of the electron donor involved in the photoreduction of the radical was further investigated by flash photolysis of the Enz-FADH° under aerobic conditions.

Nanosecond Laser Flash Photolysis. Laser flash photolysis of photolyase with 532 nm generates an initial transient at $\lambda_{max} = 420$ nm ($\tau = 1$ μsec) which has been attributed to that of the excited quartet state (see below) of the flavin radical.[38,40] This excited state reacts within several microseconds to yield a relatively stable-reduced flavin. Thus difference spectra taken after $t >$ msec are mirror images of the FADH° spectrum. In contrast, spectra taken within several microsecond intervals are likely to be composite spectra of (FADH°-$FADH^-$) difference spectrum plus the difference

[37] G. Payne, P. F. Heelis, and A. Sancar, *Biochemistry* **26,** 7121 (1987).
[38] P. F. Heelis, G. Payne, and A. Sancar, *Biochemistry* **26,** 4634 (1987).
[39] Y. F. Li, P. F. Heelis, and A. Sancar, *Biochemistry* **30,** 6322 (1991).
[40] P. F. Heelis, T. Okamura, and A. Sancar, *Biochemistry* **29,** 5694 (1990).

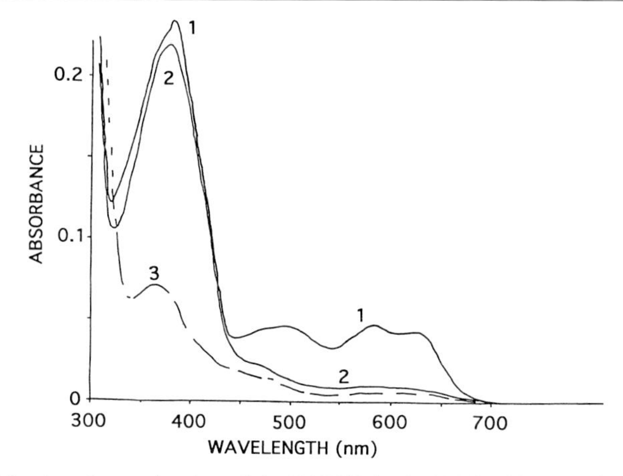

FIG. 4. Selective photoreduction of the FADH° in the E-MTHF-FADH° form of *E. coli* photolyase by camera flash photolysis. The samples were in 50 m*M* Tris–HCl, pH 7.4, 50 m*M* NaCl, 20% glycerol under aerobic conditions. Curve 1: before camera flash photolysis. Curve 2: after 40 flashes with filtered light ($\lambda > 520$ nm). The flavin radical is converted to a two-electron reduced form (photoreduction). Curve 3: after 100 flashes with white light. Short wavelength irradiation (350–400 nm) causes photoreduction of flavin and "photodecomposition" of the MTHF to species which do not absorb in the 300- to 500-nm range. From Heelis *et al.*[38]

spectrum of the electron donor (DH-DH$^{\circ+}$ or DH-D°). Indeed, although a difference spectrum taken 1 min after the flash looked like a mirror image of FADH°, an absorption spectrum taken 4 μsec after the flash was distinctively different, indicating an additional component (Fig. 5).[40] The spectrum of the additional component DH$^{\circ+}$ (or D°) was simply obtained by carrying out $\Delta\Delta A = \Delta A$ (4 μsec)-ΔA (1 min). Then the identity of DH$^{\circ+}$ or D° could be made by comparisons of the $\Delta\Delta A$ to the known spectra of the oxidized forms of amino acids. The possible contributors to be considered were the oxidized radicals of tryptophan, tyrosine, and histidine since other amino acid radicals do not absorb significantly at $\lambda > 350$ nm. Cysteine was also eliminated as a potential candidate because no pH dependence of photoreduction was observed in the pH range of 6–10. The comparisons of the spectra of Trp, Tyr, or His radical spectra revealed that the best match was with the Trp radical. Since photoreduction could proceed by electron or H-atom transfer, we compared the spectra ($\Delta\Delta A$) with the absorption spectra of both Trp$^{\bullet}$ and Trp$^{\bullet+}$. Data in Fig. 5 show that the best match was obtained with Trp$^{\bullet}$, suggesting H-atom transfer. However, the resolution of difference spectra was not sufficiently high to eliminate Trp$^{\bullet+}$ as the intermediate and electron transfer as the immediate reaction in

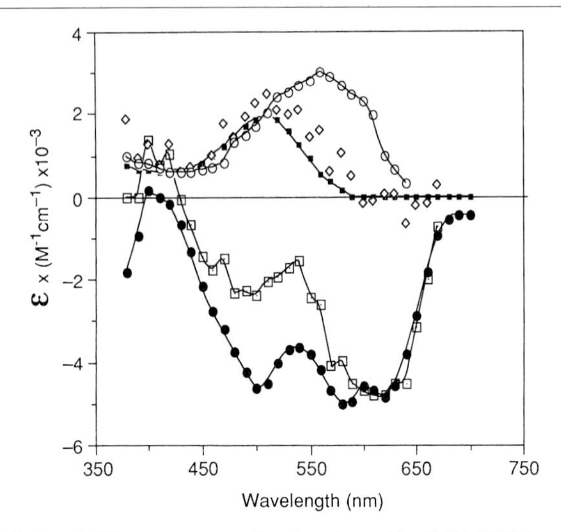

FIG. 5. Flash-induced difference spectra for photolyase (the E-FADH° form) on an absolute extinction coefficient scale. (Closed circles) Determined 1 min after a single camera flash in the presence of 10 m*M* dithiothreitol. (Open squares) Laser-induced transient spectrum determined 4 μsec after excitation, normalized (closed circles) at 620 nm. (Closed squares) Absorption spectrum of tryptophan neutral radical (Trp°). (Open circles) Absorption spectrum of tryptophan cation radical ($TrpH^{\circ +}$). (Diamonds) Difference spectrum ($\Delta\Delta A$) of ΔA (4 μsec; open squares) and ΔA (1 min; closed circles).

photoreduction. We have resolved this ambiguity of photoreduction mechanism (electron vs H-atom transfer) by the combination of site-specific mutagenesis, isotopic labeling, and time-resolved EPR studies.

Site-Specific Mutagenesis. The particular Trp responsible for photoreduction was identified by site-directed mutagenesis. Each of the 15 Trp residues were replaced by Phe in photolyase, individually, and 8 mutant proteins that were overproduced were purified and tested (Table II).[39] All but W306F were photoreducible by continuous illumination or camera flash photolysis. Thus Trp 306 was identified as the immediate electron (H-atom) donor in photoreduction.[39] When nanosecond laser flash photolysis was conducted with wild-type and mutant enzymes at 0.5 μsec, the characteristic 420-nm band was observed with both forms. With the wild type, the quartet spectrum was followed by depletion of absorption in the 450- to 650-nm region as a result of FADH° $\rightarrow$ $FADH_2$ conversion. In contrast, in the W306F mutant, the transient spectrum shows that following the initial depletion of FADH absorption, the transient decayed back to FADH° with a first-order rate constant of k $\sim 2 \times 10^6$ sec^{-1}. All these data combined are consistent with the photoreduction of the flavin-excited state quartet

TABLE II
PROPERTIES OF W → F MUTANTS OF *E. coli* PHOTOLYASE

Amino acid position photolyase	Overproduction	*In vivo* relative $\varepsilon\phi$[a]	*In vitro* relative photoreduction rate[b]
PL-WT	+	1.0	1 (0.7–1.4)
W6F	−	0	ND[c]
W41F	+	0.98	1.1
W157F	+	0.98	0.7
W271F	−	1.10	ND
W277F	+	0.96	1.0
W300F	−	1.00	ND
W306F	+	1.00	0.0
W316F	−	1.05	ND
W338F	−	1.03	ND
W359F	−	0.89	ND
W382F	−	1.14	ND
W384F	+	1.10	1.3
W418F	+	1.00	1.1
W434F	+	1.00	1.4
W436F	+	1.00	0.9

[a] The $\varepsilon\phi$ measurements at 384 nm were conducted in parallel with a strain containing the wild-type enzyme.

[b] Photoreduction was conducted in the presence of 25 m*M* dithiothreitol. The two values given for the wild type are the rates obtained in two separate experiments. The rates of the mutants are relative to the average value for the wild type.

[c] Not determined.

by Trp-306, which is in turn reduced by reducing agents such as EDTA and dithiothreitol present in the photoreduction buffer. In the absence of reductants in the buffer, the Trp-306 radical is repaired by back electron transfer from the reduced flavin. Photoreduction does occur in the absence of MTHF and thus the second chromophore is not directly involved in photoreduction. However, photoreduction per incident photon is more efficient in the presence of MTHF because a higher fraction of incident photons are absorbed and utilized to generate the excited state of the flavin radical by energy transfer from MTHF to flavin.

Picosecond Laser Flash Photolysis. In order to understand the dynamics of the excited state properties of the flavin radicals, we have performed picosecond laser photolysis on *E. coli* photolyase.[40,41] The transient absorp-

[41] T. Okamura, A. Sancar, P. F. Heelis, Y. Hirata, and N. Mataga, *J. Am. Chem. Soc.* **111,** 5967 (1989).

tion spectra of DNA photolyase containing Enz-FADH° are shown in Fig. 6.[41] In these time-dependent spectra, absorption bands that decay with a rate constant of ca. 10^{10} sec^{-1} are clearly identified in the wavelength regions less than 500 nm and more than 600 nm. We assign these bands to the lowest excited doublet state (D_1) of the Enz-FADH° which is formed by the very rapid internal conversion of the doublet manifold ($D_J \rightarrow D_1$) following laser excitation to the higher doublet state, D_J. This excited state (D_1) does not decay to the ground state (D_0) but yields another long-lived excited state of Enz-FADH°. Nanosecond flash photolysis studies have shown that the excited state of FADH° must have an intrinsic lifetime of about 1 μsec. Therefore, it was proposed that the long-lived excited state of Enz-FADH° is the lowest excited quartet state (Q_1) generated by a doublet–quartet intersystem crossing of the flavin radical. In the case of the flavin radical in DNA photolyase, excitation at the longest wavelength band of 500–700 nm almost certainly involves $\pi\pi^*$ transitions. The energy of the (Q_1) ($n\pi^*$) state is expected to be lower than that of the D_1 ($\pi\pi^*$) state because it corresponds to a transition of an electron from a nonbonding

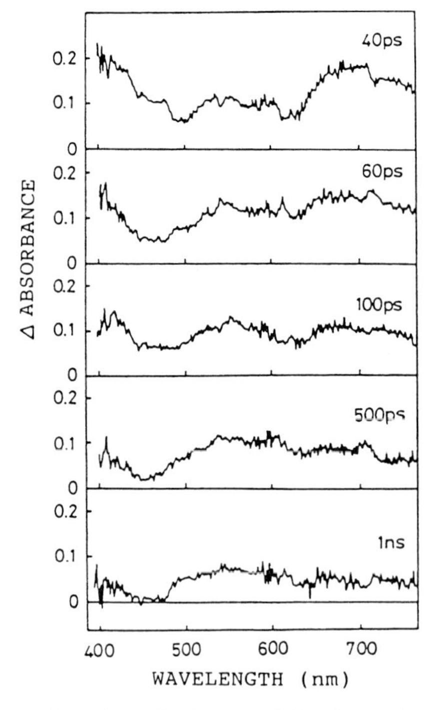

FIG. 6. Picosecond transient absorption spectra of *E. coli* photolyase containing only Enz-FADH°. Delay times from the exciting pulse are indicated in the figure.

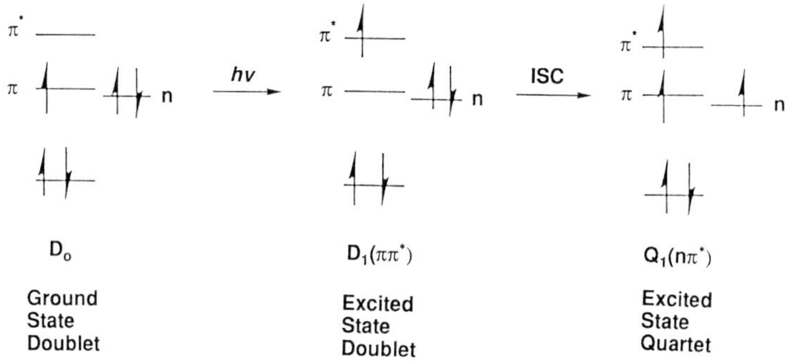

FIG. 7. Possible electronic configuration of $\pi\pi^*$ doublet and $n\pi^*$ quartet states.

orbital to the lower empty orbital. Thus, the presence of this low-lying quartet state of $n\pi^*$ character provides the route for intersystem crossing (Fig. 7). The strong spin–orbit coupling (El-Sayed's rule) of $\pi\pi^*$ and $n\pi^*$ would enhance the possibility of intersystem crossing competing with internal conversion. The much longer quartet lifetime of more than 1 μsec is consistent with the spin-forbidden nature of the intersystem crossing from Q_1 to D_0. Finally, the Q_1 reacts with an amino acid residue in photolyase apoenzyme via an electron transfer.

To summarize, our laser photolysis studies of the Enz-FADH° form of the DNA photolyase have identified the primary excited state present 40 psec after excitation as the first excited doublet state of the flavin radical (Enz-^{2}FADH°*). Intersystem crossing to the lowest excited quartet state (Enz-^{4}FADH°*) occurs within 100 psec. The intermediate Enz-^{4}FADH°* is then quenched by abstracting an electron from a nearby amino acid residue. This primary reduction process is reversed over a period of tens of milliseconds or in the presence of exogeneous electron donors, and the back reaction is prevented by the reduction of the oxidized amino acid leading to the fully reduced Enz-$FADH^-$ form of the enzyme.

Time-Resolved EPR. The results of flash photolysis and site-specific mutagenesis studies of DNA photolyase suggested that the intrinsic reductant was a tryptophanyl residue. However, these experiments did not have the resolution to differentiate between electron and proton transfer. Therefore, isotopic labeling, flash photolysis, and time-resolved EPR were conducted in order to generate and identify the radical intermediate(s) involved in the reduction of FADH°[42,43] and thus complement the flash photolysis (absorption) studies.

[42] C. Essenmacher, S. T. Kim, M. Atamian, G. T. Babcock, and A. Sancar, *J. Am. Chem. Soc.* **115,** 1602 (1993).

[43] S. T. Kim, A. Sancar, C. Essenmacher, and G. T. Babcock, *Proc. Natl. Acad. Sci. U.S.A.* **90,** 8023 (1993).

Figure 8A[42] shows a steady-state EPR spectrum of the Enz-FADH° form of photolyase isolated from a tryptophan auxotroph grown in medium supplemented with tryptophan. The spectrum of Enz-FADH° from bacteria grown on specifically labeled tryptophan showed line widths and Δ values identical to those in Fig. 8A. Flash photolysis of Enz-FADH° produced an intense transient whose amplitude and sense of signal changed with applied magnetic field. Figure 8B[42] shows an example of such a kinetic trace at the indicated field value relative to the Enz-FADH° spectrum in Fig. 8A. The directions of the transient spikes observed at different field positions were reproducible and were not consistent with the FADH° spectrum. Therefore, the signals must arise from the light-induced generation of another radical(s) in the $\gamma = 2$ region. Using gated integration (48-μsec integration

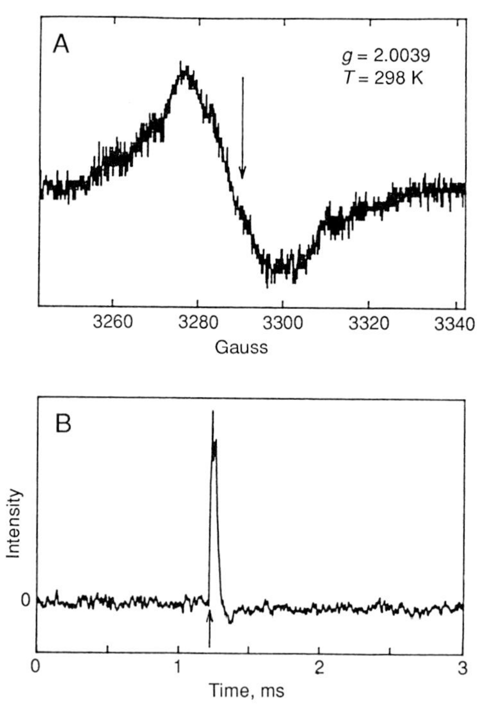

FIG. 8. EPR detection of light-induced radical in *E. coli* DNA photolyase. (A) The dark-stable EPR spectrum of E-FADH° and (B) the kinetic transient of E-FADH° are shown. Arrow in A is the field position at which the kinetic transient is taken in each scan. Arrow in B is the time at which the lamp was fired in each scan. For the kinetic trace in B, 1000 flashes were averaged. Instrument conditions: microwave frequency, 9.22226 GHz; microwave power, 6.3 mW; modulation amplitude, 2.8 G. Enzyme (0.1 m*M*) was in 50 m*M* potassium phosphate, pH 7.0, 50 m*M* NaCl, 5 m*M* $K_3Fe(CN)_6$, 1 m*M* $K_4Fe(CN)_6$, and 10% glycerol. All experiments were performed at 298 K under a gentle stream of cold nitrogen gas. From Kim *et al.*[43]

window and 4-μsec delay following flashlamp trigger) and ac coupling ($v > 10$ Hz) to discriminate against stable signals (FADH°), we obtained the spectrum of the transient in Enz-FADH° induced by the light flash (Fig. 9). The spectrum has emissive/absorptive patterns, which indicates that the radical is born in a spin polarized state, and consists of three well-resolved components with splitting of 15–16 G as indicated by the three solid upward arrows in spectrum A. Superimposed on each of the three resolved components is a partially resolved fine structure that reflects an additional coupling of = 5 G, as indicated by dashed downward arrows on the low-field, resolved transitions in spectrum A. On deuteration of the tryptophan indole ring α protons (tryptophan-d_5), the partially resolved hyper-

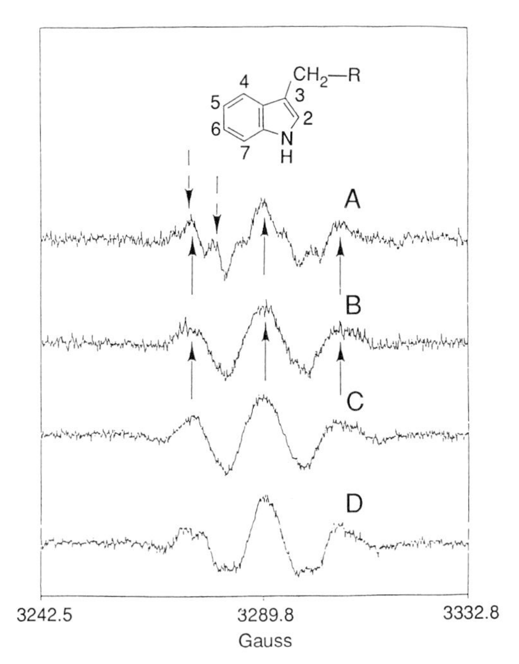

FIG. 9. Time-resolved EPR detection of the light-induced tryptophan in *E. coli* DNA photolyase. All spectra are in the first-derivative mode. Spectrum A: E-FADH° in wild type; 5 scans were averaged to obtain this spectrum. Spectrum B: FADH° labeled with [indole-$^2H_5$]tryptophan (indole-d_5); 20 scans were averaged to obtain the spectrum. Spectrum C: E-FADH° labeled with [indole-2,5-$^2H_5$]tryptophan (indole-2,5-d_2); 20 scans were averaged to obtain the spectrum. Spectrum D: E-FADH° labeled with [indole-^{15}N]tryptophan; 20 scans were averaged to obtain the spectrum. The solid upward arrows in spectra A and B show the smaller coupling that is superimposed on this large coupling. The numbering scheme for tryptophan is shown at the top. Instrument conditions: microwave frequency, 9.22226 GHz; microwave power, 6.3 mW; modulation amplitude, 2.8 G; time constant, 35 μsec. A 4-μsec delay and a 48-μsec aperture were used for the integration window. The enzyme concentration was 0.1 m*M*. From Kim *et al.*[43]

fine structure collapses, but the three major components are still observable, as indicated by the solid arrow in spectrum of Fig. 9.[43] From the data we concluded that the radical indeed arose from a tryptophan side chain.

To identify the particular residue in *E. coli* photolyase that is responsible for photoreduction of the dark-stable flavin radical, we obtained transient EPR spectra of mutant photolyases (Fig. 10).[43] All mutant proteins tested gave absorption and steady-state EPR spectra identical to the wild type but behaved differently with regard to the transient EPR spectrum. No transient EPR signal was detected with W306F and W306Y (Fig. 10, spectra B and C). In contrast, W157F (Fig. 10, spectrum D) and W418F (Fig. 10, spectrum E) had no effect on the transient spectrum. Since chemical reduction of FADH° in the W306F and W306Y mutants restores the cata-

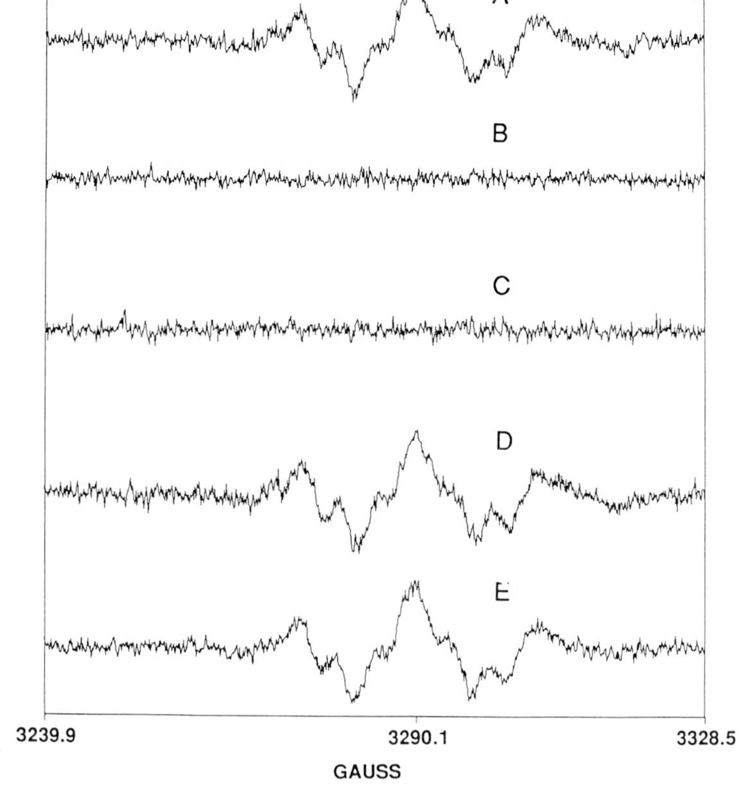

FIG. 10. Effect of Trp → Phe replacement on time-resolved EPR spectrum of the light-induced tryptophan radical in *E. coli* DNA photolyase (Enz-FADH°). Experimental conditions are the same as in Fig. 9 except that, for all spectra, five scans were averaged to obtain the final spectrum. From Kim *et al.*[43]

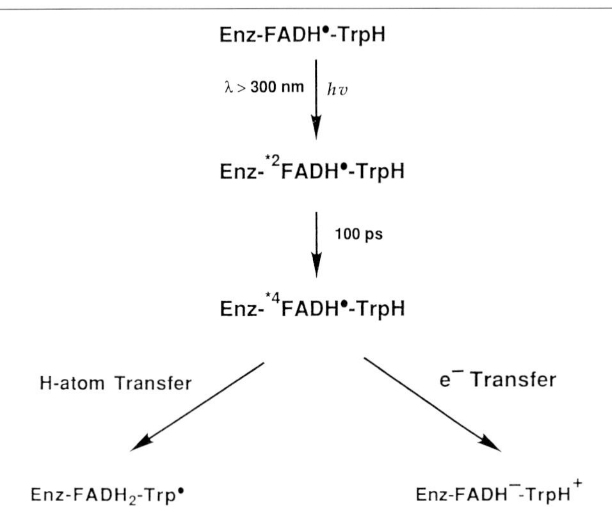

FIG. 11. Two proposed photoreduction mechanisms of Enz-FADH° in DNA photolyase: H-atom transfer vs electron transfer.[43]

lytic activity identical to that of wild-type enzyme, the absence of a transient EPR signal in these mutants is not due to conformational change but is the consequence of eliminating the photoreductant. These data and the earlier optical data conclusively show that Trp-306 is the immediate electron donor in photoreduction in FADH°.

The results presented so far, combined with those obtained from laser flash photolysis, lead to the general FADH° photoreduction scheme presented in Fig. 11.[43] Photoexcitation of FADH° generates the excited doublet state of the flavin radical, which then decays by intersystem crossing to yield the quartet state within 100 psec. The long-lived quartet state of FADH° is then quenched either by a H-atom transfer, generating the tryptophan neutral radical and $FADH_2$, or by an electron transfer to generate the tryptophan cation radical and $FADH^-$. Time-resolved EPR in combination with specifically labeled photolyase can be used to test these two proposals. Huckel-McLachlan molecular orbital and intermediate neglect of differential overlap calculations have revealed that the spin density of tryptophan radical is localized primarily at the C-2 and C-3 positions in the cation radical and at C-3 and N-1 in the neutral radical (Fig. 12).[43,44] These qualitatively different patterns of spin density distribution predicted for the tryptophan radicals by the theoretical calculations indicate that it is possible to differentiate between the two mechanisms (electron vs H-atom transfer) experimentally by examining the transient EPR spectra of the

[44] B. M. Hoffman, J. E. Roberts, C. H. Kang, and E. Margoliash, *J. Biol. Chem.* **256,** 6556 (1981).

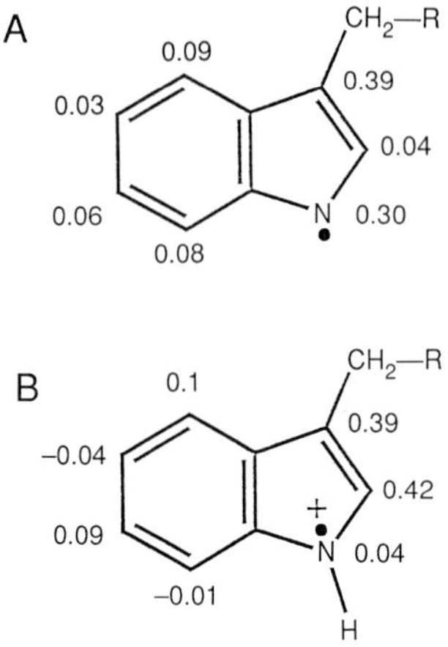

FIG. 12. The spin density distribution in tryptophan. Neutral (A) and cation (B) radicals were calculated by the Hückel–McLachlan technique.[43] Values are from Hoffman *et al.*[44]

protein samples containing indole-^{15}N, indole-2,5-d_2, and indole-d_5 tryptophans. If the unpaired electron in the transient species was primarily localized on the C-2 and C-3 positions, as is the case for the tryptophan cation radical, the spectrum of the tryptophan-2,5-d_2 sample is expected to be equivalent or similar to that of the tryptophan-d_5 sample. Spectra B and C of Fig. 8 show that the two spectra are nearly identical; this observation is consistent with a cation radical species and shows that a significant fraction of the free electron is localized at either C-2 or C-5. Significant spin density at C-5 is not expected because the MO calculations (Fig. 12) show that the spin density is small for both the cation and the neutral radical at this position. Therefore, we conclude that the coupling to the C-5 position is insignificant and that a large portion of the unpaired electron spin in the radical resides in the C-2 position, which is in agreement with the spin density distribution predicted for the cation radical, but not with that predicted for the neutral radical (Fig. 12A). To ascertain whether the ^{14}N nucleus was responsible for the major splitting, as would be expected for a neutral radical, the time-resolved EPR spectrum of ^{15}N-labeled photolyase was taken under the same experimental conditions. The spectrum obtained with the ^{15}N-labeled sample (Fig. 9, spectrum D) has the same major three-line splitting as the protonated sample (Fig. 9, spectrum A), indicating that the hyperfine splitting contribution of the ^{14}N nucleus to the spectrum is negligible. The absence of large coupling due to nitrogen further argues against the transient species being a neutral radical. This leads to the conclusion that the major three-line splitting observed with the protonated sample (Fig. 9, spectrum A) arises from hyperfine coupling to the two methylene protons at the C-3 position. The results in Fig. 9, which show that the unpaired electron is localized primarily at C-3 and C-2, together with MO calculations summarized in Fig. 12, demonstrate that the transient species associated with the major EPR signal ($\gamma = 2.0039$) is a tryptophan cation radical that results from electron abstraction by photoexcited FADH°. Therefore, of the two alternative pathways for photoactivation in photolyase shown in Fig. 11, the EPR data provide strong evidence in favor of the electron cation radical route.

To summarize, two Trp residues (W277 and W306) in *E. coli* photolyase act as electron donors during *in vitro* photoreactivation. W306 donates an electron to the excited state flavin radical to convert it into the catalytically active two-electron reduced form. W277 is involved in binding DNA and, in addition, upon direct excitation with far UV splits the cyclobutane ring by single electron transfer to the photodimer. In both instances, transient Trp cation radicals are generated which can be detected by a variety of methods.

Acknowledgments

The work reported in this paper was conducted in collaboration with Drs. G. T. Babcock (Michigan State University) and T. Okamura (Tokyo Zokei University). The work in our laboratories was supported by NIH Grants GM31082 (A. S.) and ESo5557 (S. T. K.) and by a grant from NATO (P. F. H.)

[23] Glycyl Free Radical in Pyruvate Formate-Lyase: Synthesis, Structure Characteristics, and Involvement in Catalysis

By JOACHIM KNAPPE and A. F. VOLKER WAGNER

The discovery of a tyrosyl radical in ribonucleotide reductase in the laboratories of Reichard and Ehrenberg[1] represents a hallmark in enzymology, showing the occurrence of enzymes that contain stable protein-based free radicals as coenzyme-like entities for performing catalysis of metabolic reactions. Pyruvate formate-lyase is a more recent example of free radical

[1] P. Reichard and A. Ehrenberg, *Science* **221,** 514 (1983).

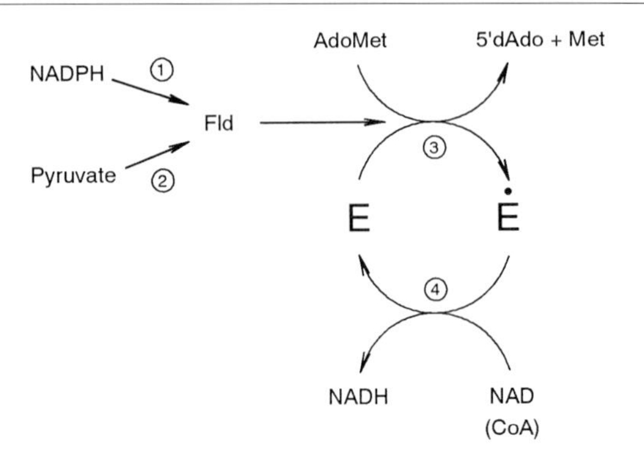

FIG. 1. Post-translational interconversion of pyruvate formate-lyase between nonradical (E) and radical forms (Ė) in *E. coli.*[4] Proteins involved are flavodoxin (Fld) as electron carrier, NADPH-Fld oxidoreductase (1), pyruvate-Fld oxidoreductase (2), pyruvate formate-lyase-activating enzyme (3), and pyruvate formate-lyase-deactivating enzyme/AdhE (4). AdoMet, *S*-adenosylmethionine; 5′-dAdo, 5′-deoxyadenosine.

enzymes. Its radical content, detected in 1984,[2] was identified in 1992 as a glycyl radical in which the spin is located on the polypeptide backbone.[3]

Pyruvate formate-lyase promotes the thiolytic cleavage of pyruvate into acetyl-CoA and formate which is the key step of the glucose fermentation route characteristically performed by *Escherichia coli* cells and related microorganisms:[4]

$$\text{Pyruvate} + \text{CoA} \rightleftharpoons \text{acetyl-CoA} + \text{formate}. \tag{1}$$

In *E. coli,* a facultative anaerobic bacterium, an elaborate regulation at both the transcriptional and the post-translational level exists to allow for the oxygen sensitivity of the glycyl radical and the exclusive appearance of pyruvate formate-lyase activity at anaerobiosis. Anaerobic growth conditions induce the protein expression about 10-fold and trigger the reaction of a specific converter enzyme, pyruvate formate-lyase-activating enzyme (activase), that generates the radical by hydrogen atom abstraction from glycine-734:

$$\text{E-H} + \text{AdoMet} + \text{FlH}_2 \rightarrow \dot{\text{E}} + 5'\text{-deoxyadenosine} + \text{Met} + \dot{\text{F}}\text{lH}. \tag{2}$$

Positive redox potentials (aerobic conditions) in turn trigger the reduction of the glycyl radical by pyruvate formate-lyase-deactivating enzyme (Fig. 1). This activity is an extra catalytic function of the AdhE protein of *E. coli*

[2] J. Knappe, F. A. Neugebauer, H. P. Blaschkowski, and M. Gänzler, *Proc. Natl. Acad. Sci. U.S.A.* **81,** 1332 (1984).

[3] A. F. V. Wagner, M. Frey, F. A. Neugebauer, W. Schäfer, and J. Knappe, *Proc. Natl. Acad. Sci. U.S.A.* **89,** 996 (1992).

[4] J. Knappe and G. Sawers, *FEMS Microbiol. Rev.* **75,** 383 (1990).

catalyzing normally the reduction of acetyl-CoA to ethanol during glucose fermentation.

The glycyl radical serves as a catalytic element in achieving the cleavage of the carbon–carbon bond in pyruvate by a radical chemical mechanism. This reaction pattern of pyruvate formate-lyase, characterized in recent years,[5] is unique among enzymatic reactions in the main routes of cellular metabolism, which are dominated by polar processes. The purpose of this chapter is to describe the principal assays for studying the catalytic cycle and an experimental procedure for obtaining homogenous preparations of the radical enzyme. This chapter also describes the spectroscopic and chemical properties of the glycyl radical and the mode of its synthesis.

Preparation of the Radical Form of Pyruvate Formate-Lyase

Purification of the radical enzyme ($\dot{E}$) directly from anaerobically grown bacteria would require elaborate laboratory equipment for strict oxygen exclusion during all workup and has not been carried out in our laboratory. Instead, we routinely obtain the radical form by subjecting the purified nonradical form (E) *in vitro* to the conversion reaction with activase. Both proteins are oxygen insensitive and can be isolated readily from overproducing *E. coli* strains by standard protein purification techniques. Photoreduced 5-deazaflavin is employed as an electron donor cosubstrate, replacing the physiological flavodoxin system.

Pyruvate Formate-Lyase (Nonradical Form)

For this protein we use *E. coli* strain 234M1 transformed with the expression vector p153E1 (containing pfl^+ and Km^r).[3] This host lacks the activase (by Cm^r insertion into the chromosomal *act* gene) which guarantees that the homogenous enzyme will be isolated free of contamination from oxygenolytic fragments of the radical form. Bacteria are grown on minimal medium with 50 m*M* glucose, 0.1 m*M* thiamin, 20 μg/ml of chloramphenicol, and 50 μg/ml of kanamycine sulfate under aeration and pH adjustment to 6.7–7.5. Fractionation of the cell extract comprises nucleic acid precipitation with Polymin G35 (BASF), gel chromatography on AcA44 (LKB), and DEAE-cellulose chromatography (Whatman) as described.[6] Maximal quality, suitable for protein crystallization, is obtained by a final ammonium sulfate gradient fractionation [Sepharose CL-4B column with 1.7 to 0.7 *M* $(NH_4)_2SO_4$ in 0.1 *M* phosphate, pH 6.5, containing 100 m*M* KCl, 5 m*M*

[5] J. Knappe, S. Elbert, M. Frey, and A. F. V. Wagner, *Biochem. Soc. Trans.* **21,** 731 (1993).

[6] H. Conradt, M. Hohmann-Berger, H.-P. Hohmann, H. P. Blaschkowski, and J. Knappe, *Arch. Biochem. Biophys.* **228,** 133 (1984).

EDTA, and 1 m*M* dithiothreitol]. The overall yield is 4 mg from 1 g of cell paste. A_{280} of a solution of 1 mg/ml is 0.98. Optimal storage conditions are 25 m*M* MOPS–KOH (pH 7.2) containing 100 m*M* KCl, 2 m*M* dithiothreitol, and 1 m*M* EDTA.

Pyruvate formate-lyase has been characterized as a homodimer composed of 85.1 kDa subunits (759 amino acid residues).[6,7] Secondary structure prediction suggests that Gly-734 is located in a β-turn segment (VSGY) and the covalent catalytic Cys-418/Cys-419 pair in a hydrophobic β-sheet element. The protein can be cleaved specifically at Arg-624 and four sites downstream with trypsin (1 : 60) applied at 0°, yielding a core (2×68 kDa) and a 8.9-kDa polypeptide (682 to 759) that are quite stable toward further degradation, indicating that Gly-734 resides in an autonomous folding domain.[7a] The limited proteolysis pattern, which is analyzed by SDS–polyacrylamide gelelectrophoresis (15% gel), is a useful means for testing the structural integrity of mutant proteins.

Pyruvate Formate-Lyase-Activating Enzyme (Activase)

For this protein we use *E. coli* K12 transformed with pKE-1 (act^+ and Amp^r)[8]; cell growth is on glucose/minimal medium (pH 7) containing 10 μg/ml of ampicillin. The cell extract (12% activase) is carried through nucleic acid precipitation with Polymin G35 and chromatography on Ultrogel AcA44.[6] The activase fraction (7 mg from 1 g of cells) is stored in 50 m*M* MOPS–KOH (pH 7.5) containing 100 m*M* KCl, 5 m*M* dithiothreitol, and 0.1 m*M* EDTA.

Activase has been characterized as a monomer of 28 kDa (245 amino acid residues).[6,7] As isolated, the purified enzyme usually shows a charge transfer absorption between 320 and 500 nm which is due to the content of Fe^{2+}. For maximal catalytic activity in the pyruvate formate-lyase conversion reaction, free Fe^{2+} must be present.

5-Deazaflavin as Electron Donor

The natural one electron donor for the activation of pyruvate formate-lyase is dihydroflavodoxin ($FlH_2/\dot{F}lH$, −455 mV), which is generated *in vivo* by two alternative oxidoreductase reactions (see Fig. 1).[9] This protein

[7] W. Rödel, W. Plaga, R. Frank, and J. Knappe, *Eur. J. Biochem.* **177,** 151 (1988).

[7a] A. Pühlhofer and J. Knappe, unpublished results (1992), see Knappe *et al.*[5]

[8] M. Frey, M. Rothe, A. F. V. Wagner, and J. Knappe, *J. Biol. Chem.* **269,** 12432 (1994).

[9] H. P. Blaschkowski, G. Neuer, M. Ludwig-Festl, and J. Knappe, *Eur. J. Biochem.* **123,** 563 (1982).

system can be replaced by various low molecular weight reductants such as Ti^{3+} or reduced methylviologen. The reagent that we use routinely in our work[6] is photoreduced 5-deazariboflavin (dF) described by Massey and Hemmerich.[10] The powerful dFH radical reductant ($d\dot{F}H/dF$, −0.65 V) is generated *in situ* by illuminating (from a common slide projector) dF-containing reaction mixtures made up with Tris, MOPS, or imidazole buffer, which serves as a photosubstrate. In the dark, $d\dot{F}H$ will recombine to the inert covalent dimer $[(dFH)_2]$, thus halting any reduction reaction. If required, the photoreduction system may be supplemented with catalytic amounts of riboflavin and catalase for the purpose of oxygen scavenging (autoxidation of dihydroriboflavin).

Enzyme Conversion Procedure[6]

The conversion reaction is performed strictly anaerobically under argon using a glass tube equipped with a ground-glass stopper with an outlet for gas exchange. The standard reaction mixture, in a final volume of 0.5 ml, contains 0.15 *M* Tris–HCl, pH 7.5, 0.1 *M* KCl, 5 m*M* dithiothreitol, 0.2 m*M* $Fe(NH_4)_2(SO_4)_2$, 50 μM 5-deazaflavin, 10 m*M* potassium oxamate,[11] 0.5 m*M* adenosylmethionine (up to 3 mg), and pyruvate formate-lyase activase (up to 0.1 mg). Mix all components except $Fe(NH_4)_2(SO_4)_2$ and deaerate the solution by evacuating several times with an oil pump and flushing with argon; then add Fe^{2+} (10 μl of 10 m*M*) under argon and close the tube. Preincubate in a water bath (30°) for 10 min, then illuminate with the slide projector (150 W/24 V halogen lamp) and run the reaction to completion; the usual time period is 10 to 30 min. Place the sample on ice.

In most cases (and all assays described in this chapter), the presence of components/coproducts of the activation reaction will not interfere with studies of the radical enzyme, so the sample may be utilized as obtained. If required, gel filtration on Sephadex G-25 or AcA44 (for separation from activase) may be performed, with the column operated under argon and with anaerobic buffer (50 m*M* MOPS or Tris, pH 7.5) containing 5 m*M* dithiothreitol and 0.1 m*M* Fe $(NH_4)_2(SO_4)_2$. The enzyme fraction is collected into argon-flushed tubes.

[10] V. Massay and P. Hemmerich, *Biochemistry* **17,** 9 (1978). 5-Deazariboflavin is not commercially available; for its synthesis, see M. Janda and P. Hemmerich, *Angew. Chem.* **88,** 475 (1976).

[11] Oxamate replaces pyruvate as an effector compound required for pyruvate formate-lyase activation.[12] When pyruvate is utilized, the reaction mixture will yield the radical enzyme form that contains acetyl on Cys-419 (Ė-acetyl intermediate).

[12] J. Knappe, H. P. Blaschkowski, P. Gröbner, and T. Schmitt, *Eur. J. Biochem.* **50,** 253 (1974).

Comments on Handling and Quantification of the Radical Form

For storage and any handling (dilutions etc.) of the radical enzyme, we routinely use dithiothreitol containing anaerobic buffer media that are supplemented with Fe^{2+} (0.1 m*M*), making use of the Fe^{2+}-catalyzed autoxidation of thiols to remove residual oxygen. The transfers of enzyme can be made with conventional automatic pipettes, but gas-tight syringes are also useful.

The enzyme is stable for several days when stored at 0°. At 30°, activity and radical content are usually fully retained for 1 to 2 hr, but become slowly lost on prolonged storage. Our limited experience indicates that the enzyme is quite stable in 40% (v/v) dimethyl sulfoxide or dimethyl formamide solutions.

The radical form as produced *in vitro* contains one glycyl radical and one active site per dimer as estimated by spin quantification and ^{14}C acetyl charging (see below).[13] (The reason for the apparent "half of the sites" property is unresolved.) Enzyme quantities in any sample can be determined most sensitively and accurately by catalytic activity measurement. The correlation is that 1 nmol (0.17 mg) of $\dot{E}$ displays 35 U, where 1 U affords 1 μmol pyruvate conversion to acetyl-CoA per min at pH 8 (30°) in the standard optical assay.[12] The specific activity value is 200 U/mg.

Studies of Hydrogen Abstraction from Gly-734[8]

Production of the radical in pyruvate formate-lyase occurs by stereospecific abstraction of the *proS* (H_{Si}) hydrogen atom from C-2 of the Gly-734 residue, which is incorporated into the methyl group of the 5′-deoxyadenosine coproduct (Fig. 2). A 5′-deoxyadenosine-5′-yl-free radical is suggested as the immediate H atom abstractor, which would be generated via $1e^-$ reduction of adenosylmethionine in the active site of activase when the converter enzyme interacts with the protein substrate. This mechanism has been concluded from studies of [2-^2H]glycine-substituted enzyme and of peptide models homologous to the Gly-734 site as follows.

Pyruvate formate-lyase was expressed in p153E1-transformed glycine auxotrophic *(glyA6) E. coli* strain WFG501, grown with glucose and [U-^2H]glycine (98%), which yielded ^2H-labeled enzyme with a ^2H enrichment of about 25% in the glycine C-2 hydrogens. When this preparation was carried through the conversion reaction, the 5′-deoxyadenosine produced was found, by ^1H-NMR and mass spectroscopy, to contain 50% ^2H enrichment in the 5′ methyl group.[8]

[13] V. Unkrig, F. A. Neugebauer, and J. Knappe, *Eur. J. Biochem.* **184,** 723 (1989).

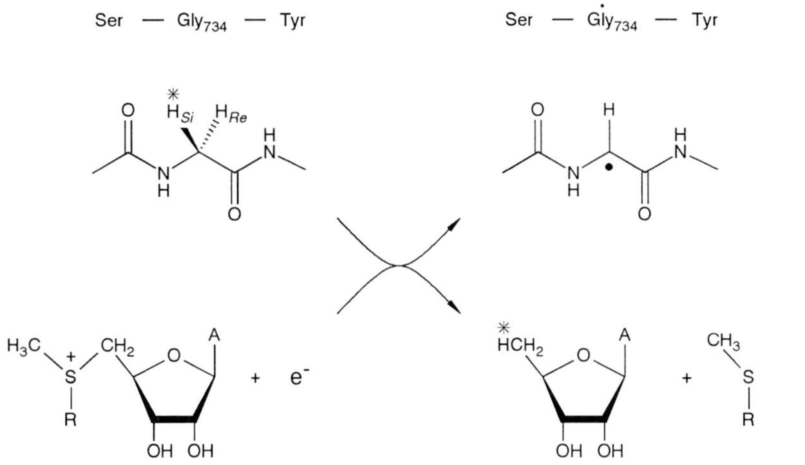

Ser — Gly734 — Tyr
Ser — Ġly734 — Tyr

FIG. 2. Synthesis of the glycyl radical in pyruvate formate-lyase (R–S–CH_3 denotes methionine).

Hydrogen abstraction from synthetic peptides was examined by applying [8-^{14}C-adenine]AdoMet to the standard conversion procedure and assaying the formation of ^{14}C-labeled 5′-deoxyadenosine with a radiochromatographic setup. The 7-mer peptide Arg–Val–Ser–Gly–Tyr–Ala–Val corresponding to the sequence stretch 731–737 of pyruvate formate-lyase and the peptide variant with a D-Ala in the 734 position were both active in this system, showing K_m values of 0.22 and 0.05 mM, respectively (K_m for pyruvate formate-lyase is 1.4 μM). However, the 7-mer peptide variant with a normal Ala substitution was totally inactive.[8]

The proS hydrogen assignment inferred from these studies is consistent with the inconvertability to a radical form of the pyruvate formate-lyase mutant 734 Gly → Ala.[5]

Characteristics of the Glycyl Radical Site

EPR Spectroscopic Properties[3,13]

EPR spectra of pyruvate formate-lyase may be obtained at room temperature on enzyme solutions contained in a flat quartz cell (0.2 mm) or at low temperatures (we routinely use −20°) on frozen samples in 0.2-cm quartz tubes. Filling is made under argon and final sealing with a plastic film. As an organic-free radical noninteracting with a metal, the EPR signal of the enzymes' glycyl radical (g = 2.0037) is readily microwave power saturated, which will blur the hyperfine structure. Saturation effects start with 0.3 and 0.05 mW at −20 and −166°, respectively. For fully activated enzymes, a quantity of 1 ± 0.04 spin per protein dimer was determined by double integration of the first derivative signal, using a Cu^{2+} standard (1 m*M*/10 m*M* EDTA) for calibration.[13]

Typical spectra, taken on 10 to 40 μ*M* enzyme samples, are presented in Fig. 3; free (Ė-SH) and acetylated (Ė-S-acetyl) enzyme states display identical signals. The [2-^{13}C]glycine-substituted enzyme was obtained through growth of *E. coli* strain WFG501 (see above) with ^{13}C-labeled glycine (99%).[3] Its anisotropic EPR multiplet spectrum shows the large hyperfine splitting of 4.9 mT assigned to parallel coupling ($A_{\parallel}$) of the unpaired electron to the central ^{13}C-2 (I = 1/2) nucleus, which is the principal evidence for the radical structure. The hyperfine splitting of 1.5 mT which dominates the EPR signal of normal (unsubstituted) enzymes (and is readily recognized also in the multiplet of [2-^{13}C]glycine-substituted en-

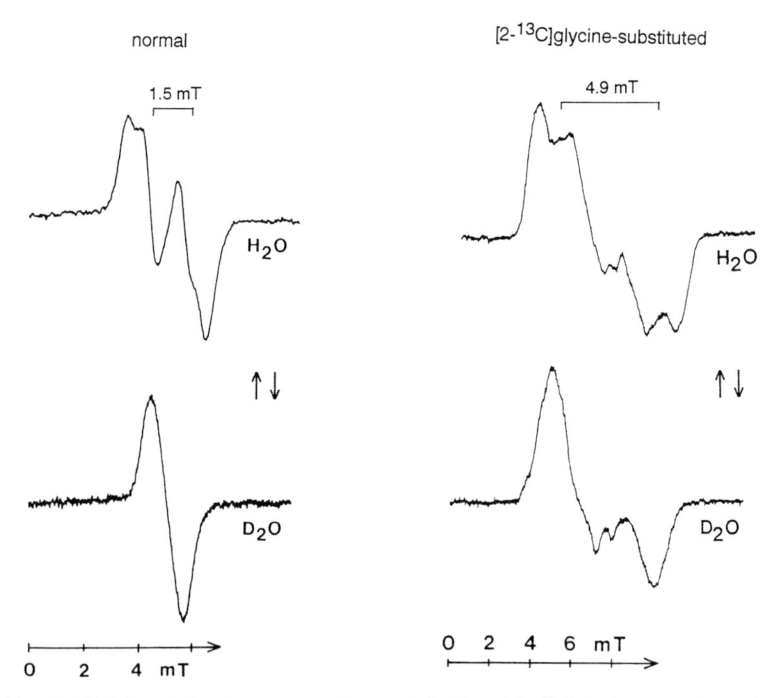

FIG. 3. EPR first derivative spectra of normal (left) and [2-^{13}C]glycine-substituted (right) pyruvate formate-lyase, each measured in $^{1}H_2O$ or $^{2}H_2O$ buffer medium as indicated. Instrument conditions: Modulation amplitude, 0.14 mT; modulation frequency, 100 kHz (Bruker ESP 300 spectrometer, X band); temperature, 253 K. Coupling constants (mT) are $A_{iso}(^{1}H_{\alpha})$, 1.5; $A_{iso}(^{1}H)$, 0.6; $A_{iso}(^{1}H)$, 0.45; $A_{\parallel}(^{13}C_z)$, 4.9. (Reproduced from Wagner *et al.*[3]).

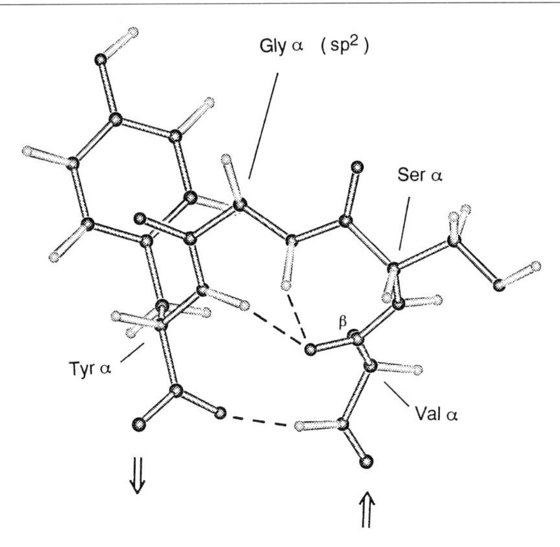

FIG. 4. Proposed β-turn structure of the Gly-734 site (Val–Ser–Gly–Tyr) in the radical form. Notable features are the extra hydrogen bond from Gly–NH to Val–CO and the placement of the $-\dot{C}H-$ group in plane with the adjacent peptide units. (Reprinted with permission from Frey *et al.*[8])

zyme) results from α coupling to the single C-2 hydrogen. Further ^1H splittings (0.6 and 0.45 mT) are yet unassigned but should be from the adjacent NH and/or protons on the adjacent amino acid residues.

A distinctive chemical property of the C-2 hydrogen of the glycyl radical is its ready exchangeability with solvent protons.[13] The 1.5-mT splitting is absent when the enzyme sample has been gel filtered into D_2O buffer medium and will reappear on reverse transfer into H_2O medium.[14] The half-time of the exchange process was determined to be about 5 min at 0° and 0.5 min at 20°. This property made it impossible to disclose the pyruvate formate-lyase radical identity by ^2H labeling, which is the common (and less expensive) approach for identifying amino acid radicals in proteins via isotope substitutions.

The spin density on glycine C-2 is estimated by the α hydrogen coupling constant to be about 0.55; the remainder is delocalized over the flanking carboxamide groups. According to a structure model of the putative β turn with the sp^2-hybridized carbon (Fig. 4), the $-\dot{C}H-$ group could lie almost

[14] The asymmetric doublet EPR signal of normal enzyme is retained in pH 10 buffer medium and room temperature, which is not indicative of a proton dissociation. A possible mechanism of hydrogen exchange is equilibration of the glycyl-NH with D_2O, followed by an intraradical hydrogen transfer ($-ND-\dot{C}H-CO- \rightarrow -NH-\dot{C}D-CO-$) as reported for glycine crystals [L. L. Gautney, Jr., and I. Miyagawa, *Radiat. Res.* **62,** 12 (1975)].

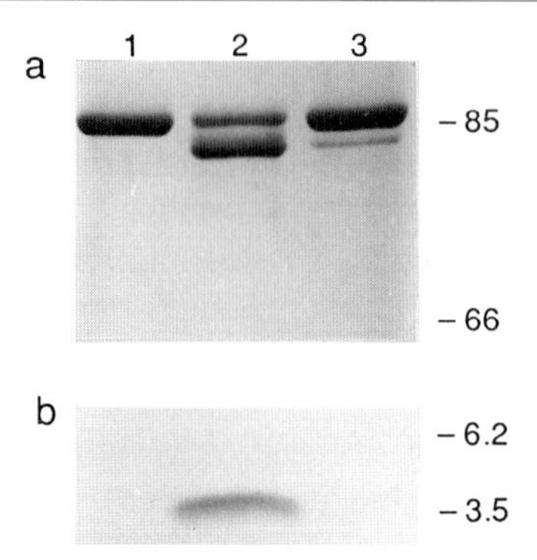

FIG. 5. Oxygen fragmentation of the glycyl radical-bearing polypeptide. Enzyme samples were resolved by SDS/polyacrylamide gel electrophoresis using an 8% gel for the high molecular weight range (a) and a 16% gel for the low molecular weight range (b); marker protein migration is shown in kilodaltons. (1) Nonradical enzyme; (2) oxygen-inactivated radical enzyme; (3) oxygen-treated radical enzyme that was previously transformed to the 1-hydroxy-1-phosphoryl-ethyl radical intermediate; see Assays of Enzyme Modifications. (Reproduced from Wagner *et al.*[3])

perfectly in plane with the adjacent peptide units, enabling optimal delocalization of the electron and thus thermal stabilization of the π radical.

UV Spectrum[13]

A further spectroscopic signal characterizing the glycyl radical in pyruvate formate-lyase is the electronic absorption band at 365 nm with an absorption coefficient of 8 mM^{-1} cm^{-1}.

Oxygen Fragmentation of the Polypeptide Chain[3]

The glycyl radical in pyruvate formate-lyase is very sensitive toward destruction by oxygen; in air-saturated buffer at 0°, the half-life was estimated (via loss of catalytic activity) to be about 10 sec. This reaction was found to cleave the polypeptide chain at the Gly-734 site into 82- and 3-kDa fragments (Fig. 5). The glycyl residue is converted into an oxalyl residue which constitutes the N terminus of the small fragment:[15]

[15] The chemical modification by O_2 probably occurs via formation of a C-2 peroxyl radical which undergoes elimination and hydrolysis of the imine, followed by oxidation of the putative glyoxylyl group:

$$O_2 + \text{R–CONH–CH–CONH–R}' \rightarrow \text{R–CONH–CH(OO}\bullet\text{)–CONH–R}' \rightarrow$$
$$\text{R–CON=CH–CONH–R}' \rightarrow \text{R–CONH}_2 + \text{HCO–CONH–R}'(\rightarrow \text{HOOC–CONHR}').$$

Protein fragmentation by γ-radiolysis follows a similar mechanism [W. M. Garrison, *Chem. Rev.* **87,** 381 (1987)].

$$\text{R–CONH–}\dot{\text{C}}\text{H–CONH–R}' \xrightarrow{O_2} \text{R–CONH}_2 + \text{HOOC–CONH–R}'. \quad (3)$$

We routinely use this protein fragmentation, which can be readily analyzed by gel electrophoresis, for sensitive determination of the glycyl radical content in pyruvate formate-lyase preparations, replacing EPR spectroscopic analysis which requires high concentrations. It can be advantageously applied to enzyme mutants where the radical content is not deducible from activity assays. Briefly, the enzyme solution, placed on ice, is adjusted to 10 m*M* EDTA (anaerobically added), then gassed with air for 1 min. Aliquots are carried through sodium dodecyl sulfate–polyacrylamide gel electrophoresis directly or after concentration with trichloroacetic acid using a 8% gel[16] to resolve the 82-kDa fragment from the parental 85-kDa polypeptide. An *in vitro*- activated enzyme with 1 spin/dimer yields a 1 : 1 pattern of these bands. Coomassie blue staining, with optional quantitation by gel scanning, readily detects 1.7 μg of the 82-kDa band which is equivalent to an original spin amount of 10 pmol. Immunostaining with PFL antiserum allows detection of low levels of radical and makes the method applicable to crude enzyme preparations of cell extracts. This method has been used for determining the radical status of pyruvate formate-lyase in intact *E. coli* cells.[17]

AdhE Protein-Catalyzed Reduction of the Glycyl Radical[17]

EPR signal and catalytic activity are lost spontaneously during long-term storage of the enzyme (see previous section). This quenching is probably due to the slow reaction of the glycyl radical with thiols such as dithiothreitol contained in the anaerobic buffers. The radical form can be recovered, usually with a high yield, after reapplication of the activation reaction.

We have not studied systematically whether the glycyl radical could be reduced in a controlled way by chemical 1e^- donor agents, but Ti^{3+}, reduced viologens, or the 5-deazaflavin radical are totally inert—a prerequisite for use of these compounds in the radical generation process. With dithionite (1 m*M*; 30°) the EPR signal becomes quenched within 1 min, which, however, results in irreversible inactivation of the enzyme (i.e., nonreactivatability).

Reversible conversion of pyruvate formate-lyase to the nonradical form can be achieved enzymatically by the AdhE protein of *E. coli,* which is readily purified from an overproducing strain.[17] This protein, a homopolymer of 100-kDa subunits, functions chiefly as acetaldehyde/CoA and etha-

[16] U. K. Laemmli, *Nature* (*London*) **227,** 680 (1970).

[17] D. Kessler, W. Herth, and J. Knappe, *J. Biol. Chem.* **267,** 18073 (1992).

nol dehydrogenase but can also deactivate pyruvate formate-lyase by using Fe^{2+}, NAD, and CoA as cofactors or cosubstrates:[18]

$$\dot{E} \xrightarrow{\text{NAD,CoA,Fe}^{2+}} E. \qquad (4)$$

In the standard reaction mixture described,[17] 1 μg of AdhE promotes total conversion of 0.1 nmol of pyruvate formate-lyase in about 10 min. It should be noted that pyruvate inhibits the reaction.

Catalytic Properties of Pyruvate Formate-Lyase

Pyruvate conversion to acetyl-CoA and formate (1) is a fully reversible reaction (K_{eq} = 750).[12] Each direction can be followed continuously with a coupled optical assay which monitors generation or consumption of NADH: Acetyl-CoA production (forward reaction) is linked to malate dehydrogenase/citrate synthase reactions; pyruvate production (reverse reaction) is linked to the lactate dehydrogenase reaction, with continuous (re)generation of acetyl-CoA via acetyl phosphate/phosphotransacetylase. The assays are carried out in anaerobic cuvettes; details of their operation have been described.[12,19]

Catalytic parameters (for 30° and pH 8.1) determined with the optical assays are as follows[12]: K_m = 2 mM, 6.8 μM, 51 μM, and 24 mM for pyruvate, CoA, acetyl-CoA, and formate, respectively; and k_{cat} = 770 (sec^{-1}) and 260 (sec^{-1}) for the forward and reverse reactions, respectively. The initial steady-state kinetics display a ping-pong reaction pattern in accordance with pyruvate formate-lyase operating via an indirect acetyl group transfer mechanism:

$$\dot{E}\text{-SH} + \text{pyruvate} \rightleftharpoons \dot{E}\text{-S-acetyl} + \text{formate} \qquad (1a)$$
$$\dot{E}\text{-S-acetyl} + \text{CoA} \rightleftharpoons \dot{E}\text{-SH} + \text{acetyl-CoA}. \qquad (1b)$$

The half-reactions may be assayed individually, again using catalytic amounts of enzyme, by the [^{14}C]formate pyruvate and [^{14}C]CoA–acetyl-CoA isotope exchange reactions, respectively. For the assay of (1a), ^{14}C incorporation into pyruvate is measured traditionally via the phenylhydrazone derivative,[20] whereas the assay of (1b) comprises ^{14}C determination

[18] The mechanism or redox stoichiometry of this process, which yields NADH as a coproduct, is not resolved; it could additionally involve dithiothreitol contained in the buffer medium.

[19] J. Knappe and H. P. Blaschkowski, this series, Vol. 41, p. 508.

[20] This classical assay procedure for pyruvate formate-lyase is described[19]; replace dinitrophenylhydrazine by phenylhydrazine, and determine the specific radioactivity of pyruvate phenylhydrazone by liquid scintillation counting and spectral quantification (E_{320} nm = 19.5 mM^{-1}cm^{-1}).

on acetyl-CoA which is separated from CoA chromatographically, preferably by HPLC (C_{18} column; CH_3CN gradient in 0.1% trifluoroacetic acid). The chemical equilibrium constants of the half-reactions were estimated to be 50 (1a) and 15 (1b).[12]

Reaction Mechanism

Three amino acid residues in pyruvate formate-lyase have been identified as being involved in the catalytic cycle of the enzyme: Cys-418, Cys-419, and Gly-734.[5] The glycyl radical is required for half-reaction (1a) of carbon–carbon bond cleavage/synthesis and is important also for (1b). (Measurement on the nonradical form, performed with the direct protein modification assay described below, found the rate of acetyl/CoA transfer half-reaction to be diminished 10^4- to 10^5-fold.) The Cys-419 SH group constitutes the central acetyl carrier, to which the acetyl group is linked, as thioester, in the intermediary isolatable $\dot{E}$-S-acetyl.[21] Finally, the SH group of the Cys-418 residue is a functional element also required for each part of the catalytic cycle. 418 Ala or Ser mutants are inactive in assays of the overall reaction as well as the half-reactions.

The Cys-418 residue has been proposed to participate in (1a) in the form of its *thiyl* radical (Fig. 6).[5] This would be generated by oxidation of the SH group by the glycyl radical when pyruvate is bound to the active site (thiohemiketal adduct to Cys-419); the reverse process, with regeneration of the glycyl radical, would occur when the formate product has been released. Thus the 418 thiyl is suggested to serve actually as the protein radical that allows processing of pyruvate, by an initial nucleophilic addition to the carboxyl.[22] (This notion is supported particularly by the formation of the substrate–analog radical described in the last section of this chapter.) In the second half-reaction (1b), Cys-418 participates in its SH form as a relay for acetyl transfer to CoA. [This is substantiated by acetylation of the Cys-418–SH group of the 419 Ser mutant (see below)].

Assays of Enzyme Modifications by Substrates and Analogs

As has been mentioned earlier, pyruvate formate-lyase undergoes covalent modification by its substrates or substrate–analogs (hypophosphite, acetylphosphinate) which can be monitored using radioactive-labeled compounds and may be analyzed further by site-mapping experiments. These

[21] W. Plaga, R. Frank, and J. Knappe, *Eur. J. Biochem.* **178,** 445 (1988).

[22] Previous proposals suggested a hydrogen atom abstraction from the substrate (pyruvate-thiohemiketal or formate) by the protein radical as the initial substrate conversion step.[21,23]

[23] E. J. Brush, K. A. Lipsett, and J. W. Kozarich, *Biochemistry* **27,** 2217 (1988).

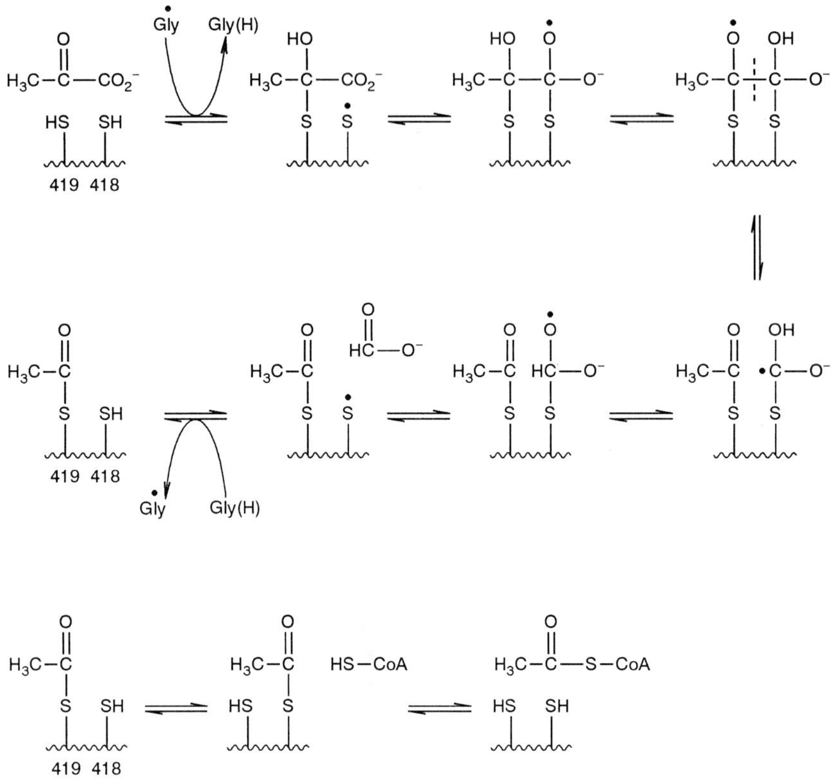

FIG. 6. Proposed reaction mechanism of pyruvate formate-lyase.[5] Numbered SH groups refer to Cys-418 and Cys-419 residues.

assays are outlined here. They have been applied as yet to normal enzyme and several mutants, which has been instrumental for the functional assignment of 418/419 cysteine residues and for indicating the significance of the glycyl radical. They should be useful for investigating effects of other amino acid replacements (or that of any protein-engineered enzyme) on individual steps in the catalytic cycle.

Acetyl-Enzyme[12,21]

Charging of the enzyme protein (radical form) on Cys-419 with ^{14}C-labeled acetyl and in this way quantitating the amount of active sites in any sample can be performed with [2-^{14}C]pyruvate (Amersham) or [1-^{14}C]acetyl-CoA (Amersham product; or enzymatically prepared from [^{14}C]pyruvate and CoA with the pyruvate formate-lyase reaction and

final purification by HPLC). Specific radioactivities required are ≥1000 dpm/nmol.

To the enzyme (0.05 to 0.3 mg) contained in anaerobic buffer, pH 7.2 to 7.5 (0.2 ml) and cooled on ice, admix 5–20 μl substrate sample containing 10 nmol of [^{14}C]pyruvate or 100 nmol of [^{14}C]acetyl-CoA. Quench the reaction mixture after 30 sec with 0.8 ml TCA mix (0.18 *M* trichloroacetic acid/20 m*M* $Na_4P_2O_7$/10 m*M* pyruvate). Allow the mixture to stand at 0° for 30 min, then transfer the precipitated protein onto a Whatman glass fiber GF/A filter (2.4 cm) and wash thoroughly with at least 20 ml of the TCA mix. Place the filter in a scintillation vial, add 0.5 ml 0.2 *M* NaOH and 50 μl 10% SDS, then incubate at 60° for 10 min to dissolve the protein. After cooling, add 0.1 ml 1 *M* acetic acid and 10 ml scintillation cocktail and determine the radioactivity. A stoichiometry of 1 nmol of ^{14}C-acetyl bound to 0.17 mg (1 nmol; 35 U) has been determined with homogenous enzyme. The acetyl location was mapped to Cys-419 by radiosequencing of a BrCN peptide.[21]

Stoichiometric acetylation by acetyl-CoA (but not pyruvate) can be achieved as well with the nonradical form, which, however, takes 30 min at a temperature of 30° for completion.[24] With the mutant enzyme (radical form) where the 419-SH is replaced by OH (C419S), acetyl-CoA acetylates the Cys-418 residue whereas pyruvate is inert. Mutant enzymes lacking the 418-SH (C418S or C418A) are inactive with either substrate.[5]

Pyruvate Thiohemiketal-Enzyme[5,24]

The adduct of the pyruvate carbonyl to the central Cys-419 SH, which is postulated as the initial step of the catalytic cycle, can be measured by using the *nonradical* form. Pyruvate binding to the protein is detected via gel filtration which must be performed at 0 to 4°. At room temperature, the covalent adduct dissociates readily (and immediately on protein precipitation with TCA).

Incubate the protein sample (0.5 mg) in 0.15 *M* Tris–HCl (pH 7.2)/5 m*M* dithiothreitol with 1 m*M* [^{14}C]pyruvate (final concentration) for 20 min at 30° (0.2 ml). Then cool on ice and carry the solution in the cold room (4°) through a Sephadex G-25 column (2 ml) using 0.1 *M* potassium phosphate, pH 7. Measure radioactivity of the protein fraction (between about 0.4 and 0.65 column volumes). The stoichiometry is usually 0.8 to 1 nmol of [^{14}C]pyruvate bound per protein dimer (0.17 mg).

Mutant protein C419S is inert with this assay whereas C418S or C418A mutants have virtually normal binding properties.

[24] S. Elbert and J. Knappe, unpublished results (1992), see Knappe *et al.*[5].

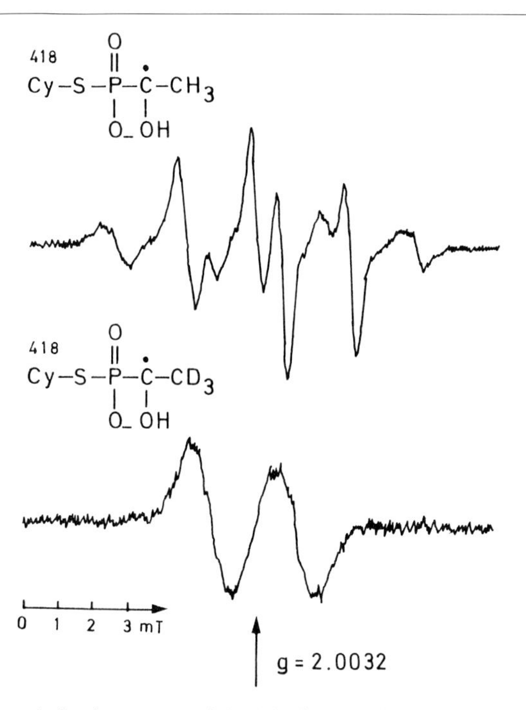

FIG. 7. EPR first derivative spectra of the 1-hydroxy-1-phosphoryl ethyl radical produced from hypophosphite and Ė-acetyl or Ė-[2-$^2H_3$]acetyl. (Reprinted from Unkrig *et al.*[13]) The identical signal is obtained from acetylphosphinate and free enzyme (Ė).[3,25] Instrument conditions: modulation amplitude, 0.25 mT; modulation frequency, 100 kHz; X band; temperature, 253 K.

1-Hydroxyethylphosphoryl Thioester of Cys-418[13,21]

Suicide reactions of pyruvate formate-lyase with phosphorus-containing substrate analogs can specifically modify the Cys-418 residue. The reactions of hypophosphite (formate analog) with acetylated enzyme (Ė-acetyl) and of acetylphosphinate (pyruvate analog) with free enzyme (Ė) yield 1-hydroxyethylphosphonate that is linked as a thioester to Cys-418 SH. The glycyl radical is reduced during this modification process. When incubated at a low temperature, the reaction mixtures will accumulate the α-phosphonyl radical intermediate (Fig. 7). Hence, the substrate–analogs can be envisaged as operating as spin-trap reagents of the putative Cys-418 thiyl radical. The mechanistic rationale is presented in Fig. 8.

[25] M. Frey, Ph.D. Thesis, University of Heidelberg (1993).

[26] O. A. Swanepoel and N. J. J. Van Rensburg, *Photochem. Photobiol.* **4,** 833 (1965).

^{32}P-labeled hypophosphite was used in the original experiment that signaled covalent fixation of the analog.[2] But the labeling method of choice for assaying the Cys-418 modification is using normal hypophosphite and ^{14}C-acetylated enzyme.[27] Typically, 2.5 mg (15 nmol) of enzyme (radical form) in 1.5 ml anaerobic buffer (pH 7.6) is reacted with 0.5 μmol of [2-^{14}C]pyruvate (2000 dpm/nmol) at 0° for acetyl charging (30 sec), then 30 μmol of sodium hypophosphite is admixed under argon. Incubate at 30° for 10 min, then store on ice (oxygen exclusion is not required). Protein-bound radioactivity is determined on an aliquot of the sample as described earlier (Acetyl-Enzyme). Since the linkage is a phosphoryl thioester, it is resistent to hydroxylamine, in contrast to the acetyl thioester. This property can be used to check the product identity. Add 0.5 ml of 1 *M* NH_2OH (pH 6.5) to a TCA precipitated protein aliquot (pelleted in a Eppendorf tube), suspend by vortexing, and allow to stand for 15 min at room temperature; then add TCA mix and proceed as before to obtain protein-bound radioactivity. The modification yield is usually $\geq$90%.

The 1-hydroxyethylphosphoryl group has been localized to Cys-418 by radiosequencing of a chymotryptic peptide.[21] The free hydroxyethylphosphonate can be released by Br_2 oxidation or by digestion with phosphodiesterase. Chemical shift values (δ) of ^{31}P resonances are +44 ppm for the thioester and +19 ppm for the free phosphonate.

To accumulate the intermediary radical species of the modification process and obtain its EPR spectrum (Fig. 7), it suffices to reduce the reaction temperature to 0° (and apply high concentrations of the analogs). Typically, 50 m*M* hypophosphite or 2 m*M* acetylphosphinate (Ciba-Geigy, Basel) is admixed to the enzyme (3 mg/ml) that is contained in anaerobic buffer with 20 m*M* pyruvate ($\dot{E}$-acetyl) or without pyruvate ($\dot{E}$), respectively. The reaction mixture is incubated for 2 min, and is then frozen into EPR tubes for spectroscopy. For the experiment with $\dot{E}$-[2-$^2H_3$]acetyl and hypophosphite (Fig. 7), [3-$^2H_3$]pyruvate was utilized, which was prepared from oxaloacetic acid by equilibration with and thermal decarboxylation in ^2H_2O.

The 1-hydroxy-1-phosphoryl-ethyl radical as contained in the native protein structure has no light absorption at wavelengths $\geq$310 nm. Its half-life of persistence at 0° is about 20 min.[13]

[27] Hypophosphite can also react with free enzyme ($\dot{E}$), presumably at the Cys-418(thiyl); the rate, determined via enzyme inactivation and glycyl radical reduction, is about 1/100 of the reaction rate with $\dot{E}$-acetyl.[2,13] The kinetics of hypophosphite- and acetylphosphinate-mediated inactivation have been described by Kozarich and co-workers.[23,28]

[28] L. Ulissi-DeMario, E. J. Brush, and J. W. Kozarich, *J. Am. Chem. Soc.* **113,** 4341 (1991).

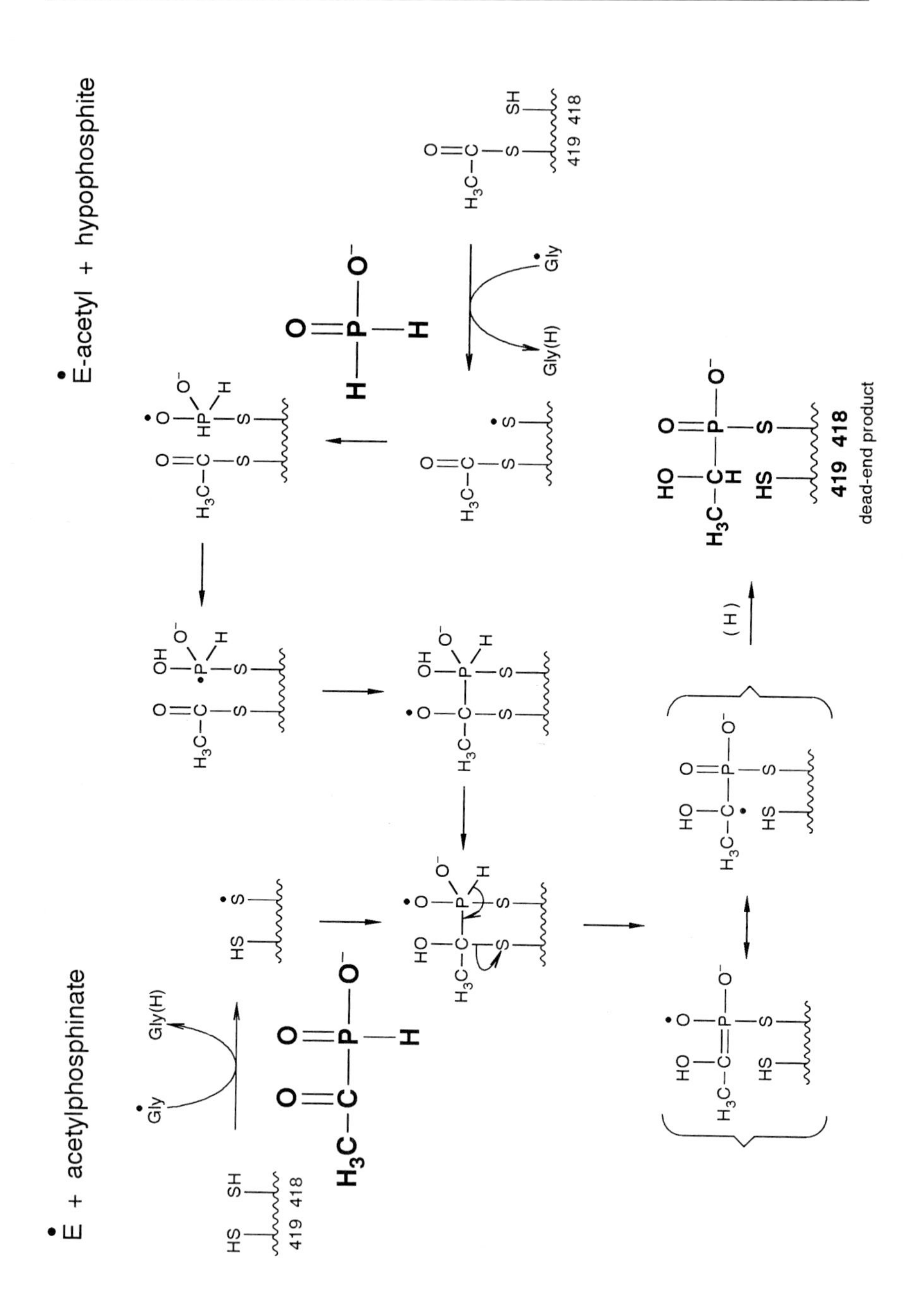

•E + acetylphosphinate
•E-acetyl + hypophosphite
Gly
Gly(H)
419 418
(H)
dead-end product

Conclusions

Data obtained with the suicide substrates and mutant variants of the enzyme consistently point to a thiyl radical (of Cys-418) as the protein-based radical actually performing the substrate processing. The nucleophilic addition of thiyl to the carboxyl of pyruvate thiohemiketal (on Cys-419) or, in the reverse reaction, to the carboxyl of formate would generate the crucial substrate radical than can undergo homolysis or afford radical annealing, respectively, of the carbon–carbon bond. In the resting state of the enzyme, however, the unpaired electron is centered at the α-carbon of the glycine residue 734, where it is resonance stabilized. From chemical model systems, glycine is known to be favored over other (aliphatic) amino acid radicals.[29] Separate sites of spin storage and spin utilization, which are connected mutually by electron transfer reactions, may constitute a general organization principle of free radical enzymes. This is clearly documented for the prototypical ribonucleodide reductase, where the tyrosyl radical iron center and the active site (with a putative Cys–thiyl intermediate) are harbored on different protein subunits.[30] (EPR spectroscopic evidence of protein thiyls has not been accomplished yet for any radical enzyme, and the delineation of the electron transfer route through the protein matrix between the resting state site and the business end is an intriguing issue to be resolved.)

It should be noted that the anaerobic (class III) ribonucletoide reductase (nrdD) of *E. coli* has been suggested as a further glycyl radical enzyme.[31] The glycyl residue is presumably in a 5 amino acid stretch (RVCGY) that is homologous to the Gly-734 stretch of pyruvate formate-lyase (RVSGY); otherwise there is no sequence similarity between the two enzymes. The databases (searched in 1993) indicate putative proteins of 14 kDa (ORF X in the *phlA* region of *Serratia liquifaciens*[32]; ORF 63.1 of bacteriophage

[29] C. J. Easton and M. P. Hay, *J. Chem. Soc., Chem. Commun.*, p. 55 (1986).

[30] For recent reviews, see S. Brooker, J. Broderick, and J. Stubbe, *Biochem. Soc. Trans.* **21,** 727 (1993); P. Nordlund, A. Åberg, U. Uhlin, and H. Eklund, *ibid.*, p. 735.

[31] E. Mulliez, M. Fontecave, J. Gaillard, and P. Reichard, *J. Biol. Chem.* **268,** 2296 (1993).

[32] M. Givskov and S. Molin, *Mol. Microbiol.* **6,** 1363 (1992).

FIG. 8. Proposed routes of the mechanism-based Cys-418 modification by acetylphosphinate and hypophosphite.[5] The extra hydrogen (H) in the last step is exogeneous. Reaction of [^{3}H]hypophosphite with Ė-acetyl releases the ^{3}H label totally to the solvent,[2] proving that glycyl radical reduction through this analog is indirect. Reaction of hypophosphite with thiyl radicals is known from chemical studies.[26]

T4[33]) showing carboxy-terminal stretches of 60 amino acid residues that are identical to the carboxyl terminus of pyruvate formate-lyase (700–759) to an extent of 80%. This could suggest that the glycine-734 domain has evolved as an autonomous protein module.

[33] K. Valerie, J. Stevens, M. Lynch, E. E. Henderson, and J. K. de Riel, *Nucleic Acids Res.* **14,** 8637 (1986).

[24] Characterization of a Radical Intermediate in the Lysine 2,3-Aminomutase Reaction

By George H. Reed and Marcus D. Ballinger

Introduction

The first step in the metabolism of lysine by clostridia is an internal redox reaction: the conversion of L-lysine to β-lysine.[1]

This mutation is catalyzed by a homohexameric enzyme, lysine 2,3-aminomutase.[2] As isolated, the enzyme contains three cofactors: pyridoxal phosphate (PLP), Fe–S centers, and cobalt or zinc.[3] Activity also depends on an extrinsic cofactor, *S*-adenosylmethionine. The Fe–S centers have properties consistent with cubane (4Fe–4S) type centers, and the cobalt that is detected by electron paramagnetic resonance (EPR) is a high-spin Co^{2+}.[4] This internal rearrangement reaction, or 1,2 shift of a hydrogen atom and a substituent, fits the pattern normally associated with adenosylcobalamin (coenzyme B_{12})-dependent enzymes. However, lysine 2,3-aminomutase does not contain and is not activated by coenzyme B_{12}. Rather, *S*-adenosylmethionine is used in place of coenzyme B_{12} as a source of the 5′-deoxyadenosine-5′-yl radical which initiates the rearrangement by way of radical intermediates.[5]

[1] T. C. Stadtman, *Adv. Enzymol. Relat. Areas Mol. Biol.* **38,** 413 (1973).
[2] T. P. Chirpich, V. Zappia, R. N. Costilow, and H. A. Barker, *J. Biol. Chem.* **245,** 1778 (1970).
[3] R. M. Petrovich, F. J. Ruzicka, G. H. Reed, and P. A. Frey, *J. Biol. Chem.* **266,** 7656 (1991).
[4] R. M. Petrovich, F. J. Ruzicka, G. H. Reed, and P. A. Frey, *Biochemistry* **31,** 10774 (1992).
[5] M. L. Moss and P. A. Frey, *J. Biol. Chem.* **262,** 14859 (1987).

METHODS IN ENZYMOLOGY, VOL. 258

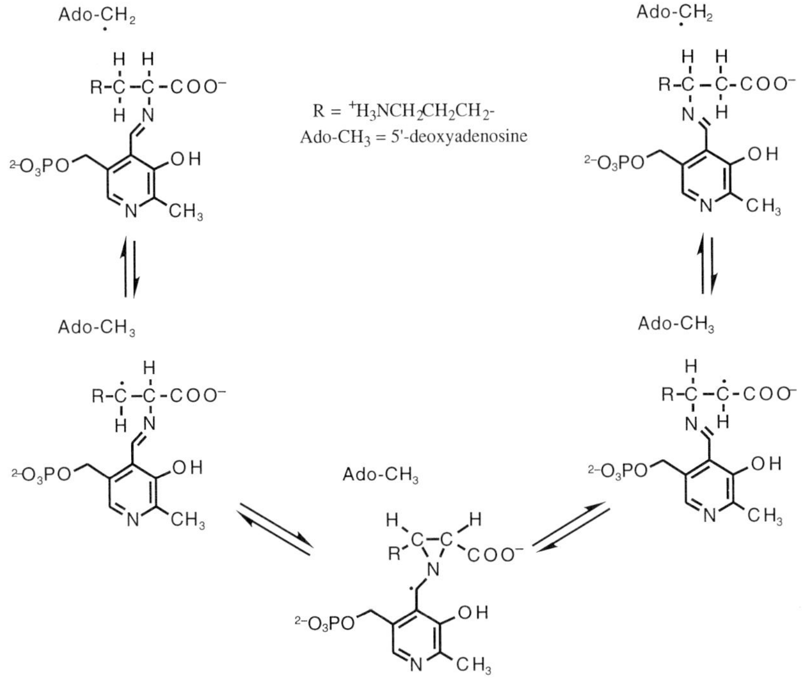

$R = {}^{+}H_3NCH_2CH_2CH_2$-
Ado-CH_3 = 5'-deoxyadenosine

SCHEME I

The mechanism proposed for this rearrangement is shown in Scheme I.[5] The pathway for the generation of the 5′-deoxyadenosine-5′-yl radical is not yet clear but likely involves reaction of *S*-adenosylmethionine with either or both types of the inorganic cofactor.[4]

The absence of coenzyme B_{12} simplifies the region of the EPR spectrum where organic radicals absorb because coupling between radical intermediates and the low spin Co^{2+} in B_{12}(r) that dominates the EPR spectra in coenzyme B_{12}-dependent reactions[6] is avoided. The reduced Fe–S centers are EPR silent, and the only other paramagnet in the sample is a form of Co^{2+} that relaxes rapidly at 77 K and does not give rise to interfering resonance signals. A window is therefore open for detection and characterization of radical intermediates whose concentrations reach the threshold for EPR detection ($\sim\mu M$).

Organic radicals are, in many respects, ideal intermediates for experimental investigation. The intrinsic sensitivity of EPR enhances chances

[6] J. R. Pilbrow, *in* "B12" (D. Dolphin, ed.), Vol. 1, p. 431. Wiley (Interscience), New York, 1982.

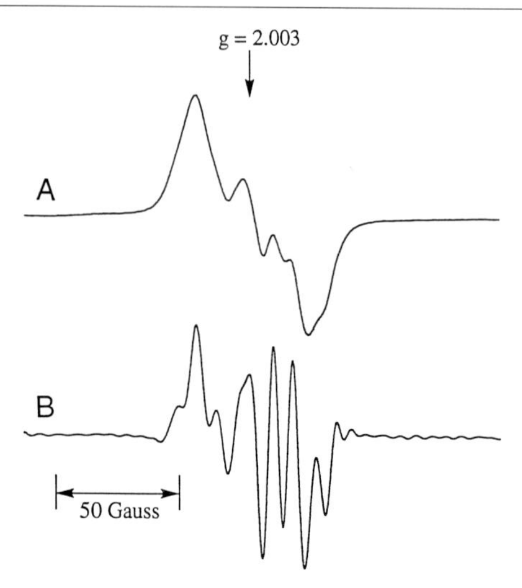

FIG. 1. EPR spectrum (77 K) of the substrate radical intermediate in the lysine 2,3-aminomutase reaction (A) and the resolution-enhanced version (B) of the same spectrum. To prepare the sample, the enzyme was first reductively incubated in a solution containing 60 m*M* Tris–H_2SO_4 at pH 8.0, 1.5 m*M* sodium dithionite, 0.4 m*M* pyridoxal phosphate, 1 m*M* ferric ammonium citrate, and 7 m*M* dihydrolipoate for 4 hr at 37°. This solution was then combined with reaction components to give the following final concentrations: enzyme (13 mg/ml), *S*-adenosylmethionine (1.2 m*M*), Tris–H_2SO_4 at pH 8.0 (80 m*M*), sodium dithionite (2.0 m*M*), and L-lysine · H_2SO_4 (150 m*M*); the sample was frozen in liquid nitrogen after 45 sec. The microwave frequency was 9.044 GHz. The details of the resolution enhancement are described in the legend of Fig. 4.

for detection of such intermediates, and the EPR signals are gravid with information pertaining to the structure of the host molecule and its immediate surroundings. When reaction mixtures of lysine 2,3-aminomutase are frozen during turnover in either direction, the samples exhibit a characteristic EPR signal[7,8] (Fig. 1) centered at *g* 2.00. The methods used to identify the source of this EPR signal and to determine the structure of the radical from which it derives are described in this chapter.

Generation of and Trapping of Intermediates

The greatest experimental challenges in any study of radical intermediates are encountered right at the starting gate—namely, finding an EPR signal that stems from an "authentic" radical component of the reaction

[7] M. D. Ballinger, G. H. Reed, and P. A. Frey, *Biochemistry* **31,** 949 (1992).
[8] M. D. Ballinger, P. A. Frey, and G. H. Reed, *Biochemistry* **31,** 10782 (1992).

and avoiding adventitious radicals that may arise from side reactions. In this endeavor, experience with isolation, handling, and assaying of the enzyme will provide clues to conditions that will maximize chances for success.

A widely accepted method for trapping of intermediates for subsequent EPR measurements is the rapid-mix/freeze-quench procedure described early on by Bray.[9] Rapid freezing is accomplished by injection of the reaction mixture into a chilled hydrocarbon reservoir, and the resulting sample snow is packed into an EPR tube appended to the bottom of the reservoir. These techniques provide a means to capture radicals in the steady state or pre-steady state (time scale in the low millisecond region) of enzymatic reactions. The rapid-mix/freeze-quench approach remains the ultimate method for establishing the "kinetic competence" of any radical intermediates that might be discovered by other means. The specialized equipment required in this procedure is available commercially, and the experimental procedures have been described in detail elsewhere.[10] Enzymatic reactions that have a cofactor, such as coenzyme B_{12} or *S*-adenosylmethionine, which initiates free radical chemistry, may exhibit a "complex" activation period that may also require the presence of substrate. The requirements of the activation period obviously need to be considered in judging the kinetic competence of radical intermediates.[11]

If the steady state of the reaction can be maintained for $\geq$20 sec at concentrations of enzyme conducive to EPR detection of intermediates, then one can simply mix activated enzyme and substrate in an EPR sample tube. The mixture is then frozen by immersing the tube in liquid N_2 or in a chilled hydrocarbon bath. In this procedure, two competing demands are encountered. The intermediates may be present at only a small fraction of the total concentration of enzyme. Hence, high concentrations of enzyme maximize chances for spectroscopic detection at the micromolar threshold. At the same time, high concentrations of enzyme bring the reaction to equilibrium rapidly. Consequently, high concentrations of substrate are required to keep the reaction in the steady state during the several seconds that are required for mixing and freezing. Another strategy that has been employed successfully with coenzyme B_{12}-dependent enzymes[11,12] is to use a "slow substrate" to retard the approach to equilibrium. For the lysine 2,3-aminomutase reaction, even with the decided handicap of mixing and handling samples in an anaerobic chamber, reaction times as short as 20

[9] R. C. Bray, *Biochem. J.* **89,** 189 (1961).

[10] H. Beinert, R. E. Hanson, and C. R. Hartzell, *Biochim. Biophys. Acta* **423,** 339 (1976).

[11] O. C. Wallis, R. C. Bray, S. Gutteridge, and M. R. Holloway, *Eur. J. Biochem.* **125,** 299 (1982).

[12] B. M. Babior, T. H. Moss, W. H. Orme-Johnson, and H. Beinert, *J. Biol. Chem.* **249,** 4537 (1974).

sec are achieved routinely. The advantages of this "steady state" method are: (1) no special equipment is needed; (2) normally less enzyme is required than in the rapid-mix/freeze-quench experiment because no enzyme is left behind in the "plumbing;" and (3) quantitative comparison of the intensities of EPR signals from separate samples is straightforward because there is no variability in packing of the samples. In the absence of a rate measurement as a test for kinetic competence of radical signals detected in this procedure, other experiments can be carried out to determine whether the properties of the radical are consistent with that of a genuine intermediate. For example, the concentration of the radical should correlate with enzyme activity present in the sample.[7] Moreover, the EPR spectrum should change when isotopic substitutions are introduced into the appropriate precursor species. These two screens—especially changes in the hyperfine structure in the spectrum of the radical on isotopic substitutions in the substrate, lysine—provided strong evidence that the radical in the lysine 2,3-aminomutase mixtures came from a normal intermediate in the reaction.[7,8]

A third approach to finding intermediates that is feasible for reversible reactions is to examine the reaction mixture at equilibrium. In principle, such measurements could be carried out on samples in the liquid phase. There are, however, some drawbacks to the liquid phase tactic. For example, the volume of aqueous samples that can be placed in a microwave cavity (or other resonant devise used for EPR) is about 5–10 times smaller than for frozen samples because of dielectric losses associated with liquid water in a conventional EPR cavity. Furthermore, increased incubation times in liquid solution greatly enhance the opportunity for generation of artifacts from unwanted side reactions. Finally, if the radicals are bound in the active site of their enzymic host, their EPR spectra are likely to have powder pattern (or solid state) characteristics anyway because the protein rotates too slowly to average the anisotropies in the g factor and hyperfine interactions. Thus, no simplifications are apt to result from measurements in the liquid phase. The problems that can arise on prolonged incubations are exemplified by measurements on reactions mixtures of lysine 2,3-aminomutase. Reactions that are incubated for a period long enough to reach equilibrium, starting from either direction, exhibit variable amounts of a signal from a second radical.[7] EPR spectra of samples initiated with isotopically labeled lysine show that this second radical is not hosted on the carbon skeleton of the substrate. Signals for this second radical are not present when substrate and product are premixed, at an equilibrium ratio, and added to the enzyme. Hence, this second radical appears to be a side product that arises on prolonged incubation of reaction mixtures.

Characterization of EPR Signals

All organic radicals exhibit EPR signals in a narrow region of the spectrum near *g* 2.00. A relatively narrow signal occurring in this region of the spectrum at 77 K is a signature of an organic radical because signals from transition metal ions are typically broader and span a wider region of the magnetic field than those of organic radicals. The absence of "*g* value dispersion" in the spectra of organic radicals means that the positions of signals in the spectrum are not especially useful in distinguishing one radical from another. The key to positive identification of the radical and determination of its structure is the hyperfine structure which appears in the EPR spectrum. The hyperfine structure stems from a magnetic coupling between the unpaired electron spin and nuclear spins in the molecule. Analysis of the hyperfine splitting parameters yields structural information such as the distribution of unpaired electron spin and the conformation of the radical.

Free radicals are relatively easy to generate in reaction mixtures that would support enzymatic reactions which have radical intermediates. In particular, traces of molecular oxygen in the presence of one electron reductants will almost certainly give rise to detectable radical species. Hence, unless molecular oxygen is a substrate in the reaction, it must be rigorously excluded from contact with the samples. Lysine 2,3-aminomutase is very susceptible to inactivation by molecular oxygen, and purification steps as well as sample preparation are carried out in an anaerobic chamber. Control experiments should be also applied to ensure that the radical under observation is related to the enzymatic reaction. For example, obvious controls such as the dependence of the appearance of the radical on the presence of enzyme, substrate, and cofactors should be established. Other control experiments, such as isotopic substitutions within the suspected precursor molecule, will also aid in subsequent identification and characterization of the radical.

The powder pattern EPR spectrum of an organic radical intermediate in the active site of an enzyme will not resemble the sharp multiline patterns that are typical of organic radicals undergoing rapid rotation in solution. Rather, the hyperfine structure, if present, is likely to be poorly resolved. Anisotropies in the hyperfine interactions and, to a lesser extent, the *g* factor conspire to prevent analysis of the spectrum by inspection, even for radicals that have a relatively simple hyperfine structure. The most powerful experimental tool in "sorting out" the spectrum is isotopic substitution, e.g., ^{13}C for ^{12}C, ^{2}H for ^{1}H, ^{15}N for ^{14}N. A detectable change in the EPR pattern coincident with an isotopic substitution indicates that the unpaired electron spin is coupled to the nuclear spin at this site. Hyperfine coupling is normally strongest for nuclei that are closest to the unpaired electron, and it is thus possible to "home in" on the center of unpaired spin through a logical sequence of isotopic substitutions. The spectra for samples prepared with isotopically substituted forms of lysine (shown in Fig. 2) illustrate how the α-radical of β-lysine was identified.[8] In particular, the collapse of the splitting in the spectrum (Fig. 2C) of the sample prepared with [2-^{2}H]lysine and the dramatic increase in the width of the spectrum (Fig. 2E) in the sample prepared with [2-^{13}C]lysine show that the radical is centered at C-2 of a lysine fragment.

One caveat regarding substitution of ^{2}H for ^{1}H should be remembered from studies of organic radicals in crystals. Hydrogens on carbons bearing the unpaired electron may undergo exchange with hydrogens from neigh-

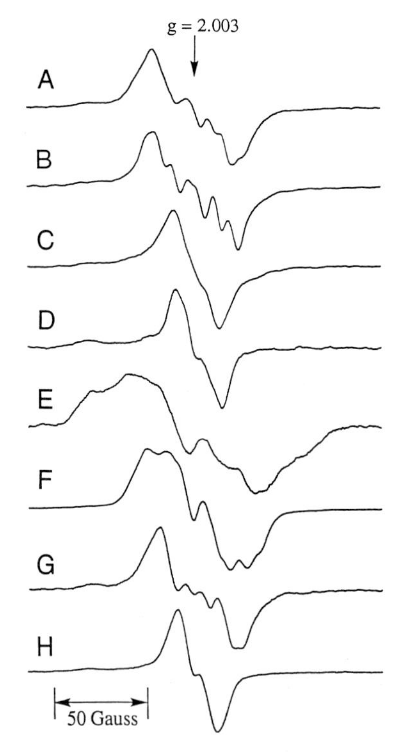

FIG. 2. The effects of isotopic substitutions in the substrate on the radical EPR spectrum. Reaction conditions were similar to those described in the legend for Fig. 1, and the following substrates were used: (A) unlabeled lysine; (B) [3,3,4,4,5,5,6,6-$^{2}H_{8}$]lysine; (C) [2-^{2}H]lysine; (D) [$^{2}H_{9}$]lysine; (E) [2-^{13}C]lysine; (F) [1-^{13}C]lysine; (G) [α-^{15}N]lysine; (H) [2-^{2}H, α-^{15}N]lysine. The microwave frequency was 9.044 GHz.

boring molecules.[13] For this reason, ^{2}H substitutions have not been as useful in studies of organic radicals as one might have anticipated. In one enzymatic system, pyruvate formate-lyase, the glycine radical of the protein backbone, exchanges its hydrogen with solvent.[14] The mechanism of such exchanges is likely to be a reversible process wherein the radical center effectively migrates to another atom via capture of a hydrogen atom from a neighbor. If this hydrogen is in an exchangeable site, e.g., an RN–H or RO–H, then exchange with a solvent can occur. Migration of the radical center might easily get out of control and the radical center could end up in a position that would not be conducive to continued support of the enzymatic reaction. We therefore expect that most enzymes would hold these reactive intermediates in a manner that would minimize chances for side reactions that would be detrimental to maintenance of catalytic turnover. Fortunately, in the lysine 2,3-aminomutase reaction, the effects of ^{2}H substitutions were entirely consistent with the proposed structure of the radical, an no evidence for exchange appeared in any of the samples with ^{2}H substitutions in the substrate or solvent.[8]

Results of the isotopic substitutions in the substrate lysine (see Fig. 2) show that the EPR spectrum from reaction mixtures of lysine 2,3-aminomutase originates from a radical centered on the α-carbon of lysine. This information is sufficient to support the rearrangement mechanism (Scheme I) that was proposed for the enzymatic reaction. The hyperfine splittings in the EPR spectra, however, contain more information about the structure and conformation of the radical. In order to reap this bounty, it is necessary to extract the hyperfine splitting parameters from the spectra. The analysis of such EPR spectra will normally require an iterative process of computer simulations in an effort to match the experimental spectra both for the unlabeled and isotopically substituted samples and determine the hyperfine splitting parameters. At this stage of the analysis, one normally has at least a tentative structural model of the radical, and there is an extensive literature documenting hyperfine splitting parameters for organic radicals to aid in modeling.[13,15] With modern computers, including PCs, simulations of powder spectra no longer present economic or time barriers in the analysis. We have also found that resolution enhancement methods provide spectra that are more exacting targets for simulations such that the hyperfine splitting parameters can be extracted with greater certainty.

[13] J. R. Morton, *Chem. Rev.* **64,** 453 (1964).

[14] A. F. V. Wagner, M. Frey, F. A. Neugebauer, W. Schafer, and J. Knappe, *Proc. Natl. Acad. Sci. U.S.A.* **89,** 996 (1992).

[15] H. Fischer, *in* "Free Radicals" (J. K. Kochi, ed.), Vol. 2, p. 435. Wiley, New York, 1973.

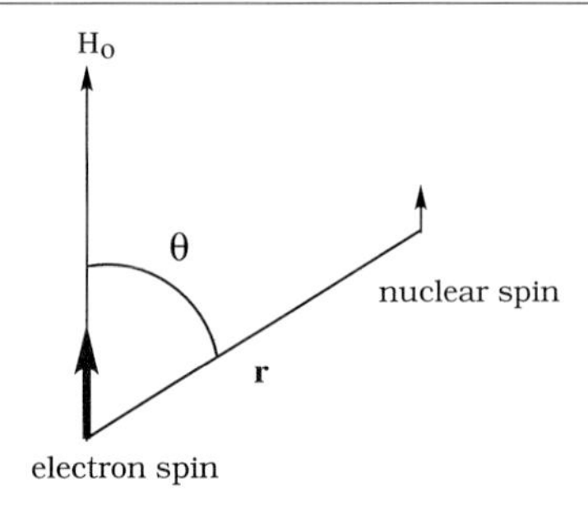

SCHEME II

Properties of π-Radicals

The α-radical of β-lysine and many other carbon-centered radicals are called π-radicals because the unpaired electron occupies a *p* orbital on a sp^2 carbon. α-Radicals of carboxylic acids have been thoroughly investigated in single crystals, and their EPR properties are well documented in the chemical literature.[13,15] Hyperfine splitting in the EPR spectra of these π-radicals comes primarily from coupling between the unpaired electron and the α-proton and the nuclear spins of atoms attached directly to the β-carbon(s). For the lysine radical, the sources of the major hyperfine splitting are the α-proton, the β-proton, and the β-nitrogen.

Hyperfine splitting from the α-proton is normally the dominant feature in EPR spectra of π-alkyl radicals. The collapse of the splitting pattern in the spectrum obtained with $[2\text{-}^2H]$lysine (Fig. 2C) confirms that the α-proton is the source of the largest hyperfine splitting in the spectrum of the radical derived from lysine. For the α-proton (or any other nuclear spin), the hyperfine interaction has two sources: a through bond, isotropic interaction, a_o; and a through space, anisotropic dipole–dipole interaction, *T*. The net hyperfine splitting at any orientation is the algebraic sum of these two interactions. The sign and the magnitude of the through space coupling depend on the orientation of the molecule in the laboratory magnetic field and on the inverse cube of the distance between the unpaired electron and the nucleus. Splitting of the EPR signals from this interaction varies as $(3\cos^2\theta - 1)/r^3$, where θ is the angle made by the magnetic field vector, H_o, and the position vector, r, between the unpaired electron spin and the nuclear spin (Scheme II). The anisotropic hyperfine interaction is expressed as a Cartesian tensor with principal values T_{xx}, T_{yy}, and T_{zz}. The net hyperfine splitting A_{ii} along the principal axes is $T_{ii} + a_o$. The principal axes of the π-radical are shown in Scheme III.

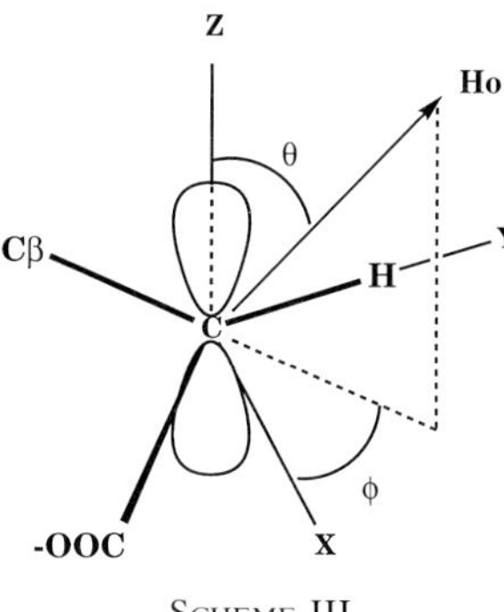

SCHEME III

The unpaired electron is "delocalized" in a molecular orbital so there is some ambiguity in assigning to it an "average" position for the purpose of calculating the splitting at each angle (i.e., the α-proton is too close for the "point–dipole approximation"). The "theoretical" α-proton hyperfine tensor has been computed.[16] Moreover, α-proton hyperfine tensors have been determined experimentally for many π-radicals, and principal values close to those given below are found to be diagnostic for an α-proton as the source of hyperfine splitting. Variations in the principal values of the tensor for an α-proton are due to variations in unpaired spin density on the α-carbon.[15]

The anisotropic hyperfine tensor, T, and the hyperfine tensor (with the isotropic coupling included), A, for the prototypical malonic acid radical are[17]

$$T = \begin{bmatrix} -11.4 & 0 & 0 \\ 0 & 10.7 & 0 \\ 0 & 0 & 0.4 \end{bmatrix} (\text{G}); A = \begin{bmatrix} -32.5 & 0 & 0 \\ 0 & -10.3 & 0 \\ 0 & 0 & -21.7 \end{bmatrix} (\text{G})$$

In a single crystal a doublet splitting with the values in A would be measured along the x, y, and z axes, respectively. The doublets would be centered at slightly different positions in the scan due to anisotropy in the g factor. The g factor anisotropy is relatively slight in organic radicals, and consequently it is difficult to determine in powder spectra at X-band frequencies. Principal values of g tensors of several classes of organic radical have been determined from single crystal studies, and a reasonable strategy to follow here is to pick a g tensor from a known radical whose structure is similar

[16] H. M. McConnell and J. Strathdee, *Mol. Phys.* **2,** 129 (1959).

[17] H. M. McConnell, C. Heller, T. Cole, and R. W. Fessenden, *J. Am. Chem. Soc.* **82,** 766 (1960).

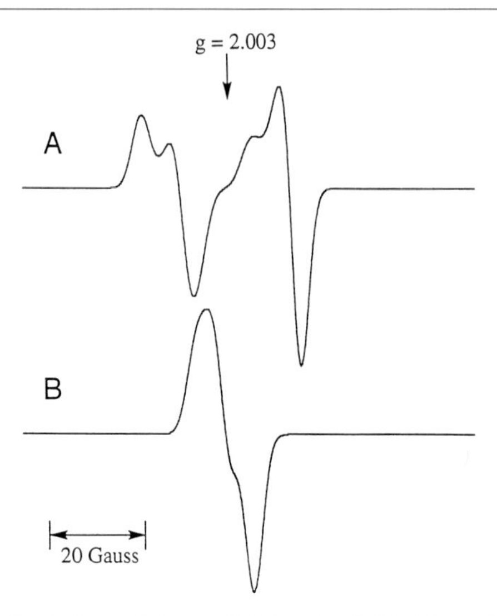

FIG. 3. Computer simulations of the α-^{1}H (A) and α-^{2}H (B) hyperfine splittings in π-type alkyl radicals. The simulations were obtained using $g_{xx} = 2.0043$; $g_{yy} = 2.0037$; $g_{zz} = 2.0021$; for α-^{1}H (A), $A_{xx} = 32.6$ G, $A_{yy} = 10.1$ G, $A_{zz} = 22.5$ G; for α-^{2}H, $A_{xx} = 5.0$ G, $A_{yy} = 1.6$ G, $A_{zz} = 3.5$ G; and line widths of 2.2 G. Crystal orientations (2025) were sampled in the simulations.

to the radical you are attempting to model.[15] The *g* factor anisotropy influences the hyperfine splitting patterns and therefore needs to be included for faithful simulations. In a powder spectrum, the α-proton splitting gives rise to the pattern shown in Fig. 3A.

The hyperfine splitting parameters for ^{2}H in place of ^{1}H may be calculated from the values of the latter because the splitting scales as $\gamma^2\mathrm{H}/\gamma^1\mathrm{H} = 0.1535$, where γ is the magnetogyric ratio of the respective isotope. Hence substitution of ^{2}H for ^{1}H changes the doublet splitting to a 1:1:1 triplet ($I = 1$ for ^{2}H) where the splitting is ~one-sixth that of ^{1}H. Positions that exhibit a substantial ^{1}H splitting require inclusion of the ^{2}H hyperfine splitting parameters on deuterium substitution for accurate simulations (Fig. 3B). The magnitude of the isotropic splitting, a_o, of the α-proton is an indicator of the spin density, ρ, in the *p* orbital on the α-carbon through the McConnell relationship: $a_o = Q\,\rho$ where Q (~27 G) is an empirically determined constant.[15]

In alkyl radicals, hyperfine interactions with β-substituents are largely isotropic because they originate primarily from hyperconjugation.[18] Hyperconjugation mixes a small amount of the 1*s* orbital of the β-hydrogen into

[18] J. E. Wertz and J. R. Bolton, "Electron Spin Resonance," p. 164. Chapman & Hall, New York, 1986.

the molecular orbital that contains the unpaired electron. If geometry favors such an admixture, protons on the β-carbon can give rise to hyperfine splitting comparable to that from the α-proton. Thus, hyperfine splitting from β-protons is of interest because of its dependence on the rotational conformation about the C_α–C_β bond. Hyperconjugation is strongest when the β-substituent is eclipsed with the half-filled *p* orbital on C_α. Hyperfine splitting from β-protons follows the empirical relationship: $a_{\beta H} = C_1 + C_2 \cos^2\chi$ (2), where C_1 (0.92 G) and C_2 (42.6 G) are empirically determined constants and χ is the dihedral angle between the plane containing the axis of the *p* orbital and the plane defined by both carbons and the β-proton (Scheme IV).[15]

Although data are not as extensive, hyperfine splittings from nitrogens on β-carbons appear to follow a dependence on the dihedral angle similar to that of β-protons.[15] The β-proton is sufficiently close to the unpaired electron that the through space coupling can be expected to produce slight anisotropy in the hyperfine interaction. This anisotropy can be detected in well-resolved spectra. On the other hand, anisotropy in splitting from β-nitrogens is normally insignificant in CW spectra because of the small magnetogyric ratio, γ, of ^{14}N and the distance of the β-nitrogen from the center of unpaired spin.

Resolution Enhancement

In all forms of spectroscopy, resolution is a limiting factor in analysis of the experimental data. The detection method used in EPR spectrometers

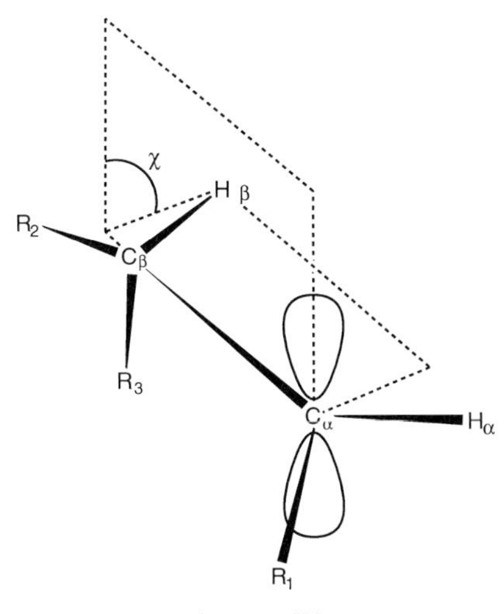

SCHEME IV

(field modulation and detection of the modulation frequency in the microwaves) produces signals that correspond to the first derivative of the absorption. Thus, EPR spectra come from the spectrometer with one form of resolution enhancement (RE), the first derivative, already applied. Various methods for further resolution enhancement have been described.[19] A Fourier transform method of RE was demonstrated for EPR early on by Hedberg and Ehrenberg.[20] Similar methods are in common use in NMR spectroscopy and in FTIR spectroscopy.[21] The underlying principle in these methods is that a spectrum, in the frequency domain, is a convolution of a set of impulses (at discrete frequencies) with a line shape. If we are more interested in the frequencies (positions) of signals in the spectrum than in their line widths, then we can partially deconvolute the frequencies from the line shape to achieve increased resolution. Deconvolution of the line shape from the frequencies involves simple division (or multiplication) in the time domain. Fast Fourier transform routines are widely available,[22] so an inverse Fourier transformation (i.e., frequency domain to time domain) is the first step in the RE process. This first step differs from RE in FTIR or NMR where the experimental data are normally recorded as interferograms or as free induction decays (FIDs) in the time domain. The (inverse) Fourier transform of an EPR spectrum yields real and imaginary transforms (EPR FIDs). The real and imaginary parts are treated in the same way with regard to deconvolution and apodization prior to transformation. The frequency information in the time domain of a spectrum is contained in the periodicity or "wiggles" of the FID whereas the line shape information is contained in the decay or "damping" of the pattern (see Fig. 4b). An exponential decay in the time domain is a characteristic of Lorentzian line shapes, and a decay of this type is a reasonable starting point in RE. Moreover, an assumption of exponential decay for non-Lorentzian line shapes is not restrictive.[20] Dividing the FID by the exponential decay function effectively erases the original line shape or broadening of the signals. The trade-off here is an artificial enhancement of noise, which is comparable to signal strength at the end of the FID. Another problem is the finite length of the data set, where an abrupt cutoff or truncation generates severe side lobes when the transformation back to the frequency domain is performed. To minimize these two problems, an apodization or windowing function is

[19] R. S. Alger, "Electron Paramagnetic Resonance: Techniques and Applications," p. 91. Wiley (Interscience), New York, 1968.

[20] A. Hedberg and A. Ehrenberg, *J. Chem. Phys.* **48,** 4822 (1968).

[21] J. K. Kauppinen, D. J. Moffatt, H. H. Mantsch, and D. G. Cameron, *Appl. Spectrosc.* **35,** 271 (1981).

[22] E. O. Brigham, "The Fast Fourier Transform and Its Applications." Prentice-Hall, Englewood Cliffs, NJ, 1988.

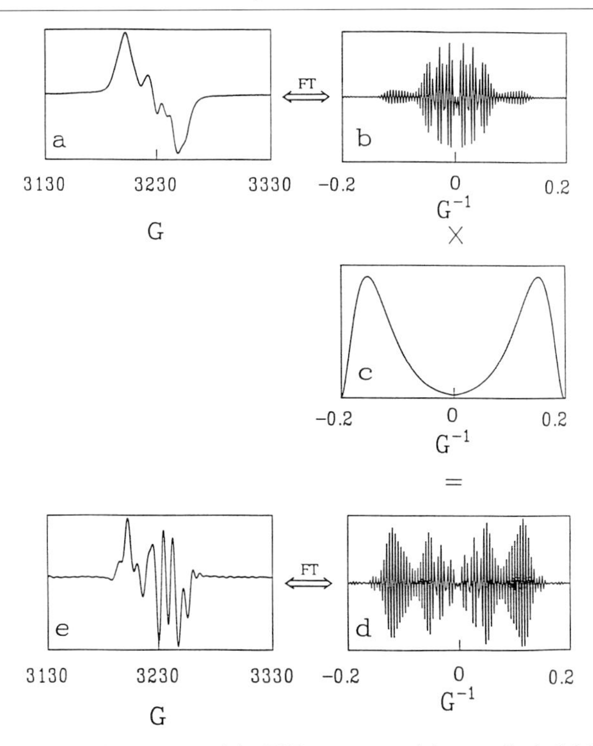

FIG. 4. Resolution enhancement of the EPR spectrum of the α-radical of β-lysine. (a) The spectrum obtained with unlabeled lysine as a substrate, as described in the legend for Fig. 1. (b) The real part of the inverse Fourier transform of the spectrum in a. (c) The combined deconvolution and windowing function. The exponential has a line width, σ, of 8 G and the squared Bartlett window function has a cutoff at $L = 0.2\ G^{-1}$. (d) The result of multiplication of data in b and c. (e) The resulting resolution-enhanced spectrum obtained by Fourier transformation of the deconvoluted real (d) and imaginary (not shown) parts of the time domain data.

applied prior to the final Fourier transformation. In practice the deconvolution and apodization functions are combined and are applied in the same operation (Fig. 4c). Signal theory has provided many different choices for apodization functions. In practice, a more gradual "cutoff" will result in smaller side lobes but less sharpening of the signals. In our experience, the choice of apodization function is not critical in most applications, and the squared Barlett function $[1 - (|x|/L)]^2$ described by Kauppinen *et al.*[21] works well for EPR applications.[23] The apodization function does, however, dictate the line shape of the resolution-enhanced spectrum. In addition to

[23] D. G. Latwesen, M. Poe, J. S. Leigh, and G. H. Reed, *Biochemistry* **31,** 4946 (1992).

the choice of apodization function, there are two variables that influence the final result. The first variable is the line width, σ, in the Lorentzian deconvolution function, $\exp(2\pi\sigma|t|)$. Larger values of σ correspond to more aggressive RE. In principle, setting σ to the width of the narrowest signal in the spectrum is the recommended limit for RE. In practice, σ values ~one-half the width of the narrowest signal provide a reasonable starting point. The other variable is the cutoff or truncation value, L, in the time domain. Larger values for L enhance both resolution and noise. In selecting a value for L, one examines the time domain spectrum (FID), and a reasonable starting value for L is the point at which the signal has decayed to the level of noise. These two variables, σ and L, are then adjusted in an iterative manner to achieve a balance between signal to noise (S/N) and RE. The major limitation to RE is S/N in the original experimental spectrum. The "rule of thumb" is that the factor by which signals can be narrowed is $\log_{10}$ (S/N).[21] Application of RE to the spectrum of the α-radical of β-lysine is illustrated in Fig. 4.

Spectral Simulations

As noted above, the EPR powder patterns of free radicals are difficult (if not impossible) to analyze without the aid of spectral simulations. One can view a powder as a collection of tiny crystals wherein all orientations of the crystals with respect to the external magnetic field are equally probable. However, even in a random orientation of crystals, the directions of the magnetic field vector with respect to principal axes of hyperfine and g tensors are not equally probable. Consequently, the spectral intensity is not distributed evenly across the pattern, and "turning points" in the powder spectrum correspond to principal values of the tensors. It is therefore possible to determine the principal values of tensors from the positions of extrema in powder spectra. The first derivative, dF/dH, of a powder spectrum is computed by numerical integration of the following expression[24]:

$$\frac{dF}{dH} = \int_0^{2\pi} \int_0^{\pi} \frac{dG}{dH} \sin\theta \, d\theta \, d\phi,$$

where G is a Gaussian line shape function centered at a field position, H_r, and θ and ϕ are, respectively, the polar and azimuthal angles of the magnetic field vector in the principal coordinate system (see Scheme III). At each set of angles, θ and ϕ, there will be a hyperfine multiplet centered at $H_c = h\nu/g(\theta, \phi)\beta$, where $g(\theta, \phi) = (g_{xx}{}^2 \sin^2\theta \cos^2\phi + g_{yy}{}^2 \sin^2\theta \sin^2\phi + g_{zz}{}^2 \cos^2\theta)^{1/2}$, h is Planck's constant, and ν is the spectrometer frequency.

[24] H. G. Hecht, "Magnetic Resonance Spectroscopy," p. 129. Wiley, New York, 1967.

The number of lines in the hyperfine multiplet for n nuclear spins is $\Pi_{i=1}^{n}(2I_i + 1)$, where I is the spin of the ith nucleus.

The offsets, ΔH's, of the hyperfine lines for n-coupled nuclear spins are obtained from the expression

$$\Delta H_{\mathrm{M}_{I1}} \ldots \mathrm{M}_{In} = [\mathrm{K}_1(\theta, \phi)\mathrm{M}_{I1} + \ldots + \mathrm{K}_n(\theta, \phi)\mathrm{M}_{In}]/\mathrm{g}(\theta, \phi)\beta$$

where

$$\mathrm{K}_k(\theta, \phi) = \frac{1}{\mathrm{g}(\theta, \phi)}$$

$$(\mathrm{A}_{xx}{}^2\mathrm{g}_{xx}{}^2 \sin^2\theta \cos^2\phi + \mathrm{A}_{yy}{}^2\mathrm{g}_{yy}{}^2 \sin^2\theta \sin^2\phi + \mathrm{A}_{zz}{}^2\mathrm{g}_{zz}{}^2 \cos^2\theta)^{1/2}$$

and M_{Ii} is the nuclear spin quantum number of the ith nucleus; the values of M_{Ii} cover all $2I + 1$ values for each nucleus. The resonant field, $\mathrm{H_r}$, for each component of the multiplet is given by $\mathrm{H_r} = \mathrm{H_c} - \Delta\mathrm{H}$. A line shape is then fitted about these field positions, and the calculations are repeated for a new set of θ and ϕ's. This process is equivalent to summing single crystal spectra over a grid which covers the $\sin\theta\, d\theta$ and $d\phi$ integrals, respectively. Treatment of anisotropic splitting from nuclei other than the α-proton or from a ^{13}C at the radical center is usually more complicated because the principal axes of the tensors will generally not be coincident. Expressions dealing with noncoincident tensors have been provided by Rieger.[25]

For the α-radical of β-lysine, the hyperfine splitting comes from three nuclei: α-^{1}H, β-^{1}H, and β-^{14}N. Spectra from samples with isotopic substitutions at each of these three positions and at other positions in lysine were obtained. The spectrum of the sample prepared with [3,3,4,4,5,5,6,6-^{2}H]lysine replaced splittings from the β-^{1}H with the much smaller splitting from ^{2}H. Fitting of this spectrum, first, provided close approximations for the principal values of the α-^{1}H hyperfine tensor and for the β-^{14}N splitting. The g tensor and hyperfine splitting parameters, or derivatives thereof appropriate for isotopic substitutions, were then used to simulate spectra from samples prepared with other isotopically labeled forms of lysine, in an iterative manner, until a self-consistent set of parameters that gave the best overall fit to all of the spectra was obtained.[8] The degree of difficulty encountered in achieving a simultaneous fit to all of the spectra gave confidence in the uniqueness of the ultimate solution. The resolution-enhanced versions of the experimental spectra were used to guide the simulations. Contributions of the hyperfine splittings from α-^{1}H, β-^{14}N, and β-^{1}H are illustrated in a cumulative manner in Fig. 5.

[25] P. H. Rieger, *J. Magn. Reson.* **50,** 485 (1982).

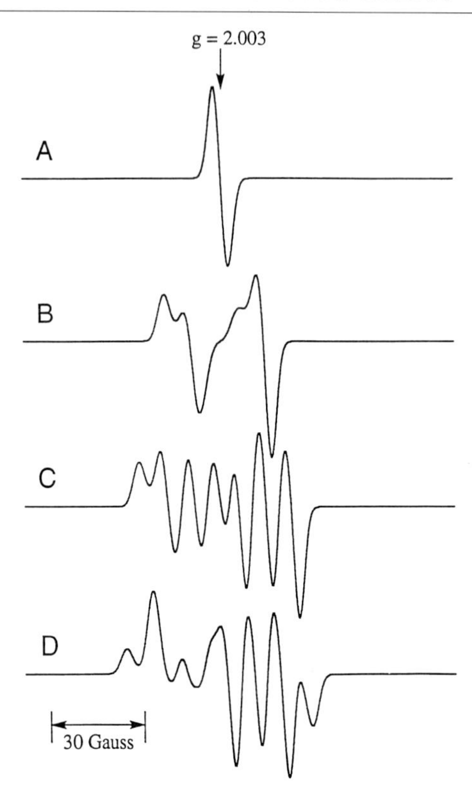

FIG. 5. Simulated EPR spectra illustrating the progression in hyperfine structure as the individual splittings composing the α-radical of β-lysine signal are included. In the simulations, the following interactions were sequentially integrated into the spectrum: (A) The g tensor, $g_{xx} = 2.0043$, $g_{yy} = 2.0037$, $g_{zz} = 2.0021$, (B), the α-^{1}H hyperfine splitting, $A_{xx} = 32.6$ G, $A_{yy} = 10.1$ G, $A_{zz} = 22.5$ G; (C) the β-^{14}N splitting, $a^{14}{}_{N\beta} = 8.3$ G; and (D) the β-^{1}H hyperfine splitting, $A_{xx} = A_{yy} = 3.6$ G, $A_{zz} = 8.8$ G, and Euler angles $\alpha = 3°$ and $\beta = -80°$. A line width of 2.2 G was used, and 2025 crystal orientations were sampled in the simulations.

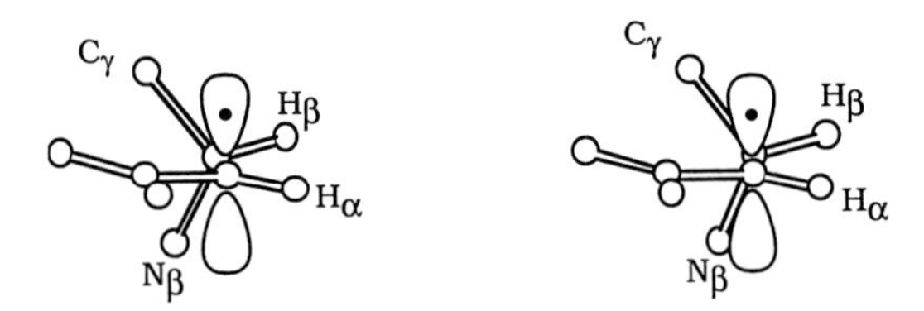

FIG. 6. Stereo view of a ball and stick model of the radical center of the β-lysine radical intermediate in the lysine 2,3-aminomutase reaction. The conformation about the Cα–Cβ bond was obtained from the hyperfine splitting of the β-^{1}H and the β-^{14}N.

Structure

The hyperfine splittings from the α-^{1}H, β-^{1}H, and β-^{14}N indicate a structure for the radical. In this structure, the spin density, ρ, at the α-carbon is ~80%, as calculated from the isotropic part of the α-proton hyperfine splitting used in the simulations. The dihedral angle, $\chi \cong 70°$ (see Scheme IV), was derived from the hyperfine splitting of the β-^{1}H.[8] A schematic view of the conformation of the radical center is shown in Fig. 6.

Acknowledgments

We gratefully acknowledge the collaboration of Dr. Perry A. Frey in the research on lysine 2,3-aminomutase. We also thank Dr. Russell R. Poyner for preparation of Fig. 4. The research described was supported by research NIH Grants GM35752 (G. H. R.) and DK28607 (P. A. F.) and by an NRSA Fellowship (M. D. B.) from NIH Training Grant GM08293.

[25] Role of Oxidized Amino Acids in Protein Breakdown and Stability

By EARL R. STADTMAN

Since the early 1980s, it has become evident that the free radical-mediated modification of protein is associated with a variety of physiological disorders, including aging, atherosclerosis, arthritis, diabetes, cataractogenesis, muscular dystrophy, pulmonary dysfunction, and a number of neurological disorders.[1] The kinds of protein modification include fragmentation of the polypeptide chain, the generation of protein–protein cross-linkage, and the oxidation of amino acid side chains. The possible relevance of such modifications to studies of quinoproteins derives from the consideration that some radical-mediated modifications lead to quinoprotein-like derivatives and also the possibility that principles utilized in these modification reactions are involved also in the biosynthesis of biologically active quinoproteins and in the mechanisms of their catalytic activities.

The types of protein modification that have been observed and the kinds of "active-oxygen" generating systems employed are so varied that detailed descriptions of the numerous methods used in their study cannot be achieved within the space allocated. Moreover, those methods unique to the studies carried out in the author's laboratory have already been

[1] E. R. Stadtman and C. N. Oliver, *J. Biol. Chem.* **266,** 2005 (1991).

published.[2,3] The purpose of this contribution is to provide an overview of studies on oxidative protein modifications with appropriate references to the methodology that has been used in their determinations.

The oxidation of proteins is mediated by interactions with one or more reactive oxygen species ($\dot{O}H$, $O_2\cdot^-$, H_2O_2, O_3, $^1O_2$, ferryl, perferryl), which may be generated by ionizing radiation, metal ion-catalyzed reactions, photochemical processes, or as by-products of enzyme-catalyzed redox reactions. Elucidation of the basic principles that underlie these modifications comes largely from the pioneering studies of Garrison,[4,5] Swallow,[6,7] Schuessler,[8,9] and their collaborators. These workers took advantage of the fact that if the radiolysis of water is carried out in the presence of N_2O (anaerobic), O_2 or a mixture of O_2 and formate, then as a consequence of secondary reactions, the only major oxygen radicals formed are $\dot{O}H$ only, $\dot{O}H$ plus $O_2\cdot^-$ or $O_2\cdot^-$ only, respectively, according to the overall reactions 1–3:

$$H_2O + N_2O \rightarrow 2\,\dot{O}H + N_2 \quad (1)$$
$$H_2O + O_2 \rightarrow O_2\cdot^- + \dot{O}H + H^+ \quad (2)$$
$$2\,O_2 + HCOO^- \rightarrow CO_2 \rightarrow 2\,O_2\cdot^- + H^+. \quad (3)$$

Therefore, the modification of protein under these three conditions of radiolysis could be attributed to the action of the specific radical species generated and also O_2, when present. From the results of these studies, it became evident that the modification of proteins is initiated by interactions of $\dot{O}H$ with the main chain α-CH- groups, the hydrocarbon side chains, and the functional groups of amino acid residues. This leads to scission of the polypeptide chain, the conversion of amino acid residue side chains to hydroxy or carbonyl derivatives, and the formation of protein–protein cross-linkages.

Oxygen Radical-Mediated Cleavage of the Polypeptide Chain

In the presence of O_2, radiolysis leads to cleavage of the polypeptide main chain by either of two pathways depicted in Fig. 1. Plausible mecha-

[2] R. L. Levine, D. Garland, C. N. Oliver, A. Amici, I. Climent, A.-G. Lenz, B.-W. Ahn, S. Shaltiel, and E. R. Stadtman, this series, Vol. 186, p. 464.

[3] R. L. Levine, J. A. Williams, E. R. Stadtman, and E. Shacter, this series, Vol. 133, p. 346.

[4] W. M. Garrison, *Chem. Rev.* **87,** 381 (1987).

[5] W. M. Garrison, M. E. Jayko, and W. Bennett, *Radiat. Res.* **16,** 487 (1962).

[6] A. J. Swallow, *in* "Radiation Chemistry of Organic Compounds" (A. J. Swallow, ed.), p. 211. Pergamon, New York, 1960.

[7] R. C. Armstrong and A. J. Swallow, *Radiat. Res.* **41,** 563 (1969).

[8] H. Schuessler and A. Herget, *Int. J. Radiat. Biol.* **37,** 71 (1980).

[9] H. Schuessler and K. Schilling, *Int. J. Radiat. Biol.* **45,** 267 (1984).

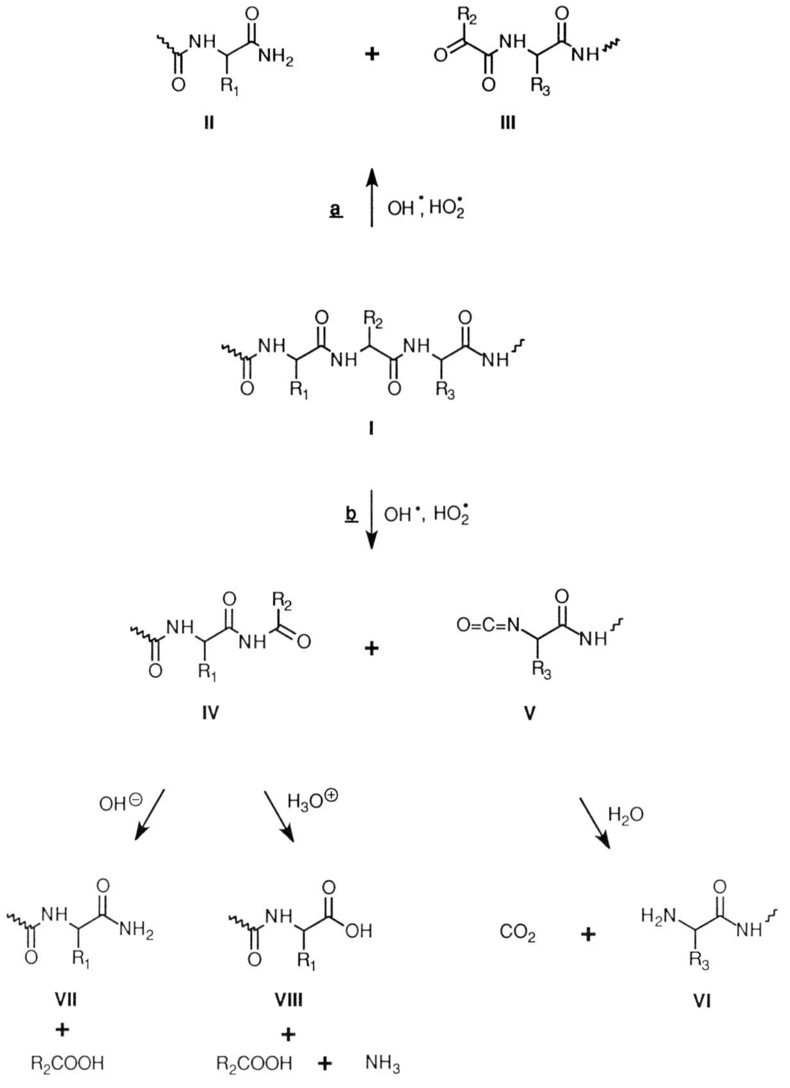

FIG. 1. Oxidative cleavage of the polypeptide chain.

nisms have been discussed by Garrison.[4] In both pathways (a and b, Fig. 1), the primary reaction with $\dot{O}H$ is assumed to involve α-hydrogen abstraction from the R_2 residue to form a carbon-centered radical, followed by transient, sequential reactions with O_2 and $HO_2^{\cdot}$ to produce alkylperoxy radical and alky peroxide intermediates, and, in the case of pathway b, to the further formation of an alkoxy radical intermediate. Subsequent decomposition of these intermediates leads finally to the cleavage products illustrated in Fig. 1.[4,10,11]

It is evident from the scheme shown in Fig. 1 that the C-terminal amino acid residues of peptide fragment II obtained by pathway a will exist as the amide derivative and that the N-terminal amino acid residue of peptide fragment III will exist as an α-ketoacyl derivative. Therefore, on acid hydrolysis, NH_3 will be released from fragment II and an α-ketoacid will be generated from fragment III. The resistance of fragment III to Edmond degradation is therefore diagnostic for cleavage by the α-amidation pathway a, and its occurrence can be confirmed and quantified by conversion of the α-ketoacyl moiety to its 2,4-dinitrophenylhydrazone derivative, either before[1,5] or after[5] acid hydrolysis.

Peptide fragment IV produced in pathway b (Fig. 1) possesses a diamide structure at the C-terminal position. On acid hydrolysis, this yields fragment VIII in which the R_1-amino acid residue occupies the C-terminal position, and also NH_3 and a carboxylic acid (R_2COOH), which is derived from the R_2-amino acid residue. On mild alkaline hydrolysis, fragment IV is converted to R_2COOH plus fragment VII in which the C-terminal amino acid residue is the amide derivative of the R_1 residue. Formation of the carboxylic acid R_2COOH is therefore diagnostic for occurrence of pathway b. Significantly, fragment V, produced as a primary product of the cleavage reaction by pathway b, possesses an isocyanate group at the N-terminal position; this may undergo rapid hydrolysis to form peptide fragment VI (Fig. 1), which contains the unblocked R_3-amino acid residue at the N-terminal position.[4,10]

Site-Specific Cleavage Reactions

In addition to the more general mechanisms depicted in Fig. 1, site-specific cleavage of the polypeptide chain can occur as a result of the oxidation of proline and glutamic acid (or aspartic acid) residues. Based on the observation that number of peptide fragments obtained when bovine serum albumin (BSA) is exposed to radiolysis is approximately equal to the number of proline residues in the protein, and also the consideration that tertiary amide bonds are more susceptible to oxidation than secondary amide bonds, Schuessler and Schilling[9] proposed that proline residues are preferred sites of peptide bond cleavage. This proposition was subsequently

[10] I. A. Platis, M. R. Ermacora, and R. O. Fox, *Biochemistry* **32,** 12761 (1993).

[11] R. C. Bateman, Jr., W. W. Youngblood, W. H. Busby, Jr., and J. S. Kizer, *J. Biol. Chem.* **260,** 9088 (1985).

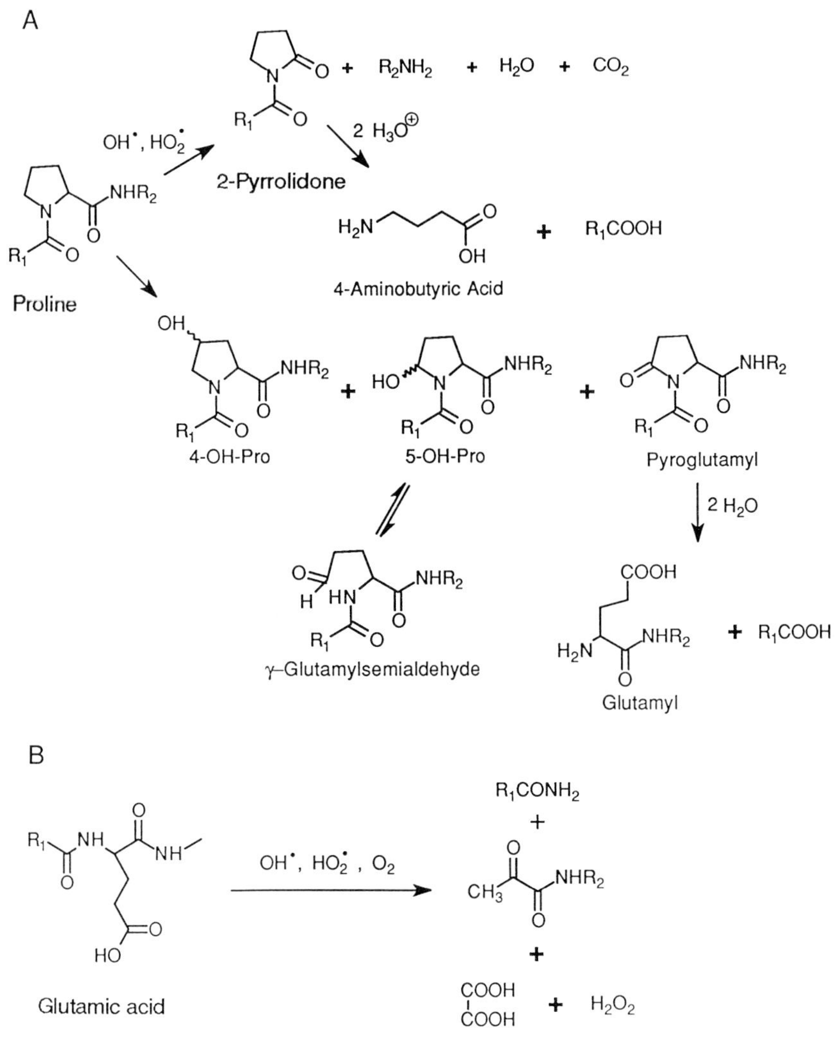

A

O + R_2NH_2 + H_2O + CO_2

R_1 O

$OH^\bullet$, $HO_2^\bullet$

2 $H_3O^\oplus$

2-Pyrrolidone

NHR_2

N O

R_1 O

Proline

H_2N O OH + R_1COOH

4-Aminobutyric Acid

OH

NHR_2 + HO NHR_2 + O NHR_2

N O N O N O

R_1 O R_1 O R_1 O

4-OH-Pro 5-OH-Pro Pyroglutamyl

2 H_2O

O NHR_2

H HN O

R_1 O

γ-Glutamylsemialdehyde

COOH

H_2N NHR_2 + R_1COOH

O

Glutamyl

B

R_1 NH O NH

O

O

HO

Glutamic acid

$OH^\bullet$, $HO_2^\bullet$, O_2

R_1CONH_2

+

O

CH_3 NHR_2

O

+

COOH

COOH + H_2O_2

FIG. 2. Site-specific cleavage of the polypeptide chain mediated by oxidation of prolyl (A) and glutamyl (B) residues.

verified by the studies of Uchida *et al.*[12,13] showing that the metal-catalyzed oxidation (MCO) of protein residues in collagen and proline-containing peptides leads to the generation of 2-pyrrolidone residues and concomitant cleavage of the peptide bond (Fig. 2A). Because acid hydrolysis of 2-

[12] K. Uchida, Y. Kato, and S. Kawakishi, *Biochem. Biophys. Res. Commun.* **169,** 265 (1990).
[13] Y. Kato, K. Uchida, and S. Kawakishi, *J. Biol. Chem.* **267,** 23646 (1992).

pyrrolidone leads to the formation of 4-aminobutyric acid, the amount of this compound in acid hydrolysates may be used as a presumptive measure of peptide bond cleavage by this pathway.[12]

In addition to 2-pyrrolidone, 4-hydroxyproline,[14] γ-glutamylsemialdehyde,[15] and pyroglutamic acid[13,15] have been identified among the oxidation products of proline residues. Moreover, glutamic acid, likely derived from pyroglutamyl residues, has been identified as a major derivative of proline oxidation in acid hydrolysates of proline-rich proteins.[13,15,16] The possibility that the conversion of pyroglutamyl residues to glutamyl residues might be a major route of peptide bond cleavage associated with protein oxidation has also been considered[17] (see Fig. 2A); however, there is no direct evidence that this reaction occurs under physiological conditions.

Site-specific cleavage may arise also from the oxidation of glutamyl residues, according to the overall reaction shown in Fig. 2B. A reasonable mechanism involving the $\dot{O}H$-dependent abstraction of hydrogen from the γ-carbon of the glutamyl side chain, followed by reactions with O_2 and $HO_2^{\cdot}$ to yield alkylperoxy radical and alkoxy radical intermediates, has been described by Garrison.[4] Whatever the mechanism, this is a unique example of peptide bond cleavage since it involves H-abstraction from a side chain carbon of an amino acid residue instead of from the α-hydrogen of a main chain carbon atom. An analogous site-specific process occurs also with the aspartyl acid residues of proteins. Notably, in the case of glutamyl residues, this kind of cleavage results in the production of oxalic acid and peptide fragment II (derived from the N-terminal portion of the protein) in which the C-terminal residue exists as the amide derivative and peptide fragment III (derived from the C-terminal portion of the protein) in which the N terminus is blocked by a pyruvyl group (Fig. 2B).

Side Chain Modifications

Aromatic Amino Acid Residues

The aromatic amino acid residues of proteins are particularly sensitive to oxidation by ozone, singlet oxygen, radiolysis, and metal ion-catalyzed reactions. As in the case of many active oxygen-mediated reactions, the detailed mechanisms of the modification reactions are not known; however,

[14] J. M. Poston, *Fed. Proc., Fed. Am. Soc. Exp. Biol.* **46,** 1979 abstr. (1988).
[15] A. Amici, R. L. Levine, L. Tsai, and E. R. Stadtman, *J. Biol. Chem.* **264,** 3341 (1989).
[16] B. Cooper, J. M. Creeth, and A. S. R. Donald, *Biochem. J.* **228,** 615 (1985).
[17] S. P. Wolfe, A. G. Garner, and R. T. Dean, *Trends Biochem. Sci.* **11,** 27 (1986).

in some cases, the oxidation products have been identified and reasonable explanations for their formation have been proposed.

Histidine Residues. Histidine residues are particularly susceptible to oxidative modification by all forms of active oxygen. As illustrated in Fig. 3, histidine residues are converted to asparagine, aspartic acid,[18,19] and 2-oxohistidine[20] residues. By means of reverse-phase, high-performance liquid chromatography and electrochemical detection, Uchida and Kawakishi[21] developed a method for the quantitation of 2-oxohistidine. In contrast to MCO systems (see below), the modification of histidine residues by ozone and radiolysis is a more or less random process; however, the rate of oxidation is a function of the primary, secondary, and quaternary structure. For example, although glutamine synthetase and BSA have nearly identical numbers of histidine residues per subunit (50–60 kDa), the rate of histidine destruction in BSA by ozone is more than two times greater than in glutamine synthetase.[18] Furthermore, whereas the rates of histidine destruction by ozone in tripeptides containing two alanine and one histidine residues are independent of the amino acid sequence, the yields of aspartate and/or asparagine are 32, 75, and 100% for the peptides in which histidine occupies the N-terminal, C-terminal, and central positions, respectively.[18]

Phenylalanine Residues. When proteins are exposed to hydroxyl radical-generating systems, the phenylalanine residues are converted to their 2-, 3-, and 4-hydroxyphenylalanine (tyrosine) derivatives (Fig. 3).[22–24] By means of selected ion-monitoring gas chromatography/mass spectrometry and by using deutero-*o*-tyrosine (d_4-*o*-tyr) as an internal standard, Baynes and his collaborators[23,25] demonstrated that *o*-tyrosine is the major hydroxyphenylalanine derivative produced when proteins are exposed to a MCO system [H_2O_2/Cu(II)] or to radiolytic oxidation. They suggest that the formation of *o*-tyrosine is a reliable indicator of radical-mediated protein damage even though the yield of this compound under their experimental conditions is quite low (10–50 nmol)/mole of phenylalanine residue).

[18] B. S. Berlett and E. R. Stadtman, unpublished data.

[19] J. M. Farber and R. L. Levine, *J. Biol. Chem.* **261,** 4574 (1986).

[20] K. Uchida and S. Kawakishi, *Bioorg. Chem.* **17,** 330 (1989).

[21] K. Uchida and S. Kawakishi, *J. Biol. Chem.* **269,** 2405 (1994).

[22] Z. Maskos, J. D. Rush, and W. H. Koppenol, *Arch. Biochem. Biophys.* **296,** 521 (1992).

[23] M. C. Wells-Knecht, T. G. Huggins, D. G. Dyer, S. R. Thorpe, and J. W. Baynes, *J. Biol. Chem.* **269,** 12348 (1993).

[24] I. Balakrishnan and M. P. Reddy, *J. Phys. Chem.* **74,** 850 (1970).

[25] T. G. Huggins, M. C. Wells-Knecht, N. A. Detorie, J. W. Baynes, and S. R. Thorpe, *J. Biol. Chem.* **268,** 12341 (1993).

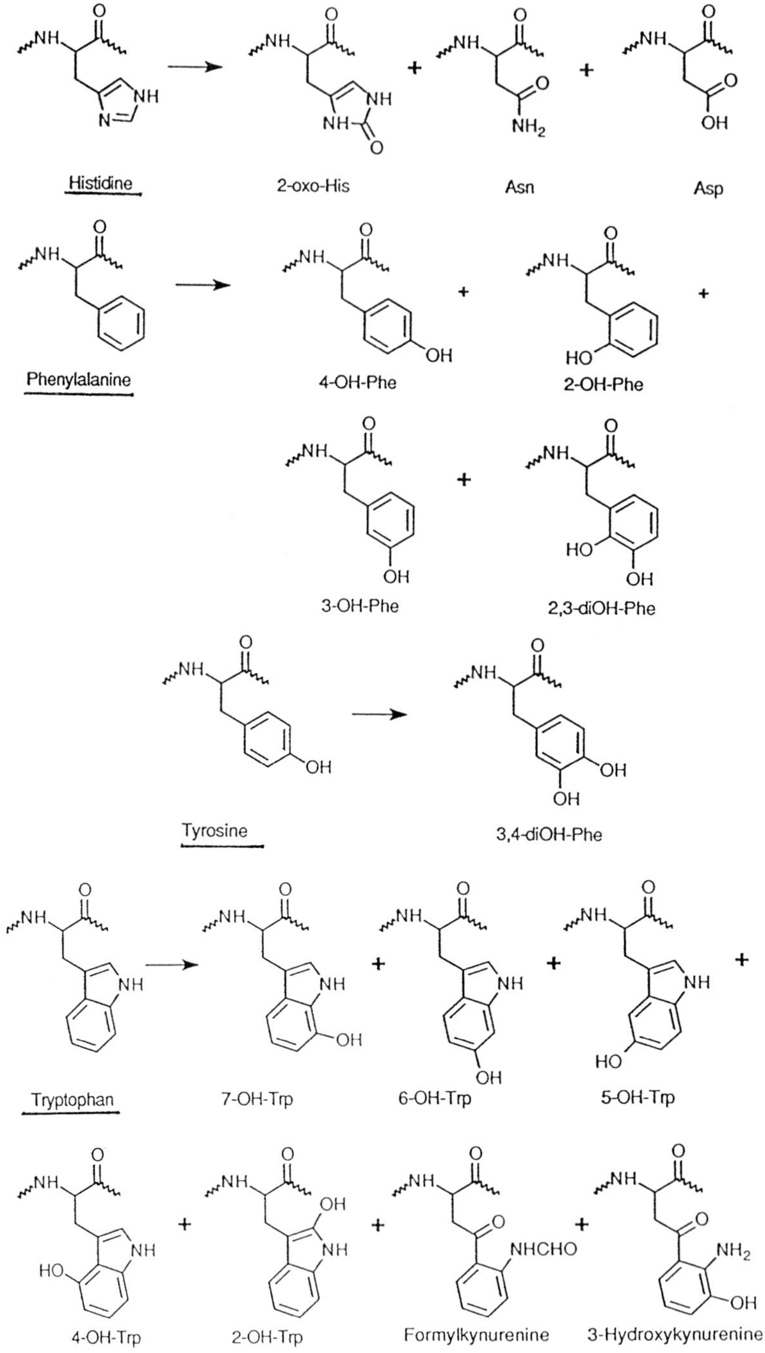

Histidine → 2-oxo-His + Asn + Asp

Phenylalanine → 4-OH-Phe + 2-OH-Phe + 3-OH-Phe + 2,3-diOH-Phe

Tyrosine → 3,4-diOH-Phe

Tryptophan → 7-OH-Trp + 6-OH-Trp + 5-OH-Trp + 4-OH-Trp + 2-OH-Trp + Formylkynurenine + 3-Hydroxykynurenine

FIG. 3. Products formed in the oxidation of aromatic amino acid residues.

Tyrosine Residues. Radiolytic and metal-catalyzed oxidation of tyrosine residues leads mainly to the production of 3,4-dihydroxyphenylalanine (dopa)[22,26–29] and to the generation of tyrosine–tyrosine cross-linkages (Fig. 3).[23,25,30–32]

Fluorescence spectroscopy has been generally used to measure the extent of dityrosine (DT) cross-linkages in oxidized proteins. This method is based on the fact that, upon irradiation at 325 nm, DT exhibits strong fluorescence at 410–420 nm. Unfortunately, estimates of the dityrosine content of proteins based solely on fluorescence measurements are invalid because several other protein modifications (viz., the conversion of tryptophan to *N*-formylkynurenine, the conjugation of 4-hydroxynonenal with lysine residues[33]; the formation of retinoic acid adducts with proteins[34]) also yield products that exhibit similar fluorescence characteristics. A more reliable procedure has been described by Guilivi and Davies,[31] which involves fluorescence measurements of DT after its separation from other amino acids in proteolytic (pronase) digests by means of reverse-phase high-performance liquid chromatography. Details of this method have been described.[35]

In another procedure described by Wells-Knecht *et al.,*[23] deuterium-labeled DT (d_6-DT) containing three deuterium atoms in each of the phenyl rings was added as an internal standard to protein hydrolysates. Then, following enrichment of the DT by means of a solid-phase Supelclean LC-18 extraction tube, the amino acids were converted to their *N,O*-pentafluropropyl isopropyl esters and the mixture was subjected to selected ion monitoring–gas chromatography/mass spectrometry, and the concentrations of major ions *m/z* 865 derived from DT and *m/z* 871 derived from d_6-DT were measured. The amount of DT was then calculated from the ratio of the two ion species and the known concentration of d_6-DT that was added as an internal standard. The content of DT in eye lens proteins determined by this highly specific procedure was less than 1% of that determined by the protein fluorescence technique. This emphasizes the

[26] G. L. Fletcher and S. Okada, *Radiat. Res.* **15,** 349 (1961).

[27] S. P. Gieseg, J. A. Simpson, T. S. Charlton, M. W. Duncan, and R. T. Dean, *Biochemistry* **32,** 4780 (1993).

[28] K. J. A. Davies, M. E. Delsignore, and S. W. Lin, *J. Biol. Chem.* **262,** 9902 (1987).

[29] R. T. Dean, S. Gieseg, and M. J. Davies, *Trends Biochem. Sci.* **18,** 437 (1993).

[30] G. Boguta and A. Dancewicz, *Int. J. Radiat. Biol.* **39,** 163 (1981).

[31] C. Guilivi and K. J. A. Davies, *J. Biol. Chem.* **268,** 8752 (1993).

[32] J. W. Heinecke, W. Li, H. L. Deehnke, III, and J. A. Goldstein, *J. Biol. Chem.* **268,** 4069 (1993).

[33] L. Szweda, B. Friguet, and E. R. Stadtman, unpublished data (1994).

[34] L. Szweda, *J. Biol. Chem.* (in press) (1994).

[35] C. Guilivi and K. J. A. Davies, this series, Vol. 233, p. 363.

need for caution in the interpretation of results based on fluorescence measurements only.

It has been known since the early 1960s that dopa derivatives of tyrosine residues are produced when proteins are subjected to radiolysis. Renewed interest in the conversion of tyrosine residues to dopa was provoked by the studies in Dean's laboratory[27] showing that exposure of proteins to ionizing radiation and to MCO systems leads to the production of low levels of reducing activity as measured by the ability to reduce cytochrome *c*, Fe(III), and Cu(II). This activity was partly attributed to the conversion of some tyrosine residues to dopa derivatives. Because protein dopa derivatives are capable of redox recycling, it was suggested[36] that they might have a catalytic role in the progression of aging and age-related diseases by virtue of their ability to reduce Fe(III) and Cu(II), and thereby facilitate the generation of hydroxyl radicals from hydrogen peroxides via Fenton chemistry. This is an interesting suggestion; however, the physiological significance of nonspecific, radiation-induced dopa generation remains to be established. The γ-irradiation dose used in these radiolysis studies are orders of magnitude greater than that to which normal individuals would be exposed, and the levels of uncomplexed Cu(II), 15 m*M*, required to generate dopa derivatives[36] (even in the presence of very high H_2O_2 concentrations, 5 m*M*) are much greater than that normally present *in vivo*. The ability of quinones to catalyze the reduction of nitroblue tetrazolium (NBT) in the presence of a high concentration of glycine is the basis of a general method developed by Paz *et al.*[37] for the estimation of quinoproteins (see article [11], this volume). When applied to some proteins in crude tissue extracts, the method may give low values because the reduced dye forms insoluble complexes with some proteins. This problem is partly resolved by extracting the reduced NBT from reaction mixtures with *n*-butanol, as follows[18]:

Reagents

2 *M* KGLY buffer: prepared by adjusting a glycine solution (2 *M* final concentration) with KOH

0.24 m*M* NBT: prepared just before use by dissolving NBT in the 2 *M* KGLY buffer.

[36] R. T. Dean, J. Gebicki, S. G. Gieseg, A. J. Grant, and J. A. Simpson, *Mutat. Res.* **275,** 387 (1992).

[37] M. A. Paz, R. Fluckiger, A. Boak, H. M. Kagan, and P. M. Gallop, *J. Biol. Chem.* **266,** 689 (1991).

Procedure

In a 12 × 75-mm test tube, place the protein sample (1–2 mg) and add 2 *M* KGLY buffer to a final volume of 0.4 ml. Then add 1.0 ml of the 0.24 m*M* NBT reagent, vortex, and incubate in a 37° waterbath for 20 min in the dark. If the reaction mixtures are turbid after incubation, add 1.4 ml of *n*-butanol, vortex, and centrifuge at 3000 rpm. Siphon off the butanol phase and measure absorbance at 530 nm. Compare the absorbance values with a standard curve generated by the same procedure using L-dopa (0–1 nmol) as the reference standard.

The *n*-butanol extraction serves two purposes: it facilitates dissociation of reduced NBT from the protein and it leads to separation of the reduced NBT from turbid reaction mixtures. Using this procedure, it was determined that rat hepatocytes contain about 0.4 nmol of quinoprotein-like material per mg of total cellular protein and that the concentration of NBT-reactive protein in 3-month-old Sprague–Dawley rats is the same as that in 24-month-old animals.[18]

Tryptophan Residues. The benzenoid and indole rings of tryptophan residues are both susceptible to attack by ozone, radiolysis, and singlet oxygen. As shown in Fig. 3, radiolysis leads to hydroxylation of the C-2 position of the indole ring[7,38] and the 4, 5, 6, and 7 carbon atoms of the benzenoid ring.[38,39] Upon exposure to radiolysis, ozone, metal ion-catalyzed reactions, or to singlet oxygen, the indole ring of the tryptophan residue is cleaved with the formation of *N*-formylkynurenine.[40–42]

Aliphatic Amino Acid Residues

All aliphatic amino acid residues of proteins are potential targets for modification by radiolysis.[4] The radiolytic attack of glycine and alanine residues occurs almost exclusively at the α-carbon atom of the polypeptide backbone and leads to peptide bond cleavage as described in Fig. 1.[4] Radiolysis of aliphatic amino acid residues possessing long hydrocarbon side chains may involve preferential hydrogen abstraction from the side chain carbon atoms with the ultimate formation of hydroxy and carbonyl derivatives. Thus, as shown in Fig. 4, oxidation of valine residues yields 3-hydroxyvaline and oxidation of leucine residues gives rise to 3-hydroxy- and 4-

[38] Z. Maskos, J. D. Rush, and W. H. Koppenol, *Arch. Biochem. Biophys.* **296,** 514 (1992).
[39] S. Solar, *Radiat. Phys. Chem.* **26,** 103 (1985).
[40] R. V. Winchester and K. R. Lynn, *Int. J. Radiat. Biol.* **17,** 541 (1970).
[41] P. Guptasarma and D. Balasubramanian, *Biochemistry* **31,** 4296 (1992).
[42] W. A. Pryor and R. M. Uppu, *J. Biol. Chem.* **268,** 3120 (1993).

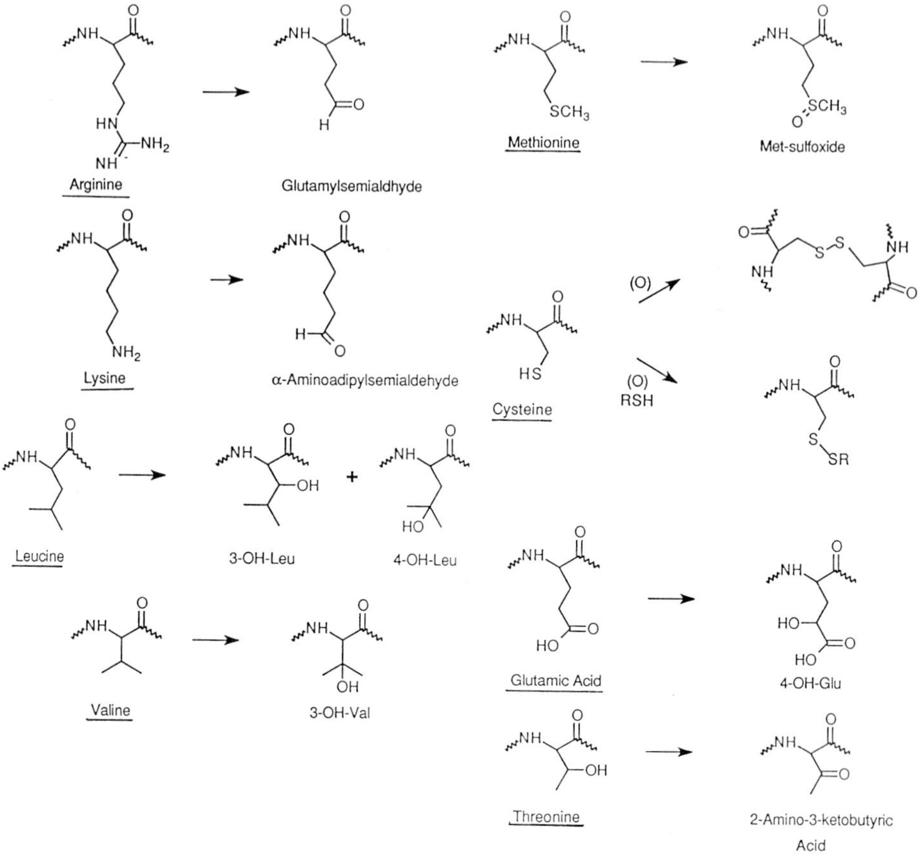

FIG. 4. Products formed in the oxidation of aliphatic amino acid residues.

hydroxyleucine.[4] Neither of these hydroxy compounds has been detected in metal ion-catalyzed reactions or as products of ozonolysis.

Aliphatic acid residues possessing functional side chain groups are all quite susceptible to free radical attack. As summarized in Fig. 4, lysine residues are converted to adipic semialdehyde derivatives[43]; methionine residues are converted to methionine sulfoxide and methionine sulfonyl derivatives[5,44]; cysteine residues are converted to protein–protein disulfide cross-linkages and to mixed disulfides[5,44]; arginine residues are converted

[43] D. G. Miller and E. R. Stadtman, unpublished observations.

[44] A. J. Swallow, *in* "Radiation Chemistry of Organic Compounds" (A. J. Swallow, ed.), p. 211. Pergamon, New York, 1960.

to glutamic semialdehyde residues[15,45]; threonine residues are converted to 2-amino-3-ketobutyric acid residues[46]; and the glutamic acid residues are converted to 4-hydroxyglutamic acid residues.[4] In addition, non-heme iron centers of proteins are destroyed.[47]

Metal Ion-Catalyzed Reactions

As noted earlier, the modifications of proteins by radiolysis and by high concentrations of Fenton's reagent [Cu(II) or Fe(III) and H_2O_2] are very similar. With both systems, all amino acid residues, especially the aromatic amino acid residues, and sulfur-containing amino acid residues are prime targets. With high concentrations of Fenton's reagent, large quantities of $\dot{O}H$ are generated in solution according to reactions 4–7:

$$H_2O_2 + Cu(II) \rightarrow Cu(I) + HO_2^{\cdot} + H^+ \quad (4)$$
$$HO_2^{\cdot} + Cu(II) \rightarrow Cu(I) + O_2 = H^+ \quad (5)$$
$$2\ H_2O_2 + 2\ Cu(I) \rightarrow 2\ \dot{O}H + 2\ OH^- + O_2 \quad (6)$$
$$\text{SUM}\ 3\ H_2O_2 \rightarrow 2\ \dot{O}H + 2\ H_2O + O_2. \quad (7)$$

It is, therefore, not surprising that reactions mediated by high concentrations of Fenton's reagent are very similar to those produced by radiolysis under conditions where $\dot{O}H$ is the major product (cf. reactions 1–3). However, distinction must be made between the reactions catalyzed by nonphysiological concentrations of Fenton's reagent and the MCO systems in which the reactive oxygen species are generated by reduction of O_2 and Fe(III) or Cu(II) by enzyme-catalyzed and nonenzymatic redox reactions.

The enzymes involved in the latter class of MCO systems include a number of flavoproteins, such as the NADH and NADPH oxidases, NADH quinone reductase, cytochrome P450 reductase/oxidase, and xanthine oxidase.[1,48,49] Nonenzymic MCO systems include the ascorbate/Fe(III)/O_2[49] and the RSH/Fe(III)/O_2[50] systems. A common feature of all these systems is their ability to reduce Fe(III) and Cu(II) to Fe(II) and Cu(I), respectively, and to reduce O_2 to $O_2^{\cdot -}$ and/or H_2O_2. In contrast to the classical Fenton systems, the modification of proteins by these MCO systems is much more limited and selective. Characteristically, only one or a few amino acid residues at metal-binding sites on a protein are the preferred targets. This

[45] I. Climent, L. Tsai, and R. L. Levine, *Anal. Biochem.* **182,** 226 (1989).
[46] G. Taborsky, *Biochemistry* **12,** 1341 (1973).
[47] P. R. Gardner and I. Fridovich, *J. Biol. Chem.* **267,** 8757 (1992).
[48] E. R. Stadtman, *Free Radical Biol. Med.* **9,** 315 (1990).
[49] R. L. Levine, C. N. Oliver, R. M. Fulks, and E. R. Stadtman, *Proc. Natl. Acad. Sci. U.S.A.* **78,** 2120 (1981).
[50] K. Kim, S. G. Rhee, and E. R. Stadtman, *J. Biol. Chem.* **260,** 15394 (1985).

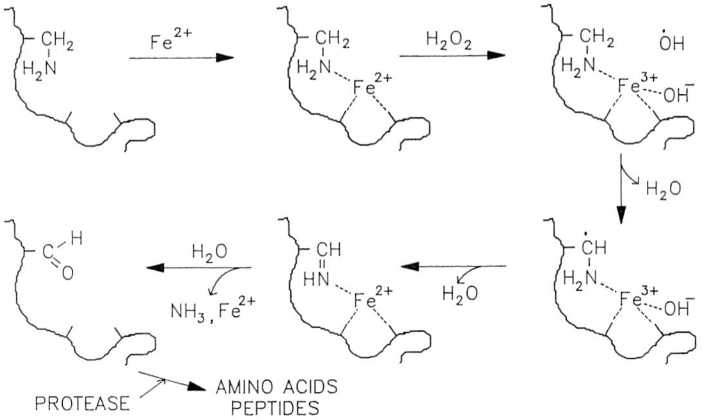

FIG. 5. Possible mechanism for site-specific metal-catalyzed oxidation of amino acid residues in proteins. In this model, it is assumed that a lysyl residue ($-CH_2NH_2$) is one of several ligands to which Fe(II) binds. Interaction of the protein-bound Fe(II) with H_2O_2 leads to the production of $\dot{O}H$, OH^-, and Fe(III). Abstraction of a hydrogen atom from the $-CH_2NH_2$ carbon leads to formation of a carbon-centered radical ($-\dot{C}HNH_2$). Transfer of the lone electron from the alkyl radical to Fe(III) leads to regeneration of Fe(II) and formation of an immino derivative, which undergoes rapid hydrolysis to form the adipic semialdehyde derivative of the lysyl side chain, and conversion of the protein to a form that is susceptible to proteolytic degradation. Although $\dot{O}H$ is depicted as the reactive oxygen species produced by the interaction of Fe(II) with H_2O_2, other reactive oxygen species, such as ferryl or perferryl, might be involved as noted.

prompted the proposal that Fe(II) or Cu(I) and H_2O_2 generated by these MCO systems interact mainly at metal-binding sites on proteins to yield $\dot{O}H$ or some other form of active oxygen (ferryl; perferryl ion), which preferentially attack the side chains of proximal amino acid residues. One possible mechanism is illustrated in Fig. 5. These reactions are thus visualized as "caged" reactions, accounting for their high specificity and the fact that the modification reactions are not very sensitive to inhibition by free radical scavengers. The site-specific nature of the reaction is supported by the results of detailed studies on the oxidative modification of *Escherichia coli* glutamine synthetase,[45,51] which demonstrate that the two histidines and one arginine residue that are modified by the ascorbate/Fe(III)/O_2 MCO system are located at metal-binding sites on the protein. Significantly, in contrast to radiolysis and high concentrations of the Fenton reagent, tryptophan, tyrosine, and phenylalanine are not major targets for modification by this class of MCO systems. This is presumably because these aromatic acids are not commonly present at metal-binding sites of proteins.

[51] J. A. Sahakian, B. D. Shames, and R. L. Levine, *FASEB J.* **5,** A1177 (1991).

Instead, histidine, lysine, and arginine residues are major targets of these MCO systems, and these amino acids are among those most commonly found at metal-binding sites.[19,45]

Formation of Carbonyl Groups

It is noteworthy that the oxidative cleavage of the peptide main chain (Figs. 1 and 2) and oxidation of the side chains of glutamyl, aspartyl, lysyl, arginyl, prolyl, and threonyl residues leads to the generation of carbonyl derivatives. Because the carbonyl content of protein is a putative measure of oxidative damage, several highly sensitive methods have been developed for the quantitation of protein carbonyl groups. Detailed procedures for these methods have been published previously.[2,3] By these techniques, it has been established that there is an exponential increase in the level of oxidized protein in various tissues as a function of animal age and that protein oxidation is associated with a number of biological proceses and pathological conditions.[1,52]

Role of Oxidative Modifications of Proteins in Protein Turnover

The proposition that oxidative modification targets proteins for degradation was highlighted by the results of studies with *E. coli* showing that metal ion-catalyzed oxidation predisposes glutamine synthetase and several other enzymes to proteolytic degradation. This led to clear enunciation of the principle that oxidative modification "marks" proteins for degradation.[49,53] In the meantime, this concept gained support from studies showing that the oxidized forms of proteins are more susceptible than their native forms to degradation by purified preparations of various proteases and that the degradation of endogenous proteins is accelerated when intact cells are exposed to "oxidative stress."[54–56]

[52] E. R. Stadtman, *Science* **257,** 1220 (1992).

[53] C. N. Oliver, R. L. Levine, and E. R. Stadtman, *in* "Metabolic Interconversion of Enzymes" (H. Holzer, ed.), p. 259. Springer-Verlag, Berlin, 1980.

[54] K. J. A. Davies, *J. Free Radicals Biol. Med.* **2,** 155 (1986).

[55] A. J. Rivett, *Curr. Top. Cell. Regul.* **28,** 291 (1986).

[56] E. R. Stadtman, *Biochemistry* **29,** 6323 (1990).

Author Index

Numbers in parentheses are footnote reference numbers and indicate that an author's work is referred to although the name is not cited in the text.

A

Abeles, R. H., 74, 82
Åberg, A., 2, 361
Aberth, W., 92
Abrams, W., 127
Achiwa, K., 53, 58(3)
Adachi, O., 21, 70, 124
Adams, G. W., 91, 108(1), 116, 117(14), 120(14), 121
Adams, R. N., 54
Adamsons, K., 89
Agostinelli, E., 70
Ahlberg, P., 56
Ahn, B.-W., 380, 393(2)
Alger, R. S., 374
Alper, C. A., 115, 117(10)
Amberg, A. C., 114
Ameyama, M., 21, 124
Amici, A., 380, 384, 393(2)
Anderson, S. E., 2
Antonini, E., 70
Armstrong, R. C., 380, 389(7)
Arnow, L. E., 6
Ataka, T., 13
Atamian, M., 337, 338(42)
Atkin, C. L., 280, 295(12)
Atkins, P. W., 301

B

Babcock, G. T., 305, 306, 307(8), 312, 314, 316, 318, 320, 327, 337, 338(42, 43), 339(43), 340(43), 341(43), 342(43)
Babior, B. M., 365
Bachrach, U., 20, 122
Backes, G., 318
Bae, J.-Y., 54, 59(12), 60(12), 62(12), 65(12)
Baer, M. E., 328
Balakrishnan, I., 385
Balasubramanian, D., 389
Baldwin, M. A., 91, 92, 95, 100(10), 107(17)
Ballinger, M. D., 364, 366(7, 8), 368(8), 369(8), 377(8)
Ballou, D. P., 286, 287(15), 303(15)
Balog, J., 82
Barber, M., 92
Barbry, P., 115
Bardley, W. G., 122
Barker, H. A., 362
Barry, B. A., 304, 305, 306, 307(8), 308, 309(7), 311(35), 312, 313(7, 9, 14, 15, 35), 314, 315(9, 4, 15), 316(7, 8, 9, 14, 15), 317(34, 40), 318
Bateman, R. C., Jr., 382
Bates, J. L., 32, 70
Bax, A., 51
Baynes, J. W., 385
Bedell-Hogan, D., 127
Befani, O., 70
Behmoaras, T., 327, 329(31)
Beinert, H., 70, 302, 365
Bellelli, A., 70, 87(13, 14)
Bellew, B. F., 306
Bender, C. J., 318
Benedict, C. V., 2, 7(18), 11(28), 12(28), 14(18), 17(28), 19(28)
Bennett, W., 380, 382(5), 390(5)
Bensasson, R. V., 327
Bergethon, P. R., 54
Berlett, B. S., 385, 388(18), 389(18)
Berlingame, A. L., 92, 100(10, 13), 102(13), 107(13)
Beruldur, R. W., 84
Biemann, K., 92, 93, 95, 100(9), 107(9)
Bienert, H., 365
Bird, T. A., 124

Bixby, K. A., 305, 306(9), 308, 313(9), 315(9), 316(9), 317(34)
Blakley, R. L., 302
Blank, L., 54
Blankenship, R. E., 305
Blaschkowski, H. P., 344, 345, 346, 347, 348(12), 354(12), 355(12), 356(12), 359(2)
Bloemendal, H., 99
Blundell, T. L., 304, 319(5)
Boak, A. M., 54
Boak, A., 7, 128, 388
Boerner, R. J., 304, 305, 306(15), 308, 309(7), 311(35), 312(35), 313(7, 15, 35), 314(7), 315(15), 316(7, 15)
Boguta, G., 387
Bollinger, J. M., 306
Bollinger, J. M., Jr., 278, 280(5, 6, 7), 290(4), 297(6), 298(5, 6)
Bolton, J. R., 312, 372
Bondi, A., 80
Boor, P. J., 115
Bordoli, R. S., 92
Borner, R. J., 305, 306(9), 313(9), 315(9), 316(9)
Boska, M., 305
Boussac, A., 318
Bradford, M. M., 6
Bray, R. C., 286, 287(14), 365
Brettel, K., 305
Brigham, E. O., 374
Brito, M., 3
Brockie, I. R., 324
Broderick, J., 361
Brok, M., 305
Brooker, S., 361
Brown, D. E., 22, 31(16, 17), 32, 33(16, 17), 34, 36(2), 37(2), 38, 53, 70, 85(15), 87(15), 89(15), 115, 117(11, 12), 122, 302
Brudvig, G. W., 305, 306, 312, 318(13)
Bruice, T. C., 53
Bruinenberg, P. G., 114
Brunori, M., 70, 87(13), 87(14)
Brush, E. J., 355, 359
Brzovic, P., 52, 63, 81, 82(35), 83(35)
Buffoni, F., 20
Burlingame, A. L., 22, 31(15), 32(15), 34, 37(1), 38(1), 53, 70, 91, 93, 94, 95(15), 97(15), 100, 102, 108(1), 116, 117(14), 120(14), 121, 125, 278
Burzio, L. O., 3
Busby, W. H., Jr., 382
Buser, C. A., 306
Buttner, W., 305

C

Cai, D., 91, 108(1)
Calaman, S. D., 127
Cameron, D. G., 374, 375(21), 376(21)
Caskey, C. T., 100, 116
Cedergren, E., 304, 319(4)
Champe, M., 115
Champigny, G., 115
Chanderkar, L. P., 324
Chandrashekar, T. K., 318
Chapman, R. F., 54
Charlton, T. S., 387, 388(27)
Chase, J. W., 327
Chassande, O., 115
Chia, V. K. F., 2
Chiancone, E., 70
Chirpich, T. P., 362
Chiu, A. H., 99
Cleland, W. W., 78
Climent, I., 380, 391, 392(45), 393(2, 45)
Cochran, A. G., 329
Coffman, J. A., 92
Cole, T., 371
Collins, J. F., 126
Compton, T., 117
Conradt, H., 345, 347(6)
Cook, R. G., 100, 116
Cooper, B., 384
Cooper, R. A., 32, 34
Corbett, J. F., 54
Corey, E. J., 53, 58(3)
Costa, M. T., 70
Costilow, R. N., 362
Coté, C. E., 32, 70
Creeth, J. M., 384
Cronshaw, A. D., 126
Császár, J., 82

D

D'Sgostino, L., 115
Dancewicz, A., 387
Daniele, B., 115

Davies, K. J. A., 387, 393
Davies, M. J., 387
Davies, R., 1
De Jong, W. W., 99
de Riel, J. K., 362
Dean, R. T., 384, 387, 388
Dearborn, D. G., 75
Debus, R. J., 304, 305, 306(9, 10), 313(9), 315(9), 316(9)
Deisenhofer, J., 304, 320
Dekker, J. P., 305
Delsignore, M. E., 387
Denhardt, D. T., 121
Derulles, J., 307
Detorie, N. A., 385
Dickson, D. E., 22, 39, 54, 64(13)
Dietrich, H., 81
Diner, B. A., 305, 306, 312, 318(13)
Dodson, M. L., 328
Donald, A. S. R., 384
Donnelly, D., 304, 319(5)
Donohue, T. J., 84
Dooley, D. M., 22, 31(16, 17), 32, 33(16, 17), 34, 36(2), 37(2), 38, 53, 70, 73(16), 85(15), 87(15), 89(15), 90(16), 115, 117(11, 12), 122, 124, 302
Dorsett, L. C., 1
Dower, H. J., 13
Drewe, W. F., 82
Dreyer, W. J., 91
Driessen, H. P. C., 99
Duich, L., 124
Duine, J. A., 21, 22, 124
Duleba, A., 2
Duncan, M. W., 387, 388(27)
Dunn, M. F., 81, 82
Dusek, K. A., 2
Dyer, D. G., 385

E

Easton, C. J., 361
Ebbesen, T., 327
Edmondson, D. E., 278, 280(5, 6, 7), 290(4), 297(6), 298(5, 6)
Ehrenberg, A., 318, 343, 374
Eker, A. M. P., 319
Eklund, H., 278, 279(3a), 361
El-Deeb, M. K., 306, 314
El-Sherbini, S., 87
Elbert, S., 345, 349(5), 357, 361(5)
Elgren, T. E., 278
Elliott, G. J., 92
Emerson, G. M., 311
Erlich, H., 116
Ermacora, M. R., 382
Essenmacher, C., 320, 337, 338(42, 43), 339(43), 340(43), 341(43), 342(43)
Etienne, A. L., 318
Evans, S., 91
Evers, M., 114

F

Falick, A. M., 91, 95, 100, 105, 107(17)
Faloona, F., 116
Farber, J. M., 385, 393(19)
Farnum, M., 56
Ferrari, C., 115
Ferrari, G., 115
Fessenden, R. W., 371
Filley, J., 278, 290(4)
Filpula, D. R., 1
Finazzi-Agró, A., 70, 87, 114
Fischer, H., 369, 370(15), 372(15), 373(15)
Fletcher, G. L., 387
Floris, G., 70, 87(13), 87(14)
Flückiger, R., 128, 388
Flückinger, R., 7
Flynn, J. E., Jr., 89
Fontecave, M., 361
Fothergill-Gilmore, L. A., 126
Fox, R. O., 382
Frahn, J. L., 1
Frank, J., 21, 124
Frank, R., 346, 355, 356(21), 357(21), 359(21)
Franzblau, C., 123
Fraser, D. R., 124
Frelin, C., 115
Frey, M., 278, 344, 345, 346, 349(3, 5, 8), 350(3), 351(8), 352(3), 357(3, 5), 358, 361(5), 369
Frey, P. A., 362, 363(4), 364, 366(7, 8), 368(8), 369(8), 377(8)
Fridovich, I., 391
Friguet, B., 387
Fritsch, E. F., 121
Fulks, R. M., 391
Fuller, J. H., 34, 114

G

Gacheru, S. N., 122, 127
Gaillard, J., 361
Gallop, P. M., 7, 54, 128, 388
Gänzler, M., 344, 359(2)
Gardner, P. R., 391
Garland, D., 380, 393(2)
Garner, A. G., 384
Garrison, W. M., 352, 380, 381(4), 382(4, 5), 384(4), 389(4), 390(4, 5)
Gato, M., 57
Gebhard, J., 128
Gebicki, J., 388
Gerken, S., 305
Ghanotakis, D. F., 312
Ghetti, F., 327
Giatosi, A., 70, 87(13)
Gibbs, "R. A., 116
Gibbs, R. A., 100
Gibson, B. W., 91, 92, 95, 100, 107(17)
Gieseg, S. G., 388
Gieseg, S. P., 387, 388(27)
Gillece-Castro, B. L., 91, 92, 100(10)
Givskov, M., 361
Grant, A. J., 388
Grant, J., 70
Gräslund, A., 278
Greenaway, F. T., 122, 127
Grefen, G. J., 306
Grice, J. A., 1
Griffen, P. R., 92
Griffin, R. G., 306
Gröbner, P., 347, 348(12), 354(12), 355(12), 356(12)
Grunstein, M., 99
Guilivi, C., 387
Guptasarma, P., 389
Gutierrez, E., 3
Gutteridge, S., 365

H

Hagan, H. M., 22
Hall, J. M., 70
Hamm-Alvarez, S., 319
Hansch, C., 79
Hansen, D., 1
Hansen, R. E., 70
Hanson, R. E., 365
Hansson, O., 305, 306(12)
Hartman, C., 114, 115(1)
Hartman, R. F., 320
Hartmann, C., 52, 63, 69, 70(1), 71, 72(1), 74(1), 75, 76(23), 77(20), 79(20), 80(20), 81, 82(35), 83(35)
Hartner, F. W., 58
Hartzell, C. R., 365
Hatchard, C. G., 324
Hawkins, C. J., 1, 2, 8
Hay, M. P., 361
Hayashi, M., 21, 124
Hecht, H. G., 376
Hedberg, A., 374
Heelis, P. F., 320, 322(9), 326, 332, 333(38, 40), 334(39), 335, 336(41)
Heinecke, J. W., 387
Hélène, C., 327, 329(31), 371
Hemmerich, P., 347
Henderson, E. E., 362
Herdman, M., 307
Herget, A., 380
Herth, W., 353, 354(17)
Hessels, J. K. C., 319
Hevey, R., 74
Hewick, R. M., 91
Hewitt, N. A., 123
Hines, W. H., 121
Hines, W. M., 91, 95, 100, 107(17), 108(1), 116, 117(14), 120(14), 121(14)
Hirata, Y., 320, 335, 336(41)
Hoffman, B. M., 341, 342(44)
Hoganson, C. W., 316, 327
Hogness, D. S., 99
Hohmann, H.-P., 345, 347(6)
Hohmann-Berger, M., 345, 347(6)
Holl, S. M., 1
Holloway, M. R., 365
Hood, L. E., 91, 92
Horn, G., 116
Housley, T. J., 9, 12(29)
Hubbard, A. T., 2
Huggins, T. G., 385
Hulmes, D. J. S., 126
Hunkapiller, M. W., 91
Husain, I., 325
Husain, M., 78, 323

Huynh, B. H., 278, 280(5, 6, 7), 290(4), 297(6), 298(5, 6)
Hyde, D., 124
Hysmith, R. M., 115

I

Ikeda, T., 57
Inamasu, M., 21
Ishizaki, H., 20
Ito, N., 306
Itoh, S., 53

J

Jacobson, A. R., 54, 59(12), 60(12), 62(12), 65(12)
Janes, S. M., 22, 31(15, 16, 17), 32(15, 16, 17), 33(16, 17), 34, 36(2), 37(1, 2), 38, 39, 53, 70, 84, 85(40), 94, 95(15), 97(15), 115, 117(11, 12), 122, 125, 278
Jayko, M. E., 380, 382(5), 390(5)
Jensen, R. A., 1, 6(5), 9(5)
Jentoff, N., 75
Johnson, J. L., 319
Johnson, R. S., 93, 95
Jongejan, J. A., 21, 22,
Jonjegan, J. A., 124
Jordan, S. P., 324
Jorns, M. S., 320, 324
Juarez-García, C., 278, 295, 297(16), 298(16), 300(16)

K

Kagan, H. M., 54, 122, 123, 124, 125(6, 14), 126, 127, 128, 129, 131(27, 28), 388
Kagan, H., 7, 127
Kalb, V. F., 84
Kang, C. H., 341, 342(44)
Kano, K., 57
Kaplan, H. M., 123
Karolczak, M., 54
Kato, Y., 383, 384(12, 13)
Kauppinen, J. K., 374, 375(21), 376(21)
Kaur, S., 22, 31(15), 32(15), 34, 37(1), 38(1), 53, 70, 91, 92, 94, 95(15),97(15), 100(10), 125, 278
Kawakishi, S., 383, 384(12, 13), 385
Ke, B., 312
Keen, J. N., 306
Kelly, I., 70
Kelm, S., 102
Kessler, D., 353, 354(17)
Khorana, H. G., 320
Ki, S. T., 320
Killgore, J., 124
Kim, K., 391
Kim, S. T., 319, 320, 324, 325, 330, 331(36), 332(36), 337, 338(42, 43), 339(43), 340(43), 341(43), 342(43)
Kittler, J. M., 124, 125(14)
Kizer, J. S., 382
Klein, M., 58
Klinman, J. P., 22, 31(15, 16, 17), 32(15, 16, 17), 33(16, 17), 34, 36(2), 37(1, 2), 38, 39, 40(1), 42, 44(2, 8), 45(8), 49(2), 51(2), 52, 53, 56, 59, 60(22), 63, 70, 71, 74, 75, 76(23), 77(20), 78, 79(20, 30), 80(20), 81, 82(30, 35), 83(35), 84, 85(40), 91, 94, 95(15), 97(15), 108(1), 114, 115, 116, 117(11, 12, 14), 120(14), 121(14), 122, 125, 278
Kluetz, M. D., 89
Knappe, J., 344, 345, 346, 347, 348, 349(3, 5, 8, 13), 350(3), 351(8, 13), 352(3, 13), 353, 354(12, 17), 355, 356(12, 21), 357, 359(2, 13, 21), 361(5), 369
Knowles, P. F., 32, 34, 70, 84, 85(15), 87(15), 89(15), 122, 302, 306
Knowles, P., 70
Kochevar, G. J., 2
Koerber, S. C., 81
Kohzuma, T., 54, 57(7)
Kokil, P. B., 58
Kokil, P. K., 56
Kooiman, P., 319
Koppenol, W. H., 385, 389
Korytowski, W., 2
Kozarich, J. W., 355, 359
Krappe, J., 278
Kuipers, J., 114
Kuma, H., 54, 57(7)

L

Laemmli, U. K., 353
Latwesen, D. G., 375

Laursen, R. A., 1
Lavin, M. F., 1, 3, 6(19), 8, 9(19), 10(19)
Lazdunski, M., 115
Lee, C. C., 100, 116
Lee, F. G. H., 22, 39, 54, 64(13)
Lee, S.-M., 1
Lee, Y., 54, 59(12), 60(12), 62, 63(23), 65(12)
Leigh, J. S., 375
Lenz, A.-G., 380, 393(2)
Leo, A., 79
Levene, C. A., 124
Levene, C., 124
Levine, R. L., 380, 384, 385, 391, 392, 393
Lewinsohn, R., 115
Li, Y. F., 330, 331(36), 332, 334(39)
Lichtenthaler, H. K., 312
Lin, S. W., 387
Lindström, B., 318
Linguegliа, E., 115
Link, R. P., 1
Lipman, R. S. A., 320
Lipsett, K. A., 355
Liu, T.-Y., 14
Livingston, B. D., 102
Lloyd, R. S., 328
Lobenstein-Verbeek, C. L., 21, 124
Loehr, T. M., 318
Long, G., 280, 295(12)
Lucchelli, P., 115
Ludwig-Festl, M., 346
Lynch, J. B., 278, 295, 297(16), 298(16), 300(16)
Lynch, M., 362
Lynn, K. R., 389

M

Maassen, D., 42
MacBeath, J. R. E., 126
MacDonald, G. M., 308, 312, 317(34, 40)
MacGibbon, A. K. H., 81
Maes, P., 115
Makáry, A., 82
Malhotra, K., 319, 325
Malin, E. L., 13
Maltby, D. A., 91, 94, 95(15), 97(15)
Maltby, D., 22, 31(15), 32(15), 34, 37(1), 38(1), 53, 70, 278
Maltby, S., 125
Manabe, T., 70
Manian, A. A., 22, 54, 64(13)
Maniatis, T., 121
Mann, G. J., 278
Mann, P. J. G., 70
Mantsch, H. H., 374, 375(21), 376(21)
Markovic, R., 74
Marti, T., 320
Martin, S. A., 93, 95(14)
Maskos, Z., 385, 389
Mason, H. S., 70
Massay, V., 78, 347, 323
Mataga, N., 320, 322(9), 335, 336(41)
Matsudaira, P., 99
Matsushita, K., 21, 124
Mayer, P., 91, 108(1), 116, 117(14), 120(14), 121
Mazzacca, G., 115
McBride, M. B., 2
McConnell, H. M., 371
McCreery, R., 54
McGuirl, M. A., 32, 34, 38, 53, 70, 73(16), 85(15), 87(15), 89(15), 90(16), 122, 302
McIntire, W. S., 32, 34, 69, 70, 72(1), 74(1), 75, 85(15), 87(15), 89(15), 114, 115(1), 122, 302
McIntosh, L., 305, 306(10)
McPherson, M. J., 306
Medzihradszky, K. F., 91, 92, 93, 95, 100(10, 13), 102, 105, 107(13, 17), 108(1), 116, 117(14), 120(14), 121
Melko, M., 124
Mentasti, E., 14
Merefield, P. M., 1
Metz, J. G., 305, 306(13), 318(13)
Mewes, W., 13
Michel, H., 304, 320
Miller, A.-F., 312
Miller, D. G., 390
Minamiura, N., 20
Moffatt, D. J., 374, 375(21), 376(21)
Mogi, T., 320
Molin, S., 361
Mondovi, B., 70
Moog, R. S., 22, 124
Mori, T., 57
Morpurgo, L., 70
Morse, D. E., 1, 6(5), 9(5)
Morton, J. R., 369, 370(13)
Moss, M. L., 362
Moss, T. H., 365

Mu, D., 22, 31(15, 16), 32(15, 16), 33(16), 34, 37(1), 38, 39, 40(1), 53, 70,91, 94, 95(15), 97(15), 108(1), 114, 115, 116, 117(11, 14, 17),120(14), 121(14), 122, 125, 278
Mu, S., 53
Mueller, K. M., 1, 8(4), 9(4), 14(4), 17(4)
Mulliez, E., 361
Mullis, K., 116
Münck, E., 278, 295, 297(16), 298(16), 300(16)
Munemitsu, S., 115
Munzy, D. M., 100
Mure, M., 22, 34, 36(2), 37(2), 38(2), 39, 44(2, 8), 45(8), 49(2), 51(2), 59,60(22), 115, 117(12)
Murray, J. C., 124
Murrill, E., 54
Musso, H., 42
Mute, M., 42
Muzny, D. M., 116

N

Nakahara, A., 70
Nakamura, N., 54, 57(7)
Nanba, O., 304, 306(2)
Neuer, G., 346
Neugebauer, F. A., 344, 345(3), 348, 349(3, 13), 350(3), 351(13), 352(3, 13), 357(3, 13), 359(2, 13), 369
Neugehauer, F. A., 278
Neumann, R., 74
Nguyen, A. P., 305, 306(9), 313(9), 315(9), 316(9)
Nixon, P. J., 305, 318(13)
Noren, G. H., 305, 306(9, 13, 14), 308, 311(35), 312(35), 313(9, 14, 35), 315(14, 9), 316(9, 14)
Norlund, P., 2, 278, 279(3a), 361
Norton, J. R., 278, 290(4)
Notter, M. F. D., 19
Nugent, J. H. A., 304, 319(5)

O

O'Day, C. L., 324
O'Gara, C. Y., 122
Ochiai, E.-I., 278
Offner, G. D., 129, 131(27, 28)
Ogel, Z. B., 306
Ohshiro, Y., 53
Okada, S., 387
Okamura, T., 70, 320, 322(9), 332, 333(40), 335, 336(41)
Oliver, C. N., 379, 380, 382(1), 390(1), 391, 393
Olsson, B., 56, 81, 83(34)
Olsson, J., 56, 70, 81, 83(34)
Orme-Johnson, W. H., 302, 365
Ormö, M., 2
Ouwehand, L., 305

P

Palcic, M. M., 22, 31(17), 32(17), 33(17), 34, 36(2), 37(2), 38(2), 56, 78, 79(30), 82(30), 115, 117(12), 124
Palmer, G., 78, 286, 287(15), 303(15)
Pardo, J., 3
Pargoliash, E., 341, 342(44)
Parker, C. A., 324
Parry, D. L., 1, 8
Paulson, J. C., 102
Payne, G., 319, 323, 324(19), 325, 326(27), 329(27), 332, 333(38)
Paz, M. A., 7, 128, 388
Pederson, J. Z., 87
Pelizzetti, E., 14
Percival, A., 54
Petersson, L., 318
Petrovich, R. M., 362, 363(4)
Petruzzelli, R., 114
Pettersson, G., 56, 70, 81, 83(34), 84
Phillips, S. E. V., 306
Pignata, S., 115
Pilbrow, J. R., 363
Piotrowski, E. G., 13
Plaga, W., 346, 355, 356(21), 357(21), 359(21)
Platis, I. A., 382
Plijter, J. J., 305
Poe, M., 375
Poston, J. M., 384
Pratt, A. M., 129, 131(27, 28)
Pryor, W. A., 389
Pühlhofer, A., 346

Q

Qin, X.-X., 3, 6(19), 9(19), 10(19)
Que, L., Jr., 278, 295, 297(16), 298(16), 300(16)

R

Rajagoplan, K. V., 319
Ramsey, R. R., 78
Ravi, N., 278, 280(6, 7), 297(6), 298(6)
Reamer, R. A., 58
Reddy, M. P., 385
Reed, G. H., 362, 363(4), 364, 366(7, 8), 368(8), 369(8), 375, 377(8)
Reichard, P., 280, 295(12), 343, 361
Reiser, K., 124
Rhee, S. G., 391
Rice-Ficht, A. C., 2
Rice-Ficht, A. M., 1, 4, 5(22), 8, 9(2, 22), 10(2), 11(22), 17(22), 19(2, 22)
Rinaldi, A., 70, 87(13)
Rippka, R., 307
Rius, F. X., 84
Rivett, A. J., 393
Roberts, J. E., 341, 342(44)
Robins, S. P., 14
Rödel, W., 346
Rodriuez, E. J., 53
Rogner, M., 305, 306(13), 318(13)
Romero-Chapman, N., 124
Rose, S. D., 320, 325(16)
Rosenbloom, J., 127
Rosenthal, S. M., 20
Ross, I. L., 1, 8
Rossi, A., 114
Rothe, M., 346, 349(8), 351(8)
Rotilio, G., 70, 87
Rougee, M., 327
Rucker, R. B., 124
Ruffle, S. V., 304, 319(5)
Rush, J. D., 385, 389
Rutherford, A. W., 305, 306
Ruzicka, F. J., 362, 363(4)
Rzepecki, L. M., 1, 3, 6(19), 8(4), 9(4, 19), 10(19, 21), 14(4), 17(4)

S

Saez, C., 3
Sahakian, J. A., 392
Sahlin, M., 318
Saiki, R., 116
Saini, G., 14
Sakurai, T., 70
Salcedo, L. L., 123
Salowe, S. P., 280, 302, 318
Sambrook, J., 121
Sancar, A., 319, 320, 321, 322(9), 323, 324, 325, 326, 328, 329(27), 330, 331(36), 332, 333(38, 40), 334(39), 335, 336(41), 337, 338(42, 43), 339(43), 340(43), 341(43), 342(43)
Sancar, G. B., 319, 321, 322(9), 326(9), 328
Sanders-Loehr, J., 318
Sandusky, P. O., 306, 314
Sarna, T., 2
Satoh, K., 304, 306(2)
Sauer, K., 305, 306
Sawers, G., 344
Sayre, L. M., 54, 56, 58, 59(12), 60(12), 62, 63(23), 65(12)
Scaman, C. H., 22, 31(17), 32(17), 33(17), 34, 36(2), 37(2), 38(2), 115, 117(12), 124
Schaefer, J., 1
Schäfer, W., 344, 345(3), 349(3), 350(3), 352(3), 357(3), 369
Schagger, H., 128
Scharf, S., 116
Schilling, K., 380, 382(9)
Schlodder, E., 305
Schmitt, T., 347, 348(12), 354(12), 355(12), 356(12)
Schuessler, H., 380, 382(9)
Schultz, P. G., 329
Sedwick, R. D., 92
Senob, S., 33, 54
Shackleton, D. R., 126
Shacter, E., 380, 393(3)
Shah, M. A., 54, 122, 124
Shaltiel, S., 380, 393(2)
Shames, B. D., 392
Shater, W., 278
Shinagawa, E., 21, 124
Simpson, J. A., 387, 388
Simpson, R. J., 14
Singel, D. J., 306
Singh, M. P., 56, 58
Sithole, I., 305
Sjöberg, B.-M., 2, 278, 318
Smidt, C., 124
Smith, A. J., 22, 31(15, 16, 17), 32(15, 16, 17), 33(16, 17), 34, 36(2), 37(1, 2), 38(1, 2), 53, 70, 91, 94, 95(15), 97(15), 108(1), 115, 116, 117(11, 12, 14), 120(14), 121(14), 122, 125

Smith, A., 278
Smith, C. A., 325
Smith, F. W., 321, 322(9), 326(9)
Smith, R. A., 21
Solar, S., 389
Soriaga, M. P., 2
Spacciapoli, P., 122
Spaguolo, S., 115
Stadtman, E. R., 379, 380, 382(1), 384, 385, 387, 388(18), 389(18), 390, 391, 393
Stadtman, T. C., 362
Stanier, R., 307
Steenkamp, D. J., 78
Stevens, C., 306
Stevens, J., 362
Strathdee, J., 371
Straub, K. M., 92
Strausberg, R. L., 1
Strausberg, S. L., 1
Strickland, S., 78
Stubbe, J. A., 318
Stubbe, J., 278, 280, 290(4), 297(6), 298(5, 6), 306, 361
Styring, S., 304, 306, 319(4)
Sugasawara, R., 329
Summers, M. C., 74
Summers, M. F., 51
Suva, R. H., 82
Suzuki, S., 54, 57(7), 70
Svensson, B., 304, 319(4)
Swallow, A. J., 380, 389(7), 390
Swan, G. A., 54
Swanepoel, O. A., 358, 361(26)
Swartz, H., 2
Szweda, L., 387

T

Tabor, C. W., 20
Tabor, H., 20
Taborsky, G., 391
Tang, S. S., 129, 131(27, 28)
Tanzer, M. L., 9, 12(29)
Tartar, A., 115
Taylor, J.-S., 324, 325
Taylor, S. W., 2
Tchitchibabime, A. E., 42
Tesser, G. I., 99
Thanassi, J. W., 124, 125(14)
Thélander, L., 278, 280, 295(12)
Thiblin, A., 56
Thompson, L. K., 306
Thorpe, S. R., 385
Tinker, N., 124
Tong, W. H., 278, 280(7)
Toulme, J.-J., 327, 329(31)
Trackman, P. C., 122, 126, 127, 129, 131(27, 28)
Trackman, P., 127
Troxler, R. T., 129, 131(27, 28)
Tsai, L., 384, 391, 392(45), 393(45)
Tsugita, T., 13
Tull, R., 58
Turowski, P. N., 70, 73(16), 85(15), 87(15), 89(15), 122, 302
Tyler, A. N., 92

U

Uchida, H., 13
Uchida, K., 383, 384(12, 13), 385
Uhlin, U., 361
Ulissi-DeMario, L., 359
Ullrich, A., 115
Un, S., 306
Unkig, V., 348, 349(13), 351(13), 352(13), 357(13), 359(13)
Uno, B., 57
Uppu, R. M., 389

V

Valerie, K., 362
Van Camp, J. R., 320
van der Meer, R. A., 22, 124
van Gorkom, H. J., 305
Van Rensburg, N. J. J., 358, 361(26)
Vander Zwan, M. C., 58
Vänngård, 'T., 87
Vass, I., 304, 319(4)
Velde, J. V., 319
Venkataraman, B., 58
Ventriglia, R., 115
Vermaas, W. F. J., 305, 306(12)

W

Wagner, A. F. V., 278, 344, 345, 346, 349(3, 5, 8), 350(3), 351(8), 352(3), 357(3, 5), 361(5), 369

Waite, J. H., 1, 2, 3, 4, 5(22), 6(5, 19), 7(18), 8, 9, 10(2, 19, 21), 11(14, 22, 28), 12(28, 29), 14, 17(4, 22, 28), 19(2, 22, 28, 35)
Wallis, O. C., 365
Walls, F. C., 91, 92, 100(10), 105
Walsh, C. T., 319
Walsh, C., 323, 324(19)
Wang, B., 324
Wang, F., 54, 56, 59(12), 60(12), 62(12), 65(12)
Wang, G. H., 91
Warden, J. T., 305
Watanabe, K., 21
Waterbury, J. B., 307
Waterham, H. R., 114
Wells-Knecht, M. C., 385
Wemmer, D., 22, 31(15), 32(15), 34, 37(1), 38(1), 53, 70, 94, 95(15), 97(15), 125, 278
Wen, D. X., 102
Wertz, J. E., 312, 372
Wesselink, L. G., 2
Whittaker, J. W., 306
Whittaker, M. M., 306
Williams, J. A., 380, 393(3)
Williams, J. G. K., 307
Williams, K. R., 327
Williamson, P. R., 22, 123, 124, 125(14)
Wills, M., 323, 324(19)
Wilson, K. J., 99
Winchester, R. V., 389
Wingo, W. J., 311
Winzor, D. J., 2
Witkop, B., 33, 54
Witt, H. T., 305
Wolanski, A., 129, 131(27, 28)
Wolfe, S. P., 384
Wu, X., 100, 116

Y

Yadav, K. D. S., 306
Yadav, K. P. S., 70
Yamada, H., 20, 33(5), 70
Yamano, T., 70
Yasui, A., 319
Yasunobu, K. T., 20, 21, 33(5)
Yasunobu, K., 70
Yates, J. R., 92
Yi, Y. F., 328
Young, T., 320
Youngblood, W. W., 382
Yu, P. H., 56
Yu, Z., 92, 100(10)
Yuan, P. M., 99
Yuen, S. W., 99

Z

Zappia, V., 362
Zeller, E. A., 114
Zepperzauer, M., 81
Zhang, X., 34, 114
Zoski, C. G., 126

Subject Index

A

AdhE protein, reduction of pyruvate formate-lyase, 353–354
Alcohol dehydrogenase, cloning from bacteria, mutant complementation, 217–220
Amicyanin
electron acceptor for methylamine dehydrogenase, 151, 164–165, 177, 190, 192
–methylamine dehydrogenase complex, 192–193
X-ray crystallography, 193, 208–210
Amine oxidase, *see* Copper amine oxidase; Lysyl oxidase
Amino acid analysis, dopa-containing proteins, 13–15
Amino acid sequencing, dopa-containing proteins, 15, 17, 19
4-Amino-6-*tert*-butylresorcinol
absorption spectra, 51
Schiff base analog preparation, 45
synthesis, 44–45
4-Amino-6-[2-(pivalamido)ethyl]resorcinol, synthesis, 66–67
Arginine
metal-catalyzed oxidation, 392–393
oxidative modification in proteins, 390–391
Aromatic amine dehydrogenase
absorption properties, 182
aminoquinol formation, 183–184
dithionite reduction, 182
electron acceptors, 177
subunit structure, 178

B

Benzylamine
deamination by topa quinone pivalamide derivative
absorption spectroscopy analysis, 62–63
aerobic reaction conditions, 65
catalyst destruction, 57–58
chemical synthesis of products, 68–69
copper effects, 57
deuterium isotope effect, 56
mechanism, 60, 62
nuclear magnetic resonance analysis, 62–64, 66–68
pH effect on yield, 54, 56
rate, 54–55
substrate specificity, 56–57
oxidation by tryptophan tryptophylquinone derivative, 172–173
substrate in amine oxidase assay, 71, 74
Bovine serum amine oxidase, *see* Copper amine oxidase

C

Carbonyl group, quantitation in proteins, 393
Catechol oxidase, and dopa-containing protein, cosecretion, 1
CD, *see* Circular dichroism
Circular dichroism, *see also* Magnetic circular dichroism
galactose oxidase, 274–276
semiquinone intermediate in amine oxidase, 87
sensitivity, 274
theory, 274
Copper amine oxidase, *see also* Lysyl oxidase
ammonia
freeze–quench kinetics of release, 84
radiochemical analysis of production, 84–85
retention on enzyme, 83–85
assay
oxygen electrode assay, 71
radiochemical assay, 84–85
spectrophotometric assay, 71, 74

bovine serum amine oxidase
  amino acid sequence, 113
  cloning
    library screening, 115, 119–121
    mixed oligonucleotide primed amplification of cDNA, 116–119
    N-terminal sequencing, 108–109
    primer design, 117, 121–122
  resonance Raman spectroscopy, 137–138
  sequence analysis
    HPLC/electrospray ionization mass spectrometry, 112
    tandem mass spectrometry, 92–94, 97–100, 102–103, 105–108, 110–111
  site of synthesis, 115
carbonyl reagent sensitivity, 20, 122
copper role in catalysis, 53, 70
electron transfer, temperature-jump detection, 89–90
multiple regression analysis in Hammett value determination, 79–80
oxidative half reaction, 70, 73
peptide isolation
  thermolytic digest, 25–26, 30–31
  tryptic digest, 26–27, 91, 100
phenylhydrazine reaction, 21, 23–24, 33, 133–135
proteolytic digestion, 24–25, 30–31
reductive half reaction, 70, 72
Schiff base intermediates
  product, 83
  substrate, 74–75
sodium cyanoborohydride effects
  enzyme–substrate complex inactivation, 74–75
  free enzyme inactivation, 75–76
steady-state kinetics, 78–79
stopped-flow spectrometry
  anaerobic conditions, 77
  data analysis, 77–78, 81
  deuterium isotope effects, 82–83
  rapid-scanning analysis, 80–83
  relaxation characteristics, 81–82
subunit structure, 69–70
topa quinone chromophore
  accessibility to modifying reagents, 35
  history of attempts at identification, 20–22
  peptide analysis
    mass spectrometry, 28–29, 95, 97
    nuclear magnetic resonance, 29
    resonance Raman spectroscopy, 31–32, 34, 139
    visible absorbance spectroscopy, 31–32, 34–38
  pyrroloquinoline quinone investigations, 21–22, 32, 34
  semiquinone intermediate
    absorption spectroscopy, 88–89
    circular dichroism, 87
    electron paramagnetic resonance, 85–87
transamination mechanism, 39–40, 53

D

5-Deazariboflavin, electron donor for pyruvate formate-lyase, 347
3,5-Di-*tert*-butyl-1,2-benzoquinone, transamination reactions, 53, 58
5-Dihydroflavodoxin, electron donor for pyruvate formate-lyase, 346
3,4-Dihydroxyphenyl-L-alanine, *see also* Dopa-containing protein
  metal affinity, 2
  stability, 2
2,4-Dinitrophenylhydrazine
  modified proteins, absorption spectra 133–134
  quinoprotein derivatization, 133
DNA photolyase
  apoenzyme preparation, 323
  assays
    coupled enzyme assay, 325
    DNA repair quantum yield measurement, 326
    gel retardation, 325
    spectrophotometric assay, 325
  electron paramagnetic resonance
    instrumentation, 327
    isotope effects, 342
    time-resolved spectra, 337–343
  flash photolysis, 326–327
    camera flash photolysis, 332
    nanosecond laser flash photolysis, 332–334
    picosecond laser flash photolysis, 335–337

photosensitizing cofactors, 319–320, 329
purification from *Escherichia coli*
affinity chromatography, 322–323
ammonium sulfate precipitation, 322
cell growth, 321
cell lysis, 322
gel filtration, 323
hydroxylapatite chromatography, 323
reconstitution, 323–324
substrate synthesis, 324
tryptophans
binding contribution, 320–321, 327–329
electron transfer, 320–321, 343
flavin radical photoreduction, 332–343
photosensitized repair of pyrimidine dimers, 329–332
site-directed mutagenesis, 320–321, 328, 331–332, 334–335
Dopa-containing protein
amino acid analysis
autoanalyzer parameters, 14–15
hydrolysis, 13–14
amino acid sequencing, 15, 17, 19
and catechol oxidase, cosecretion, 1
dopa quantitation
Arnow assay, 6–7
redox cycling, 7–8
liver fluke vitelline proteins
amino acid composition, 17
extraction, 4–6
reverse-phase high-performance liquid chromatography, 9–10
Mytilus edulis foot proteins
amino acid composition, 17
dopa/protein ratio, 6
extraction, 2–4
gel filtration, 9
purification assessment, 6
reverse-phase high-performance liquid chromatography, 9–10
peptide purification, 12–13
phenyl boronate agarose chromatography, 8
proteolytic digestion, 11–12
tyrosine modification, 1–2, 140
Dopaquinone, attachment to protein thiols, 140–141
DTBQ, *see* 3,5-Di-*tert*-butyl-1,2-benzoquinone

E

Edman degradation, N-terminus sequencing of amine oxidase, 99, 108
Electron paramagnetic resonance
aminosemiquinone of methylamine dehydrogenase, 184
anisotropy, 370–373
DNA photolyase
instrumentation, 327
isotope effects, 342
time-resolved spectra, 337–343
Fourier transform, 374–375
galactose oxidase, 236–237, 254
apoenzyme, 267–268
double integration, 265
nitric oxide complex, 269
silent copper complex, 267
spin standards, 265–267
lysine 2,3-aminomutase
isotopic substitution, 367–369, 377
liquid phase measurements, 366
π-radical properties, 370–373
rapid-mix/freeze-quench trapping of intermediates, 365
signal characterization, 367–369
slow substrate trapping of intermediates, 365–366
spectral properties, 363–365, 367
structure determination, 379
photosystem II tyrosyl radicals
apparatus, 312
Ḍ, 312–314
deuteration effects, 314, 316
M^+, 306, 315–316
Ż, 312–314
pyruvate formate-lyase
data collection, 349–350
spectra, 350–351
resolution enhancement, 373–376
ribonucleotide reductase R2 protein reaction with excess Fe^{2+} and O_2, rapid freeze-quench studies
apparatus, 286–287
calibration of sampling apparatus, 287
control experiment, 288–289
quantitative analysis, 290–292, 303
rate constants, 292–293
trapping of intermediates, 286

semiquinone intermediate in amine oxidase, 85–87
sensitivity, 264–265
spectral simulation, 376–377
theory, 265
EPR, *see* Electron paramagnetic resonance

F

Fenton's reagent, protein modification, 391
Flash photolysis, DNA photolyase
camera flash photolysis, 327, 332
nanosecond laser flash photolysis, 326, 332–334
picosecond laser flash photolysis, 326–327, 335–337
Flavin adenine dinucleotide, oxidation state in DNA photolyase, 319–320
Fourier transform infrared spectroscopy, photosystem II tyrosyl radicals, 316–318
FT-IR, *see* Fourier transform infrared spectroscopy

G

Galactose oxidase
active site structure, 264
circular dichroism, 274–276
copper role, 236, 257–258, 272
electron paramagnetic resonance, 236–237, 254
apoenzyme, 267–268
double integration, 265
nitric oxide complex, 269
silent copper complex, 267
spin standards, 265–267
magnetic circular dichroism, 276–277
optical absorption
azide effects, 272
free radical identification, 273–274
ligand-to-metal charge transfer, 271–272
spectra, 270–271
titration curves, 272–273
purification from *Dactylium dendroides*, 237
reaction catalyzed, 235, 263
spectroscopic techniques, overview, 262–263
substrate specificity, 236, 260–261
thioether bond between tyrosinyl and cysteinyl residues
biosynthesis, 257
catalysis role, 257–258, 264
confirmatory evidence, 251–254
mutation effects, 256
stacking interaction of tryptophan, 254–255
X-ray crystallography, 251
X-ray crystallography
accuracy of model, 244–245
acetate ion, 243–244
copper coordination, 247–248
copper site geometry, 249–251
crystallization, 239–238
data collection, 238
heavy atom derivatives, 239–241
initial model building, 241–243
phase determination, 238–241
refinement, 243–244
sequence determination, 241–242
stacking tryptophan mutant, 255
substrate binding, 260–262
superbarrel motif, 213, 215–216, 245–247
thioether cysteine mutant, 256
Glucose dehydrogenase, cloning from bacteria
mutant complementation, 217–220
oligonucleotide probing, 220
Glutamate
oxidation and polypeptide cleavage, 382, 384
oxidative modification in proteins, 391
Glutamine synthetase
metal-catalyzed oxidation, 392
oxidation and turnover, 393
Glycine, radicals and catalysis, *see* Pyruvate formate-lyase

H

Hammett value, multiple regression analysis, 79–80
High-performance liquid chromatography
liver fluke vitelline proteins, 9–10
lysyl oxidase peptide separation, 130–131
Mytilus edulis foot proteins, 9–10
pyrroloquinoline quinone, quantitative assay, 232

Histidine
metal-catalyzed oxidation, 392–393
oxidative modification in proteins, 385
HPLC, *see* High-performance liquid chromatography
2-Hydrazinopyridine, quinoprotein derivatization, 135
2-Hydroxy-5-*tert*-butyl-1,4-benzoquinone
absorption spectra, 48–50
Schiff base analog
absorption spectra, 52
preparation, 43–44
two-dimensional nuclear magnetic resonance analysis, 45, 47–48
synthesis, 42–43
6-Hydroxydopaquinone, *see* Topa quinone
2-Hydroxy-5-(2-pivalamidoethyl)-1,4-benzoquinone
conversion to quinoneimine, 69
deamination of benzylamine
absorption spectroscopy analysis, 62–63
aerobic reaction conditions, 65
catalyst destruction, 57–58
chemical synthesis of products, 68–69
copper effects, 57
deuterium isotope effect, 56
mechanism, 60, 62
nuclear magnetic resonance analysis, 62–64, 66–68
pH effect on yield, 54, 56
rate, 54–55
substrate specificity, 56–57
synthesis, 54, 59–60, 65

I

Infrared spectroscopy, *see* Fourier transform infrared spectroscopy

L

Leucine, oxidative modification in proteins, 389–390
Ligand-to-metal charge transfer, optical absorption, 271–272
Lysine
metal-catalyzed oxidation, 392–393
oxidative modification in proteins, 390
Lysine 2,3-aminomutase
cofactors, 362
electron paramagnetic resonance
anisotropy, 370–373
Fourier transform, 374–375
isotopic substitution, 367–369, 377
liquid phase measurements, 366
π-radical properties, 370–373
rapid-mix/freeze-quench trapping of intermediates, 365
resolution enhancement, 373–376
signal characterization, 367–369
slow substrate trapping of intermediates, 365–366
spectral properties, 363–365, 367
spectral simulation, 376–377
structure determination, 379
inactivation by oxygen, 367
reaction catalyzed, 362
reaction mechanism, 363
Lysyl oxidase
active site peptide
cyanogen bromide cleavage, 128
gel electrophoresis, 128
labeling, 127–128
proteolytic digestion
digestion conditions, 130
peptides generated by, HPLC, 130–131
reductive alkylation pretreatment, 129–130
sequencing, 128–129
biological role, 123
carbonyl reagent sensitivity, 122–123
cofactor characterization, 124–125, 132
ethylenediamine inhibition, 127–128
oxidative half reaction, 124
purification from bovine aorta
extraction, 125
gel filtration, 126
hydroxyapatite batch elution, 125–126
ion-exchange chromatography, 126–127
reductive half reaction, 124

M

Magnetic circular dichroism, galactose oxidase, 276–277
Mass spectrometry
bovine serum amine oxidase, sequence analysis

HPLC/electrospray ionization mass spectrometry, 112
tandem mass spectrometry, 92–94, 97–100, 102–103, 105–108, 110–111
dityrosine quantitation in proteins, 387–399
peptides
molecular weight determination, 91–92, 110–111
sequence analysis, 91–94
topa quinone peptide analysis, 28–29, 95, 97
tryptophan tryptophylquinone peptides, 159–161
MCD, *see* Magnetic circular dichroism
Methanol dehydrogenase
calcium role, 213
cloning from bacteria, mutant complementation, 217–220
electron acceptors, 195
genes, bacterial
expression systems, 221–222
mutant construction, 224–226
pyrroloquinoline quinone cofactor, 193
subunit structure, 193
X-ray crystallography
amino acid sequence analysis, 199
cofactor identification, 200–201
crystallization conditions, 194–195
pyrroloquinoline quinone orientation, 211–213
structure analysis and refinement, 194, 196–198, 210–212
superbarrel motif, 211, 213, 215–216
Methionine, oxidative modification in proteins, 390
Methylamine dehydrogenase
absorption spectra, 152–153, 155, 163
adducts
ammonium, 185–186
hydroxide, 186
structures, 179–180
assay, 152
carbonyl reagents
effect on absorption spectra, 186–187
phenylhydrazine derivatization, 135
sensitivity, 156
cloning from bacteria
mutant complementation, 217–220
oligonucleotide probing, 220
electron acceptors, 151, 164–165, 192
genes, bacterial
expression systems, 221–222
mutant construction, 226
regulation, 226–227
sequencing, 222–224
Methylobacillus extorquens, cloning of small subunit, 162–163
reaction mechanism, 178–179
redox forms
absorption spectra, 180–182
aminoquinol formation, 182–183
aminosemiquinone formation, 184
electron paramagnetic resonance, 184
fluorescence spectra, 183
reduction with dithionite, 181–182
structures, 179–180
resonance Raman spectroscopy, 138–139
stopped-flow spectroscopy
amicyanin oxidation, 190
apparatus, 187, 189
methylamine reduction, 189–190
subunit structure, 151–152, 178, 191–192
tryptophan tryptophylquinone isolation
amino acid composition, 159
carboxypeptidase-Y treatment, 158–159
cofactor stabilization, 156
leucine aminopeptidase treatment, 158
pronase treatment, 157–158
sequence analysis, 159
subunit isolation, 157
sulfhydryl group protection, 156
X-ray crystallography, 163, 168, 178
amino acid sequence analysis, 199
cofactor identification, 200–201
crystallization conditions, 194–195
heavy subunit, 204
light subunit, 205–207
Paracoccus denitrificans enzyme
amicyanin complex, 193, 208–210
cytochrome *c*–amicyanin ternary complex, 210
structure analysis and refinement, 194, 196–198
superbarrel motif, 213, 215–216
Thiobacillus versutus enzyme, hydrazine complex, 208
tryptophan tryptophylquinone orientation, 168–169, 201, 203–207

Methylamine oxidase, absorption of phenylhydrazine derivatives, 133–134
Mössbauer spectroscopy, ribonucleotide reductase R2 protein reaction with excess Fe^{2+} and O_2, rapid freeze-quench studies
  apparatus, 294–295
  quantitative analysis, 296–297, 299, 301
  rate constants, 301
  reference sample preparation
    diferric cluster, 295
    diferric radical species, 296
    ferrous ion in HEPES buffer, 296
    ferrous-R2, 296
  reference spectra, 297–299
  sensitivity, 303
  spectra acquisition, 296
  time course of reaction, 299

N

Neuraminidase, superbarrel motif, 213, 215–216
Nitroblue tetrazolium
  dopa staining in proteins, 7–8
  pyrroloquinoline quinone protein assay
    detection in gel electroblots, 147–149
    inhibition, 142–143
    principle, 141
    quinoprotein identification in fluids and tissues, 143, 145
    reaction conditions, 146–147
  reduced form, extraction from reaction mixtures, 388–389
p-Nitrophenylhydrazine
  absorption of modified proteins, 134–135
  topa quinone reaction, 35, 38, 134–135
NMR, *see* Nuclear magnetic resonance
Nuclear magnetic resonance
  2-hydroxy-5-*tert*-butyl-1,4-benzoquinone Schiff base analogs, 45, 47–48
  pyrroloquinoline quinone
    carbon-13 study, 233–235
    sample preparation, 235
  topa quinone peptides, 29, 34
  topa quinone pivalamide derivative, 62–64, 66–68
  tryptophan tryptophylquinone peptides, 161–162

O

Oxygen radical, polypeptide cleavage, 380–382

P

PCR, *see* Polymerase chain reaction
Phenylalanine, oxidative modification in proteins, 385
Phenylhydrazine
  absorption of modified proteins, 133–134, 136
  copper amine oxidase labeling, 21, 23–24, 33, 133
Photosystem II
  electron transfer, 304
  isotopic labeling of tyrosine in cyanobacteria, 307–308
  oxygen evolution assay, 311–312
  purification from cyanobacteria
    anion-exchange chromatography, 309–311
    thylakoid membrane preparation, 308–309
  subunit structure, 304
  tyrosyl radicals
    D site, 304–305
    electron paramagnetic resonance
      apparatus, 312
      $\dot{D}$, 312–314
      deuteration effects, 314, 316
      $\dot{M}^+$, 306, 315–316
      $\dot{Z}$, 312–314
    Fourier transform infrared spectroscopy, 316–318
    M^+ site, 305, 319
    redox kinetics, 305–306
    Z site, 304–305
Polymerase chain reaction, cloning of bovine serum amine oxidase, 116–119
Proline, oxidation and polypeptide cleavage, 382–384
Pyrroloquinoline quinone
  benzylamine deamination, 53, 60, 62
  biosynthesis
    amino acid precursors, 228–229
    genes involved in, 224, 226
    radiolabeling, 228–233

distinguishing from cyclic topa quinone, 21–22, 32, 34
isolation from bacterial culture, 231
high-performance liquid chromatography, quantitative assay, 232
nuclear magnetic resonance
carbon-13 study, 233–235
sample preparation, 235
redox cycling assay in proteins
detection in gel electroblots, 147–149
inhibition, 142–143
principle, 141
quinoprotein identification in fluids and tissues, 143, 145
reaction conditions, 146–147
structure, 193
X-ray crystallography in proteins, 200–201, 211–213
Pyruvate formate-lyase
activating enzyme, 344
conversion of purified enzyme, 347
purification, 346
assay, 348, 354–355
catalytic parameters, 354
covalent intermediates
acetyl-enzyme, 356–357
1-hydroxyethylphosphoryl thioester, 358–359
pyruvate thiohemiketal-enzyme, 357
electron donors, 346–347
glycyl radical
activation, 344
AdhE protein-catalyzed reduction, 353–354
catalytic role, 345, 355, 361
electron paramagnetic resonance
data collection, 349–350
spectra, 350–351
hydrogen abstraction mechanism, 348–349
hydrogen exchange, 351
oxygen fragmentation, 352–353
stereochemistry, 348–349
induction in bacteria, 344
purification, recombinant *Escherichia coli* enzyme
cell growth, 345
chromatography, 345
reaction catalyzed, 344
reaction mechanism, 355
stability, 348, 353
subunit structure, 346

Q

Quinone tanning, mechanism, 1

R

Radiolysis
polypeptide cleavage, 380–382
protein modification, 380, 385, 387, 389
Raman spectroscopy, *see* Resonance Raman spectroscopy
Redox cycling, *see also* Nitroblue tetrazolium
dopa quantitation in proteins, 7–8
factors affecting color yield, 141–142
pyrroloquinoline quinone protein assay
detection in gel electroblots, 147–149
inhibition, 142–143
quinoprotein identification in fluids and tissues, 143, 145
reaction conditions, 146–147
quinone detection, 141
Resonance Raman spectroscopy
baseline subtraction, 140
data collection, 137–138
isotopic substitution and band assignment, 139
laser-induced changes in sample, monitoring, 137
laser selection, 136
methylamine dehydrogenase, 138–139
quinoprotein sample preparation
concentration, 135–136
derivatization, 133–135
sensitivity, 132, 137
topa quinone peptides, 31–32, 34, 139
Ribonucleotide reductase, *Escherichia coli,* diferric-tyrosyl radical cofactor assembly
glycyl radical evidence, 361–362
mechanism, 302
rapid freeze-quench electron paramagnetic resonance
apparatus, 286–287
calibration of sampling apparatus, 287

control experiment, 288–289
quantitative analysis, 290–292, 303
rate constants, 292–293
trapping of intermediates, 286
rapid freeze-quench Mössbauer spectroscopy
apparatus, 294–295
quantitative analysis, 296–297, 299, 301
rate constants, 301
reference sample preparation
diferric cluster, 295
diferric radical species, 296
ferrous ion in HEPES buffer, 296
ferrous-R2, 296
reference spectra, 297–299
sensitivity, 303
spectra acquisition, 296
time course of reaction, 299
reducing equivalents, 278–279
stopped-flow spectroscopy
absorbance changes, 282–283
apparatus, 281
mixing reactions, 281–282
rate constants, 283–285, 293
requirements, 280–281
structure, 278

S

Sodium cyanoborohydride, copper amine oxidase inactivation
enzyme–substrate complex, 74–75
free enzyme, 75–76
Stopped-flow spectroscopy
copper amine oxidase
anaerobic conditions, 77
data analysis, 77–78, 81
deuterium isotope effects, 82–83
rapid-scanning analysis, 80–83
relaxation characteristics, 81–82
methylamine dehydrogenase
amicyanin oxidation, 190
apparatus, 187, 189
methylamine reduction, 189–190
ribonucleotide reductase diferric-tyrosyl radical cofactor assembly
absorbance changes, 282–283
apparatus, 281
mixing reactions, 281–282
rate constants, 283–285, 293
requirements, 280–281

T

Thioether bond
galactose oxidase
biosynthesis, 257
catalysis role, 257–258, 264
confirmatory evidence, 251–254
mutation effects, 256
stacking interaction of tryptophan, 254–255
X-ray crystallography, 251
tyrosinase, 259–260
Threonine, oxidative modification in proteins, 391
Topa hydantoin quinone
absorption spectra, 48–50
synthesis, 39–42
Topa quinone
enzymes containing, 32
internal cyclization, 33–34
p-nitrophenylhydrazine reaction, 35, 38
peptide analysis
mass spectrometry, 28–29, 34
nuclear magnetic resonance, 29, 34
resonance Raman spectroscopy, 31–32, 34
visible absorbance spectroscopy, 31–32, 34–38
phenylhydrazine reaction, 21, 23–24, 33
pivalamide derivative
deamination of benzylamine
absorption spectroscopy analysis, 62–63
aerobic reaction conditions, 65
catalyst destruction, 57–58
chemical synthesis of products, 68–69
copper effects, 57
deuterium isotope effect, 56
mechanism, 60, 62
nuclear magnetic resonance analysis, 62–64, 66–68
pH effect on yield, 54, 56
rate, 54–55
substrate specificity, 56–57
synthesis, 54, 59–60, 65

pK of derivatized cofactor, 37–38
semiquinone intermediate in amine oxidase
absorption spectroscopy, 88–89
circular dichroism, 87
electron paramagnetic resonance, 85–87
2,4,5-Tribenzyloxyphenethylamine, synthesis, 64
N-(2,4,5-Tribenzyloxyphenethyl)pivalamide, synthesis, 64–65
2,7,9-Tricarboxy-1*H*-pyrrolo[2,3-f]quinoline-4,5-dione, *see* Pyrroloquinoline quinone
2,4,5-Trihydroxyphenylalanine, deamination of benzylamine, 57
2,4,5-Trihydroxyphenylalanine quinone, *see* Topa quinone
Tryptophan
deuteration, 314, 316, 325
electron transfer, 320–321, 343
oxidative modification in proteins, 389
Tryptophan tryptophylquinone, *see also* Aromatic amine dehydrogenase; Methylamine dehydrogenase
absorption spectra in proteins, 152–153, 163
biosynthesis mechanism, 173–176
carbonyl reagent sensitivity, 154–155
electron acceptors, 151, 154, 177
enzymes with, resonance Raman spectroscopy, 138–139, 154, 178
model compound
benzylamine oxidation, 172–173
cyclic voltammogram, 170
spectral properties, 167–168, 170, 175
synthesis, 165–167, 173–176
molecular geometry, 168–169
structure, 150, 165, 177, 191
confirmation, 163
mass spectral analysis, 159–161
nuclear magnetic resonance, 161–162
X-ray crystallography, 163, 168, 178
Tyrosine
dimer quantitation
fluorescence, 387
mass spectrometry, 387–399
oxidative modification in proteins, 387–389
pyrroloquinoline quinone precursor, 228–229
radicals, *see* Photosystem II
synthesis of radiolabeled compounds, 229–230
Tyrosine phenol-lyase, expression in *Erwinia herbicola* culture, 229–230

V

Valine, oxidative modification in proteins, 389–390

X

X-ray crystallography
galactose oxidase
accuracy of model, 244–245
acetate ion, 243–244
copper coordination, 247–248
copper site geometry, 249–251
crystallization, 239–238
data collection, 238
heavy atom derivatives, 239–241
initial model building, 241–243
phase determination, 238–241
refinement, 243–244
sequence determination, 241–242
stacking tryptophan mutant, 255
substrate binding, 260–262
superbarrel motif, 213, 215–216, 245–247
thioether cysteine mutant, 256
methanol dehydrogenase
amino acid sequencing, 199
cofactor identification, 200–201
crystallization conditions, 194–195
pyrroloquinoline quinone orientation, 211–213
structure analysis and refinement, 194, 196–198, 210–212
superbarrel motif, 211, 213, 215–216
methylamine dehydrogenase, 163, 168, 178
amino acid sequencing, 199
cofactor identification, 200–201
crystallization conditions, 194–195
heavy subunit, 204
light subunit, 205–207

Paracoccus denitrificans enzyme
  amicyanin complex, 193, 208–210
    cytochrome *c*–amicyanin ternary
      complex, 210
structure analysis and refinement, 194,
  196–198
superbarrel motif, 213, 215–216
Thiobacillus versutus enzyme, hydrazine complex, 208
  tryptophan tryptophylquinone orientation, 201, 203–207
tryptophan tryptophylquinone structure,
  163, 168, 178